Catharanthus roseus

M. Naeem • Tariq Aftab • M. Masroor A. Khan
Editors

Catharanthus roseus

Current Research and Future Prospects

Springer

Editors
M. Naeem
Plant Physiology Section
Department of Botany
Aligarh Muslim University
Aligarh, UP, India

Tariq Aftab
Plant Physiology Section
Department of Botany
Aligarh Muslim University
Aligarh, UP, India

M. Masroor A. Khan
Plant Physiology Section
Department of Botany
Aligarh Muslim University
Aligarh, UP, India

ISBN 978-3-319-51619-6 ISBN 978-3-319-51620-2 (eBook)
DOI 10.1007/978-3-319-51620-2

Library of Congress Control Number: 2017933948

Printed on acid-free paper

This Springer imprint is published by Springer Nature
The registered company is Springer International Publishing AG
The registered company address is: Gewerbestrasse 11, 6330 Cham, Switzerland

Contents

Contributors

Harshita Abul Department of Pharmaceutics, College of Pharmacy, SGIT, Ghaziabad, U.P., India

Tariq Aftab Plant Physiology Section, Department of Botany, Aligarh Muslim University, Aligarh, Uttar Pradesh, India

Bilal Ahmad Plant Physiology Section, Department of Botany, Aligarh Muslim University, Aligarh, Uttar Pradesh, India

M. Masidur Alam Plant Physiology Section, Department of Botany, Aligarh Muslim University, Aligarh, Uttar Pradesh, India

Department of Botany, Malda College, Malda, West Bengal, India

Akbar Ali Plant Physiology Section, Department of Botany, Aligarh Muslim University, Aligarh, Uttar Pradesh, India

Muzamil Ali Cellular Differentiation and Molecular Genetics Section, Department of Botany, Hamdard University, New Delhi, India

Petra Andrade-Hoyos Procesos Naturales, Ingeniería Forestal Comunitaria, Universidad Intercultural del Estado de Puebla, Calle Principal a Lipuntahuaca S/N; Lipuntahuaca, Huehuetla, Puebla, México

Matin ArefiFard Biological Department, Arak University-Islamic Azad University of Arak, Arak, Iran

S. Bala Abirami Department of Biotechnology, PRIST University, Thanjavur, Tamilnadu, India

Md. Abul Barkat Department of Pharmaceutics, Nanomedicine Research Lab, Faculty of Pharmacy, Jamia Hamdard, New Delhi, India

Department of Pharmaceutics, Faculty of Pharmacy, Integral University, Kursi Road, Lucknow, India

Hebert Jair Barrales-Cureño Procesos Naturales, Ingeniería Forestal Comunitaria, Universidad Intercultural del Estado de Puebla, Calle Principal a Lipuntahuaca S/N; Lipuntahuaca, Huehuetla, Puebla, México

Ramesh Bellamkonda Department of Food Technology, Vikramasimhapuri University, Nellore, Andhra Pradesh, India

Sasi Bhusana Rao Bongu Department of Biochemistry, Sri Krishnadevaraya University, Anantapur, Andhra Pradesh, India

Salvador Chávez-Salinas Departamento de Ingeniería en Biotecnología, Universidad Politécnica del Valle de Toluca, Santiaguito Tlalcilalcali, Almoloya de Juárez, México

Saralakumari Desireddy Department of Biochemistry, Sri Krishnadevaraya University, Anantapur, Andhra Pradesh, India

V. Devi Rajeswari Department of Biomedical Sciences, School of Biosciences and Technology, VIT University, Vellore, Tamilnadu, India

R. Elangomathavan Department of Biotechnology, PRIST University, Thanjavur, Tamilnadu, India

Sudhakara Gujjala Department of Biochemistry, Sri Krishnadevaraya University, Anantapur, Andhra Pradesh, India

Mohd. Idrees Department of Mathematics and Sciences, College of Arts and Applied Sciences, Dhofar University, Salalah, Sultanate of Oman

Hassan Jaleel Plant Physiology Section, Department of Botany, Aligarh Muslim University, Aligarh, Uttar Pradesh, India

Jagjit Kaur Department of Biotechnology, Maharishi Markandeshwar University, Mullana, Haryana, India

Amjad Khalil Department of Life Sciences, King Fahad University of Petroleum and Minerals, Dhahran, Saudi Arabia

M. Masroor A. Khan Plant Physiology Section, Department of Botany, Aligarh Muslim University, Aligarh, Uttar Pradesh, India

Shamshad Ahmad Khan Division of Biochemical Sciences, CSIR-National Chemical Laboratory, Pune, Maharashtra, India

Kuldeep Kumar Department of Biotechnology, Multani Mal Modi College, Patiala, Punjab, India

Luis Germán López-Valdez Laboratorio de Productos Naturales, Área de Química, Departamento de Preparatoria Agrícola, Universidad Autónoma Chapingo, Texcoco Estado de, México

Alfonso Luna-Cruz Catedrático CONACYT, Facultad de Ciencias Agropecuarias, Universidad Michoacana de San Nicolás de Hidalgo, Morelia, Michoacán, Mexico

Ajay Kumar Mathur Plant Biotechnology Division, Central Institute of Medicinal and Aromatic Plants (CSIR-CIMAP), Lucknow, Uttar Pradesh, India

A. Mujib Cellular Differentiation and Molecular Genetics Section, Department of Botany, Hamdard University, New Delhi, India

M. Naeem Plant Physiology Section, Department of Botany, Aligarh Muslim University, Aligarh, Uttar Pradesh, India

Qifang Pan Plant Biotechnology Research Center, SJTU-Cornell Institute of Sustainable Agriculture and Biotechnology, Fudan-SJTU-Nottingham Plant Biotechnology R&D Center, School of Agriculture and Biology, Shanghai Jiao Tong University, Shanghai, People's Republic of China

Sanjeev Pandey Department of Botany, Banwarilal Bhalotia College, Asansol, West Bengal, India

Varsha Parasharami Division of Biochemical Sciences, CSIR-National Chemical Laboratory, Pune, Maharashtra, India

Teena Pathak Department of Biotechnology, Multani Mal Modi College, Patiala, Punjab, India

S. Patharajan Department of Biotechnology, PRIST University, Thanjavur, Tamilnadu, India

Mallaiah Putakala Department of Biochemistry, Sri Krishnadevaraya University, Anantapur, Andhra Pradesh, India

Md. Akhlaquer Rahman Department of Pharmaceutics, Faculty of Pharmacy, Integral University, Lucknow, India

S. Ramesh Department of Microbiology, PRIST University, Thanjavur, Tamilnadu, India

M.V. Rao Department of Plant Science, Bharathidasan University, Tiruchirappalli, Tamil Nadu, India

Karuna Rasineni Department of Internal Medicine, University of Nebraska Medical Center, Omaha, NE, USA

Cesar Reyes-Reyes Departamento de Ingeniería en Biotecnología, Universidad Politécnica del Valle de Toluca, Santiaguito Tlalcilalcali, Almoloya de Juárez, México

Yawar Sadiq Plant Physiology Section, Department of Botany, Aligarh Muslim University, Aligarh, Uttar Pradesh, India

Saisree Sagree Department of Biochemistry, Sri Krishnadevaraya University, Anantapur, Andhra Pradesh, India

D. Sathya Prabhu Department of Biomedical Sciences, School of Biosciences and Technology, VIT University, Vellore, Tamilnadu, India

P. Senbagalakshmi Department of Plant Science, Bharathidasan University, Tiruchirappalli, Tamil Nadu, India

T. Senthil Kumar Department of Industry University Collaboration, Bharathidasan University, Tiruchirappalli, Tamil Nadu, India

Asfia Shabbir Plant Physiology Section, Department of Botany, Aligarh Muslim University, Aligarh, Uttar Pradesh, India

Sreenivasa Reddy Singareddy Biochemistry Division, National Institution of Nutrition, Hyderabad, Telangana, India

Apoorva Singh Department of Biotechnology, Multani Mal Modi College, Patiala, Punjab, India

V. Subhashini Department of Environmental Sciences, Acharya Nagarjuna University, Guntur, Andhra Pradesh, India

A.V.V.S. Swamy Department of Environmental Sciences, Acharya Nagarjuna University, Guntur, Andhra Pradesh, India

Kexuan Tang Plant Biotechnology Research Center, SJTU-Cornell Institute of Sustainable Agriculture and Biotechnology, Fudan-SJTU-Nottingham Plant Biotechnology R&D Center, School of Agriculture and Biology, Shanghai Jiao Tong University, Shanghai, PR China

Dipti Tonk Cellular Differentiation and Molecular Genetics Section, Department of Botany, Hamdard University, New Delhi, India

Moin Uddin Botany Section, Women's College, Aligarh Muslim University, Aligarh, Uttar Pradesh, India

Israr Ul Hassan Department of Mathematics and Sciences, College of Arts and Applied Sciences, Dhofar University, Salalah, Sultanate of Oman

Priyanka Verma Division of Biochemical Sciences, CSIR-National Chemical Laboratory, Pune, Maharashtra, India

Nadia Zafar Cellular Differentiation and Molecular Genetics Section, Department of Botany, Hamdard University, New Delhi, India

About the Editors

M. Naeem is an Assistant Professor in the Department of Botany at Aligarh Muslim University, India. For more than a decade, he has devoted his research to improving the yield and quality of commercially important medicinal and aromatic plants (MAPs). His research focuses on escalating the production of MAPs and their active principles using a novel and safe technique involving radiation-processed polysaccharides as well as the application of potent PGRs. To date, he has successfully run three major research projects as Principal Investigator, two of which were sanctioned by the Department of Science & Technology, New Delhi. He also completed another intriguing research project on *Catharanthus roseus* awarded by CSTUP, Lucknow. Dr. Naeem has published seven books and more than 80 research papers in reputable national and international journals. He has also participated in various national and international conferences and acquired life memberships to various scientific bodies in India and abroad. Based on his research contributions, Dr. Naeem has been awarded a Research Associateship from the Council of Scientific & Industrial Research, New Delhi; a Young Scientist Award (2011) from the State Government of Uttar Pradesh; a Fast Track Young Scientist Award from the Department of Science & Technology, India; a Young Scientist of the Year Award (2015) from the Scientific and Environmental Research Institute, Kolkata; and a Rashtriya Gaurav Award (2016) from the International Friendship Society, New Delhi.

Tariq Aftab received his Ph.D. in the Department of Botany at Aligarh Muslim University, India, and is currently an Assistant Professor there. He is the recipient of a prestigious Leibniz-DAAD fellowship from Germany, a Raman Fellowship from the Government of India, and Young Scientist Awards from the State Government of Uttar Pradesh and Government of India. He is also a member of various scientific associations from India and abroad. He has worked as Visiting Scientist at IPK, Gaterleben, Germany, and in the Department of Plant Biology, Michigan State University, USA. He has published research articles in several peer-reviewed national and international journals and is the lead editor of the book *Artemisia annua:* Pharmacology and Biotechnology. His research interests include physiological, proteomic, and molecular studies on medicinal and aromatic plants.

M. Masroor A. Khan is a Professor in the Department of Botany at Aligarh Muslim University, India. After completing his Ph.D., he worked as a postdoctoral fellow at the Ohio State University, USA (1987–1988). Then he worked as Pool Scientist for the Council of Scientific and Industrial Research and as a Research Associate for CSIR and the University Grants Commission before he joined AMU as an Assistant Professor. In his 30 years of teaching experience at the university level, he has guided 8 Ph.D., 2 M.Phil., and 27 M.Sc. students. For the convenience of his students, he has also developed study materials and launched more than 100 class notes and PowerPoint presentations on his website. Prof. Khan has run six research projects sponsored by national and international funding agencies and has published about 120 research papers in reputable journals and books. In his research, Prof. Khan is working to promote the productivity and active ingredients of medicinal and aromatic plants using different strategies such as the application of mineral nutrition, PGRs, and nanoparticles. Prof. Khan has contributed greatly toward the establishment of Radiation-Processed Polysaccharides (RPPs) as plant growth promoters. In collaboration with the Bhabha Atomic Research Centre, Mumbai, he is researching the appropriate doses of RPPs that would help farmers increase the productivity of medicinal and aromatic plants and their active constituents. Another field of his research is nanoparticles, some of which he has found can enhance the essential oil production of some aromatic plants. He has also filed a patent in this regard.

Strategies for Enhancing Alkaloids Yield in *Catharanthus roseus* Via Metabolic Engineering Approaches

Kexuan Tang and Qifang Pan

Abstract As the only source for the low-abundance antitumor agents vinblastine and vincristine, *Catharanthus roseus* is highly valued for its diversity of more than 130 monoterpenoid indole alkaloids (MIAs) and has been studied extensively as a model for medicinal plants improvement. However, the low yield increases the cost and limits the industrial production of these valuable MIAs in medical use. The biosynthesis of these MIAs is a complex multistep enzymatic network that is tightly regulated by developmental and environmental factors. Many genes encoding constitutive structural biosynthetic enzymes, transcription factors, and transporters involved in these pathways have been cloned and characterized. To improve the MIA production, a couple of approaches have been carried on the plants, hairy roots, and cell culture of *C. roseus*, as well as on heterogeneous plant (like *Nicotiana benthamiana*), including abiotic and biotic methods. The main strategies for enhancing alkaloids yield is to genetically modify the MIA pathway and enhance the metabolic flux to MIA production via metabolic engineering strategies. Here, we will review the past decades' efforts on the MIA production.

Keywords Alkaloids • *Catharanthus roseus* • Genes • MIAs pathways

1 Introduction

Monoterpenoid indole alkaloids (MIAs) are important alkaloids for their medicinal bioactivities of highly value. *Catharanthus roseus* is the main and natural source to produce these valuable MIAs, including the antitumor drugs vinblastine and

K. Tang (✉) • Q. Pan
Plant Biotechnology Research Center, SJTU-Cornell Institute of Sustainable Agriculture and Biotechnology, Fudan-SJTU-Nottingham Plant Biotechnology R&D Center, School of Agriculture and Biology, Shanghai Jiao Tong University, Shanghai 200240, PR China
e-mail: kxtang@sjtu.edu.cn; panqf@sjtu.edu.cn

M. Naeem et al. (eds.), *Catharanthus roseus*,
DOI 10.1007/978-3-319-51620-2_1

vincristine. The trace amount of useful MIAs in *C. roseus* forces great efforts to improve their production via various methods. Metabolic engineering is an approach to modify metabolic pathways and metabolites production via gene transfer technology. With the increasing knowledge of the MIA pathway and its biosynthetic genes, metabolic engineering is widely performed on MIA biosynthesis in *C. roseus* to boost the yields of targeted MIAs and their analogs.

1.1 Natural Compound-Targeted Engineering Strategies

Different types of genes cloned from the MIA pathway have been overexpressed in the cells, hairy roots, and plants of *C. roseus*, which showed different effects on MIA biosynthesis as well as at the level of the full metabolome (Table 1).

1.1.1 Structural Genes

MIA biosynthesis in *C. roseus* is a complex pathway including more than 35 coordinately regulated enzymatic steps producing at least 35 known intermediates (Fig. 1). Up to now, 30 structural genes of the key biosynthetic steps from MIA pathways have been cloned and identified. They are the first target genes used in metabolic engineering of *C. roseus*.

Genes from the Tryptophan Pathway

The expression of a more tryptophan inhibition-resistant Arabidopsis *ASα* enzyme coupled with a glucocorticoid-inducible promoter in *C. roseus* hairy roots dramatically increased tryptophan and tryptamine yields but not of MIAs, except lochnericine, after induction with 3 μM dexamethasone (Hughes et al. 2004a). Transgenic hairy roots expressing both *ASα* and *ASβ* subunits produced more tryptamine and showed a greater resistance to feedback inhibition of AS activity by tryptophan than those only expressing *ASα* (Hong et al. 2006). When fed with the terpenoid precursors 1-deoxy-D-xylulose, loganin, and secologanin, respectively, hairy roots overexpressing *ASα* or *Asβ* could increase the levels of hörhammericine, catharanthine, ajmalicine, lochneriricine, and tabersonine (Peebles et al. 2006). As a side effect, the metabolic flux into the flavonoid pathway was also transiently increased when the *AS* overexpressing hairy roots were induced by 0.2 μM dexamethasone, which caused increases of catechin and naringin in hairy roots (Chung et al. 2007). This might be due to the induction of the phenylalanine/tyrosine pathway by tryptophan (Verpoorte and Alfermann 2000). TDC overexpression in *C. roseus* transgenic calli results in increased tryptamine levels but not in increased MIA production (Goddijn et al. 1995), neither in *C. roseus* cell cultures (Whitmer et al. 2002b). On the contrary, no increase of tryptamine

Table 1 Overexpression of genes involved in MIA biosynthesis in the cell cultures, hairy roots, and plants of *Catharanthus roseus*

Varieties	Genes	Metabolites significantly affected in levels	References
Hairy roots	*GmMYBZ2*	Catharanthine	Zhao et al. (2001)
	HMGR	Campesterol, serpentine, ajmalicine, catharanthine	Ayora-talavera et al. (2002)
	ASα	Tryptophan, tryptamine, lochnericine	Hughes et al. (2004a)
	TDC	Serpentine	Hughes et al. (2004b)
	ASα+ TDC	Tryptamine	
	ASα	Tryptamine	Hong et al. (2006)
	ASα +ASβ	Tryptophan, tryptamine	
	ASα +ASβ+TDC	Tryptamine	
	ASαβ	Naringin, catechin, salicylic acid	Chung et al. (2007)
	DAT	Hörhammericine	Magnotta et al. (2007)
	ORCA2	Catharanthine, vindoline	Liu et al. (2011)
	G8O(G10H)	Catharanthine	Wang et al. (2010)
	G8O+ORCA3	Catharanthine	
Cell cultures	*ORCA3*	Serpentine, ajmalicine, tabersonine, hörhammericine	Peebles et al. (2009)
	DXS	Ajmalicine, serpentine, lochnericine, tabersonine, hörhammericine	Peebles et al. (2010)
	G8O	–	
	Asα	Tryptophan, tryptamine, lochnericine, tabersonine, hörhammericine	
	DXS +G8O	Ajmalicine, tabersonine, lochnericine, hörhammericine	
	DXS+ Asα	tryptamine, tabersonine, lochnericine, hörhammericine, tryptophan	
	TDC	Tryptamine	Canel et al. (1998); Whitmer et al. (2002b)
	STR	Strictosidine, ajmalicine, catharanthine, serpentine, tabersonine	Canel et al. (1998); Whitmer et al. (2002a)
	PRX1	Ajmalicine, serpentine, H_2O_2	Jaqqi et al. (2011)
	ORCA3	Tryptophan, tryptamine	Van der Fits and Memelink (2000)
	CYP76B6	10-hydroxy geraniol	Collu et al. (2001)
	CjMDR1	Ajmalicine, tetrahydroalstonine	Pomahacova et al. (2009)

(continued)

Table 1 (continued)

Varieties	Genes	Metabolites significantly affected in levels	References
Plant	*DAT*	Vindoline	Wang et al. (2012)
	ORCA3	Vindoline, catharanthine	Pan et al. (2012)
	ORCA3 + *G10H(G8O)*	Strictosidine, vindoline, catharanthine	

but a 129% increase of serpentine was noted on induction of 3 μM dexamethasone in hairy roots overexpressing *TDC* (Hughes et al. 2004b). Expressing *TDC* from *C. roseus* in cell cultures or plants of *Nicotiana tabacum* resulted in the formation of tryptamine up to 10 μg/g FW and 18–66 μg/g FW, respectively (Hallard et al. 1997). When co-overexpressing *AS* and *TDC* in hairy roots, an enhanced ability to produce tryptamine was observed, but only a transiently increased accumulation of tabersonine and lochnericine among all measured alkaloids (Hughes et al. 2004b; Hong et al. 2006). To study the effect of introducing MIA alkaloid biosynthetic genes in a plant normally only producing secologanin, Hallard and coworkers (Hallard 2000) introduced both the *TDC* and *STR* into *Weigelia* hairy roots. Compared to normal roots no more secologanin could be observed, whereas tryptamine, ajmalicine, and serpentine could be detected in the hairy roots. This confirmed the presence of a glucosidase able to hydrolyze strictosidine. Though the alkaloid levels were very low, it shows that MIAs can also be made in non-alkaloid-producing plants. That means alternative crops for making MIA. In that context also the production of strictosidine was achieved in yeast cells in which STR and SGD are overexpressed and which are fed with secologanin and tryptamine (Geerlings et al. 2001). The cells could produce 3 g/L of strictosidine in 3 days, many times more than ever achieved in plant cell cultures. As STR was mainly excreted to the medium, whereas SGD was in the cells, grinding the whole culture resulted in the production of cathenamine. The transgenic yeast cells could be grown on the juice pressed out of the berries of *Symphoricarpus albus* rich in sugar and secologanin from which strictosidine was made after feeding tryptamine (Geerlings et al. 2001).

Fig. 1 (Continued) 4-diphosphate synthase, *IDI* isopentenyl diphosphate isomerase, *GPPS* GPP synthase, *GES* geraniol synthase, *G8O* geraniol 8-oxidase, *8-HGO* 8-hydroxygeraniol oxidoreductase, *IS* iridoid synthase, *IO* iridoid oxidase, *7-DLGT* 7-deoxyloganetic acid-*O*-glucosyltransferase, *7DLH* 7-deoxyloganic acid hydroxylase, *LAMT* loganic acid-O-methyltransferase, *SLS* secologanin synthase, *AS* anthranilate synthase, *TDC* tryptophan decarboxylase, *STR* strictosidine synthase, *SGD*, strictosidine-β-D-glucosidase, *T16H*, tabersonine 16-hydroxylase, *16OMT* O-methyltransferase, *NMT* *N*-methyltransferase, *D4H* desacetoxyvindoline-4-hydroxylase, *DAT*, deacetylvindoline-4-*O*-acetyltransferase, *Prx1* α-3′,4′-anhydrovinbastine synthase (AVLBS)

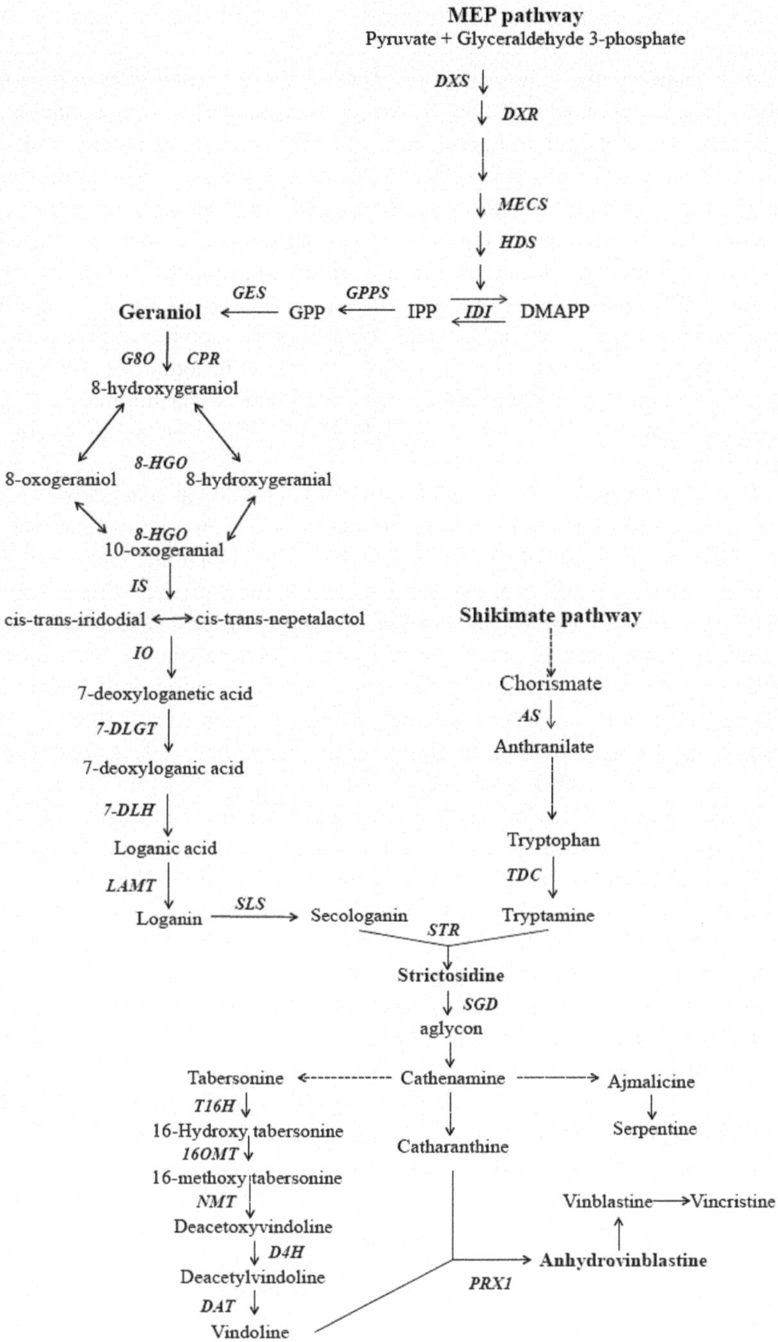

Fig. 1 The scheme of MIA biosynthesis pathways in *Catharanthus roseus*. *DXS* 1-deoxy-D-xylulose 5-phosphate synthase, *DXR* 1-deoxy-D-xylulose 5-phosphate reductoisomerase, *MECS* 2-Cmethyl-D-erythritol 2,4-cyclodiphosphate synthase, *HDS* 1-hydroxy-2-methyl-2-butenyl

Genes from the MEP and Iridoid Pathways

In the MEP pathway, the genes encoding *DXS*, *DXR*, 2-C-methyl-Derythritol-2,4-cyclodiphosphate synthase (*MECS*), hydroxymethylbutenyl-4-diphosphate (*HDS*), and *IDI1* have been cloned and characterized from *C. roseus* (Chahed et al. 2000; Veau et al. 2000; Guirimand et al. 2012). *DXS* overexpression resulted in a significant increase in ajmalicine, serpentine, and lochnericine but a significant decrease in tabersonine and hörhammericine in *C. roseus* hairy roots. In fact, overexpression of *DXS* and *DXR* has been found to increase terpenoid production in several plants. For example, *DXS* overexpression enhances the production of various isoprenoids in *Arabidopsis* (Estevez et al. 2001), and *DXR* overexpression increases essential oil yield in peppermint, carotenoid accumulation in ripening tomatoes (Mahmoud & Croteau 2001), and various isoprenoids in tobacco leaves (Hasunuma et al. 2008). There is no information about the use of *MECS* and *IDI1* in metabolic engineering on MIA production.

At the edge of primary and secondary metabolism, G8O as gatekeeper could be a carbon flux controlling step for the iridoid pathway. The encoding gene was overexpressed in the hairy roots of *C. roseus*, which resulted in a higher accumulation of catharanthine (0.063–0.107% of dry weight) than in the wild-type lines (0.019 and 0.029%) (Wang et al. 2010). When co-overexpressing *DXS* and *G8O*, the hairy roots showed a significant increase in ajmalicine by 16%, lochnericine by 31%, and tabersonine by 13% (Peebles et al. 2010). Considering their location with respect to the IPP/DMAPP branching point for terpenoid classes, it is conceivable that the overexpression of one downstream structural gene alone is unable to effect the channeling of the flux at this upstream branch point while the overexpression together with an upstream gene of the IPP/DMAPP branch point may affect the carbon fluxes by a push and pull effect toward the MIA iridoid precursor. An increased production of IPP/DMAPP and *G8O* overexpression may increase the flux in the monoterpenoid branch and from there into MIA. *DXS* and *ASα* co-overexpression displayed a significant increase in hörhammericine by 30%, lochnericine by 27%, and tabersonine by 34% in hairy roots (Peebless et al. 2010). Recent discoveries of IDI1, GPPS, GES, and iridoid synthase encoding genes provide new possibilities for the regulation and improvement of MIA production.

Genes from the Alkaloids Pathway

Cultures of *STR* transgenic cells consistently showed tenfold higher STR activity than wild-type cultures, which favored the biosynthetic flow through the pathway. Two such lines accumulated over 200 mg/L of strictosidine and strictosidine-derived MIAs, including ajmalicine, catharanthine, serpentine, and tabersonine, while maintaining wild-type levels of TDC activity (Canel et al. 1998). Whitmer et al. (2002a, b) showed that in *C. roseus* cell lines overexpressing TDC or STR have an overcapacity of indole alkaloid biosynthesis enzyme activities, as feeding of loganin resulted in a large increase of alkaloid production whereas the combination of loganin and tryptamine feeding even further increased the level of alkaloids. Apparently, the iridoid pathway is the most limiting step, but when that limitation is overcome the tryptophan pathway becomes limiting. Overcoming one limiting step immediately shows what the next limiting step is. A single structural gene overexpression will thus always have only a limited effect on the overall flux in a pathway. On the other hand, it shows that probably many biosynthetic steps are already present, and only the enzymatic machinery has to be started up by increasing the amount of the limiting substrate by feeding or genetic modification. The elucidation of the full iridoid pathway as described above is thus a major breakthrough opening new possibilities to explore for increasing MIA production.

The key gene *DAT* for the vindoline biosynthesis was introduced into *C. roseus* plants by *Agrobacterium tumefaciens*, which resulted in an increase of vindoline level in the leaves (Wang et al. 2012). However, overexpression of *DAT* in hairy roots altered their MIA profile and accumulated more hörhammericine compared to control lines (Magnotta et al. 2007). Comparative analysis revealed that MIA pathway genes have elevated expression levels in *CrPrx* overexpression transgenic hairy roots, whereas they had a significant reduction in their transcript level in *CrPrx-RNAi* transgenic hairy roots (Jaggi et al. 2011). Alkaloid analysis showed higher levels of ajmalicine and serpentine in these peroxidase overexpressing cell lines. All these transgenic lines produced higher amounts of H_2O_2 (Jaggi et al. 2011). The oxidative burst or H_2O_2 production is closely related to indole alkaloid production (Zhao et al. 2001). In leaves of *C. roseus*, PRX together with phenolic compounds were suggested to represent an important sink of excess H_2O_2, diffusing from the chloroplast under high light exposure (Ferreres et al. 2011). These results indicate a role of the *CrPrx* gene in the regulation of MIA pathway and other metabolic pathways, thus affecting the production of specific alkaloids. In order to study the role of CrPrx and CrPrx1 in plants, these two peroxidases were expressed in *Nicotiana tabacum* (Kumar et al. 2011). The transformed plants exhibited increased peroxidase activity. Increased oxidative stress tolerance was also observed in transgenics when treated with H_2O_2 under strong light conditions. However, differential tolerance to salt and dehydration stress was observed during germination of T1 transgenic seeds. Under these forms of stress, the seed germination of *CrPrx*-transformed plants and wild-type plants was clearly suppressed, whereas *CrPrx1* transgenic lines showed improved germination. *CrPrx*-transformed lines exhibited better cold tolerance than *CrPrx1*-transformed lines. These results indicate that vacuolar peroxidases play an important role in salt and dehydration stress, while cell wall-targeted peroxidases render cold stress tolerance.

1.1.2 Transporter Genes

Since MIA biosynthesis involves at least four different cell types and in each of them at least five different subcellular compartments, the trafficking of pathway intermediates from one to another compartment requires an efficient transport system. Previous research also suggested that transport is one of the potential factors in regulation of MIA biosynthesis. However, the knowledge about MIA membrane transport mechanisms is still very limited.

Transport has basically two aspects, a physicochemical and a biochemical one. In cells and in an organism diffusion will always take place. Concentration gradients make molecules to diffuse in a liquid phase. Moreover, molecules will equilibrate between the aqueous phase and lipid phase (membrane). Mass transfer factors determine the rate of the uptake in a lipophilic membrane from water and the release again to water, i.e., they affect the transport rate through a membrane. That allows calculations of the rate of diffusion of compounds between cells and cellular compartments. The complexity of this system is further increased by the pH, making that acids and bases at different pH have different solubility in the liquid phases. For example, an alkaloid in acidic conditions is poorly soluble in a lipid phase, but at basic conditions it is better lipid soluble. So at a high ratio of protonated to non-protonated alkaloids, which is at acidic conditions, transport will be slow through a membrane, at higher pH it will be the opposite. Modeling uptake in *C. roseus* vacuoles using these physical–chemical processes resulted in an ion-trap model for alkaloid uptake in vacuoles that fitted reasonably well the experimental results using isolated vacuoles. The lower pH in the vacuole than in the cytosol causes preferred accumulation of alkaloids in the vacuole if compared with the cytosol as the uptake rate on the more basic cytosolic site of the membrane is faster than on the more acidic vacuolar side. This physicochemical process requires ATP for maintaining the low vacuolar pH, so depletion of ATP will inhibit uptake, similar as in case of ABC transporters (Blom et al. 1991). On the other hand, Deus-Neumann and Zenk (1984) reported uptake kinetics for active transport for some indole alkaloids. Ajmalicine, catharanthine, and vindoline showed different rates, and all were ATP dependent. From this it was hypothesized that the vacuolar transport occurred via selective transporter proteins. Roytrakul (2004) reported a detailed study on the uptake of several *C. roseus* alkaloids and secologanin in isolated vacuoles. By adding inhibitors of the various classes of transport proteins, for each individual compound quite a different and complex picture came out. For each single compound, different transporter seems to be involved (Roytrakul and Verpoorte 2007).

The uptake into the vacuole is thus dependent on a combination of factors, first of all there is the bidirectional diffusion-driven transport. On top of that, there are multidrug-resistant associated proteins (MRP, inhibited by glibenclamide) and ATP Binding Cassette (ABC, inhibited by *ortho*-vanadate) type of transporters involved in uptake. Whereas multidrug resistant (MDR) (P-glycoproteins, inhibited by cyclosporine A and verapamil) and MDR coupled with proton symport, are responsible for extrusion. To further complicate the transport system, glutathione was found to cis-activate the MRP transport of ajmalicine into the vacuole (Roytrakul 2004; Roytrakul and Verpoorte 2007). Considering the multicompartment system involved in the MIA biosynthesis, it is clear that with the already very complex transport system into vacuoles, the model for a single-cell or multi-cell system is impossible to describe. The need for sufficient energy and co-factors in the different compartments add further to this complexity. In an attempt to calculate the rate of transport between cells by using the various available data on uptake of compounds and a number of assumptions based on observations from other plants, it became clear that at least diffusion alone would result in a biosynthetic rate more or less of what is found in the plant (Supandi et al. 2009, unpublished results). It means that the selective transport might play a role in accumulating compounds in certain cells and in some of the specific biosynthetic steps, e.g., by accumulating certain compounds in a vacuole, where they are oxidized to yield serpentine or dimeric alkaloids. In case of serpentine, this anhydronium compound is much more polar than ajmalicine from which it is formed by oxidation, thus becomes trapped into the vacuole. The fact that tobacco vacuoles excrete strictosidine, whereas *C. roseus* vacuoles store it (Hallard et al. 1997) shows at least that every plant species will have different transport systems with different selectivity. Considering that the MIAs are confined to certain cell types may also in part be due to specific transport systems in the cellular membrane(s). It means that introduction of a novel pathway in a plant may be hampered by lack of transport of intermediates.

The example of *CjMDR1,* an ABC transporter gene specific for berberine transport originally isolated from *Coptis japonica*, shows the problems one may encounter in genetically modifying transport. This gene was expressed in *C. roseus* cell cultures (Pomahacova et al. 2009). The endogenous alkaloids, ajmalicine and tetrahydroalstonine, were accumulated significantly more in *C. roseus* cells expressing *CjMDR1* in comparison with control lines after feeding these alkaloids, but transport of other alkaloids was not affected, and even no effect at all on berberine transport into the cells was observed.

A unique catharanthine ABC-transporter (*CrTPT2*) belonging to the pleiotropic drug resistance (PDR) family has been cloned and functionally characterized. It is expressed predominantly in the epidermis of young leaves (Yu and De Luca 2013). Further analysis suggested that CrTPT2 may be specific to MIA-producing plant species, where it mediates secretion of alkaloids to the leaf surface. *CrTPT2* gene expression is induced under the treatment with catharanthine, and its silencing redistributes catharanthine into the leave, causing an increase of dimeric alkaloid levels in the leaves.

Recently, strong support for active MIAs uptake by *C. roseus* mesophyll vacuoles through a specific H (+) antiport system was reported (Carqueijeiro et al. 2013). The vacuolar transport mechanism of the main MIAs accumulated in *C. roseus* leaves, vindoline, catharanthine, and α-3′,4′-anhydrovinblastine was characterized using a tonoplast vesicle system. Vindoline uptake was ATP dependent, and this transport activity was strongly inhibited by NH_4^+ and carbonyl cyanide m-chlorophenyl hydrazine and was insensitive to the ATP-binding cassette (ABC) transporter inhibitor vanadate. Spectrofluorimetric assays with a pH-sensitive fluorescent probe showed that vindoline and other MIAs indeed were able to dissipate an H^+ pre-established gradient across the tonoplast by either vacuolar H^+-ATPase or vacuolar H^+-diphosphatase. Though it was claimed that this system would be responsible for the MIA transport instead of an ion-trap mechanism or ABC transporters, it seems unlikely, as at least physicochemical-based transport will always occur and the various previous reports found alkaloid specificity for the uptake into the vacuole.

1.1.3 Transcription Factors

Transcription factors (TFs) are sequence-specific-DNA-binding proteins that interact with the promoter regions of target genes and modulate the rate of mRNA synthesis by RNA polymerase II (Gantet and Memelink 2002). They usually control the expression of more than one gene vital for normal development and functional physiology in plants. Several TFs have been found to be involved in the regulation of secondary metabolism. In *C. roseus*, MIA biosynthesis is related with plant defense and controlled by a number of signals including developmental cues, light, and biotic and abiotic stress. Regulation of MIA biosynthetic genes is coordinated by several types of TFs (Fig. 1).

ORCAs

The best-known TFs regulating MIA biosynthesis are the jasmonates-responsive ORCAs (octadecanoid-responsive *Catharanthus* AP2-domain proteins) from the plant-specific AP2/ERF (APETALA2/ethylene-responsive factor) family, i.e., ORCA2 and ORCA3, for which the regulation mechanism of the MIA biosynthetic genes in *C. roseus* is well established. *ORCAs* expression is induced by jasmonates (van der Fits and Memelink 2001), which is a major and essential signaling pathway to induce MIA biosynthesis. Jasmonates are first converted to the bioactive jasmonate isoleucine derivative (JA-Ile). Perception of JA-Ile by CrCOI1 causes the degradation of the CrJAZ proteins, derepressing the CrMYC2 protein. CrMYC2 then activates the expression of *ORCAs*, which in its turn activate the expression of MIA biosynthetic genes through binding to the JERE (jasmonate and elicitor-responsive element) in the promoter of targeted genes (Menke et al. 1999; van der Fits and Memelink 2000; Zhang et al. 2011). Ectopic expression of *ORCA3* in cell cultures of *C. roseus* increased the expression of the MIA biosynthetic genes *TDC*, *STR*, *CPR*, and *D4H*, as well as two genes encoding primary metabolic enzymes (*AS* and *DXS*) (van der Fits and Memelink 2000). This indicates that ORCA3 is a central regulator of MIA biosynthesis and positively regulates the biosynthesis of MIAs and their precursors. Nevertheless, ORCA3 does not regulate the expression of *G8O* and *DAT*. Overexpression of *ORCA3* caused an increase of ajmalicine and serpentine but a decrease in tabersonine, lochnericine, and hörhammericine in hairy roots (Peebles et al. 2009). When *ORCA3* combined with *G8O* were overexpressed in hairy roots, alkaloid accumulation level analyses showed that all transgenic clones accumulated more catharanthine, with the highest accumulation level 6.5-fold more than that of the non-expression clone (Wang et al. 2010). ORCA2 from *C. roseus* was demonstrated to regulate the expressions of *STR, TDC,* and *SGD* gene, but has no effect on the CYP-related reductase (*CPR*), which is regulated by ORCA 3 (Menke et al. 1999; Li et al. 2013). Transgenic hairy root cultures overexpressing *ORCA2* showed an average content of catharanthine that was increased up to 2.03 in comparison to the control lines, respectively. However, vinblastine could not be detected in the transgenic and control hairy root cultures by HPLC (Liu et al. 2011). Transgenic

C. roseus plants overexpressing ORCA3 alone (OR lines), or co-overexpressing G10H and ORCA3 (GO lines) were obtained by genetic modification (Pan et al. 2012). It was found that ORCA3 and G10H overexpression significantly increased the accumulation of strictosidine, vindoline, catharanthine, and ajmalicine but had limited effects on anhydrovinblastine and vinblastine levels.

ZCTs and BPF

The zinc finger-binding proteins ZCT1, ZCT2, and ZCT3 (members of the transcription factor IIIA-type zinc finger family) were found to bind to the promoters of *STR* and *TDC*. This interaction repressed the activity of *STR* and *TDC*. The binding of the ZCTs to the *STR* promoter has been suggested to counteract the activation of *STR* by ORCA2 or ORCA3 (Pauw et al. 2004).

Using an enhancer domain of the *STR* promoter as bait in a yeast one-hybrid screen resulted in the isolation of *CrBPF1*, a periwinkle homolog of the MYB-like transcription-factor BPF1 from parsley (van der Fits et al. 2000). *CrBPF1* expression is induced by elicitors but not jasmonates, which indicates that elicitors induce *STR* expression in periwinkle cells via jasmonic-acid-dependent and -independent pathways.

Sequence analysis of the *STR* and *TDC* promoters shows that they contain a G-box or G-box-like binding site. Two G-box-binding factors, CrGBF1 and CrGBF2, were subsequently identified in *C. roseus* and shown to repress the transcription of *STR* by binding to the G-box sequence (Siberil et al. 2001).

WRKYs

A *C. roseus* WRKY transcription factor, CrWRKY1, is preferentially expressed in roots and induced by the phytohormones jasmonate, gibberellic acid, and ethylene (Suttipantaa et al. 2011). Overexpression of *CrWRKY1* in *C. roseus* hairy roots upregulated several key MIA pathway genes, especially *TDC*, as well as transcriptional repressors *ZCT1*, *ZCT2*, and *ZCT3*. However, *CrWRKY1* overexpression repressed the transcriptional activators, *ORCA2*, *ORCA3*, and *CrMYC2*. Overexpression of a dominant repressive form of CrWRKY1, created by fusing the SRDX-repressor domain to CrWRKY1, resulted in down-regulation of *TDC* and *ZCTs* but up-regulation of *ORCA3* and *CrMYC2*. CrWRKY1 binds to the W-box elements of the *TDC* promoter in the electrophoretic mobility shift, yeast one-hybrid and *C. roseus* protoplast assays. Up-regulation of *TDC* increased TDC activity, tryptamine concentration and resistance to 4-methyl tryptophan inhibition of *CrWRKY1* hairy roots. Compared to control roots, *CrWRKY1* hairy roots accumulated up to threefold higher levels of serpentine. The preferential expression of *CrWRKY1* in roots and its interaction with transcription factors including ORCA3, CrMYC2, and ZCTs may play a key role in determining the root-specific accumulation of serpentine in *C. roseus* plants.

Other TFs

The root-specific MADS-box transcription factor Agamous-like 12 (Agl12) from *Arabidopsis thaliana* was expressed on the differentiation of suspension cells from *C. roseus* (Montiel et al. 2007). The expression of Agl12 is sufficient to promote an organization of suspension cells into globular parenchyma-like aggregates but is insufficient by itself to induce complete morphological root differentiation. Agl12 expression selectively increases the expression of genes encoding enzymes involved in the early biosynthetic steps of the terpenoid precursor of the alkaloids. The transgenic cell lines expressing Agl12 produced significant amounts of ajmalicine, which indicates that TFs involved in tissue or organ differentiation may constitute new metabolic engineering tools to produce specific valuable MIAs. Murata and De Luca (2005) reported that ORCA3 and an AP2/ERF type of transcription factors were expressed in all four cell types (epidermis, IPAP, laticifers, and idioblast cells).

Although different types of TFs have been reported to interact with the genes in the MIA pathway, regulation of the key enzyme genes involved in its branches still remains unclear and need to be figured out, such as the iridoid pathway, vindoline pathway, and bisindole alkaloids pathway.

1.2 Unnatural Compound-Targeted Engineering Strategies

Approaches to generate new-to-nature compounds from plant-based pathways are also developed on *C. roseus*, which modifies the structure of a natural product to improve the biological activity of the compound. Replacement of an endogenous starting material with an unnatural compound is a strategy that has been broadly applied in prokaryotic biosynthetic pathways (O'Connor 2012). Now, genetic manipulation is performed on the MIA pathway combined with precursor-directed biosynthesis and engineered enzymes to produce various unnatural products in *C. roseus*.

RNA-mediated suppression of tryptamine biosynthesis in *C. roseus* hairy root culture eliminates the production of monoterpenoid indole alkaloids derived from tryptamine and secologanin. But when an unnatural tryptamine analog, 5-fluorotryptamine 1a, was fed to both wild-type and silenced cultures, a variety of novel fluorinated alkaloids, such as fluoro-ajmalicine, fluoro-tabersonine, and fluoro-serpentine, were produced and not contaminated with the natural alkaloid counterparts in silenced lines (Runguphan et al. 2009). The flux of the unnatural substrate could be enhanced to the downstream alkaloids through some branches of the pathway when the natural, endogenous substrate is limited or unavailable. Targeted silencing of substrate biosynthesis combined with precursors feeding programs a plant alkaloid pathway to more effectively produce desirable novel products, which opens new areas of combining synthesis and biosynthesis to increase chemodiversity.

A mutant strictosidine synthase gene with reengineered substrate specificity was transformed into *Catharanthus roseus*. The resulting transgenic plant cell culture produced a variety of unnatural alkaloid compounds when cocultured with simple, achiral, commercially available precursors that the reengineered enzyme was designed to accept (Runguphan and O'Connor 2009). This work demonstrates the power of engineering new structures of complex alkaloidal natural products in plant cultures.

Another example is to validate the function of the engineered flavin-dependent halogenase RebH. In vivo, the tryptamine-specific RebH mutant (Y455W) was transformed into the alkaloid-producing plant *C. roseus*, and the de novo production of the halogenated alkaloid 12-chloro-19, 20-dihydroakuammicine was observed. The resulting tissue cultures accumulated substantial levels of 7-chlorotryptophan while wild-type (WT) RebH has been integrated into periwinkle metabolism previously. By installing chlorine onto tryptamine, the RebH Y455W mutant circumvents the bottleneck that tryptophan decarboxylase accepts 7-chlorotryptophan at only 3% of the efficiency of the native substrate tryptophan. In comparison with cultures harboring RebH and WT RebF, tissue cultures containing mutant RebH Y455W and RebF also accumulate microgram per gram fresh-weight quantities of 12-chloro-19,20-dihydroakuammicine but, in contrast, do not accumulate 7-chlorotryptophan, demonstrating the selectivity and potential utility of this mutant in metabolic engineering applications (Glenn et al. 2011).

The development of approaches to generate new-to-nature compounds from *C. roseus* MIA pathway will produce a number of MIA analogs which have to improve or alter biological activity, and will further enhance our ability to hijack the downstream MIA pathways (O'Connor 2012).

References

Ayora-Talavera T, Chappell J, Lozoya-Gloria E et al (2002) Overexpression in *Catharanthus roseus* hairy roots of a truncated hamster 3-hydroxy-3-methylglutaryl-CoA reductase gene. Appl Biochem Biotech 97:135–145

Blom T, Sierra M, Van Vliet T et al (1991) Uptake and accumulation of ajmalicine into isolated vacuoles of cultured cells of *Catharanthus roseus* (L.) G. Don. and its conversion into serpentine. Planta 183:170–177

Canel C, Lopes-Cardoso M, Whitmer S et al (1998) Effects of over-expression of strictosidine synthase and tryptophan decarboxylase on alkaloid production by cell cultures of *Catharanthus roseus*. Planta 205:414–419

Carqueijeiro I, Noronha H, Duarte P et al (2013) Vacuolar transport of the medicinal alkaloids from *Catharanthus roseus* is mediated by a proton driven antiport. Plant Physiol 162:1486–1496

Chahed K, Oudin A, Guivarc'h N et al (2000) 1-Deoxy-D-xylulose 5-phosphate synthase from periwinkle: cDNA identification and induced gene expression in terpenoid indole alkaloid-producing cells. Plant Physiol Biochem 38:559–566

Chung IM, Hong SB, Peebles CAM et al (2007) Effect of the engineered indole pathway on accumulation of phenolic compounds in *Catharanthus roseus* hairy roots. Biotechnol Prog 23:327–332

Collu G, Unver N, Peltenburg-Looman AM, Heijden RVD et al (2001) Geraniol 10-hydroxylase, a cytochrome P450 enzyme involved in terpenoid indole alkaloid biosynthesis. Febs Lett 508:215–220

Deus-Neumann B, Zenk M (1984) A highly selective alkaloid uptake system in vacuoles of higher plants. Planta 162:250–260

Estévez JM, Cantero A, Reindl A et al (2001) 1-deoxy-D-xylulose-5-phosphate synthase, a limiting enzyme for plastidic isoprenoid biosynthesis in plants. J Biol Chem 276:22901–22909

Ferreres F, Figueiredo R, Bettencourt S et al (2011) Identification of phenolic compounds in isolated vacuoles of the medicinal plant *Catharanthus roseus* and their interaction with vacuolar class III peroxidase: an H_2O_2 affair? J Exp Bot 62:2841–2854

van der Fits L, Memelink J (2000) ORCA3, a jasmonate-responsive transcriptional regulator of plant primary and secondary metabolism. Science 289:295–297

van der Fits L, Memelink J (2001) The jasmonate-inducible AP2/ERF-domain transcription factor ORCA3 activates gene expression via interaction with a jasmonate-responsive promoter element. Plant J 25:43–53

van der Fits L, Zhang H, Menke FLH et al (2000) A *Catharanthus roseus* BPF-1 homologue interacts with an elicitor-responsive region of the secondary metabolite biosynthetic gene *Str* and is induced by elicitor via a JA-independent signal transduction pathway. Plant Mol Biol 44:675–685

Gantet P, Memelink J (2002) Transcription factors: tools to engineer the production of pharmacologically active plant metabolites. Trends Pharmacol Sci 23:563–569

Geerlings A, Redondo F, Contin A et al (2001) Biotransformation of tryptamine and secologanin into plant terpenoid indole alkaloids by transgenic yeast. Appl Microbiol Biotechnol 56:420–424

Glenn WS, Nims E, O'Connor SE (2011) Reengineering a tryptophan halogenase to preferentially chlorinate a direct alkaloid precursor. J Am Chem Soc 133:19346–19349

Goddijn OJ, Pennings EJ, van der Helm P et al (1995) Overexpression of a tryptophan decarboxylase cDNA in *Catharanthus roseus* crown gall calluses results in increased tryptamine levels but not in increased terpenoid indole alkaloid production. Transgenic Res 4:315–323

Guirimand G, Guihur A, Phillips MA et al (2012) A single gene encodes isopentenyl diphosphate isomerase isoforms targeted to plastids, mitochondria and peroxisomes in *Catharanthus roseus*. Plant Mol Biol 79:443–459

Hallard DAC (2000) Transgenic plant cells for the production of indole alkaloids, Ph.D thesis, Rijksuniversiteit te Leiden

Hallard DAC, Van der Heijden R, Verpoorte R et al (1997) Suspension cultured transgenic cells of *Nicotiana tabacum* expressing tryptophan decarboxylase and strictosidine synthase cDNAs from *Catharanthus roseus* produce strictosidine upon secologanin feeding. Plant Cell Rep 17:50–54

Hasunuma T, Takeno S, Hayashi S et al (2008) Overexpression of 1-deoxy-D-xylulose-5-phosphate reductoisomerase gene in chloroplast contributes to increment of isoprenoid production. J Biosci Bioeng 105:518–526

Hong S-B, Peebles CA, Shanks JV et al (2006) Expression of the Arabidopsis feedback-insensitive anthranilate synthase holoenzyme and tryptophan decarboxylase genes in *Catharanthus roseus* hairy roots. J Biotechnol 122:28–38

Hughes EH, Hong SB, Gibson SI et al (2004a) Expression of a feedback-resistant anthranilate synthase in *Catharanthus roseus* hairy roots provides evidence for tight regulation of terpenoid indole alkaloid levels. Biotechnol Bioeng 86:718–727

Hughes EH, Hong SB, Gibson SI et al (2004b) Metabolic engineering of the indole pathway in *Catharanthus roseus* hairy roots and increased accumulation of tryptamine and serpentine. Metab Eng 6:268–276

Jaggi M, Kumar S, Sinha AK (2011) Overexpression of an apoplastic peroxidase gene *CrPrx* in transgenic hairy root lines of *Catharanthus roseus*. Appl Microbiol Biotechnol 90:1005–1016

Kumar S, Jaggi M, Sinha AK (2011) Ectopic overexpression of vacuolar and apoplastic *Catharanthus roseus* peroxidases confers differential tolerance to salt and dehydration stress in transgenic tobacco. Protoplasma 249:423–432

Li CY, Leopold AL, Sander GW et al (2013) The ORCA2 transcription factor plays a key role in regulation of the terpenoid indole alkaloid pathway. BMC Plant Biol 13:155

Liu DH, Ren WW, Cui LJ et al (2011) Enhanced accumulation of catharanthine and vindoline in *Catharanthus roseus* hairy roots by overexpression of transcriptional factor ORCA2. Afr J Biotechnol 10:3260–3268

Magnotta M, Murata J, Chen J et al (2007) Expression of deacetylvindoline-4-*O*-acetyltransferase in *Catharanthus roseus* hairy roots. Phytochemistry 68:1922–1931

Mahmoud SS, Croteau RB (2001) Metabolic engineering of essential oil yield and composition in mint by altering expression of deoxyxylulose phosphate reductoisomerase and menthofuran synthase. Proc Natl Acad Sci U S A 98:8915–8920

Menke FLH, Champion A, Kijne JW et al (1999) A novel jasmonate-and elicitor-responsive element in the periwinkle secondary metabolite biosynthetic gene STR interacts with a jasmonate- and elicitor-inducible AP2-domain transcription factor, ORCA2. EMBO J 18:4455–4463

Montiel G, Breton C, Thiersault M et al (2007) Transcription factor Agamous-like 12 from *Arabidopsis* promotes tissue-like organization and alkaloid biosynthesis in *Catharanthus roseus* suspension cells. Metab Eng 9:125–132

Murata J, De Luca V (2005) Localization of tabersonine 16-hydroxylase and 16-OH tabersonine-16-*O*-methyltransferase to leaf epidermal cells defines them as a major site of precursor biosynthesis in the vindoline pathway in *Catharanthus roseus*. Plant J 44:581–594

O'Connor SE (2012) Strategies for engineering plant natural products: the iridoid-derived monoterpene indole alkaloids of *Catharanthus roseus*. Methods Enzymol 515:189–206

Pan Q, Wang Q, Yuan F et al (2012) Overexpression of ORCA3 and G10H in *Catharanthus roseus* plants regulated alkaloid biosynthesis and metabolism revealed by NMR-metabolomics. PLoS One 7:1–14

Pauw B, Hilliou FAO, Martin VS et al (2004) Zinc finger proteins act as transcriptional repressors of alkaloid biosynthesis genes in *Catharanthus roseus*. J Biol Chem 279:52940–52948

Peebles CA, Hong SB, Gibson SI et al (2006) Effects of terpenoid precursor feeding on *Catharanthus roseus* hairy roots over-expressing the alpha or the alpha and beta subunits of anthranilate synthase. Biotechnol Bioeng 93:534–540

Peebles CAM, Hughes EH, Shanks JV et al (2009) Transcriptional response of the terpenoid indole alkaloid pathway to the overexpression of ORCA3 along with jasmonic acid elicitation of *Catharanthus roseus* hairy roots over time. Metab Eng 11:76–86

Peebles CAM, Sander GW, Hughes EH et al (2010) The expression of 1-deoxy-$_{D}$-xylulose synthase and geraniol-10-hydroxylase or anthranilate synthase increases terpenoid indole alkaloid accumulation in *Catharanthus roseus* hairy roots. Metab Eng 13:234–240

Pomahacová B, Dusek J, Dusková J et al (2009) Improved accumulation of ajmalicine and tetrahydroalstonine in *Catharanthus* cells expressing an ABC transporter. J Plant Physiol 166:1405–1412

Roytrakul S (2004) Transport of alkaloids and its precursor through the vacuolar membrane of *Catharanthus roseus*. Ph.D. Thesis. Leiden University, The Netherlands.

Roytrakul S, Verpoorte R (2007) Role of vacuolar trasporter proteins in plant secondary metabolism: *Catharanthus roseus* cell culture. Phytochem Rev 6:383–396

Sibéril Y, Benhamron S, Memelink J et al (2001) *Catharanthus roseus* G-box binding factors 1 and 2 act as repressors of strictosidine synthase gene expression in cell cultures. Plant Mol Biol 45:477–488

Supandi F, Hille S, Verpoorte R (2009) A reassessment of membrane transport and accumulation of terpenoid indole alkaloid precursors in *Catharanthus roseus* through mathematical modeling and simulation. MSc thesis, Leiden University, The Netherlands

Suttipantaa N, Pattanaika S, Kulshrestha M et al (2011) The transcription factor CrWRKY1 positively regulates the terpenoid indole alkaloids biosynthesis in *Catharanthus roseus*. Plant Physiol 157:2081–2093

Veau B, Courtois M, Oudin A et al (2000) Cloning and expression of cDNAs encoding two enzymes of the MEP pathway in *Catharanthus roseus*. Biochem Biophys Acta 1517:159–163

Verpoorte R, Alfermann AW (2000) Metabolic engineering of plant secondary metabolism. Springer, New York

Wang CT, Liu H, Gao XS et al (2010) Overexpression of *G10H* and *ORCA3* in the hairy roots of *Catharanthus roseus* improves catharanthine production. Plant Cell Rep 29:887–894

Wang Q, Xing S, Pan Q et al (2012) Development of efficient *Catharanthus roseus* regeneration and transformation system using agrobacterium tumefaciens and hypocotyls as explants. BMC Biotechnol 12:34

Whitmer S, van-der-Heijden R, Verpoorte R (2002a) Effect of precursor feeding on alkaloid accumulation by a strictosidine synthase over-expressing transgenic cell line S1 of *Catharanthus roseus*. Plant Cell Tiss Org Cult 69: 85–93

Whitmer S, van der Heijden R, Verpoorte R (2002b) Effect of precursor feeding on alkaloid accumulation by a tryptophan decarboxylase over-expressing transgenic cell line T22 of *Catharanthus roseus*. J Biotechnol 96:193–203

Yu F, De Luca V (2013) ATP-binding cassette transporter controls leaf surface secretion of anticancer drug components in *Catharanthus roseus*. Proc Natl Acad Sci U S A 24:110–139

Zhang H, Hedhili S, Montiel G et al (2011) The basic helix-loop-helix transcription factor CrMYC2 controls the jasmonate-responsive expression of the ORCA genes that regulate alkaloid biosynthesis in *Catharanthus roseus*. Plant J 67:61–71

Zhao J, Hu Q, Guo YQ et al (2001) Elicitor-induced indole alkaloid biosynthesis in *Catharanthus roseus* cell cultures is related to Ca^{2+} influx and the oxidative burst. Plant Sci 161:423–431

In Vitro Biotechnological Production and Pharmacological Studies of Antileukemic Alkaloids of *Catharanthus roseus*

Hebert Jair Barrales-Cureño, Petra Andrade-Hoyos, Alfonso Luna-Cruz, Cesar Reyes-Reyes, Salvador Chávez-Salinas, and Luis Germán López-Valdez

Abstract Different techniques of in vitro cultures of the medicinal plant *Catharanthus roseus* are available. In this regard, the plant is a source of important secondary metabolites that are compounds widely used in pharmacology. For instance, vinblastine and vincristine are alkaloids employed in the treatment of leukemia. This chapter discusses the techniques mostly used in the field of modern biotechnology, such as the in vitro culture of callus and suspension cells, as well as those related to organs, roots, and seedlings. Similarly, the chapter encompasses the types of explant cultures used, induction rates, and the culture environment, jointly with hormones and concentration employed. Also discussed is the level of production

---

H.J. Barrales-Cureño (✉) • P. Andrade-Hoyos
Procesos Naturales, Ingeniería Forestal Comunitaria,
Universidad Intercultural del Estado de Puebla, Calle Principal a Lipuntahuaca S/N;
Lipuntahuaca, Huehuetla, Puebla, C.P. 73475, México
e-mail: hebert.jair@uiep.edu.mx

A. Luna-Cruz
Catedrático CONACYT, Facultad de Ciencias Agropecuarias, Universidad Michoacana de San Nicolás de Hidalgo, Avenida Francisco J. Múgica S/N. Ciudad Universitaria, CP. 58030, Morelia, Michoacán, Mexico

C. Reyes-Reyes • S. Chávez-Salinas
Departamento de Ingeniería en Biotecnología, Universidad Politécnica del Valle de Toluca, Carretera Toluca-Almoloya de Juárez Km. 5.6 C.P. 50904, Santiaguito Tlalcilalcali, Almoloya de Juárez, Mexico
e-mail: hebert.jair@uiep.edu.mx

L.G. López-Valdez
Laboratorio de Productos Naturales, Área de Química, Departamento de Preparatoria Agrícola, Universidad Autónoma Chapingo, AP74 Oficina de correos Chapingo, Km 38.5, Carretera México-Texcoco, Texcoco Estado de, 56230, México

M. Naeem et al. (eds.), *Catharanthus roseus*,
DOI 10.1007/978-3-319-51620-2_2

of each category of alkaloids according to the type of in vitro culture. Similarly, new metabolites obtained from suspension cell cultures of *Catharanthus roseus*, along with major pharmacological studies recently conducted, are contained in the chapter.

Keywords *Catharanthus roseus* • Callus culture • Root cultures • Cell suspension cultures • Antileukemic alkaloids

1 Introduction

Biotechnology of plant in vitro cultures represents a successful tool in the production of callus and cell cultures that have the capacity to produce secondary metabolites, compounds of great interest in the pharmaceutical and medical fields (Barrales-Cureño and Ramírez 2013). Alkaloids vincristine and vinblastine, elements produced from *Catharanthus roseus,* are a good example of this. Vinblastine and vincristine are potent mitotic inhibitors that are used in chemotherapy for leukemia. They are complex, chemically synthesized structures, similar to other drugs used in the fight against cancer, such as Taxol (Barrales-Cureño and Soto 2012; Barrales-Cureño et al. 2012, 2015, 2016). In that regard, biotechnological approaches represent the best way in obtaining these compounds.

Recently, the production of vinblastine and vincristine has been induced and research carried out as it has never been over the plant in vitro cultures by means of hormone combination of auxins and cytokinins (Villa-Ruano et al. 2011). Cell potency represents the basis of in vitro culture, a term defined as the potential capacity of a single plant cell to regenerate into a whole plant. Several in vitro techniques, such as micropropagation of adventitious meristems or organs, including tissues and cell cultures, provide a large amount of material of *Catharanthus roseus* that is used in the isolation of dimeric and indole mono-type alkaloids with multi-therapeutic properties. In this regard, research has demonstrated that *Catharanthus roseus* have the potential to regenerate through somatic organogenesis during the induction of friable calluses. Likewise, in vitro cultures of multiple shoots can be induced directly. The great pharmacological importance of terpene indole alkaloid, associated with low content in plants (approximately 0.0005% of dry weight), stimulates intensive research regarding metabolic routes of the alkaloids occurring in various studies over in vitro culture. These allow determining the concentrations that occur in in vitro callus and cell suspension cultures.

In this chapter, the importance of different types and conditions of in vitro cultures in the production of vinblastine and vincristine as antileukemic alkaloids is highlighted, as well as that of other related metabolites including the main medical applications of *Catharanthus roseus* species.

2 Overview of In Vitro Culture

Generally, the process of in vitro culture implies the inoculation of a medium gelled culture (using agar, gelrite, or phytagel®) and a piece of tissue or plant organ, known as explant, previously treated to remove any unwanted body present on the surface (disinfestation). The culture is incubated under controlled environmental conditions of light, temperature, and humidity, along with other physicochemical and nutritional conditions, in order to construct an amorphous cell mass called callus; or toward differentiation, in an organized, embryo and organ-producing tissue (Calva and Pérez 2005). Organ cultures redifferentiate to complete plants (micropropagation) which are then transferred to a greenhouse, a condition known as acclimatization phase. The temperature is usually controlled and set between 25 and 28 °C, while the pH is between 5.2 and the 6.5 and the light ranges from 0 to 12.000 lux (Calva and Pérez 2005).

Several studies have investigated the effects of pH on cell growth and metabolite production in suspension cells (Morgan et al. 2000). Some studies confirmed that changes of pH made on cultures increased the release of secondary metabolite. In addition, few studies were conducted in order to examine the effects that produce buffers over the growth of crops or the metabolic pathways during secondary metabolism. Similarly, several authors have studied the effect of buffers on in vitro root cultures of *Catharanthus roseus,* in order to quantify the content of serpentine, ajmalicine, abersonine, horhammericine, and lochericine (Morgan et al. 2000).

Light, as a parameter, plays a prominent role in the in vitro production of secondary metabolites of *Catharanthus roseus.* Other factors, such as temperature, also have a significant influence on the growth of suspension cell cultures and in the production of ajmalicine, as is indicated by certain authors (Ten Hoopen et al. 2002). The optimum temperature for both processes, the growth of biomass and the production of secondary metabolites was 27.5 °C. Young or developing plants with meristem tissues showing vigorous vegetative growth are the best source of explants. Even though, both juvenile and adult growth can be found in the same plant, the first one is characterized by being active and not having reproductive structures, while adult growth is usually slower and plants feature sexual structures for their reproduction. Disinfestation of tissue to be used as a source of explants is performed with disinfectant agents, such as sodium hypochlorite or calcium. Disinfectant agent penetration in rough or hairy surfaces of plant tissue can be increased with the addition of surfactants, such as Tween 20. Meanwhile, activated carbon or citric acid is used as antioxidants.

Phytohormones and their inhibitors are substances produced by plants. By controlling their response to environmental stimuli, such as light, temperature, and humidity, they help regulate and coordinate processes essential for the development of the plants. They consist of auxins, gibberellins, cytokinins, abscisic acid, ethylene, brassinosteroids, polyamines, salicylic acid, and jasmonic acid. In particular, auxins and gibberellins promote the elongation of cells while inhibiting differentiation. Meanwhile, cytokinins stimulate the division process by which new cells are produced, and can thus avoid cellular aging. Ethylene stimulates fruit ripening mainly, while abscisic acid inhibits the action of auxin, gibberellins, and cytokinins, to serve as a natural defense system against physiological effects of stress. Cytokinins and auxins are compounds commonly used in plant cell cultures. In addition, the 2,4-Dichlorophenoxyacetic acid (2,4-D) is the most widely used plant hormone in the induction and maintenance of callus tissue due to its property to suppress organogenesis rigorously. With regard to cytokinins, compound 2-Indolaminopurine (2iP) is the more active structure. Notwithstanding the foregoing, the 6-Benzylaminopurine (BAP) and kinetin (KIN), the latter is a synthetic cytokinin affected by light in the wavelength range 300–800 nm, represent the most widely used in plant cell culture compounds.

In vitro cultures provide insight into the production of secondary metabolites. In this respect, there are various sources in which different alkaloids of *Catharanthus roseus* have been isolated in vitro. Examples include the ajmalicine alkaloid (molecular formula: $C_{21}H_{24}N_2O_3$, molar mass: 352.43 g mol^{-1}, present antispasmodic properties, depression treatment, antistress effects), which has been extracted using analysis conducted from callus, cell suspensions, shoots and hairy roots; and the alstonine, from in vitro callus; as well as anthirine, in cell suspensions. To this, we can add the cathindine in suspensions; serpentine (molecular formula: $C_{21}H_{22}N_2O_3$, molar mass: 349.40 g mol^{-1}, with antihypertensive properties, antispasmodic properties, anxiety treatment), in callus, suspensions, shoots and hairy roots; the aquamicine, in callus, suspensions and shoots; and lochericine, in callus, suspensions and hairy roots. Similarly, the horhammericine, in suspensions and shoots; the tabersonine, in callus and suspensions; the vindoline (molecular formula: $C_{25}H_{32}N_2O_6$, molar mass: 456.53 g mol^{-1}, exhibiting anti-ulcerative properties), in suspensions cell and shoots; catharanthine (molecular formula: $C_{21}H_{24}N_2O_2$, molar mass: 336.42 g mol^{-1}, with cytotoxic action on the HCT-116 colorectal carcinoma cell line), in suspensions, shoots and roots; and 3,4 anhydrovinblastine (molecular formula: $C_{46}H_{56}N_4O_8$, molar mass: 795.97 g mol^{-1}, to combat lung and cervico-uterine cancer treatment), in shoots; are part of isolation sources. Similarly, the leurosine in shoots; the catharine, in shoots; vinblastine (molecular formula: $C_{46}H_{58}N_4O_9$, molar mass: 810.97 g mol^{-1},), in callus, shoots and somatic embryos; and vincristine (molecular formula: $C_{46}H_{56}N_4O_{10}$, molar mass: 824.96 g mol^{-1}), in somatic embryos and shoots are also considered sources (Barrales-Cureño 2015).

3 Applications of In Vitro Culture

The main applications of the technique of in vitro culture of cells, tissues, and organs are found in the plant micropropagation, obtaining pathogen-free plants, the preservation of germplasm, as well as plant breeding. In this regard, the biosynthesis of secondary metabolites and basic research in disciplines, such as genetics, physiology, and biochemistry, are also implications of the technique. As regards micropropagation, embryogenesis, and organogenesis, these methods can be used in obtaining somatic clones, as well as in the regeneration of complete plants with uniform characteristics. From this, valuable cultivars free of microorganisms and difficult to obtain using traditional farming methods plants are established.

In vitro cultures can also be stored for long periods. This is accomplished by any of the methods of preservation used for microorganisms such as refrigeration, slow or reduced growth methods for preserving for months, and cryopreservation. The latter consists of crop storage in liquid nitrogen for cooling to −196 °C, which ensures storage for several years. The method eliminates problems related to physical space, excess labor, crop contamination, as well as effects resulting from genetic erosion (Osorio et al. 2011).

There are several advantages in cell and plant culture in fundamental research, micropropagation, and production of biological compounds, such as secondary metabolites, proteins, and genetically modified products activity. In this regard, studies are carried out in a greatly reduced time under controlled conditions, which compared favorably with those using traditional methods for a plant grown.

4 In Vitro Culture of Callus of *Catharanthus roseus*

A callus represents a set of friable dedifferentiated cells growing in a solid medium, and serve as starting material for the establishment and growth of suspension cells (Barrales-Cureño and Ramírez 2013). The calluses obtained may be subcultured for maintenance and propagation. Furthermore, differentiation can be induced to embryos and organs formation (organogenesis and embryogenesis, respectively). Similarly, these can be transferred to a liquid culture medium in order to obtain suspension cells and small aggregates. Table 1 presents the main in vitro cultures of callus from *Catharanthus roseus.*

4.1 In Vitro Culture of Cell Suspension from Catharanthus roseus

The in vitro cultivation of plant cells in liquid medium for suspension cell represents a potential topic of interest to the pharmaceutical industry showing all the advantages inherent in biotechnological processes (Barrales-Cureño et al. 2011). Cell culture (especially, cell suspensions) offer several advantages. These include that of a similar handling with respect to that performed with microorganisms; rapid cell multiplication (doubling time); and the ability to scale novel techniques, such as in the case of bioreactors and temporary immersion systems (Pérez-Alonso and Jiménez 2011). Notwithstanding the foregoing, not all compounds are produced in undifferentiated cells in terms of equal quantity and quality as those that are obtained from mother plants. Similarly, it should be considered the fact that many metabolites are synthesized from their integration into differentiation events. In this regard, several authors have pointed that cell lines are a vehicle for the production of metabolites in equal to or greater than

Table 1 In vitro cultures of callus from *Catharanthus roseus*

Explant type	In vitro culture type	Percentage	Medium	References
Hypocotyl	Callus induction	99%	Murashige and Skoog Medium supplemented with BAP (1.0 mg L^{-1}) + NAA (1.0 mg L^{-1})	Singh et al. (2011)
Leaf	Best callus response	95%	Murashige and Skoog medium supplemented with 2,4-D (1 mg L^{-1}) + Kin (1 mg L^{-1})	Haq et al. (2013)
Node	Callus	80%	Murashige and Skoog medium supplemented with 2,4-D (1 mg L^{-1}) + Kin (1 mg L^{-1})	Haq et al. (2013)
Fruit	Callus	60%	Murashige and Skoog medium supplemented with 2,4-D (1 mg L^{-1}) + Kin (1 mg L^{-1})	Haq et al. (2013)
Leaf	Callus growth	Enhancement of alkaloid content	0.50 mg L^{-1} of 2,4-D and 1.0 mg L^{-1} of BA 47.92 ± 2.85	Verma and Singh (2012)
Hypocotyls of in vitro germinated seeds	Embryogenic callus	2,4 D (1.0 mg L^{-1}) and Murashige and Skoog medium	The advanced cotyledonary embryos showed prominent root and shoot axis, which germinated into plantlets.	Aslam et al. (2014)
Petiole segments of seedlings	Callus roots	10-fold catharanthine, 125-fold serpentine, 0.5-fold vindoline and 0.34-fold ajmalicine were produced by new roots	Medium Murashige and Skoog containing 0.1 mg L^{-1} NAA and 0.1 mg L^{-1} Kin	Ataei-Azimi et al. (2008)
Leaf explants	Callus was induced in MS medium with plant growth regulator (PGR) 2 mg L, 2,4-D and 0.2 mg L Kinetin	MS medium with tryptophan 50–250 mg L^{-1}	Catharanthine content of *Catharanthus roseus* aggregate cells after 14 days of culture was increasing and has optimum content in treatment C (150 mg L^{-1}) that was equal to 50.96 µg diagonal g dw	Pandiangan et al. (2013)

Nodal section	Callus induction	Medium containing MS more Kinetin (1 mg L^{-1}) and NAA (2 mg L^{-1})	66% explants were responded in Murashige and Skoog medium supplemented with NAA and kinetin	Sandhya et al. (2016)
Leaf and stem segments from mature plants	Callus cultures	MS medium supplemented with 2,4-Dichlorophenoxy acetic acid (2,4-D) 1.0 μM and 6-furfurylaminopurine (kinetin) 1.0 μM was used to support the growth of callus cultures	The maximum amount of dry biomass (598.04 mg) was produced after 7 weeks of culture	Kalidass et al. (2010)
Stem and leaf explants in a modified MS liquid induction medium supplemented with 5.37 μM α-naphthaleneacetic acid and 4.65 μM kinetin	In the induction medium, most leaf explants developed into friable half-closed hollow callus clusters	The compact callus clusters could synthesize indole alkaloids 1.9 and 2.4-fold higher than the half-closed hollow callus clusters and dispersed cell cultures	The degree of compaction expressed by the ratio of fresh weight to dry weight of these suspension cultures was correlated to indole alkaloid production	Zhao et al. (2001)

the amount achieved under natural conditions. In addition, new substances have been detected that are not synthesized by plants in their natural habitat. From this, it can be said that the cultivation of cell lines reflexes a biotechnology of great importance for the development of new secondary metabolites (Pérez-Alonso and Jiménez 2011).

In vitro cell suspension cultures are maintained under the same physical and physicochemical conditions for callus induction. In the case of cell suspension cultures of *Catharanthus roseus,* all terpene indole alkaloids derived from intermediates as can be strictosidin, serpentine, catharanthine, ajmalicine, and tabersonine, as well as vincristine and vinblastine (Zhi-Gang et al. 2013). Strictosidine precursor is then hydrolyzed by strictosidine β-glucosidase producing cathenamine as the main product (El-Sayed et al. 2004). Once the cell culture has been established, it is possible to observe the presence of a continuous process of epigenetic or genetic changes, which causes the population to become heterogeneous. As a result, the selection of clones with high growth and production of metabolites of interest becomes a necessary aspect to observe. Meanwhile, cell lines are obtained by selecting several strategies, including macroscopic, enzyme and microscopic examination (cell viability, for example, using fluorescein diacetate) (Pérez-Alonso and Jiménez 2011). Table 2 presents the main in vitro cultures of cell suspension from *Catharanthus roseus.*

4.2 In Vitro Culture of Organ from *Catharanthus roseus*

Aspects associated with the accumulation of secondary metabolites imply the presence of certain types of cells and organelles, including the expression and regulation of catabolic and biosynthetic genes. Therefore, organ culture means an interesting alternative in the production of plant secondary metabolites. In this regard, the shoots and roots represent two types of bodies that are of major importance and that can be cultured on a large scale. In particular, organ culture produces substances of interest that have not been obtained from undifferentiated cultures. Notwithstanding the foregoing, shoot culture does not have the capacity to produce all compounds obtained in natural conditions in the leaves of plants. If the compound of interest is synthesized in roots, therefore shoot culture will tend to not appear. Moreover, it is important to consider that even if the compound is synthesized in the leaves, it is likely that the pattern and concentration result different from those obtained in intact plants. The main advantage indicates that organ culture is more stable in genetic terms if compared with the cultivation of suspension cell and callus (Pérez-Alonso and Jiménez 2011). Table 3 presents the main in vitro cultures of shoots from *Catharanthus roseus.*

Vindoline is a major alkaloid in vitro cultures of *Catharanthus roseus* outbreaks from which some authors reached 2 mg g^{-1} dry weight after 27 days of culture (Hernández-Domínguez et al. 2004).

Table 2 In vitro cultures of cell suspension from *Catharanthus roseus*

Explant type	In vitro culture type	Alkaloid type	Level production	References
Leaf explants	Suspension cell cultures	Indole alkaloids production	Murashige and Skoog medium containing 1 mg L^{-1} Kinetine under light condition. The highest value of mass cell cultures and indole alkaloids production were achieved with modified MS medium containing 300 mg L^{-1} of either L-glutamine for mass cell induction or L- tryptophan for enrichment of total indole alkaloids	Taha et al. (2009)
CRPP Cell suspension line	Cell suspensions	The cell lines were grown in MS or B5 medium supplemented with either 20 g L^{-1} glucose or 30 g L^{-1} sucrose in 250 mL Erlenmeyer flasks with 100 or 70 mL culture volume per flask	24 μmol g^{-1} dry weight	Zuwairi et al. (2014)
Shake flask suspension cultures of *Catharanthus roseus* cells in two-stage process	The processes for production of indole alkaloids	Both culture processes produced ~20 g dw^{-1} of biomass. Total and individual indole alkaloid production were ten times higher (740 mg L^{-1} and 25–4000 μg g^{-1} dw, respectively) for two-stage than for one-stage cultures	Zenk's alkaloid production medium (APM)	Tom et al. (1991)
Cell cultures in shake flasks and bioreactors	Ajmalicine production	The production of ajmalicine on production medium in a shake flask was not reproduced in a bioreactor	In turbine stirred bioreactor, at low oxygen concentration an intermediate from the tryptophan pathway, tryptamine, is accumulated. At high oxygen, ajmalicine is formed	Ten Hoopen et al. (1994)

Table 3 In vitro cultures of shoots from *Catharanthus roseus*

Explant type	In vitro culture type	Response	Medium	Reference
Hypocotyl	Shoot proliferation	89.2% in light and 71.6% in light, respectively	Murashige and Skoog basal medium supplemented with BAP (1.5 mg L^{-1}) + NAA (1.0 mg L^{-1}) and BAP (3.0 mg L^{-1}) + NAA (4.0 mg L^{-1})	Singh et al. (2011)
Hypocotyl calli	Shoot regeneration	10–15 shoots regenerated per calli.	Murashige and Skoog medium supplemented with BAP (1.5 mg L^{-1}) + NAA (1.0 mg L^{-1})	Singh et al. (2011)
Shoot tip	Multiple shoots	Percentage: 90%	BAP (1 mg L^{-1})	Haq et al. (2013)
Nodal portion	Multiple shoots	Percentage: 80%	BAP (1 mg L^{-1})	Haq et al. (2013)
Nodal explants	Shoot bud	Percentage: 100%	Murashige and Skoog medium supplemented with BAP (1.0 mg L^{-1})	Pandey et al. (2014)
Nodal segments	Multiple Shoots	Number of shoots/explant: 7.30 ± 0.64, shoot length (cm): 5.97 ± 0.17 and shooting response (%): 99%	Murashige and Skoog medium supplemented with 0.5 mg L^{-1} BAP ± 1 mg L^{-1} NAA	Bagum and Mathur (2014)
Juvenile explants such as shoot tip and nodal sections	Multiple shoots	57% of shoots tips were responded on medium containing BAP and kinetin	BAP 2 mg L^{-1} and kinetin 1 mg L^{-1}	Sandhya et al. (2016)
Shoots about 1.5–2 cm	Axillary buds	Results showed that adding 2 mg L^{-1} (BA) to the medium caused significantly increasing of parameters. The average of shoots number was recorded (4.75), leaves number (9.25), the fresh and dry weight were recorded (1103.75 and 112.00 mg), respectively	Murashige and Skoog medium in culture vessels supplemented with different concentrations of BA (0, 1, 2, 3, or 4) mg L^{-1} and 0.2 mg L^{-1} of NAA	Al-oubaidi and Mohammed-Amin (2014)

4.3 *In Vitro Culture of Roots from* Catharanthus roseus

The roots synthesize, accumulate, and secrete a variety of secondary metabolites, in addition to providing mechanical support and allow water and nutrients collection from the soil. In addition, it has been reported that biosynthetic activity of roots is also maintained in in vitro culture, a reason from which *Catharanthus roseus* root crops grow rapidly in a Murashige and Skoog medium. Several types of research determined that in vitro cultures have the ability to synthesize metabolites by producing roots. In that regard, cultures may serve as a biotechnology option for the production of alkaloids for future research. Table 4 presents the main in vitro cultures of roots from *Catharanthus roseus.*

Table 5 presents the main in vitro cultures of plantlets from *Catharanthus roseus* (Table 6).

5 Pharmacological Studies in *Catharanthus roseus*

Cancer is a term applied generically to a great number of different diseases. Due to its nature, it comprehends several malignant tumors found in different locations, such as leukemia, bone sarcoma, Hodgkin's disease, and non-Hodgkin's lymphoma (Barrales-Cureño 2015). In particular, six are the alterations identified in cancerous cells that determine their potential. In that respect, they: (1) show signs of very active growth; (2) evade apoptosis; (3) reflect loss of responsiveness to antigrowth signals; (4) release substances to the medium for tissue vascularization; (5) invade tissues and organs; and (6) experience unlimited replicative growth (Hanahan and Weinberg 2000).

5.1 *Antioxidant Enzyme Activity*

An experiment that involved different concentrations of sodium chloride was conducted over two varieties of *Catharanthus roseus* (the alba and rosea varieties). It was found that the enzyme activity of superoxide dismutase levels increased to 50 mM of sodium chloride, which contributes to a higher level of this enzyme with antioxidant value (Abdul 2009).

5.2 *Antiviral Activity*

Ozcelik et al. (2011) indicated the antiviral effect of *Catharanthus roseus* in herpes simplex virus (type I) with an effect of cytopathogenicity at 0.8 μg mL^{-1}. The catharoseumine, a monoterpene indole alkaloid, which has a single peroxy, was identified as a potential inhibitor against falcipain-2 protozoan parasites that cause malaria,

Table 4 Cultivos in vitro de raíces de *Catharanthus roseus*

Explant type	In vitro culture type	Response	Medium	Reference
Hypocotyl	Root	Best rooting response with quality roots	Half strength Murashige and Skoog medium supplemented with IBA (2.5 mg L^{-1}) + NAA (0.5 mg L^{-1})	Singh et al. (2011)
Nodal segments	Maximum rooting	Number of roots/explant: 3.60 ± 0.51, root length (cm): 1.68 ± 0.32 and rooting response (%): 90%	Murashige and Skoog medium supplemented with 5.0 mg L^{-1} IBA	Bagum and Mathur (2014)
The Plant Growth Regulators used were paclobutrazol (PBZ), gibberellic acid (GA_3) and *Pseudomonas fluorescens* elicitors (PF Elicitors). The estimated alkaloids were ajmalicine, catharanthine, tabersonine, serpentine, and vindoline	Roots	The root vindoline contents increased with PBZ and PF Elicitors treatments but the decreased under GA_3 treatments when compared to control plants. In roots, the ajmalicine content increased significantly under all the treatments on all sampling days. The catharanthine contents increased with the age in control and growth regulator treatments, but the increase was not prominent and significant in PGR treatments when compared to controls. The serpentine contents of the plant increased with PGR treatments, but the increase was more prominent in PBZ treatments when compared to other treatments	The increase was in the order PBZ > PF Elicitors > GA_3	Jaleel et al. (2009)

showing an IC50 value of 4.06 μM (Wang et al. 2012). Meanwhile, vinblastine and vincristine showed antiparasitic effects against *Trypanosoma*, which causes trypanosomiasis in humans, inhibiting its mitosis and affecting cell shape in a dose-dependent manner (Grellier et al. 1999). Also, the use of 15 μM of vinblastine and 50 μM of vincristine inhibited cytokinesis and nuclear division. Consequently, the compounds affected the cell morphology, whereas the effect of 3 μM of vinblastine and 10 μM of vincristine inhibited the cytokinesis without affecting cell cycle progression.

Table 5 In vitro cultures of plantlets from *Catharanthus roseus*

Explant type	In vitro culture type	Response	Level production	References
Nodal explants of stem through axillary shoot proliferation	Shootlets	MS medium supplemented with 0.5 mg L^{-1} BAP ± 1 mg L^{-1} NAA	99% of shooting response	Mehta et al. (2013)
Multiple shoot cultures	Production of ajmalicine in shake flasks	Effect of different concentrations of IAA and BA in the production of ajmalicine	Murashige and Skoog medium supplemented with IAA at a low concentration and BA at a low concentration, accumulated high levels of ajmalicine	Satdive et al. (2003)
(a) Hypocotyl sections	(a) In vitro callus grown seedlings	MS supplemented with naphthaleneacetic acid (NAA 2 mg L^{-1}), 6-benzyl-aminopurine (BAP, 5 mg L^{-1}), casein hydrolysate (CH, 1000 mg L^{-1}), and asparagine (100 mg L^{-1}) for callus induction	Vinblastine Yield:(a) 0.9 μg g^{-1}	Datta and Srivastava (1997)
(b) Cotyledonary leaves	(b) In vitro callus grown seedlings		(b) 0.1 μg g^{-1}	
(c) Hypocotyl sections	(c) In vitro callus grown seedlings		(c) 1.6 μg g^{-1}	
(d) Immature fruits	(d) Mature plant		(d) 0.2 μg g^{-1}	

Table 6 New metabolites obtained from in vitro culture of *Catharanthus roseus* cells

Metabolite	Function	Type of culture	References
Phosphatidyl kinase	Phospholipid metabolite enzyme	Plasmatic membranes of *Catharanthus roseus* cell suspension cultures	Wissing et al. (1994)
Trichoselin	Antibiotic	Dual cultures of *Thrichoderma harzianum* and *Catharanthus roseus* callus	Marfori et al. (2002)
Phytic acid	Phosphorous storage, mRNA cellular export, chromatin remodeling	Cell suspension cultures	Mitsuhashi et al. (2005)
Cathachunine	The bisindole alkaloid cathachunine which lost C-18′ and C-19′ was isolated from *Catharanthus roseus*. It exerted a potent antitumor effect toward human leukemia cells through the induction of apoptosis via an intrinsic pathway	Dried whole plants	Xiao-Dong et al. (2016)

5.3 Hypoglycemic Activity

Several animal studies showed that ethanolic extracts of leaves and flowers decreased the levels of glucose in the blood (Ghosh and Gupta 1982; Chattopadhyay et al. 1991). In particular, the aqueous extract decreased the glucose levels in diabetic rats by 20%. This ratio is compared with the reduction of 49–58% in glucose (Singh et al. 2001), which was related to the dichloromethane and ethanolic extracts. Meanwhile, the hypoglycemic effects derived from an increased use of glucose in the liver (Chattopadhyay 1991).

Currently, research works have been conducted on new alkaloids in *Catharanthus roseus*. Some examples include that of the vindogentianine, a hypoglycemic metabolite extracted from the leaves of the plant. Works showed hypoglycemic activity in β-TC_6 and C2C12 cells by a higher glucose consumption, as well as significant in vitro inhibition. This suggests that the hypoglycemic activity of vindogentianine derives from the increased consumption of glucose, including the PTP-1B-type inhibiting effect, a potential therapeutic agent that fights diabetes mellitus type 2 (Huat et al. 2015).

5.4 Antidiarrheal Activity In Vivo

In vivo antidiarrheal activity, produced by ethanolic extracts of leaves and using castor oil as an inducing agent of experimental diarrhea, was tested in Wistar rats. For the same purpose, loperamide, and atropine were used as standard drugs. The antidiarrheal effect caused by ethanolic extract showed a dose-dependent inhibition of castor oil, which induced diarrhea at doses of 200 and 500 mg kg^{-1} (Kyakulaga et al. 2011).

5.5 Antimicrobial Activity

The antimicrobial activity of the leaf extracts was tested against various microorganisms such as *Pseudomonas*, *Salmonella,* and *Staphylococcus*, these extracts emerge as prophylactic agents in the treatment of various diseases (Patil and Ghosh 2010). Ramya et al. (2008) evaluated the in vitro antibacterial activity by use of crude extracts of *Catharanthus roseus*. The results indicated that the leaf extract that was prepared exhibited better antibacterial activity if compared with that of the extracts from other parts of the plant. Based on the above, the aqueous extracts of leaves, stems, roots, and flowers showed low growth of microorganisms, for example, *Escherichia coli*, *Pseudomonas aeruginosa*, *Serratia marcescens*, *Salmonella typhimurium*, *Staphylococcus aureus*, *Streptococcus pyogenes*, *Bacillus cereus* and *Bacillus subtilis* (Ramya et al. 2008).

Kumari and Gupta (2013) tested *Catharanthus roseus* leaf extracts (rosea variety), which showed excellent activity against *Aspergillus*. Meanwhile, extracts from alba variety stems showed maximum inhibitory activity against *Bacillus*, while rosea variety flowers indicated superior activity against *Bacillus* in the methanolic extract. In particular, the Minimum Inhibitory Concentration of the extract against microorganisms tested was found at the range between 100 and 20 mg mL^{-1}.

Acetonic, ethanolic, and chloroformic, which has been used for leaf extracts against pathogenic microorganisms were evaluated in a different study, which was to determine their antimicrobial potential. Thus, the ethanolic extract deployed maximum antibacterial activity when compared with acetone and chloroform extracts. Thus, it proved that *Staphylococcus* was the prominent, susceptible microorganism, being followed by *Escherichia coli*, *Pseudomonas* sp. and *Streptococcus* sp. (Shanmugaraju and Bhakyaraj 2016).

5.6 Antineoplastic Effect

Catharanthus roseus plants contain various dimeric indole alkaloids, which show antitumor activity significantly. In this regard, it was found that these alkaloids have apoptosis-inducing activity against in vitro and in vivo tumor cells. This is measured by the nuclear factor kappa enhancer of activated B cells, and from c-Jun N-terminal kinase routes, where damage on DNA and mitochondrial dysfunctions are present significantly.

The nuclear factor kappa B was discovered about 20 years ago. It was identified as a protein that has the ability to bind in order to improve the immunoglobulin light chain in B cells. The factor belongs to the family of transcription factors NF-kB, which is ubiquitous and participates in the inflammatory and immune response in the formation, development, progression, and apoptosis of tumors (Echeverri and Mockus 2008). Meanwhile, kinases c-Jun N-terminal bind and phosphor c-Jun protein in Ser-63 and -73 residues were also present in the transcriptional activation domain. These activated kinases, such as cytokines, react to stress stimulus, including UV irradiation and thermal and osmotic shocks. They participate both in the differentiation of T cells, and apoptotic processes (Echeverri and Mockus 2008).

Furthermore, different percentages of crude methanolic extracts have been identified. Metrics indicate the presence of significant anticancer activity against numerous cell types under in vitro conditions (Ueda et al. 2002), versus different types of multidrug-resistant tumors (Wang et al. 2004). Moreover, Ruskin and Aruna (2014), showed that the ethanolic extract of *Catharanthus roseus* has in vivo antitumor activity in the Ehrlich tumor model. In contrast, the in vitro study of ethanolic extract exhibited significant antitumor activity.

The aqueous extract obtained from the aerial parts caused a hypoglycemic activity in rabbits intragastrically at a dose of 10 g kg^{-1}, contrary to that produced by an aqueous-ethanolic extract at a dose of 5 g kg^{-1}. The hypoglycemic action was also observed in rats exposed to aqueous and ethanol extracts of the leaf of the plant. Vinblastine molecule is reported to be an effective component of certain cancer chemotherapy regimens, specifically when used with bleomycin, and methotrexate in VBM chemotherapy for Stage IA or IIA Hodgkin lymphomas. The inclusion of vinblastine allows for lower doses of bleomycin and reduced overall toxicity with larger resting periods between chemotherapy cycles (Gobbi et al. 2003).

5.7 *Toxicity*

Acute toxicity studies were performed in mice. Research indicated that the median lethal dose of intravenous vinblastine was 9.5 mg kg^{-1}, and vincristine intraperitoneally was 5.2 mg kg^{-1}. In this regard, 20% of mice died when they were administered the total fraction of alkaloids, which was obtained from the root of the plant subcutaneously, and provided in doses of 50 mg kg^{-1}. The ethanolic extract of the leaves was orally administered in daily doses of 75 mg kg^{-1} for 24 days. The application provoked a marked reduction in the weight of the animal's testicles and prostate on the 25th day as the autopsy revealed. Similarly, the action indicated antispermatogenic activity in rats derived from the fraction of total alkaloids of plant intraperitoneally administered.

References

Abdul JC (2009) Soil salinity regimes alters antioxidant enzyme activities in two varieties of *Catharanthus roseus*. Bot Res Int 2:64–68

Al-oubaidi HKM, Mohammed-Amin AS (2014) Effect of benzyl adenine on multiplication of *Catharanthus roseus* L. in vitro. World J Pharm Sci 3:2101–2107

Aslam J, Mujib A, Prasad SM (2014) Somatic embryos in *Catharanthus roseus*: a scanning electron microscopic study. Not Sci Biol 6:167–172

Ataei-Azimi A, Delnavaz HB, Ebrahimzadeh H et al (2008) High in vitro production of antcanceric indole alkaloids from periwinkle (*Catharanthus roseus*) tissue culture. Afr J Biotechnol 7:2834–2839

Bagum T, Mathur M (2014) In vitro regeneration of *Catharanthus roseus* and *Bacopa monnieri* and their survey around Jaipur district. Int J Pure App Biosci 2:210–221

Barrales-Cureño HJ (2015) Aplicaciones farmacológicas y producción biotecnológica in vitro de los alcaloides anticancerígenos de *Catharanthus roseus*. Biotecnol Appl 32:1101–1110

Barrales-Cureño HJ, Ramírez Sepúlveda MF (2013) Una revisión sobre la producción de taxoides anticancerígenos en cultivos in vitro de callos y células de *Taxus* spp. Rev Colomb Biotecnol 15:67–177

Barrales-Cureño HJ, Soto HRM (2012) Taxoides: metabolitos secundarios del árbol del tejo (*Taxus* spp.). Rev Chapingo Ser Cie 18:207–218

Barrales-Cureño HJ, HRM S, VAC R et al (2011) Extracción y cuantificación de taxoides por HPLC en hojas in situ y en callos inducidos in vitro de *Taxus globosa* Schlecht. Span J Rural Dev 2:103–114

Barrales-Cureño HJ, de la Rosa MCR, Villegas OS (2012) Hacia una genética celular del cáncer. Revista Científica y Tecnológica de la Universidad Veracruzana La Ciencia y el Hombre 25:1–4

Barrales-Cureño HJ, Castillo HFJ, Barros GLC (2015) El Paclitaxel. Bol Soc Quim Peru 29:7–10

Barrales-Cureño HJ, Farrera RA, Reyes RC et al (2016) Generalidades del fármaco Taxol: una revisión sistemática. Rev Méd Universidad Veracruzana 16:75–91

Calva CG, Pérez VJ (2005) Cultivo de células y tejidos vegetales: fuente de alimentos para el futuro. Rev Digit Universit 6:2–16

Chattopadhyay RR (1991) A comparative evaluation of some blood sugar lowering agents of plant origin. J Ethnopharmacol 67:367–372

Chattopadhyay RR, Sarkar SK, Ganguli S (1991) Hypoglycemic and antihyperglycemic effect of leaves of *Vinca rosea* Linn. Indian J Physiol Pharmacol 35:145–151

Datta A, Srivastava PS (1997) Variation in vinblastine production by *Catharanthus roseus* during in vivo and in vitro differentiation. Phytochemistry 46:135–137

Echeverri RNP, Mockus SI (2008) Factor nuclear κB (NF-κB): signalosoma y su importancia en enfermedades inflamatorias y cáncer. Rev Fac Med 56:133–146

El-Sayed M, Choi YH, Frédérich M et al (2004) Alkaloid accumulation in *Catharanthus roseus* cell suspension cultures fed with stemmadenine. Biotechnol Lett 26:793–798

Ghosh RK, Gupta I (1982) Effect of *Vinca rosea* and *Ficus racemososus* on hyperglycemia in rats. Indian J Anim Health 19:145–148

Gobbi MDPG, Broglia MDC, Merli F et al (2003) Vinblastine, bleomycin, and methotrexate chemotherapy plus irradiation for patients with early stage, favorable Hodgkin lymphoma. Cancer 98:2393–2401

Grellier P, Sinou V, Garrea de Loubresse N et al (1999) Selective and reversible effects of vinca alkaloids on *Trypanosoma cruzi* epimastigote forms: blockage of cytokinesis without inhibition of the organelle duplication. Cell Motil Cytoskeleton 42:36–47

Hanahan D, Weinberg RA (2000) The hallmarks of cancer. Cell 100:57–70

Haq R, Naz S, Aslam F et al (2013) Comparison of in vitro response of micropropagation and callogenesis of medicinal plant, *Vinca rosea* L. J Agric Res 51:9–17

Hernández-Domínguez E, Campos-Tamayo F, Vázquez-Flota F (2004) Vindoline synthesis in in vitro shoot cultures of *Catharanthus roseus*. Biotechnol Lett 26:671–674

Huat TS, Yeng LC, Arya A et al (2015) Vindogentianine, a hypoglycemic alkaloid from *Catharanthus roseus* (L.) G. Don (Apocynaceae). Fitoterapia 102:182–188

Jaleel CA, Gopi R, Gomathinayagam M et al (2009) Traditional and non-traditional plant growth regulators alters phytochemical constituents in *Catharanthus roseus*. Process Biochem 44:205–209

Kalidass C, Ramasamy MV, Daniel A (2010) Effect of auxin and cytokinin on vincristine production by callus cultures of *Catharanthus roseus* L. (Apocynaceae). Trop Subtrop Agroecosyst 12:283–288

Kumari K, Gupta S (2013) Phytopotential of *Catharanthus roseus* L. (G.) Don. var. Rosea and Alba against various pathogenic microbes in vitro. Int J Res Pure Appl Microbiol 3:77–82

Kyakulaga A, Hassan AT, Brenda VP (2011) In vivo antidiarrheal activity of the ethanolic leaf extract of *Catharanthus roseus* Linn. (Apocynaceae) in Wistar rats. Afr J Pharm Pharmacol 5:1797–1800

Marfori EC, Kajiyama S, Fukusaki E et al (2002) Trichosetin, a novel tetramic acid antibiotic produced in dual culture of *Trichoderma harzianum* and *Catharanthus roseus* Callus. Z Naturforsch C 57:465–470

Mehta J, Upadhyay D, Paras P et al (2013) Multiple shoots regeneration of (anticáncer plant) *Catharanthus roseus*—an importante medicinal plant. Am J Pharmtech Res 3:785–793

Mitsuhashi N, Ohnishi M, Sekiguchi Y et al (2005) Phytic acid synthesis and vacuolar accumulation in suspension-cultured cells of *Catharanthus roseus* induced by high concentration of inorganic phosphate and cations. Plant Physiol 138:1607–1614

Morgan JA, Barney CS, Penn AH et al (2000) Effects of buffered media upon growth and alkaloid production of *Catharanthus roseus* hairy roots. Appl Microbiol Biotechnol 53:262–265

Osorio SA, Mascorro GJO, Rodríguez de la O et al (2011) Crioconservación de ápices de crisantemo (*Dendranthema grandiflorum* Kitam) por encapsulación-deshidratación y por vitrificación. Rev Chapingo 17:33–43

Ozcelik B, Kartal M, Orhan I (2011) Cytotoxicity, antiviral and antimicrobial activities of alkaloids, flavonoids, and phenolics acids. Pharm Biol 49:396–402

Pandey S, Bahadur AN, Kanungo VK et al (2014) *In-vitro* propagation of a medicinal plant *Catharanthus roseus* L. (G.) Don. Indian J Life Sci 4:125–128

Pandiangan D, Tilaar W, Nainggolan N et al (2013) Relations between catharanthine content enhancement with the other associated secondary metabolites in *Catharanthus roseus* cell culture that treated tryptophan. Int J Sci Res 4:2208–2212

Patil PJ, Ghosh JS (2010) Antimicrobial activity of *Catharanthus roseus*—a detailed study. Br J Pharmacol Toxicol 1:40–44

Pérez-Alonso N, Jiménez E (2011) Producción de metabolitos secundarios de plantas mediante el cultivo in vitro. Biotecnología Vegetal 11:195–211

Ramya S, Govindaraji V, Navaneetha K et al (2008) In vitro evaluation of antibacterial activity using crude extracts of *Catharanthus roseus* L. (G.) Don. J Ethnobot Leaflets 12:1067–1072

Ruskin RS, Aruna SR (2014) *In-vitro* and *in-vivo* antitumor activity of *Catharanthus roseus*. Int Res J Pharm App Sci 4:1–4

Sandhya M, Deepti L, Bhakti D et al (2016) Effect of growth regulator combination on in-vitro regeneration of *Catharanthus roseus*. Int J Life Sci 6:1–4

Satdive RK, Fulzele DP, Eapen S (2003) Studies on production of ajmalicine in shake flasks by multiple shoot cultures of *Catharanthus roseus*. Biotechnol Prog 19:1071–1075

Shanmugaraju V, Bhakyaraj R (2016) Antimicrobial potential activity of leaf extracts of *Catharanthus roseus* against human pathogens under laboratory conditions. Int J Curr Res Biol Med 1:35–51

Singh R, Kharb P, Rani K (2011) Rapid micropropagation and callus induction of *Catharanthus roseus* in vitro using different explants. World J Agr Sci 7:699–704

Singh SN, Vats P, Suri S (2001) Effect of an antidiabetic extract of *Catharanthus roseus* on enzymic activities in streptozotocin induced diabetic rats. J Ethnopharmacol 76:269–277

Taha HS, El Bahr MK, Seif-El-Nasr MM (2009) In vitro studies on Egyptian *Catharanthus roseus* (L.) G. Don. IV: manipulation of some amino acids as precursors for enhanced of indole alkaloids production in suspension cultures. Aust J Basic Appl Sci 3:3137–3144

Ten Hoopen HJG, Van Gulik WM, Schlatmann JE, Moreno RHP, Vinke JL, Heijnen JJ, Verpoorte R (1994) Ajmalicine production by cell cultures of *Catharanthus roseus*: from shake flask to bioreactor. Plant Cell Tiss Org Cult 38:85–91

Ten Hoopen HJG, Vinke JL, Moreno PRH, Verpoorte R, Heijnen JJ (2002) Influence of temperature on growth and ajmalicine production by *Catharantus roseus* suspension cultures. Enzyme and Microbial Technology 30:56–65

Tom R, Jardin B, Chavarie C et al (1991) Effect of culture process on alkaloid production by *Catharanthus roseus* cells: I suspension cultures. J Biotechnol 21:1–19

Ueda JY, Tezuka Y, Banskota AH (2002) Antiproliferative activity of Vietnamese medicinal plants. Biol Pharm Bull 25:753–760

Verma AK, Singh RR, Singh S (2012) Improved alkaloid content in callus culture of *Catharanthus roseus*. Bot Serb 36:123–130

Villa-Ruano N, Pacheco-Hernández Y, Lara-Zaragoza EB et al (2011) Biotecnologia de plantas medicinales: generando fármacos de un futuro tornado presente. Temas de ciencia y tecnología 15:13–20

Wang L, He HP, Di YT (2012) Catharoseumine, a new monoterpenoid indole alkaloid possessing a peroxy bridge from *Catharanthus roseus*. Tetrahedron Lett 53:1576–1578

Wang S, Zheng Z, Weng Y (2004) Angiogenesis and anti-angiogenesis activity of Chinese medicinal herbal extracts. Life Sci 74:2467–2478

Wissing JB, Kornak B, Funke A (1994) Phosphatidate kinase. A novel enzyme in phospholipid metabolism (characterization of the enzyme from suspension-cultured *Catharanthus roseus* cells). Plant Physiol 105:903–909

Xiao-Dong W, Chen-Yang L, Miao-Miao J et al (2016) Induction of apoptosis in human leukemia cells through an intrinsic pathway by cathachunine, a unique alkaloid isolated from *Catharanthus roseus*. Phytomedicine 23:641–653

Zhao J, Wei-Hua Z, Hu Q et al (2001) Compact callus cluster suspension cultures of *Catharanthus roseus* with enhanced indole alkaloid biosynthesis. In Vitro Cell Dev Biol 37:68–72

Zhi-Gang G, Liu Y, Mei-Zheng G et al (2013) Regulation of vinblastine biosynthesis in cell suspension cultures of *Catharanthus roseus*. Plant Cell Tiss Org Cult 112:43–54

Zuwairi SM, Rianika MN, Pomahocová B et al (2014) Analysis of metabolites in the terpenoid pathway of *Catharanthus roseus* cell suspension. Plant Cell Tiss Org Cult 117:225–239

Plant Efficacy and Alkaloids Production in Sadabahar (*Catharanthus roseus* L.): Role of Potent PGRs and Mineral Nutrients

M. Naeem, Tariq Aftab, Mohd. Idrees, M. Masidur Alam, M. Masroor A. Khan, and Moin Uddin

Abstract *Catharanthus roseus* (L.) G. Don (Apocynaceae) is a medicinal plant that produces indole alkaloids used in cancer chemotherapy. *C. roseus* has been the most commonly used model plant for the study of the biosynthetic pathways regarding indole alkaloids. Most important anticancer alkaloids of *C. roseus* are vincristine and vinblastine. These alkaloids are extracted commercially from large amount of leaf-biomass of *C. roseus* plants. The concentration of these alkaloids is very low (about 0.005%) in this plant and the artificial synthesis is very expensive and cumbersome. At present, India is the third largest manufacturer of vinblastine and vincristine in the world and is exporting these alkaloids to European and other countries. High demand and low yield of these alkaloids in the plant has led to conduct researches in order to explore for alternative means of their production. Great efforts have been made to produce these alkaloids at large scale by cultures of plant cell suspensions and diverse tissues (such as hairy roots). With the same aim, the role of plant growth regulators (PGRs) in regulation of *C. roseus* indole alkaloids biosynthesis has been extensively researched. In fact, there is immense need of enhancing the production of these medicinally important alkaloids in view of their massive demand worldwide.

The present review provides information regarding the role of potent PGRs [namely, gibberellic acid (GA_3), epibrassinolide (EBL), kinetin (Kn), and triaconta-

M. Naeem (✉) • T. Aftab • M.M.A. Khan
Plant Physiology Section, Department of Botany, Aligarh Muslim University, Aligarh, Uttar Pradesh 202002, India
e-mail: naeemgaur@gmail.com

M. Idrees
Department of Mathematics and Sciences, College of Arts and Applied Sciences, Dhofar University, P. O. Box 2509, Salalah 211, Sultanate of Oman

M.M. Alam
Department of Botany, Malda College, Malda, West Begal, India

Plant Physiology Section, Department of Botany, Aligarh Muslim University, Aligarh, Uttar Pradesh 202002, India

M. Uddin
Botany Section, Women's College, Aligarh Muslim University, Aligarh, Uttar Pradesh 202002, India

M. Naeem et al. (eds.), *Catharanthus roseus*,
DOI 10.1007/978-3-319-51620-2_3

nol (TRIA)], mineral nutrients (such as N, P, and K) and radiation-processed polysaccharides (sodium alginate and carrageenan) in boosting up the growth, metabolism, and other plant processes, particularly the production of anticancer alkaloids (vinblastine and vincristine) in *C. roseus* plants.

Keywords Alkaloids • *Catharanthus roseus* • Mineral nutrients • PGRs • Radiation-processed polysaccharides • Vinblastine • Vincristine

1 Introduction

Periwinkle [*Catharanthus roseus* (L.) G. Don], called as Sadabahar, is a prominent medicinal plant of Apocynaceae family that produces indole alkaloids used in chemotherapy. Commercially important antineoplastic alkaloids, namely, vinblastine and vincristine, are mainly present in the leaves of *C. roseus*. Periwinkle is a perennial tropical plant that produces more than 100 monoterpenoid indole alkaloids including the commercially important antitumor alkaloids, viz. vinblastine and vincristine; these alkaloids are used in cancer chemotherapy (Magnotta et al. 2006). In addition to cancer chemotherapy, the alkaloids vinblastine and vincristine are used in the treatment of various types of lymphoma and leukemia ailments (Svoboda and Blake 1975). These alkaloids are also used for the treatment of other malignant as well as nonmalignant diseases and disorders associated with platelets (Nwafor et al. 2001). *C. roseus* produces a number of biologically active alkaloids in different plant parts using complex, multi-step and strictly regulated terpenoid indole alkaloid (TIA) biosynthetic pathways (Vazquez-Flota et al. 1997; Gantet and Memelink 2002). Distribution and accumulation of alkaloids vary in roots, stems, and leaves of *C. roseus* plant (Misra and Kumar 2000).

It was reported that a number of sugars, amino acids, and organic acids may act as signals for alkaloid production that possibly cause the biotransformations (Aerts et al. 1996). Production of secondary metabolites, in general, is a highly ordered process with respect to plant development and involves a wide array of gene expression which, in turn, controls the regulation of biochemical pathways within the organs, specific cell types, and specific organelles. De Luca and Laflamme (2001) reported that alkaloid biosynthesis in particular is regulated primarily at the level of gene expression, which is controlled by the expression of transcription factors. Moreover, the activity of regulator/transcription factors is determined by their abundance at the level of gene also. The biosynthesis of these alkaloids is regulated by diverse biomolecules such as plant signaling molecules, plant growth regulators, prenylated proteins, and transcription factors (Zhou et al. 2010).

Nowadays, agronomists are trying to develop additional ways to break the present yield limits of the field crop to fill the gap between productivity and demand. DiCosmo and Misawa (1995) and Aslam et al. (2010) suggested that cultivation of this plant on scientific lines is the only commercial source for the production of drugs. *C. roseus,* as a tropical plant, needs relatively high temperature during the period of growth and development. In general, it is very sensitive even to a short-term drop of temperature (e.g., at night). The decrease of temperature caused flower bud (in spring) or fruit (silique) drop in autumn (Narkiewicz and Sadowska 1991). In fact, at 10 °C the inhibition of plant growth was also reported (Mitrev 1976). The air temperature should be about 25 °C at the beginning of growth. So, to increase fresh herb in the temperate climatic zone, the economically most efficient way is to grow it either in an open field or in unheated plastic tunnels (Sadowska et al. 1987). Removal of flower buds, flower, and apices of a plant at 2 cm below the top, at the phase of 50% of full bloom increased both root yield and alkaloid content (Pareek et al. 1985, 1991). Remarkable results were obtained by applying hydroponic technique under greenhouse conditions; the yield of alkaloids was several times higher, as compared to plants grown in the open ground (Dedio 1983; Babakhanyan 1991). However, this method is not economically attractive for the large-scale production.

Lovkova et al. (2005) studied the effect of various mineral elements, natural and synthetic auxins, cytokinins, and gibberellins on biosynthesis and accumulation of indole alkaloids in Madagascar periwinkle-seedlings; Zn and Auxin were shown to modulate various stages in the biosynthesis of monomeric indole alkaloids (catharanthine and vindoline). Foliar application of GA treatment (1000 g m^{-3}) on *C. roseus* leaf produced negative phenotypic response in total biomass production but it showed positive response in the content of total alkaloids in leaf, stem, and roots (Srivastava and Srivastava 2007; Alam et al. 2012). Furthermore, Jaleel et al. (2007) reported GA_3 to enhance ajmalicine accumulation in periwinkle roots. Root vindoline content was significantly increased with paclobutrazol (PBZ) and *Pseudomonas fluorescens* (PF) elicitor's treatment but decreased under GA_3 treatment when compared to control plants (Jaleel et al. 2009). However, higher concentration of salicylic acid (SA) induced vindoline accumulation in *C. roseus* seedlings (El-Sayed and Verpoorte 2004).

Due to paramount pharmaceutical importance and the low content of useful alkaloids in the plant, *C. roseus* became an important model plant for biotechnological studies on plant secondary metabolism. Exogenous application of plant growth regulators (PGRs) and mineral nutrients have vital impact on the growth and productivity of *C. roseus*. In view of the desired production of alkaloids, extensive work has been carried out at Aligarh, Western Uttar Pradesh, with an aim to study the effect of potent PGRs on *C. roseus* (Table 1). Concludingly, the present review is focused on the effect of several potent PGRs and mineral nutrients with regard to growth, metabolism, and other plant processes including alkaloid production of *C. roseus.*

2 Role of Plant Growth Regulators (PGRs) in Plants

A plant growth regulator is an organic compound, either natural or synthetic, that modifies or controls one or more specific physiological processes within a plant. If the compound is produced within the plant, it is called a plant hormone. Plant

Table 1 Effect of potent PGRs on growth attributes, physiological activities, and content and yield of alkaloids of *C. roseus*

PGRs	Growth attributes	Physiological activities	Content and yield of alkaloids	Reference
Gibberellic Acid (GA_3)	Fresh and dry weights of shoot and root, LAI	Total content of Chl and carotenoids, photosynthetic rate (P_N), stomatal conductance (gs), NR and CA activities, antioxidative enzymes	Contents and yield of total alkaloids, vincristine and vinblastine	Idrees et al. (2011), Alam et al. (2012), Alam (2013)
Triacontanol (TRIA)	Fresh and dry weights of shoot and root, LAI, leaf- and herbage-yield	P_N, gs, NR and CA activities, and antioxidative enzymes	Contents and yield of total alkaloids, vincristine and vinblastine	Idrees et al. (2011), Alam et al. (2012), Alam (2013), Naeem et al. (2017)
Salicylic acid (SA)	Plant height, fresh and dry weights of shoot and root, and LAI	P_N, gs, NR and CA activities and antioxidative enzymes.	Contents and yield of total alkaloids, vincristine and vinblastine	Idrees et al. (2011)
Epibrassinolide (EBL)	Average leaf area, LAI, dry weights of leaf, stem and root	Total content of Chl and carotenoids, NR and CA activities, leaf-N, -P, and -K contents	Total content and yield of alkaloids	Alam et al. (2016)
Kinetin (Kn)	Average leaf area, LAI, dry weights of leaf, stem and root	Total content of Chl and carotenoids, NR and CA activities, leaf-N, -P, and -K contents	Total content and yield of alkaloids	Alam (2013)

Chl chlorophyll, *NR* nitrate reductase, *CA* carbonic anhydrase

hormones or phytohormones are other than nutrients, synthesized at specific site/s and transported to other tissues, wherein very low concentrations they evoke specific biochemical, physiological, and/or morphological responses. They are active both at the site of their synthesis and at remote places.

The plant hormones play extremely important role in the integration of developmental activities. They are also very much concerned with the response of plants to the external physical environment. These factors often exert inductive effects by evoking changes in hormonal metabolism and/or their distribution within the plant (Moore 1989). Earlier, major lines of investigations led to the characterization of the following five groups of classical plant hormones (auxins, gibberellins, cytokinins, abscisic acid, and ethylene). It is now superseded by a view that various other molecules, remotely related with the above hormones, have varied important roles in the regulation of plant activities; these molecules include brassinosteroids, salicylates, jasmonates, etc. The phytohormones are employed to improve the efficiency of plants. They are known to affect the chlorophyll content and the photosynthetic rate of the plants and regulate the partitioning of assimilates resulting in an increase or decrease of the plant size (Arteca 1997). However, the response in terms of chlorophyll and/or photosynthesis varies from a group of plants to another group. The regulatory role of phytohormones in selective uptake and distribution of ions in plants, through their effect on membrane properties and transport of assimilates, has been shown in earlier studies (Steveninck 1976).

2.1 *Gibberellins*

Gibberellins are defined as compounds having an ent-gibberellane skeleton. They show biological activity in stimulating cell division or cell elongation or both, or such other biological activity as may be specifically associated with this type of naturally occurring substance (Paleg 1965). Of the 136 naturally occurring GAs (MacMillan 2002), that have been identified to date, in plants or fungi, relatively few are thought to possess intrinsic biological activity. GA_3 is the first widely available active form of commercial gibberellins. However, the other recognized bioactive gibberellins are GA_1, GA_4, and GA_7.

2.1.1 Effect of GA_3

GA_3 plays an important role in modulating diverse physiological processes throughout the period of plant growth and development. It is known to improve the photosynthetic efficiency through its influence on photosynthetic enzymes, leaf-area index, light interception, and nutrient-use efficiency of plants (Khan et al. 2007a). Besides, GA_3 is the best known growth hormone for seed germination, leaf expansion, stem elongation, flower and trichome initiation, and flower and fruit development (Yamaguchi 2008).

As per different studies, GA_3 increased the use efficiency of nutrients. Eid and Abou-Leila (2006) reported that GA_3 treatment increased the N, P, K, Mg, Fe, Zn, Mn, and Cu content, thereby increasing the mineral nutrient status of the plant. In fact, an adequate supply of mineral nutrients (N, P, and K) in the initial stage of plant growth and development plays a pivotal role in nutrient uptake by crops, which is improved by GA_3 application as well (Shah et al. 2006). An increase in membrane permeability by GA_3 (Wood and Paleg 1972; Crozier and Turnbull 1984; Al-Wakeel et al. 1994) might be ascribed to GA_3-facilitated absorption and utilization of mineral nutrients and transport of assimilates in plants (Khan et al. 2002; Al-Rumaih et al. 2003). The increased nutrient content might, in turn, enhance the photosynthetic potential of leaves, i.e., the strength of source. GA_3 has been reported to increase the photosynthetic rate in plant leaves (Ashraf et al.

2002). It was suggested that GA_3 treatment could lead to changes in plastid development and chloroplast structure too (Wellburn et al. 1973). Augmentation of leaf-chlorophyll content due to GA_3 application was also observed in *Lemna trisulca* L. (Bata and Nešković 1974), mustard (Afroz et al. 2005; Shah 2007a), *Stevia rebaudiana* Bertoni (Modi Arpan et al. 2011), and *Hibiscus sabdariffa* L. (Ali et al. 2012).

Nitrate reducing power of the plants, as depicted by nitrate reductase (NR) activity, is one of the important factors determining the growth. However, the process of nitrate reduction depends on three important factors (a) substrate (NO_3) level in the cytoplasm, (b) the level of functional NR, and/or (c) the activity level of functional NR. Each of these factors is directly or indirectly dependent on the metabolic sensors and/or signal transducers (Campbell 1999). Moreover, the major rate limiting step in the whole process of nitrate reduction is the reduction of nitrate to nitrite (Salisbury and Ross 1992), which is catalyzed by NR. The level of the enzyme is dependent on a number of factors born within or outside the plant. GA_3 was proved as inducer of NR activity in earlier findings (Goupil et al. 1998; Shah et al. 2006; Aftab et al. 2010a). Another enzyme, carbonic anhydrase (CA), catalyzes the reversible hydration of carbon dioxide and maintains its constant supply to RuBPCase at the level of chloroplast-grana (Majeau and Coleman 1994; Price et al. 1994). Moreover, CA is hypothesized to be involved in photosynthetic electron transport system (Stemler 1997) and is considered to maintain chloroplast pH during the rapid changes in light intensity (Reed and Graham 1981).

It is in general agreement that GA_3 enhances the metabolic activity within the pathways leading to accumulation of secondary metabolites, e.g., steroids (Ohlsson and Björk 1988), anthocyanin (Weiss et al. 1992), essential oil (terpenoid) (Singh et al. 1999), artemisinin (sesquiterpene) (Aftab et al. 2010b), and alkaloids in opium poppy (Khan et al. 2007b) and *Balanites aegyptiaca* plants (Mostafa and Alhamd 2011).

GA_3 may influence the source–sink relationship by affecting various plant processes (Marschner 1995). In higher plants, gibberellins appear to de-repress the signaling pathway by inducing proteolysis of GA-signaling repressors (DELLA proteins). It is suggested that the proteolysis of repressor proteins is an important hormone signal transduction mechanism in plants. However unlike most epigenetic signaling events, such as a phosphotransfer, proteolysis is irreversible and therefore enforces directionality on a system. Such directionality is important in controlling the mitotic cell cycle (Patton et al. 1998). By analogy, it could be argued that changes in gibberellins concentration in the levels during plant development should be followed by irreversible development of plants via the cell division and expansion. In plants, the activities of these growth determinants (gibberellins) are maximal at a crucial growth stage when the requirement of photosynthates is too high. When the plants grow old, the biomass yet increases even with decreasing activities of various cellular processes. The fact was revealed by the studied growth observations after foliar supplementation with GA_3.

2.2 Effect of Epibrassinolide

Brassinosteroids (BRs) are cholestane derivatives with significant growth-promoting activity. Brassinolide (BL) is considered to be the end-product of biosynthetic pathway of BRs and shows the highest biological activity among BRs (Grove et al. 1979). BRs are considered as hormones with pleiotropic effects, as they influence various developmental processes like plant growth, germination of seeds, rhizogenesis, flowering and senescence, etc. (Rao et al. 2002). Besides, BRs elicit various physiological responses and are essential for male fertility and xylem differentiation (Müssig and Altmann 1999; Altmann 1999). Their growth-promoting effect results primarily from the stimulation of cell elongation and includes induction of the expression of genes encoding proteins such as xyloglucan endotransglycosylases (XETs) (Xu et al. 1995), which are probably involved in cell-wall metabolism and loosening. Moreover, effects such as cell-wall space acidification appear to contribute to BRs-induced growth stimulation. Effects of BRs on cell division are less clear; however, the induction of CycD3 transcription by epibrassinolide (EBL) might represent a mechanism by which BRs can drive cell division (Hu et al. 2004). Though our understanding of the molecular mechanism of action of brassinosteroids is still in its infancy stage, few years ago, a leucine-rich protein (BRI1) from *Arabidopsis thaliana* has been identified, which is considered as the receptor of BRs (Li and Chory 1997). Unlike in animal system, where receptors for steroid hormones are intracellular, the receptor of BRs (BRI1) is located in the plasma membrane. It functions at the cell surface and transduces extracellular signals (Clouse and Sasse 1998). Further, the binding of BR molecule to the receptor causes activation of the kinase domain and the subsequent phosphorylation of additional kinases and/or phosphatases (Chow and McCourt 2006). There are evidences, that BRs do not undergo long-distance transport (Symons et al. 2006), however, they may influence long-distance signaling by altering auxin transport (Paponov et al. 2005; Wiśniewska et al. 2006). Probably, after perception in the leaves, the signals of BRs might interfere with the physiological activities within the leaves of the plant. The positive effects of EBL were also suggested in the cases of *Pinus banksiana* (Rajasekaran and Blake 1998), *Arachis hypogaea* (Vardhini and Rao 1998), pepper plants (Houimli et al. 2010), *Phaseolus vulgaris* L. (Rady 2011), *Pisum sativum* L. (Shahid et al. 2011), and *Brassica juncea* L. (Arora et al. 2012). The biosynthesis of alkaloids is an integration of several metabolic pathways, which requires linking of several steps such as continuous production of precursors, their transport and translocation to the active site of synthesis, and finally the transport of alkaloids to accumulation site. This sequence of steps depends on normal functioning of associated metabolic pathways. Any disorder in normal metabolic pathways also affects the sequence of steps in biosynthesis. Accordingly, a plant may alter/adopt its specific metabolic pathway in response to particular effect, such as nutrient imbalance and hormone application. BL and EBL are potent growth regulators that have proved efficient in altering the metabolic pathway in the case of several medicinal plants such as lavender (Youssef and Talaat 1998), mint (Maia et al. 2004; Naeem et al. 2012b), and geranium (Swamy and Rao 2009).

2.3 *Effect of Kinetin (Kn)*

Cytokinins play a central role in the regulation of plant cell division and numerous developmental processes. A variety of additional activities of the hormone were described, including the capability to induce the formation of shoots from unorganized callus tissue (Skoog and Miller 1957), to retard leaf senescence (Richmond and Lang 1957), to stimulate pigment accumulation (Bamberger and Mayer 1960), and to support plastid development (Stetler and Laetsch 1965). Generally, natural cytokinins are N6-substituted adenine derivatives. The first cytokinin discovered was N^6-furfuryladenine (kinetin) as a degradation product of DNA. In plants, cytokinins are perceived by histidine kinases and transduced by a two-component signaling system. In *Arabidopsis* cytokinin signal transduction pathway, hybrid histidine protein kinases (AHKs) serve as cytokinin receptors and histidine phosphotransfer proteins (AHPs) transmit the signal from AHKs to nuclear response regulator (ARRs), which can activate or repress transcription (Inoue et al. 2001). Histidine phosphotransfer proteins act as signaling shuttles between the cytoplasm and nucleus in a cytokinin-dependent manner. However, the identification of CRE1 (a histidine kinase identical to AHK4 and WOL) as the cytokinin receptor of *Arabidopsis* is a landmark in cytokinin research (Schmülling 2001). Moreover, the cytokinin receptor genes have been identified in numerous other flowering plant species including *Catharanthus roseus* (Papon et al. 2002).

Kinetin (Kn) is an artificial cytokinin which has been frequently used in plant research. Growth-promoting effect of Kn application on various growth attributes including leaf expansion and organ biomass production was reported by various researchers (Dey and Srivastava 2006a; Singh et al. 2006; Shah 2007b). van Staden et al. (1988) proposed that cytokinins boosted the general metabolism of the chloroplast through their action on the related processes operative at the level of the nucleus and/or cytoplasm; the cytokinins facilitated an increase in chloroplast DNA, augmented the rate of protein synthesis in the chloroplast, maintained sufficient pigment level and promoted the formation of chloroplast-grana.

In fact, the partitioning of the *C. roseus* alkaloids in various organs is a complex and organ-specific process (Misra and Kumar 2000). The growth regulators may exert their effect on alkaloid accumulation through improvement in the alkaloid precursor pool and/or through an enhancement of enzyme activities taking part in alkaloid biosynthesis (Srivastava and Srivastava 2011). The precursors of root-specific alkaloids or the enzymes involved in the biosynthesis of root alkaloids might anyway be influenced by the cytokinin signaling system (Mujib et al. 2012).

3 Effect of PGRs on the Growth and Productivity of *C. roseus*

3.1 *Growth responses*

The research work regarding GA_3 application on performance of *C. roseus* L. has been reported by various researchers (Srivastava and Srivastava 2007; Jaleel et al. 2009; Misra et al. 2009; Alam et al. 2012). Jaleel et al. (2008) investigated the effect of different plant growth regulators (PGRs) and fungicide treatments on the growth characteristics of *C. roseus*. They used PGRs like paclobutrazol (PBZ) and gibberellic acid (GA_3) through soil drenching on 38, 53, 68, and 83 days after planting (DAP). The total height of the *C. roseus* plants increased with the age in the control and gibberellic acid, but it decreased significantly under PBZ treatments.

The effect of two well-known plant growth regulators, namely, GA_3 and TRIA on growth and biochemical parameters of *C. roseus* was investigated by Idrees et al. (2011c) under normal and salt stress conditions. Two pot culture experiments were conducted to find out the effect of foliar spray of GA_3 and TRIA on crop productivity and production of indole alkaloids (vinblastine and vincristine contents) of *C. roseus* under salt stress. In another experiment, foliar spray of GA_3 and TRIA exhibited significant effect to improve the growth parameters (fresh and dry weights of shoot and root) of C. *roseus*. Alam et al. (2012) reported the significant effect of various PGRs, viz. indole-3-acetic acid (IAA), indole-3-butyric acid (IBA), naphthalene acetic acid (NAA), 6-benzylaminopurine (BAP), kinetin (Kn), thidiazuron (TDZ), gibberellic acid (GA3), salicylic acid (SA), homobrassinosteroids (HBR), and triacontanol (TRIA) at the rate of 10^{-7} M on growth parameters of two *C. roseus* varieties (Alba and Rosea). Alam (2013) did extensive work in his research study on the effect of potent PGRs (GA_3, TRIA, EBL, and Kn) on the productivity of *C. roseus*. Naeem et al. (2017) in his study also reported promotive effect of TRIA on growth attributes of *C. roseus*.

3.2 Alkaloids Production

The effects of plant growth regulators (PGRs) on the terpenoid indole alkaloids (TIAs) of *C. roseus* has been extensively studied (El-Sayed and Verpoorte 2007; Zhao and Verpoorte 2007; Pan et al. 2010). PGRs, such as methyl jasmonate (MeJA) and jasmonate (El-Sayed and Verpoorte 2005; Ruiz-May et al. 2008; Peebles et al. 2005), abscisic acid (ABA), salicylic acid (SA) (Bulgakov et al. 2002; Mustafa et al. 2009), and gibberellic acid (GA_3) (Verpoorte et al. 1999; Srivastava and Srivastava 2007; Amini et al. 2009; Alam et al. 2012) showed significant influence on TIAs production and enzymes activities of the biosynthesis pathways in *C. roseus* cell suspensions cultures, hairy roots, and seedlings (El-Sayed and Verpoorte 2004; Ruiz-May et al. 2008). Previous researches about the effects of PGRs on TIAs in *C. roseus* were mostly focused on the production of ajmalicine, serpentine, tabersonine, ajmaline, vindoline, and catharanthine (El-Sayed and Verpoorte 2005; Zhao and Verpoorte 2007; Amini et al. 2009; Mustafa et al. 2009; Peebles et al. 2005). Moreover, Satdive et al. (2003) studied the effect of phytohormones (IAA, BA, NAA, and Kinetin,) on growth rate and ajmalicine production by multiple shoot cultures of *C. roseus*. However, there are few reports about the effect of PGRs on production of vinblastine (Pan et al. 2010; Xing et al. 2011).

3.3 Work Done at Aligarh on *C. roseus* in Response to PGRs

In a study of Idrees et al. (2011a), foliar application of SA (10^{-5} M) reduced the damaging effect of salinity on plant growth, thus accelerating the restoration of growth processes. SA not only improved the growth parameters (plant height, leaf-area index, fresh and dry weights of shoot and root) and physiological attributes (photosynthetic parameters, activities of nitrate reductase and carbonic anhydrase, ascorbic acid, and antioxidant enzymes) but also reversed the effects of salinity. Total alkaloids content was improved by SA application both in unstressed and stressed plants. According to their study, the highest level of total alkaloids content recorded in leaves of SA-treated stressed plants was 11.1%. Foliar spray of SA overcame the adverse effect of salinity by improving the content of vincristine (14%) and vinblastine (15%) in plants treated with 100 M NaCl.

Idrees et al. (2011b) conducted another pot experiment to find out the effect of foliar spray of triacontanol (TRIA) in order to ameliorate the adverse effects of salt stress on periwinkle. Foliar application of TRIA (10^{-5} M) reduced the damaging effect of stress on plant growth attributes including plant height, leaf-area index, shoot and root fresh weights, shoot and root dry weights, thus accelerating the restoration of growth processes. It not only improved the growth parameters but also partially reversed the effects of salt stress. Total alkaloids content was improved by TRIA application both in stressed and normal plants. Foliar spray of TRIA not only overcame the adverse effect of stress but also improved the content of vinblastine and vincristine in salt-depressed plants.

Further, Idrees et al. (2013) conducted an experiment on periwinkle (*C. roseus* L.). In their study, 30-days-old plants of *C. roseus* were supplied with different treatments comprising basal application of nickel (0, 50, 100, and 150 mg kg^{-1}) and foliar application of SA (0 and 10^{-5} M). Total alkaloid content was significantly declined in Ni-treated plants. Foliar application of SA reduced the deleterious effects of Ni on plant growth, accelerating the restoration of growth and physiological processes. They reported that SA also improved the total alkaloid content under normal as well as adverse conditions. Foliar spray of SA significantly improved the content of anticancer alkaloids vincristine (by 22%) and vinblastine (by 50%) in plants treated with 150 mg kg^{-1} of Ni.

According to Alam et al. (2012), the plants of two cultivars (Rosea and Alba) of *C. roseus* were sprayed with various PGRs, viz. IAA, IBA, NAA, BAP, KIN, TDZ, GA_3, SA, HBR, and TRIA at the rate of 10^{-7} M at 60 days after planting (DAP). Cultivar Rosea gave higher yield of foliage and roots and that of alkaloids compared to Alba (Idrees et al. 2010). Their results showed that HBR significantly improved most of the growth attributes. Application of HBR, KIN, and GA_3 resulted in the ameliorative effects on plant productivity as well as on physiological and biochemical parameters as compared to the unsprayed (control) plants. However, the effect of TDZ was not significantly different than the control. GA_3 application significantly increased the vincristine content (7.3%) while TDZ exhibited reduced the vincristine content. The effect of other PGRs was insignificant regarding vincristine and vinblastine contents. The response of Rosea toward exogenous PGRs application was higher than Alba in terms of crop productivity, physiological and biochemical parameters, and alkaloid production.

Foliar application of GA_3 stimulated various growth characteristics, viz. number of leaves, average leaf area, leaf area index, fresh and dry weights of leaves and stems and roots per plant at 6 and 9 MAP (months after planting) (Alam 2013). Also EBL-treated plants showed considerable improvement in growth attributes. EBL also improved the growth attributes of the plants in comparison to those treated with GA_3 and kinetin (Alam 2013). Application of GA_3 also increased the contents of total chlorophyll and carotenoids of *C. roseus* plants significantly at both the stages (Alam 2013). Further, he also reported that the activities of NR and CA in the leaves were significantly increased as a result of foliar application of GA_3. Moreover, it was revealed that foliar application of EBL influenced most of the biochemical parameter significantly. However, EBL application did not affect the content of total carotenoids and phosphorus in the leaves (at 6 months after planting). Rest of the parameters, viz. leaf chlorophyll content, NR and CA activity and the content of leaf-N and -P were notably enhanced by EBL application at both the growth stages (6 and 9 MAP). The foliar application of kinetin was effective to increase content of leaf-N, -P, and -K as compared to the untreated plants. Further, in his investigation, the leaves supplemented with kinetin possessed a larger quantity of chlorophyll than those sprayed with water (control). Besides, the carotenoids were also enhanced due to Kn application. The level of enzyme activity like CA activity in leaves was significantly affected while leaf-NR activity was affected only at 90 MAP by Kn application.

Additionally, Alam (2013) studied the effect of GA_3 on the leaf content of *C. roseus* alkaloids; in addition, he noted the positive effect of GA_3 spray on the yield of different plant parts like leaf, stem, and root. The content of total alkaloids in all plant parts was significantly increased by the application of GA_3 treatments. The yield of total alkaloids in the plant parts was augmented accordingly owing to significant increase in total herbage-yield of the plant. Further, he examined that EBL application positively influenced the total content and yield of leaf and root alkaloids significantly.

Kinetin-treated plants showed no significant influence in leaf and stem alkaloid content compared to the control plants (Alam 2013). However, Kn supplementation enhanced the root alkaloids significantly. However, the yield of leaf, stem, and root alkaloids were augmented significantly through Kn application at both the growth stages (6 and 9 MAP). Various EBL concentrations positively influenced the leaf and root alkaloid contents significantly. The yield of leaf and root alkaloids per plant was also affected significantly by EBL application.

Alam et al. (2016) discovered the promotive effect of EBL on plant growth, physiological activities, and production of total alkaloids in leaves, stems, and roots in *C. roseus* at 6 and 9 months after planting (MAP). Five concentrations of EBL (10^{-0}, 10^{-8}, 10^{-7}, 10^{-6}, and 10^{-5} M) were applied on *C. roseus*. They found out that EBL at 10^{-7} M significantly increased several growth, physiological and biochemical parameter at 6 and 9 MAP. This concentration also increased the leaf and root alkaloid content/yield at both the stages.

Naeem et al. (2017) carried out a pot culture experiment to explore the effect of triacontanol (TRIA) on plant growth, physiology, and production of total alkaloids (including anticancer alkaloids) in *C. roseus* at 120 and 150 days after planting. Four concentrations of TRIA (10^{-0}, 10^{-7}, 10^{-6}, and 10^{-5} M) were tested through foliar spray. TRIA at 10^{-6} M significantly increased the attributes related to plant growth, physiological and biochemical processes and yield. As compared to the control (10^{-0} M), leaf-applied TRIA at 10^{-6} M improved the production (yield) of anti-cancerous alkaloids vinblastine (+71.6%) and vincristine (+73.1%) and caused to maintain the highest content and yield of vindoline.

3.4 Radiation-Processed Natural Polysaccharides

Sodium alginate is a natural polysaccharide that is largely obtained from the brown algae (*Sargassum* sp.). Commercial varieties of alginate are extracted from seaweeds, including the giant kelp *Macrocystis pyrifera*, *Ascophyllum nodosum*, *Sargassum sinicola*, and various species of *Laminaria* (Day 1998). The chemical compound, sodium alginate is the sodium salt of alginic acid. Its empirical formula is $NaC_6H_7O_6$. It has combined features of linear copolymers of L-guluronic acid and D-mannuronic acid units (Xu et al. 2006). Carrageenan is sulfated anionic polymer of red seaweed that comprises the main structural polysaccharides (Rhodophyceae). They are composed of d-galactose units linked alternately with α-1,4 and β-1,3 linkages. They are mixtures of water-soluble, linear, and sulfated galactans. Recent researches show that radiation-processed polysaccharides (RPPs) proved potent plant growth promoting activity.

Gamma rays irradiation is employed to degrade and lower down the molecular weight of some natural polysaccharides like aliginate, chitosan, and carrageenan into small sized oligomers of comparatively low molecular weight. These oligomers, when applied to plants in the form of foliar sprays, elicited various kinds of biological and physiological activities, including promotion of plant growth, seed germination, shoot elongation, root growth, flower production, suppression of heavy metal stress, etc. Furthermore, application of these oligomers can shorten the harvesting period of various crops and helps in reducing the use of insecticides and chemical fertilizers (Ohta et al. 1999; Hien et al. 2000; Kume et al. 2002; Hafeez et al. 2003; Luan et al. 2003). It has been used as a wonderful growth-promoting substance in its depolymerized form for a number of medicinal and agricultural plants (Idrees et al. 2011c; Aftab et al. 2011; Naeem et al. 2015a, b, Naeem et al. 2017). Application of radiolytically (using gamma rays) degraded oligomers in the form of foliar sprays is also reported to have been used as plant growth promoters (Kume 2006; Mollah et al. 2009; Khan et al. 2011; Idrees et al. 2010; Sarfaraz et al. 2011; Naeem et al. 2012a, b, Naeem et al. 2014) regarding several medicinal and other plants.

3.4.1 Effect of Gamma-Irradiated Sodium Alginate and Carrageenan on *C. roseus*

Idrees et al. (2011c) studied the effects of various concentrations of γ-irradiated sodium alginate (ISA), viz. deionized water (control), 20, 40, 60, 80, and 100 mg L^{-1} on the performance of *C. roseus* L. (var. Rosea). Their work revealed that ISA applied as leaf-sprays at 80 mg L^{-1} significantly enhanced the plant growth, photosynthesis, physiological activities, and alkaloid production in *C. roseus*. In another study, Naeem et al. (2015a) applied ISA on *C. roseus* plants. They studied various parameters including plant growth, physiological activities, and production of anticancer alkaloids (vinblastine and vincristine) at 120 and 150 DAP. 80 mg L^{-1} of ISA enhanced the leaf-yield by 25 and 30% and the herbage-yield by 29 and 34% at 120 and 150 DAP, respectively, as compared to the control. The spray of ISA at 80 mg L^{-1} improved the yield of vinblastine by 68 and 71% and that of vincristine by 68 and 76% at 120 and 150 DAPS, respectively, over the control. Further, as compared to control, the application of ISA at 80 mg L^{-1} resulted in the maximum swell in the content and yield of vindoline, increasing them by 19 and 29% and by 89 and 87% at 120 and 150 DAP, respectively.

Considering the importance of irradiated carrageenan (ICR) as a promoter of plant growth and alkaloids production in *C. roseus*, a pot experiment was carried out by Naeem et al. (2015b) to explore the effect of ICR on the plant growth, physiological activities, and production of anticancer alkaloids in *C. roseus* at 120 and 150 DAP. Foliar application of ICR (at 0, 20, 40, 60, 80, and 100 mg L^{-1}) significantly improved the performance of *C. roseus*. 80 mg L^{-1} of ICR enhanced the leaf-yield by 29 and 35% and the herbage-yield by 32 and 37% at 120 and 150 DAP, respectively, over the control. Foliar spray of ICR at 80 mg L^{-1} increased the yield of vinblastine by 64 and 65% and of vincristine by 75 and 77% at 120 and 150 DAP, respectively, as compared to the control.

Rasheed et al. (2016) examined the effect of single and combined application of methyl jasmonate (MeJA) and irradiated sodium alginate (ISA) on growth parameters, physiological attributes, and production of anticancer alkaloids (vincristine and vinblastine) of periwinkle. They examined that foliar application of ISA at 80 mg L^{-1} improved the growth and other physiological parameters, while MeJA at 40 mg L^{-1} enhanced the content and yield of vincristine and vinblastine at 6 and 9 MAP. Combined application of these treatments (ISA 80 mg L^{-1} + MeJA 40 mg L^{-1}) synergistically enhanced the total content of alkaloids and vincristine.

3.5 Effect of Mineral Nutrients on C. roseus

Several factors, viz. climatic factors, agricultural measures (including propagation), sowing and harvest dates, planting methods, nutrient requirement and irrigation, are responsible for the optimum yield and performance of *C. roseus*. Of the above factors, nutrients play pivotal role in the cultivation of plant growth. Jana and Varghese (1996) observed that adequate mineral nutrition could increase the yield and the alkaloid content in periwinkle. In fact, an adequate mineral nutrition in the initial phase of plant growth and development is most important. van Iersel et al. (1998) reported that shoot-N concentration was linearly correlated to the shoot dry mass of *Catharanthus* seedlings. Nitrogen, followed by other macro elements, is the main yield determining factor under field conditions. According to Pareek et al. (1985), higher N were required under irrigation than under rainfall conditions for better performance of periwinkle; in addition, applications of farmyard manure at 15 t/ha was profitable for superior yield of the herb. Yanishevskii and Dzhaparidze (1990) reported an advantageous effect of K fertilizer in pot and field experiments conducted on *C. roseus*. According to researchers, the effect of foliar fertilization on the yield of dry matter of both leaves and roots, as well as on the total alkaloid content was observed by the applications of amide nitrogen [$Co(NH_2)_2$] (Łata and Sadowska 1996), phosphorus (Chandra 1981), boron and zinc (Vasuki et al. 1980). Nitrogen plays an important role in the biosynthesis and accumulation of alkaloids in plants. Effects of different levels of nitrogen application rates (from 50 to 150 kg/ha) on plant yield have been tested, with 150 kg/ha giving the best response (Shylaja et al. 1996; Sreevalli et al. 2004; Gholamhosseinpour et al. 2011). Root and shoot dry weights were greatest when high nitrate-N to ammonium-N ratio were applied, with high levels of ammonium-N showing an adverse effect on Periwinkle (Thomas and Latimer 1996). Abdolzadeh et al. (2006) studied to evaluate of the effects of varied sources of nitrogen (2.75, 5.5, 11, 22, and 32 mM) on growth and the total alkaloids content of *C. roseus* including that of vincristine and vinblastine. According to their study, the highest content of amino acids, proteins, total nitrogen, total alkaloids, vincristine and vinblastine were noticed in plant supplied with nitrate plus ammonium.

Łata (2007) reported enormous information on cultivation, climatic and soil requirements, agricultural measures, dates of sowing and harvest, planting methods, fertilization, and irrigation requirements/methods for *Catharanthus roseus*. According to Łata (2007), researchers emphasized the significance of N application on the crop, but the recommended doses varied considerably, viz. 50 kg/ha (Rajeswara Rao and Singh 1990), 80–100 kg/ha (Pareek et al. 1985) and 150 kg N ha^{-1} (Shylaja et al. 1996). Application of farmyard manure at 15 t/ha was also profitable for higher yield of the herb. The other macronutrients, such as P and K, had a smaller effect on plant yield. However, beneficial effect of K fertilizer in pot and field experiments was noted by Yanishevskii and Dzhaparidze (1990). Most significant effect of N, P, and K fertilization that increased the vinblastine content in *C. roseus* leaves was recorded when potassium was not applied in the chloride form. An increased N supply stimulated the uptake of K and P and their translocation to the leaves was proven by the results of the chemical analysis (Łata and Sadowska 1996). According to Dovrat and Goldschmidt (1978) and Chandra (1981), P had the greatest influence on alkaloid synthesis of *C. roseus*. Boldyreva and Velichko (2003) studied the path of biosynthesis of alkaloids in callus tissues of *C. roseus*, applying different minerals to the nutrient medium. Alkaloid biosynthesis in callus tissue was enhanced by 33% when Mg was increased to two-fold and a 35% increase was obtained with the application of Co, which was applied in a five times higher concentration than that present in the original nutrient medium. In another study of Hassan et al. (2009), effect of N and K on growth, yield, and alkaloid content in *C. roseus* was studied under field conditions. Significant improvement in all the parameters was recorded to the highest extent in plants fertilized with 150 Kg/fed of N and 25 Kg/fed of K.

Misra and Gupta (2006) investigated that the activities of various antioxidant enzymes and accumulation of various alkaloid in different leaf pairs (apical, middle, basal) and in roots of *C. roseus* seedlings under different doses and sources of nitrogen (20 mM KNO_3 and 2 mM NH_4Cl) and in the absence (nonsaline control) or presence (100 mM NaCl) of salinity, as managed with the nutrient solution supplied. As per the data recorded, the biomass production of ammonium-fed plants was lower than that of nitrate-fed plants. Further, they noticed that higher accumulation of alkaloid was found in all leaf pairs, as well as in roots of *C. roseus* plants when supplied with NO_3^- as compared to $NH4^+$. Jaleel et al. (2007) grew *C. roseus* plants with NaCl and $CaCl_2$ with the aim of finding out the ameliorative effect of $CaCl_2$ on NaCl-induced oxidative stress in terms of lipid peroxidation (TBARS content), H_2O_2 content, osmolytes concentration, proline-metabolizing enzymes, antioxidant enzyme activities, and indole alkaloids accumulation. The antioxidant enzymes such as SOD, POX, and CAT were increased under salinity stress; the enhancement effect was further progressed due to application of $CaCl_2$. They claimed that plants treated with NaCl and $CaCl_2$ showed an increase in total indole alkaloids content in shoots and roots.

Singh and Agrawal (2015) found out that growth and metabolism of *C. roseus* plant were affected differently by elevated CO_2 and availability of N. Nitrogen applied at recommended dose was more favorable for alkaloids production than that with no N dose or with double dose of recommended N. Cartmill et al. (2016) studied to quantify the optimum rates of water-soluble phosphorus (P) on the growth of *C. roseus* var. "Pacifica White" in soilless media in recirculating subirrigation and top-watering systems. In this study, N, P, K, Mg, Mn, Zn, and Cu, determined in the shoot, were greater in the subirrigated plants when compared to top-watered plants. However, shoot-measured Ca, S, Fe, Al, and B were greater in top-watered plants when compared to subirrigated plants.

Freitas et al. (2016) studied to verify the effect of macronutrient deficiencies (N, P, K, Ca, Mg, S) and (B), over ajmalicine bioproduction in roots of *Vinca* (*C. roseus*). Potassium deficiency resulted in 19% increment of ajmalicine within roots, while deficiencies of N, P, Mg, and S reduced ajmalicine concentration in 55%, 33%, 22%, and 26%, respectively, when compared with complete nutrient treatment. Deficiencies in Ca and B had no significant effect in ajmalicine concentration within the plant roots.

4 Conclusion

Several scientific strategies have recently been tried to improve alkaloid production of *C. roseus*. In this regard, significant work has been done to enhance the content of all and/or specific alkaloid/s of periwinkle. Plant researchers should focus more to alter biosynthetic pathway at molecular level in order to enhance the yield of alkaloids. In this review, considerable information has been covered about the effect of potent PGRs such as GA_3, EBL, Kn, and TRIA on various processes of plants, particularly *C. roseus*. In addition, mineral nutrients (N, P and K) play a pivotal role in improving the productivity as well as biosynthesis of indole alkaloids in *C. roseus* plants. It may be concluded that foliar application of PGRs and RPPs in combination with mineral fertilizers may further boost the desired production of periwinkle alkaloids.

References

Abdolzadeh A, Hosseinian F, Aghdasi M, Sadgipoor H (2006) Effects of nitrogen sources and levels on growth and alkaloid content of periwinkle. Asian J Plant Sci 5:271–276

Aerts RJ, Schäfer A, Hesse M, Baumann TW, Slusarenko A (1996) Signalling molecules and the synthesis of alkaloids in *Catharanthus roseus* seedlings. Phytochemistry 42:417–422

Afroz S, Mohammad F, Hayat S, Siddiqui MH (2005) Exogenous application of gibberellic acid counteracts the ill effect of sodium chloride in mustard. Turk J Biol 29:233–236

Aftab T, Khan MMA, Idrees M, Naeem M, Singh M, Ram M (2010a) Stimulation of crop productivity, photosynthesis and artemisinin production in *Artemisia annua* L. by triacontanol and gibberellic acid application. J Plant Interact 5:273–281

Aftab T, Khan MMA, Idrees M, Naeem M, Hashmi N (2010b) Effects of gibberellic acid on growth, photosynthetic efficiency and artemisinin content of *Artemisia annua* L. Med Arom Plant Sci Biotechnol 5:25–29

Aftab T, Khan MMA, Naeem M, Idrees M, Varshney L (2011) Enhancing the growth, photosynthetic capacity and artemisinin concentration in *Artemisia annua* L. by irradiated sodium alginate. Radiat Physic Chem 80:833–836

Alam (2013) Influence of GA3, epibrassinolide and kinetin on the performance of *Catharanthus roseus* L. with special reference to alkaloid production. PhD Thesis, AMU, Aligarh, India

Alam MM, Naeem M, Idrees M, Khan MMA, Moinuddin (2012) Augmentation of photosynthesis, crop productivity, enzyme activities and alkaloids production in Sadabahar (*Catharanthus roseus* L.) through application of diverse plant growth regulators. J Crop Sci Biotechnol 15:117–129

Alam MM, Naeem M, Khan MMA (2016) Exploiting the epibrassinolide as a plant growth promoter for augmenting the growth, physiological activities and alkaloids production in *Catharanthus roseus* L. J Med Plant Stud 4:88–93

Ali HM, Siddiqui MH, Basalah MO, Al-Whaibi MH, Sakran AM, Al-Amri A (2012) Effects of gibberellic acid on growth and photosynthetic pigments of *Hibiscus sabdariffa* L. under salt stress. Afr J Biotechnol 11:800–804

Al-Rumaih MM, Rushdy SS, Warsy AS (2003) Effect of cadmium stress in the presence and absence of gibberellic acid on mineral nutrition of cowpea (*Vigna unguiculata* L.) during ontogenesis. Agric Sci 15:141–151

Altmann T (1999) Molecular physiology of brassinosteroids revealed by the analysis of mutants. Planta 208:1–11

Al-Wakeel SAM, Dadoura SS, Hamed AA (1994) Interactive effects of water stress and gibberellic acid on mineral composition of fenugreek plant. Egypt J Physiol Sci 18:269–282

Amini A, Glevarec G, Andreu R, Rideau M, Creche J (2009) Low levels of gibberellic acid control the biosynthesis of ajmalicine in *Catharanthus roseus* cell suspension cultures. Planta Med 75(2):187–191

Arora P, Bhardwaj R, Kanwar MK (2012) Effect of 24-epibrassinolide on growth, protein content and antioxidative defense system of *Brassica juncea* L. subjected to cobalt ion toxicity. Acta Physiol Planta 34:2007–2017

Arteca RN (1997) Plant growth substances: principles and applications. CBS Publishers and Distributors, New Delhi, India

Ashraf M, Karim F, Rasul E (2002) Interactive effects of gibberellic acid (GA_3) and salt stress on growth, ion accumulation and photosynthetic capacity of two spring wheat (*Triticum aestivum* L.) cultivars differing in salt tolerance. Plant Growth Regul 36:49–59

Aslam J, Khan SH, Siddiqui ZH, Fatima Z, Maqsood M, Bhat MA, Nasim SA, Ilah A, Ahmad IZ, Khan SA, Mujib A, Sharma MP (2010) *Catharanthus roseus* (L.) G. Don. An Important Drug: it's applications and production. *Pharm Glob* Int J Compr Pharm 1:1–16

Babakhanyan MA (1991) Introduction of *Catharanthus roseus* G. Don in open-air hydroponics in the Ararat Valley in Armenia. Herba Hung 30:76–81

Bamberger E, Mayer AM (1960) Effect of kinetin on formation of red pigment in seedlings of *Amaranthus retroflexus*. Science 131:1094–1095

Bata J, Nešković M (1974) The effect of gibberellic acid and kinetin on chlorophyll retention in *Lemna trisulca* L. Zeitschrift für Pflanzenphysiol 73:86–88

Boldyreva YA, Velichko NA (2003) Effect of the compounds of mineral nutrition on the *Catharanthus roseus* callus tissue growth and synthesis of alkaloids. Biotekhnologiya 19:53–62

Bulgakov VP, Tchernoded GK, Mischenko NP, Khodakovskaya MV, Glazunov VP, Radchenko SV (2002) Effect of salicylic acid, methyl jasmonate, ethephon and cantharidin on anthraquinone production by *Rubia cordifolia* callus cultures transformed with the rolB and rolC genes. J Biotechnol 97:213–221

Campbell WH (1999) Nitrate reductase structure, function and regulation: bridging the gap between biochemistry and physiology. Annu Rev Plant Physiol Plant Mol Biol 50:277–303

Cartmill AD, Cartmill DL, Ballweg DL, Valdez-Aguilar LA (2016) Optimum phosphorus concentration for growth of *Catharanthus roseus* (L.) G. Don 'Pacifica White' in a subirrigation and top watering system. Commun Soil Sci Plant Anal 47:52–64

Chandra V (1981) Nutrient requirements for some medicinal and aromatic plants. Biol Mem 6:81–96

Chow B, McCourt P (2006) Plant hormone receptors: perception is everything. Genes Dev 20:1998–2008

Clouse SD, Sasse JM (1998) Brassinosteroids: essential regulators of plant growth and development. Annu Rev Plant Physiol Plant Mol Biol 49:427–451

Crozier A, Turnbull CGN (1984) Gibberellins: biochemistry and action in extension growth. What's New Plant Physiol 15:9–12

Day DF (1998) Alginate. In Kaplan DL (Ed) Biopolymers from Renewable Resources. Springer, Berlin, Heidelberg, New York pp 119–143

De Luca V, Laflamme P (2001) The expanding universe of alkaloid biosynthesis. Curr Opin Plant Biol 4:225–233

Dedio I (1983) Barwinek różowy w uprawie hydroponicznej (*Vinca rosea* in hydroponic cultivation). Wiadomos'ci Zielarskie 10/11: 12–13

Dey B, Srivastava RC (2006a) Effect of kinetin on soluble protein, leghaemoglobin content and nitrogen metabolizing enzymes in *Vigna radiata* (L.) Wilczek. Vegetos 19:53–57

DiCosmo F, Misawa M (1995) Plant cell and tissue culture: alternatives for metabolite production. Biotechnol Adv 13:425–453

Dovrat A, Goldschmidt J (1978) Cultivation aspects of *Catharanthus roseus* for roots. Acta Hort 73:263–268

Eid RA, Abou-Leila BH (2006) Response of croton plants to gibberellic acid, benzyl adenine and ascorbic acid application. World J Agric Sci 2:174–179

El-Sayed M, Verpoorte R (2004) Growth, metabolic profiling and enzymes activities of *Catharanthus roseus* seedlings treated with plant growth regulators. Plant Growth Regul 44:53–58

El-Sayed M, Verpoorte R (2005) Methyljasmonate accelerates catabolism of monoterpenoid indole alkaloids in *Catharanthus roseus* during leaf processing. Fitoterapia 76:83–90

El-Sayed M, Verpoorte R (2007) Catharanthus terpenoid indole alkaloids: biosynthesis and regulation. Phytochem Rev 6:277–305

Freitas MSM, Gama MC, Monnerat PH, De Carvalho AJC, Lima TC, Vieira IJC (2016) Induced nutrient deficiencies in *Catharanthus roseus* impact ajmalicine bioproduction. J Plant Nutr 39:835–841

Gantet P, Memelink J (2002) Transcription factors: tools to engineer the production of pharmacologically active plant metabolites. Trends Pharmacol Sci 23:563–569

Gholamhosseinpour Z, Hemati K, Dorodian H, Bashir-Sadr Z (2011) Effect of nitrogen fertilizer on yield and amount of alkaloids in periwinkle and determination of vinblastine and vincristine by HPLC and TLC. Plant Sci Res 3:4–9

Goupil P, Loncle D, Druart N, Bellettre A, Rambour S (1998) Influence of ABA on nitrate reductase activity and carbohydrate metabolism in chicory roots (*Cichorium intybus* L.). J Exp Bot 49:1855–1862

Grove MD, Spencer GF, Rohwedder WK, Mandava N, Worley JF, Warthen JD, Steffens GL, Flippen-Anderson JL, Cook JC (1979) Brassinolide, a plant growth-promoting steroid isolated from *Brassica napus* pollen. Nature 281:216–217

Hassan RA, Habib AA, El-Din E (2009) Effect of nitrogen and potassium fertilization on growth, yield and alkaloidal content of periwinkle (*Catharanthus roseus* G. Don). Med Arom Plant Sci Biotechnol 3:24–26

Hafeez M, Muhamad 0, Muhammad SY, Ahmad MS (2003) Irradiated chitosan as plant promoter. Prosiding Simposium Biologic Gunaan Ke-7, Mines Beach Resort & Spa, Seri Kumbangan, Selangor, 3–4 June

Hien NQ, Nagasawa N, Tham LX, Yoshii F, Dang VH, Mitomo H, Makuuchi K, Kume T (2000) Growth promotion of plants with depolymerised alginates by irradiation. Radiat Phys Chem 59:97–101

Houimli SIM, Denden M, Mouhandes BD (2010) Effects of 24-epibrassinolide on growth, chlorophyll, electrolyte leakage and proline by pepper plants under NaCl-stress. EurAsian J BioSci 4:96–104

Hu X, Jiang X, Hwang H, Liu S, Guan H (2004) Promotive effects of alginate-derived oligosaccharide on maize seed germination. J Appl Phycol 16:73–76

Idrees M, Naeem M, Khan MMA (2010) The superiority of *cv*. 'rosea' over *cv*. 'alba' of periwinkle (*Catharanthus roseus* L.) in alkaloid production and other physiological attributes. Turk J Biol 32:81–88

Idrees M, Naeem M, Aftab T, Khan MMA, Moinuddin (2011a) Salicylic acid mitigates salinity stress by improving antioxidant defence system and enhances vincristine and vinblastine alkaloids production in periwinkle [*Catharanthus roseus* (L.) G. Don]. Acta Physiol Plant 33:987–999

Idrees M, Khan MMA, Naeem M (2011b) Anticancerous Alkaloids of *Catharanthus roseus* - Enhanced Production through PGRs. LAP LAMBERT Publishing GmbH & Co. KG, Germany

Idrees M, Naeem M, Alam M, Aftab T, Hashmi N, Khan MMA, Moinuddin VL (2011c) Utilizing the gamma irradiated sodium alginate as a plant growth promoter for enhancing the growth, physiological activities and alkaloids production in *Catharanthus roseus* L. Agric Sci China 10:1213–1221

Idrees M, Naeem M, Aftab T, Khan MMA, Moinuddin (2013) Salicylic acid restrains nickel toxicity, improves antioxidant defence system and enhances the production of anticancer alkaloids in *Catharanthus roseus* (L.). J Hazard Material 252–253:367–374

Inoue T, Higuchi M, Hashimoto Y, Seki M, Kobayashi M, Kato T, Tabata S, Shinozaki K, Kakimoto T (2001) Identification of CRE1 as a cytokinin receptor from *Arabidopsis*. Nature 409:1060–1063

Jaleel CA, Gopi R, Manivannan P, Sankar B, Kishorekumar A, Panneerselvam R (2007) Antioxidant potentials and ajmalicine accumulation in *Catharanthus roseus* after treatment with gibberellic acid. Colloids Surf B Biointerfaces 60:195–200

Jaleel CA, Gopi R, Chang-Xing Z, Azooz MM, Panneerselvam R (2008) Plant growth regulators and fungicides alters growth characteristics in *Catharanthus roseus*; comparative study. Global J Mol Sci 3:93–99

Jaleel CA, Gopi R, Gomathinayagam M, Panneerselvam R (2009) Traditional and non-traditional plant growth regulators alters phytochemical constituents in *Catharanthus roseus*. Process Biochem 44:205–209

Jana BK, Varghese B (1996) Effect of mineral nutrition on growth and alkaloid content of *Catharanthus roseus*. Indian Agric 40:93–99

Khan NA, Mir R, Khan M, Javid S, Samiullah (2002) Effects of gibberellic acid spray on nitrogen yield efficiency of mustard grown with different nitrogen levels. Plant Growth Regul 38:243–247

Khan NA, Singh S, Nazar R, Lone PM (2007a) The source–sink relationship in mustard. The Asian Australas J Plant Sci Biotechnol 1:10–18

Khan R, Khan MMA, Singh M, Nasir S, Naeem M, Siddiqui MH, Mohammad F (2007b) Gibberellic acid and triacontanol can ameliorate the opium yield and morphine production in opium poppy (*Papaver somniferum* L.). Acta Agric Scand 756:289–298

Khan ZH, Khan MMA, Aftab T, Idrees M, Naeem M (2011) Influence of alginate oligosaccharides on growth, yield and alkaloid production of opium poppy (*Papaver somniferum* L.). Front Agric China 5:122–127

Kume T, Nagasawa N, Yoshii F (2002) Utilization of carbohydrates by radiation processing. Radiat Phys Chem 63:625–627

Kume T (2006) Application of radiation in agriculture. Proceedings of International Workshop on Biotechnology in Agriculture. Takasaki Advanced Radiation Research Institute, Japan Atomic Energy Agency (JAEA). 1233 Watanuki, Takasaki, Gunma 370-1292, Japan

Łata B (2007) Cultivation, mineral nutrition and seed production of *Catharanthus roseus* (L.) G. Don in the temperate climate zone. Phytochem Rev 6:403–411

Łata B, Sadowska A (1996) Effect of nitrogen level in the substrate on yield and alkaloid concentration in *Catharanthus roseus* L. (G.) Don. Folia Hortic 8:59–69

Li J, Chory J (1997) A putative leucine-rich repeat receptor kinase involved in brassinosteroid signal transduction. Cell 90:929–938

Lovkova MY, Buzuk GN, Sokolova SM, Buzuk LN (2005) Role of elements and physiologically active compounds in the regulation of synthesis and accumulation of indole alkaloids in *Catharanthus roseus* L. Appl Biochem Microbiol 41:299–305

Luan LQ, Hien NQ, Nagasawa N, Kume T, Yoshii F, Nakanishi TM (2003) Biological effect of radiation-degraded alginate on flower plants in tissue culture. Biotechnol Appl Biochem 38:283–288

MacMillan J (2002) Occurrence of gibberellins in vascular plants, fungi, and bacteria. J Plant Growth Regul 20:387–442

Magnotta M, Murata J, Chen J, De Luca V (2006) Identification of a low vindoline accumulating cultivar of *Catharanthus roseus* (L.) G. Don by alkaloid and enzymatic profiling. Phytochemistry 67:1758–1764

Maia NB, Bovi OA, Zullo MAT, Perecin MB, Granja NP, Carmello QAC, Robaina C, Coll F (2004) Hydroponic cultivation of mint and vetiver with spirostane analogues of brassinosteroids. Acta Hortic 644:55–59

Majeau N, Coleman JR (1994) Correlation of carbonic anhydrase and ribulose-1,5-bisphosphate carboxylase/oxygenase expression in pea. Plant Physiol 104:1393–1399

Marschner H (1995) Mineral nutrition of higher plants. Academic, New York, pp 269–340

Misra N, Gupta AK (2006) Effect of salinity and different nitrogen sources on the activity of antioxidant enzymes and indole alkaloid content in *Catharanthus roseus* seedlings. J Plant Physiol 163:11–18

Misra P, Kumar S (2000) Emergence of periwinkle *Catharanthus roseus* as a model system for molecular biology of alkaloids: phytochemistry, pharmacology, plant biology and *in vivo* and *in vitro* cultivation. J Med Arom Plant Sci 22:306–337

Misra A, Srivastava NK, Srivastava AK, Khan A (2009) Influence of etherel and gibberellic acid on carbon metabolism, growth, and alkaloids accumulation in *Catharanthus roseus* L. African J Pharm Pharmacol 3:515–520

Mitrev A (1976) Laboratory investigations on the biological properties of *Vinca rosea* L. seeds produced in Bulgaria. Rasteniev Dni Nauki 13:85–90

Modi Arpan R, Shukla YM, Litoriya NS, Patel NJ, Narayan S (2011) Effect of gibberellic acid foliar spray on growth parameters and stevioside content of ex vitro grown plants of *Stevia rebaudiana* Bertoni. Med Plant - Internat J Phytomed Related Indus 3:157–160

Mollah MZI, Khan MA, Khan RA (2009) Effect of gamma irradiated sodium alginate on red amaranth (*Amaranthus cruentus* L.) as growth promoter. Radiat Phys Chem 78:61–64

Moore TC (1989) Biochemistry and physiology of plant hormones. Springer, New York

Mostafa GG, Alhamd MFA (2011) Effect of gibberellic acid and indole 3-acetic acid on improving growth and accumulation of phytochemical composition in *Balanites aegyptiaca* plants. Am J Plant Physiol 6:36–43

Mujib A, Ilah A, Aslam J, Fatima S, Siddiqui ZH, Maqsood M (2012) *Catharanthus roseus* alkaloids: application of biotechnology for improving yield. Plant Growth Regul 68:111–127

Müssig C, Altmann T (1999) Physiology and molecular mode of action of brassinosteroids. Plant Physiol Biochem 37:363–372

Mustafa NR, Kim HK, Choi YH, Erkelens C, Lefeber AWM, Spijksma G, Van der Heijden R, Verpoorte R (2009) Biosynthesis of salicylic acid in fungus elicited *Catharanthus roseus* cells. Phytochemistry 70:532–539

Naeem M, Idrees M, Aftab T, Khan MMA, Uddin M, Varshney L (2012a) Irradiated sodium alginate improves plant growth, physiological activities, and active constituents in *Mentha arvensis* L. J Appl Pharm Sci 2:28–35

Naeem M, Idrees M, Alam MM, Aftab T, Khan MMA, Uddin M (2012b) Brassinosteroid-mediated enrichment in yield attributes, active constituents and essential oil production in *Mentha arvensis* L. Russ Agric Sci 38:106–113

Naeem M, Idrees M, Aftab T, Khan MMA, Uddin M, Varshney L (2012c) Depolymerised carrageenan enhances physiological activities and menthol production in *Mentha arvensis* L. Carbohydr Polym 87:1211–1218

Naeem M, Idrees M, Aftab T, Khan MMA, Uddin M, Varshney L (2014) Employing depolymerised sodium alginate, triacontanol and 28-homobrassinolide in enhancing physiological activities, production of essential oil and active components in *Mentha arvensis* L. Ind Crop Prod 55:272–279

Naeem M, Aftab T, Ansari AA, Idrees M, Ali A, Khan MMA, Uddin M, Varshney L (2015a) Radiolytically degraded sodium alginate enhances plant growth, physiological activities and alkaloids production in *Catharanthus roseus* L. J Radiat Res Appl Sci 8:606–616

Naeem M, Idrees M, Aftab T, Alam MM, Khan MMA, Uddin M and Varshney L (2015b) Radiation processed carrageenan improves plant growth, physiological activities and alkaloids production in *Catharanthus roseus* L. Adv Bot 2015:1–11. Article ID 150474

Naeem M, Aftab T, Idrees M, Singh M, Ali A, Khan MMA, Uddin M and Varshney L (2017) Modulation of physiological activities, active constituents and essential oil production of *Mentha arvensis* L. by concomitant application of depolymerised carrageenan, triacontanol and 28-homobrassinolide. J Essent Oil Res 29:179–188

Narkiewicz M, Sadowska A (1991) Possibility of cultivation of rosy periwinkle (*Catharanthus roseus* (L.) G. Don) for seed production under conditions of Poland. Biuletyn IHAR 177:179–184

Nwafor SV, Akah PA, Okoli CO (2001) Potentials of plant products as anticancer agents. J Nat Remed 1:75–88

Ohlsson AB, Björk L (1988) Effects of gibberellic acid on cardenolide accumulation by *Digitalis lanata* tissue cultures grown in light and darkness. J Plant Physiol 133:535–538

Ohta K, Zaniguhi A, Konishi N, Idosoki T (1999) Chitosan treatment affects plants growth and flower quality in *Eustoma grandiflorum*. Hort Sci 34:233–234

Paleg LG (1965) Physiological effects of gibberellins. Annu Rev Plant Physiol 16:291–322

Pan Q, Chen Y, Wang Q, Yuan F, Xing S, Tian Y, Zhao J, Sun X, Tang K (2010) Effect of plant growth regulators on the biosynthesis of vinblastine, vindoline and catharanthine in *Catharanthus roseus*. Plant Growth Regul 60:133–141

Papon N, Clastre M, Andreu F, Gantet P, Rideau M, Crèche J (2002) Expression analysis in plant and cell suspensions of *CrCKR1*, a cDNA encoding a histidine kinase receptor homologue in *Catharanthus roseus* (L.) G. Don. J Exp Bot 53:1989–1990

Paponov IA, Teale WD, Trebar M, Blilou I, Palme K (2005) The PIN auxin efflux facilitators: evolutionary and functional perspectives. Trends Plant Sci 10:170–177

Pareek SK, Maheshwari ML, Gupta R (1985) Cultivation of periwinkle in North India. Indian Hortic 30:9–12

Pareek SK, Maheshwari ML, Gupta R (1991) In: Raychoudhuri SP (ed) Chemical weed control and water management in periwinkle, vetiver and palmarosa oil grass. Today and Tomorrow Printers and Publishers, New Delhi, pp 167–170

Patton EE, Willems AR, Tyers M (1998) Combinatorial control in ubiquitin-dependent proteolysis: don't skp the F-box hypothesis. Trends Genet 14:236–243

Peebles CAM, Hong SB, Gibson SI, Shanks JV, San KY (2005) Transient effects of over-expressing anthranilate synthase α and ß subunits in *Catharanthus roseus* hairy roots. Biotechnol Prog 21:1572–1576

Price GD, Caemmerer S, Evans JR, Yu J-W, Lloyd J, Oja V, Kell P, Harrison K, Gallagher A, Badger MR (1994) Specific reduction of chloroplast carbonic anhydrase activity by antisense RNA in transgenic tobacco plants has a minor effect on photosynthetic CO_2 assimilation. Planta 193:331–340

Rady MM (2011) Effect of 24-epibrassinolide on growth, yield, antioxidant system and cadmium content of bean (*Phaseolus vulgaris* L.) plants under salinity and cadmium stress. Scientia Hort 129:232–237

Rajasekaran LR, Blake TJ (1998) Early growth invigoration of jack pine seedlings by natural plant growth regulators. Trees 12:420–423

Rajeswara Rao BR, Singh M (1990) Effect of NPK fertilizers and spacing on periwinkle (*Catharanthus roseus*) under irrigated and rained conditions. Herba Hung 29:1–2

Rao SSR, Vardhini BV, Sujatha E, Anuradha S (2002) Brassinosteroids – a new class of phytohormones. Curr Sci 82:1239–1245

Rasheed F, Naeem M, Uddin M, Khan MMA (2016) Synergistic effect of irradiated sodium alginate and methyl jasmonate on anticancer alkaloids production in periwinkle (*Catharanthus roseus* L.) (submitted)

Reed ML, Graham D (1981) Carbonic anhydrase in plants: distribution, properties and possible physiological roles. In: Progress in phytochemistry. 7. Pergamon Press, Oxford, New York, pp 47–94

Richmond AE, Lang A (1957) Effect of kinetin on protein content and survival of detached Xanthium leaves. Science 125: 650–651

Ruiz-May E, Galaz-Avalos RM, Loyola-Vargas VM (2008) Differential secretion and accumulation of terpene indolie alkaloids in hairy roots of *Catharanthus roseus* treated with methyl jasmonate. Mol Biotechnol 41:278–285

Sadowska A, Grysiak A, Racka M, Strumińska E (1987) Wstępne oszacowanie kosztów produkcji barwinka różowego (*Catharanthus roseus* (L.) G. Don) w Polsce. Biuletyn IHAR 177:171–178

Shahid MA, Pervez MA, Balal RM, Mattson NS, Rashid A, Ahmad R, Ayyub CM, Abbas T (2011) Brassinosteroid (24-epibrassinolide) enhances growth and alleviates the deleterious effects induced by salt stress in pea (*Pisum sativum* L.). Australian J Crop Sci 5:500–510

Salisbury FB, Ross CW (1992) Plant physiology, Fourth edn. Wadsworth Publishing Company, Belmont, California, USA

Sarfaraz A, Naeem M, Nasir S, Idrees M, Aftab T, Hashmi N, Khan MMA, Uddin M, Varshney L (2011) An evaluation of the effects of irradiated sodium alginate on the growth, physiological activities and essential oil production of fennel (*Foeniculum vulgare* Mill.). J Med Plant Res 5:15–21

Satdive RK, Fulzele DP, Eapen S (2003) Studies on production of ajmalicine in shake flasks by multiple shoot cultures of *Catharanthus roseus*. Biotechnol Prog 19:1071–1075

Schmülling T (2001) CREam of cytokinin signalling: receptor identified. Trends Plant Sci 6:281–284

Shah SH (2007a) Effects of salt stress on mustard as affected by gibberellic acid application. Gen Appl Plant Physiol 33:97–106

Shah SH (2007b) Effect of kinetin spray on growth and productivity of black cumin plants. Russ J Plant Physiol 54:702–705

Shah SH, Ahmad I, Samiullah (2006) Effect of gibberellic acid spray on growth, nutrient uptake and yield attributes during various growth stages of black cumin (*Nigella sativa* L.). Asian J Plant Sci 5:881–884

Shylaja MR, Sankar MA, Nair GS, Mercy KA (1996) Response of *Catharanthus roseus* L. (G. Don) to nitrogen phosphorus and potassium fertilization. Indian Cocoa Arec Spice J 20(3):83–88

Singh A, Agrawal M (2015) Effects of ambient and elevated CO2 on growth, chlorophyll fluorescence, photosynthetic pigments, antioxidants, and secondary metabolites of *Catharanthus roseus* (L.) G Don. grown under three different soil N levels. Environ Sci Pollut Res 22:3936–3946

Singh SP, Singh U, Singh AK, Singh VN, Chaturvedi GS (2006) Kinetin reversal of salinity effect on growth and nodulation in mungbean (*Vigna radiata* L. Wilczek). Internat J Plant Sci 1:236–239

Singh P, Srivastava NK, Mishra A, Sharma S (1999) Influence of etherel and gibberellic acid on carbon metabolism, growth, and essential oil accumulation in spearmint (*Mentha spicata*). Photosynthetica 36:509–517

Skoog F, Miller CO (1957) Chemical regulation of growth and organ formation in plant tissue cultured *in vitro*. Symp Soc Exp Biol 11:118–131

Sreevalli Y, Kulkarni RN, Baskaran K, Chandrashekara RS (2004) Increasing the content of leaf and root alkaloids of high alkaloid content mutants of periwinkle through nitrogen fertilization. Ind Crop Prod 19(2):191–195

Srivastava NK, Srivastava AK (2007) Influence of gibberellic acid on $^{14}CO_2$ metabolism, growth, and production of alkaloids in *Catharanthus roseus*. Photosynthetica 45:156–160

Srivastava NK, Srivastava AK (2011) Allocation of different ^{14}C substrates into primary metabolites in relation to total alkaloid accumulation into *Catharanthus roseus* plants. J Med Plant Res 5:4747–4753

Stemler AJ (1997) The case for chloroplast thylakoid carbonic anhydrase. Physiol Plant 99:348–353

Stetler DA, Laetsch WM (1965) Kinetin-induced chloroplast maturation in cultures of tobacco tissue. Science 149:1387–1388

Steveninck RFMV (1976) Effect of hormones and related substances on transport. In: Lüttge U, Pitman MG (eds) Encyclopedia of plant physiology. Springer, Berlin, pp 307–342

Svoboda GH, Blake DA (1975) The phyto-chemistry and pharmacology of *Catharanthus roseus* (L.) G. Don. Inc. In: Taylor WJ, Farnsworth NR (eds) The *Catharanthus* alkaloids. Marcel Decker, New York, pp 45–84

Swamy KN, Rao SSR (2009) Effect of 24-epibrassinolide on growth, photosynthesis, and essential oil content of *Pelargonium graveolens* (L.) Herit. Russ J Plant Physiol 56:616–620

Symons GM, Davies C, Shavrukov Y, Dry IB, Reid JB, Thomas MR (2006) Grapes on steroids. Brassinosteroids are involved in grape berry ripening. Plant Physiol 140:150–158

Thomas PA, Latimer JG (1996) Growth of vinca as affected by form of nitrogen, presence of bark, and type of micronutrients. J Plant Nutr 18:2127–2134

Van Iersel MW, Beverly RB, Thomas PA, Latimer JG, Mills HA (1998) Fertilizer effects on the growth of impatiens, petunia, salvia, and vinca plug seedlings. HortScience 33:678–682

van Staden J, Cook EL, Noodén LD (1988) Cytokinins and senescence. In: Noodén LD, Leopold AC (eds) Senescence and aging in plants. Academic, San Diego, pp 281–328

Vardhini BV, Rao SSR (1998) Effect of brassinosteroids on growth, metabolic content and yield on *Arachis hypogaea*. Phytochemistry 48:927–930

Vasuki KPS, Rao VS, Rao KVN (1980) Effect of micronutrients and their interactions on growth and alkaloid production in *Catharanthus roseus* (L.) G. Don (Proc Indian Acad Aci). Plant Sci 89:197–201

Vazquez-Flota F, De Carolis E, Alarco AM, De Luca V (1997) Molecular cloning and characterization of desacetoxyvindoline-4-hydroxylase, a 2-oxoglutarate dependent- dioxygenase involved in the biosynthesis of vindoline in *Catharanthus roseus* (L.) G. Don. Plant Mol Biol 34:935–948

Verpoorte R, van der Heijden R, ten Hoopen HJG, Memlink J (1999) Metabolic engineering of plant secondary metabolic pathways for the production of fine chemicals. Biotechnol Lett 21:467–479

Weiss D, van Blokland R, Kooter JM, Mol JNM, van Tunen AJ (1992) Gibberellic acid regulates chalcone synthase gene transcription in the corolla of *Petunia hybrida*. Plant Physiol 98:191–197

Wellburn FAM, Welburn AR, Stoddart JL, Treharne KJ (1973) Influence of gibberellic acid and abscisic acid and the growth retardant, CCC, upon plastid development. Planta 111:337–346

Wiśniewska J, Xu J, Seifartová D, Brewer PB, Růžička K, Blilou L, Rouquié D, Benková E, Scheres B, Friml J (2006) Polar PIN localization directs auxin flow in plants. Science 312:883

Wood A, Paleg LG (1972) The influence of gibberellic acid on the permeability of model membrane systems. Plant Physiol 50:103–108

Xu W, Purugganan MM, Polisensky DH, Antosiewicz DM, Fry SC, Braam J (1995) Arabidopsis *TCH4*, regulated by hormones and the environment, encodes a xyloglucan endotransglycosylase. Plant Cell 7:1555–1567

Xu XQ, Shen H, Xu JR, Xie MQ, Li XJ (2006) The colloidal stability and coroshell structure of magnetite nanoparticle coated with alginate. App SurfSci 253:2158–2164

Xing S, Pan Q, Tian Y, Wang Q, Liu P, Zhao J, Wang G, Sun X, Tang K (2011) Effect of plant growth regulator combinations on the biosynthesis of terpenoid indole alkaloids in *Catharanthus roseus*. J Med Plant Res 5:1692–1700

Yamaguchi S (2008) Gibberellin metabolism and its regulation. Annu Rev Plant Biol 59:225–251

Yanishevskii FV, Dzhaparidze NM (1990) The effect of potassium fertilizer form on *Catharanthus roseus*. Subtrop Kul'tury 2:131–135

Youssef AA, Talaat IM (1998) Physiological effect of brassinosteroid and kinetin on the growth and chemical constituents of lavender plant. Ann Agric Sci 43:261–272

Zhou Mei-L, Hou Hong-L, Zhu Xue-M, Shao Ji-R, Wu Yan-M, Tang Yi-X (2010) Molecular regulation of terpenoid indole alkaloids pathway in the medicinal plant, *Catharanthus roseus*. J Med Plant Res 4:2760–2772

Zhao J, Verpoorte R (2007) Manipulating indole alkaloid production by *Catharanthus roseus* cell cultures in bioreactors: from biochemical processing to metabolic engineering. Phytochem Rev 6:435–457

Role of Biotechnology in Alkaloids Production

Amjad Khalil

Abstract Medicinal plants contain special chemical compounds in their leaves, stems, and roots and were used thousands of years ago for the purpose of some therapeutic uses. Medicinal plants serve as a huge reservoir for a large variety of bioactive compounds, which considered of very high value as pharmaceutical products. Many plants produce very important pharmaceutical compounds known as alkaloids. Alkaloids usually produced by variety of plants showing a great potential to be purified in relatively large quantities using new advanced techniques in the area of plant biotechnology. Alkaloids are chemical compounds that are characterized by having nitrogen atom in their structures. They are large group of chemicals divided into several categories, such as true alkaloids, protoalkaloid, polyamine alkaloids, peptide and cyclopeptide alkaloids, and pseudoalkaloids.

One of the main challenges in the area of alkaloid investigation is how to be produced on the commercial scale and their characterization to be used as very useful therapeutic and pharmaceutical compounds. In order to achieve this, researchers turned to concentrate on the role of new biotechnology methods for synthesizing and characterizing of pharmaceutically important alkaloids. The great development that has happened in recent years in the field of biotechnology and molecular biology made it easy and convenient for transferring and propagating genes in plant cells for the purpose of commercially produce alkaloid related drugs. Number of techniques were developed and proved to be very useful and practical in synthesizing pharmaceutical alkaloids from plants. This chapter provides very important information about pharmaceutical alkaloids and discusses the importance of alkaloids as pharmaceutical and the role of new biotechnology approaches in synthesizing and producing alkaloids in commercial and industrial scale. This chapter is also discussed the new advanced techniques of plant biotechnology, including tissue culture of plant cell, genetic transformation, metabolic engineering, and microbial synthesis of alkaloids.

Keywords Alkaloids biosynthesis • Biotechnology • Protoalkaloids • Polyamine alkaloids • Peptide alkaloids • Pseudoalkaloids

---

A. Khalil (✉)
Department of Life Sciences, King Fahad University of Petroleum and Minerals, Dhahran, Saudi Arabia
e-mail: amjadb@kfupm.edu.sa

M. Naeem et al. (eds.), *Catharanthus roseus*,
DOI 10.1007/978-3-319-51620-2_4

1 Introduction

Medicinal plants known as a very important source of various chemical compounds produce alkaloids as therapeutic and pharmaceutical compounds. Alkaloids are very large group of plant secondary metabolites that contains basic nitrogen atom in their structure. Alkaloids also include some neutral compounds and some weak acidic properties (Manske 1965). Some synthetic compounds of related structure are also termed alkaloids (Lewis 1998). Some alkaloids contain oxygen, sulfur and, more rarely, other elements such as chlorine, bromine, and phosphorus in addition to carbon, hydrogen, and nitrogen (Chemical Encyclopedia: alkaloids). Many different groups of organisms such as plants, animals, bacteria, and fungi are capable of producing alkaloids (Table 1).

Alkaloids have a large spectrum of therapeutic and pharmacological applications including anticancer (homoharringtonine), antimalarial (quinine), antiasthma (ephedrine) (Kittakoop et al. 2014), cholinomimetic (glutamine) (Russo et al. 2013), antiarrhythmic (quinidine), vasodilatory (vincamine), analgesic (morphine) (Raymond et al. 2010), and antibacterial (chelerythrine) (Cushnie et al. 2014). The origin of the name alkaloid comes from a German word—Alkaloide. It was first presented in 1819 by the German Chemist Carl Friedrich and derived from Latin word alkali, which comes from the Arabic word al-qalawi—plant ashes (Luch 2009). The breakthrough and large-scale usage of alkaloids came after the review publication by Oscar Jacobsen in the chemical dictionary of Albert Ladenburg in the 1880s (Jacobsen 1882). Human being has used plants containing alkaloids thousands of years ago for therapeutic and recreational uses. These plants were identified and known in the Mesopotamia around 2000 BC (Aniszewski 2007).

An important achievement in the chemistry of alkaloids was done by the French researcher Pierre Joseph Pelletier and Joseph Bienaime Caventou, who discovered quinine (year 1820) and strychnine (year 1818). Many other alkaloids such as xanthine (year 1817), atropine (year 1819), caffeine (year 1820), nicotine (year 1828), colchicine (year 1833), sparteine (year 1851), and cocaine (year 1886) were also discovered (Hesse 2002d).

Alkaloid synthesis was conducted in the year 1886 by the German chemist Albert Ladenburg (Hesse 2002e). The emergence of spectroscopic and chromatographic methods in the twentieth century has accelerated the development of the chemistry of alkaloids. By the year 2008 more than 12,000 alkaloids had been identified (Begley 2009).

Alkaloids (in comparison with other compounds of natural chemicals) are characterized by many different and complex structures which explain that the classification of alkaloids is not clear (Hesse 2002f). The new classifications of alkaloids are based on the presence of the carbon skeleton (e.g., indole and isoquinoline) or biochemical precursor (ornithine, lysine, tyrosine, tryptophan, etc.) (Chemical Encyclopedia: alkaloids).

Table 1 Selected major groups of monomeric alkaloids are listed in the table below

Class	Major groups	Examples
Alkaloids with nitrogen heterocycles (true alkaloids)		
Pyrrolidine derivatives	–	Cuscohygrine, hygrine, hygroline, stachydrine
Tropane derivatives	Atropine group	Atropine, scopolamine, hyoscyamine
	Substitution in positions 3, 6, or 7	
	Cocaine group	Cocaine, ecgonine
	Substitution in positions 2 and 3	
Pyrrolizidine derivatives	Non-esters	Retronecine, heliotridine, laburnine
	Complex esters of monocarboxylic acids	Indicine, lindelophin, sarracine
	Macrocyclic diesters	Platyphylline, trichodesmine
Quinolizidine derivatives	Lupinine group	Lupinine, nupharidin
	Cytisine group	Cytisine
	Sparteine group	Sparteine, lupanine, anahygrine
Pyridine derivatives	Simple derivatives of pyridine	Trigonelline, ricinine, arecoline
	Polycyclic noncondensing pyridine derivatives	Nicotine, nornicotine, anabasine, anatabine
	Sesquiterpene pyridine derivatives	Evonine, hippocrateine, triptonine
Isoquinoline derivatives and related alkaloids	Cularine group	Cularine, yagonine
	Pavines and isopavines	Argemonine, amurensine
	Benzopyrrocolines	Cryptaustoline
	Aporphines	Glaucine, coridine, liriodenine
Quinazoline derivatives	3,4-Dihydro-4-quinazolone derivatives	Febrifugine
	1,4-Dihydro-4-quinazolone derivatives	Glycorine, arborine, glycosminine
	Pyrrolidine and piperidine quinazoline derivatives	Vazicine (peganine)
Quinoline derivatives	Tricyclic terpenoids	Flindersine
	Furanoquinoline derivatives	Dictamnine, fagarine, skimmianine
	Quinines	Quinine, quinidine, cinchonine, cinchonidine
Alkaloids with nitrogen in the side chain (protoalkaloids)		
β-Phenylethylamine derivatives	–	Tyramine, ephedrine, pseudoephedrine, mescaline, cathinone, catecholamines (adrenaline, noradrenaline, dopamine)
Colchicine alkaloids	–	Colchicine, colchamine

(continued)

Table 1 (continued)

Class	Major groups	Examples
Muscarine	–	Muscarine, allomuscarine, epimuscarine, epiallomuscarine
Benzylamine	–	Capsaicin, dihydrocapsaicin, nordihydrocapsaicin, vanillylamine
Polyamines alkaloids		
Putrescine derivatives	–	Paucine
Spermidine derivatives	–	Lunarine, codonocarpine
Spermine derivatives	–	Verbascenine, aphelandrine
Peptide (cyclopeptide) alkaloids		
Peptide alkaloids with a 13-membered cycle	Nummularine C type	Nummularine C, Nummularine S
	Ziziphine type	Ziziphine A, sativanine H
Peptide alkaloids with a 14-membered cycle	Frangulanine type	Frangulanine, scutianine J
	Scutianine A type	Scutianine A
	Integerrine type	Integerrine, discarine D
Peptide alkaloids with a 15-membered cycle	Mucronine A type	Mucronine A
Pseudoalkaloids (terpenes and steroids)		
Diterpenes	Lycoctonine type	Aconitine, delphinine
Steroids	–	Solasodine, solanidine, veralkamine, batrachotoxin

This table was adopted and modified from the link. https://en.wikipedia.org/wiki/Alkaloid

2 Groups of Alkaloids: (Hesse 2002g)

1. True alkaloids, which contain nitrogen in the heterocycle and their precursors from amino acids, such as atropine, nicotine, and morphine (Plemenkov 2001a). This group also includes piperidine alkaloids coniine and coniceine (Hesse 2002h) although they do not originate from amino acids (Dewick 2002b)
2. Protoalkaloids, which contain nitrogen and also originate from amino acids (Plemenkov 2001a) such as ephedrine, adrenaline, and mescaline
3. Polyamine alkaloids—derivatives of putrescine, spermidine, and spermine
4. Peptide and cyclopeptide alkaloids
5. Pseudoalkaloids—alkaloid-like compounds that do not originate from amino acids, such as terpene-like and steroid-like alkaloids (Plemenkov 2001b)

3 Alkaloids, Chemical and Physical Aspects

Chemically, many alkaloids have oxygen in their structure; those compounds are usually colorless crystals at ambient conditions. Nicotine and coniine are oxygen-free alkaloids, usually volatile, colorless, oily liquids. Other alkaloids are yellow, such as berberine, and orange like sanguinarine (Grinkevich and Safronich 1983). Most alkaloids dissolve weakly in water and freely dissolve in organic solvents, for example, chloroform or 1,2-dichloroethane and diethyl ether. Cocaine, codeine, caffeine, and nicotine are slightly soluble in water. Alkaloids and acids form salts of different strengths. These salts are easily soluble in water and ethanol and ailing soluble in many organic solvents (Grinkevich and Safronich 1983). The production of alkaloids in plants seemed to have developed in response to nourishing by herbivorous animals; however, some animals have developed the ability to detoxify alkaloids (Fattorusso and Taglialatela-Scafati 2008).

4 Biotechnology and Alkaloid Production

In the last two decades, biotechnology approaches for the secondary metabolite production were rapidly grown. Synthesis of secondary metabolites from plants and other organisms by large-scale culture in bioreactors is technically achievable and promising (Verpoorte et al. 2002). Alkaloids are very important secondary metabolites known to play a significant role in many pharmaceutical applications leading to an increased commercial importance recently (Ahmad et al. 2013). The biotechnology approaches are very practical to select, improve, and characterize medicinal plants. Plant cell culture techniques form a potential renewable source of valuable medicinal plant compounds, which cannot be produced by microbial cells or chemical synthesis (Sihasar et al. 2011).

The use of plant cell culture as a major source of biosynthetic enzymes and well-known application of molecular techniques to the alkaloid ground have expedited the identification of several genes involved in many alkaloid biosynthesis. The followings are some examples on these enzymes which are produced by *Catharanthus roseus*: Desacetoxyvindoline 4-hydroxylase, Desacetoxyvindoline acetyltransferase, Glutamine synthetase, Peroxidase, Tabersonine 16-hydroxylase, and β-Glucosidase (Facchinin 2001).

4.1 Biotechnology Approaches for Alkaloid Biosynthesis

4.1.1 Tissue Culture of Plant Cell

Plant tissue culture is usually used for the in vitro sterile cells culture, organs, tissues, and their constituents under distinct conditions. Plant tissue culture is an advanced field of applied plant technology, including plant biotechnology. The tissue culture technique is effective because almost all the plant cells can be produced from stem cells. These cells carry all genetic materials and cellular mechanism necessary to generate a full organism. This technology can be used to produce a large number of plants that are genetically similar to a parent plant as well as to another.

Table 2 The table below summarizes number of successful attempts to produce some very important pharmaceuticals in commercial quantities using plant cell culture technique

S. number	Drug name	Source of production (plant)	Treatment
1	Taxol (Plaxitaxol)	Bark of Taxus tree	Anticancer
	Diterpene alkaloid)		
2	Morphine and Codeine	*Papaver somniferum*	Pain killer
3	Ginsenosides	Root of *Panax ginseng*	Tonic and highly prized medicine
4	Berberine (isoquino alkaloid)	Root of *Coptis japonica*	Antibacterial, anti-inflammatory, and immune-enhancing properties
5	Diosegnin	*Dioscorea deltoidea*	Combined oral contraceptive pills
6	Vinblastine and Vincristine	*Catharanthus roseus*	Anticancer and antitumor activity against leukemia and solid tumors

Recent advances in the area of plant tissue culture have resulted in the manufacturing of many therapeutic pharmaceutical compounds (Table 2). Wide variety of pharmaceutical drugs was produced using this approach, such as alkaloids, terpenoids, steroids, saponin, phenolics, and amino acids (Sihasar et al. 2011).

4.1.2 Genetic Transformation

The great development that has happened in recent years in the field of biotechnology and molecular biology made it easy and convenient for transferring and propagating genes in plant cells for the purpose of commercially produce alkaloid related drugs. Research in the area of plant biotechnology has resulted in the production of many pharmaceutical drugs. The introduction of new genes with new phenotype into the plants represents one of the important developments in recent advances of plant biotechnology including production of therapeutic compounds in large quantities, and clear the way for the production of many biologically active natural compounds such as artemisinin, paclitaxel, scopolamine, etc. A plant pathogenic bacteria known as *Agrobacterium tumefaciens* and *Agrobacterium rhizogenes* have been proved to be effective and highly useful vector for the production of genes into the plant genome resulting in the transfer and integration of genes of the plasmids from the bacteria into the plant DNA, transformed neoplastic tissues, crown galls, and hairy roots (Ahmad et al. 2013).

Applying the genetic transformation approach, the plants of the Papaver genus were able to synthesize morphinan alkaloids. The plant was transformed with codeinone reductase gene and constitutively expressed cDNA. Most transgenic cells revealed great increases in the alkaloid content in the field trials. This resulted in approximately tenfold higher levels of this product compared with non-transgenic plant (Larkin et al. 2007).

4.1.3 Metabolic Engineering for Alkaloid Production

Metabolic engineering is a tool for enhancing genetic and regulatory processes within cells to raise the cells' production of a certain metabolite. Such practices are chemical systems that use a chain of biochemical reactions which allow cells to convert raw materials into compounds needed by the cells (Yang et al. 1998). Large number of pharmaceutical products (drugs) originate from plant sources. One of the main challenges for large-scale drug production is the very low concentrations of some bioactive chemicals in plant cells. Metabolic engineering technique has eased the development of modified plant cell as other production platforms that can be used for this purpose (Leonard et al. 2009).

In recent studies, metabolic engineering approach was used to produce several types of plant isoquinoline alkaloids (PIAs) which possess a very important therapeutics and biotechnological properties. One major obstacle in production of these alkaloids is their low concentration in nature. Metabolic engineering offers opportunity to resolve issues related to divergence, limited productivity and availability of plant alkaloids. Plant tissue culture, plant cells and microbial cell cultures can perform as biofactories by offering their genetic information and metabolic machinery for the purpose of enhancing the conditions and increasing the yield of certain alkaloid (Diamond and Desgagne-Penix 2016).

Many alkaloids which synthesized by plants now used as therapeutic agents include morphine and codeine as analgesics, vinblastine and taxol as anticancer agents, and colchicines as gout suppressant. The powerful biological activity of alkaloids has led to their industrial exploitation as therapeutic, stimulant, narcotic, and poisons (Table 3) (Smita and Ashok 2013). Plant-derived alkaloid are now widely used as pharmacological and therapeutic drugs. Wide range of plants demonstrates their ability to contain such very important and significant chemical compounds. Table 3 below summarizes selected plants, their alkaloids, and its medicinal properties.

Applying plant tissue culture technique (*in vitro*) to produce different types of alkaloids showed that different types of alkaloids can be produced by various cell culture techniques, such as cell suspension, callus, hairy root culture, root culture and biotransformation (Table 4).

4.1.4 Microbial Synthesis of Alkaloids

It is well known and acknowledged that plant secondary metabolites are very important source of natural products with therapeutic properties. Pharmaceutical companies recently were put great efforts in screening of natural products as potential pharmaceutical drugs on a commercial scales. Production of plant metabolites in microorganisms offers an opportunity to improve the drug discovery development. Microbes showed very successful examples on the industrial production of some compounds such as antimalarial artemisinic acid which produced in yeast. In addition, microbial production expedites the possibility to synthesize a new

Table 3 Pharmacological applications of some important plant-derived alkaloids

Pharmacological applications of some important plant-derived alkaloids		
Alkaloid	Plant source	Medicinal properties
Ajmaline	*Rauvolfia serpentina*	Antiarrhythmic, antihypertensive
Berberine	*Berberis vulgaris*	Antimicrobial
Caffeine	*Coffee arabica*	Stimulant, Insecticide
Camptothecin	*Camptotheca acuminate*	Antineoplastic
Cocaine	*Erythroxylon coca*	Analgesic, narcotic, local anesthetic
Codeine	*Papaver somniferum*	Analgesic, antitussive
Emetine	*Uragoga ipecacuanha*	Antiamoebic, expectorant, emetic
Hyoscyamine	*Atropa belladonna* and others	Anticholinergic
Irinotecan	Semisynthetic derivative of camptothecin	Chemotherapeutics
Morphine	*Papaver somniferum*	Analgesic, narcotic
Nicotine	*Nicotiana tabacum*	Stimulant
Noscapine	*Papaver somniferum*	Analgesic, antitussive
Oxycodone	Semisynthetic derivative	Analgesic
Oxymorphone	Semisynthetic derivative	Analgesic
Papaverine	*Papaver somniferum*	Vasodilator
Pilocarpine	*Pilocarpus jaborandi*	Cholinergic
Quinidine	*Cinchona* spp.	Antiarrhythmic
Quinine	*Cinchona* spp.	Antimalarial
Reserpine	*Rauvolfia serpentina*	Tranquilizer, antihypertensive
Sanguinarine	*Sanguinaria canadensis*	Antibacterial
Scopolamine	*Hyoscyamus niger and others*	Sedative, anticholinergic
Strychnine	*Strychnos nux-vomica*	Stimulant, poison
Taxol	*Taxus brevifolia*	Antineoplastic
Topotecan	Semisynthetic derivative	Chemotherapeutics
Vinblastine and vincristine	*Catharanthus roseus*	Antineoplastic, chemotherapeutics
Vindesine	Semisynthetic derivative	Chemotherapeutics
Vinflunine	Semisynthetic derivative	Chemotherapeutics
Vinorelbine	Semisynthetic derivative	Chemotherapeutics
Yohimbine	*Pausinystalia yohimbe*	Erectile dysfunction treatment

This table adapted from reference (Smita and Ashok 2013)

compound with new activity or advance the medical properties of current drugs (Narcross et al. 2016).

Another example on microbial synthesis of alkaloids was achieved by Sato and Kumagai (2013). In an effort to solve the problems associated with secondary metabolite production in plants, they developed a microbial procedure for the production of isoquinoline alkaloids which involve the combination of the microbial and plant metabolic pathways into a single system (Sato and Kumagai 2013).

Table 4 Summarizes the in vitro plant cell/tissue cultivation for alkaloid production

In vitro plant cell/tissue cultivation for alkaloid production		
Plant source	Type of alkaloid	Alkaloid production by plant cell culture
Ailanthus altissima	Alkaloids	Cell suspension
Ailanthus altissima	Cathinone alkaloids	Cell suspension
Brucea javanica (L.) Merr.	–	–
Catharanthus roseus	Indole alkaloids	Cell suspension
Choisya ternata	Furoquinoline alkaloids	Cell suspension
Cinchona L.	Alkaloids	Cell suspension
Corydalis ophiocarpa	Isoquinoline alkaloids	Callus
Fumaria capreolata	–	–
Duboisia leichhardtii	Tropane alkaloids	Callus
Hyoscyamus niger	Tropane alkaloids	Callus
Nandina domestica	Alkaloids	Callus
Nicotiana rustica	Alkaloids	Callus
Nothapodytes foetida	Camptothecin	Callus
Ophiorrhiza pumila	Camptothecin related alkaloids	Callus
Peganum harmala L.	β-Carboline alkaloids	Cell suspension
Ptelea trifoliata L.	Dihydrofuro [2,3-b] quinolinium alkaloids	Callus
Rauwolfia sellowii	Alkaloids	Cell suspension
Thalictrum minus	Berberine	Cell suspension
Catharanthus roseus	Catharanthine	Cell suspension
Ephedra spp.	L-Ephedrine	Cell suspension
–	D-Pseudoephedrine	–
Cinchona ledgeriana	Quinoline alkaloids	Hairy root culture
Neotyphodium uncinatum	Loline alkaloids	Cell suspension
Catharanthus roseus	Ajmalicine	Cell suspension
Rauvolfia sp.	–	–
Catharanthus roseus	Vinblastine	Cell suspension
Catharanthus roseus	Vincristine	Cell suspension
Rauvolfia sp.	Reserpine	Cell suspension
Vinca sp.	Vincamine	Cell suspension
Cinchona sp.	Quinine	Cell suspension
Cinchona sp.	Quinidine	Cell suspension
Ochrosia elliptica	Ellipticine	Cell suspension
Rauvolfia sp.	Rescinnamine	No data available
Camptotheca acuminala	Camptothecine	Cell suspension
Cephaelis ipecacuanha	Emetine	Root culture
Coffea, Thea	Caffeine	Cell suspension
Theobroma	Theobromine	Cell suspension

(continued)

Table 4 (continued)

In vitro plant cell/tissue cultivation for alkaloid production		
Plant source	Type of alkaloid	Alkaloid production by plant cell culture
Atropa belladonna	Atropine	Hairy root culture
Atropa belladonna	Scopolamine	Hairy root culture
Duboisia leichhardtii	Scopolamine	Hairy root culture
Coptis japonica	Berberine	Cell suspension
Papaver somniferum	Morphine	Cell suspension
Papaver somniferum	Codeine	Cell suspension
Nicotiana sp.	Nicotine	Cell suspension
Colchicum autumnale	Colchicine	Callus culture
Dicentra peregrina	Alkaloids	Shoot culture
Calystegia sepium	Tropane alkaloids	Root culture
Hyoscyamus albus	Hyoscyamine	Root culture
Hyoscyamus muticus	Hyoscyamine	Hairy root culture
B. vulgaris	Betalains	Root culture
Papaver somniferum	Codeine	Biotransformation
Spirulina platensis	Morphine	Biotransformation

This table adapted from reference (Smita and Ashok 2013)

Nakagawa et al. (2011) reported the construction of a bacterial platform for the production of plant alkaloid from simple carbon sources. In these procedures, they build up a tailor-made alkaloid biosynthetic pathway from L-tyrosine using *E. coli* cells. The genetically modified strain has the ability to produce 46.0 mg L^{-1} of (*S*)-reticuline from glycerol, suggesting that the fermentation procedures would allow low-cost production of several various alkaloids (Nakagawa et al. 2011).

5 Conclusion

The recent advances in plant biotechnology techniques make easy and feasible to produce alkaloids on the commercial and industrial scales. Knowing more details about the genetic information and metabolic machinery of the plants facilitate using these techniques on various types of alkaloids-producing plants. Techniques, such as plant cell culture, genetic transformation, metabolic engineering, and using microbes, clearly demonstrate the high potential to produce certain types of alkaloids in significant and promising quantities. Plant biotechnology approaches (techniques) which discussed in this chapter have proven that medicinal plants can be used efficiently to produce many types of pharmaceutical alkaloids for therapeutic applications.

Using metabolic engineering strategies, which involved gene transfer, showed number of fruitful and successful achievements in terms of increasing alkaloid production. For example, the transformation of *Nicotiana tabacum* with gene from *Catharanthus roseus* resulted in high-level production of tryptamine (Smita and Ashok 2013). Taking into consideration the degree of complexity in bacteria and other microorganism systems, and their ability to work as factories for large scale of production, microbes are more accessible than plant cells and tissue culture to achieve more efficient processes for alkaloid production.

In conclusion, the above discussed plant biotechnology approaches offer a great opportunity in order to produce pharmaceutically important alkaloids on a large scale. Eventually, for plant and microbial systems, large-scale production, understanding the genetic manipulation of alkaloid genes, and very efficient extraction methods will be crucial factors for the production of alkaloid on commercial level. Finally, a better understanding of genetic manipulation tools as well as the metabolic regulations of genes producing alkaloids will clear the way for more achievements in the area of plant biotechnology for alkaloid production.

References

Ahmad S, Garg M, Tamboli ET, Abdin MZ, Ansari SH (2013) *In vitro* production of alkaloids: factors, approaches, challenges and prospects. Pharmacogn Rev 7(13):27–33

Aniszewski T (2007) Alkaloids—secrets of life. Alkaloids chemistry, biological significance application and ecological role. Elsevier, Amsterdam

Begley TP (2009) Encyclopedia of chemical biology. Wiley

Cushnie TP, Cushnie B, Lamb AJ (2014) Alkaloids: an overview of their antibacterial, antibiotic-enhancing and antivirulence activities. Int J Antimicrob Agents 44(5):377–386

Dewick PM (2002b) Medicinal natural products. A biosynthetic approach, 2nd edn. Wiley, New York, p 381

Diamond A, Desgagne-Penix I (2016) Metabolic engineering for the production of plant isoquinoline alkaloids. Plant Biotechnol J 14:1319–1328

Chemical Encyclopedia: alkaloids. http://www.xumuk.ru/encyklopedia/119.html

Facchinin PJ (2001) Alkaloid biosynthesis in plants: biochemistry, cell biology, molecular regulation, and metabolic engineering applications. Annu Rev Plant Physiol Plant Mol Biol 52:29–66

Fattorusso E, Taglialatela-Scafati O (2008) Modern alkaloids: structure, isolation, synthesis and biology. Wiley, Weinheim

Grinkevich NI, Safronich LN (1983) The chemical analysis of medicinal plants: Proc. allowance for pharmaceutical universities. M.P 131

Hesse M (2002d) Alkaloids: nature's curse or blessing? Wiley, Weinheim, p 413

Hesse M (2002e) Alkaloids: nature's curse or blessing? Wiley, Weinheim, p 204

Hesse M (2002f) Alkaloids: nature's curse or blessing? Wiley, Weinheim, p 11

Hesse M (2002g) Alkaloids: nature's curse or blessing? Wiley, Weinheim, p 12

Hesse M (2002h) Alkaloids: nature's curse or blessing? Wiley, Weinheim, p 31

Jacobsen O (1882) Alkaloide. In: Ladenburg, Handwörterbuch der Chemie, vol 1. Eduard Trewendt, Breslau, pp 213–422

Kittakoop P, Mahidol C, Ruchirawat S (2014) Alkaloids as important scaffolds in therapeutic drugs for the treatments of cancer, tuberculosis, and smoking cessation. Curr Top Med Chem 14(2):239–252

Larkin PJ, Miller JA, Allen RS, Chitty JA, Gerlach WL, Frick S, Kutchan TM, Fist AJ (2007) Increasing morphinan alkaloid production by over-expressing codeinone reductase in transgenic Papaver somniferum. Plant Biotechnol J 5(1):26–37
Leonard E, Runguphan W, O'Connor S, Prather KJ (2009) Opportunities in metabolic engineering to facilitate scalable alkaloid production. Nat Chem Biol 5:292–300
Lewis RA (1998) Lewis' dictionary of toxicology. CRC Press, Boca Raton, p 51
Luch A Molecular, clinical and environmental toxicology. Springer, Basel, p 20
Manske RHF (1965) The alkaloids. Chemistry and physiology, vol VIII. Academic, New York, p 673
Nakagawa A, Minami H, Kim J, Koyanagi T, Katayama T, Sato F, Kumagai H (2011) A bacterial platform for fermentative production of plant alkaloids. Nat Commun 2 Article number 326
Narcross L, Fossati E, Bourgeois L, Dueber JE, Martin VJJ (2016) Microbial factories for the production of benzylisoquinoline alkaloids. Trends Biotechnol 34 No. 3
Plemenkov VV (2001a) Introduction to the chemistry of natural compounds. Kazan, p 242
Plemenkov VV (2001b) Introduction to the chemistry of natural compounds. Kazan, p 246
Raymond SS, Jonathan SJ, Watkins-Pitchford JM (2010) The essence of analgesia and analgesics. Cambridge University Press, Cambridge, pp 82–90
Russo P, Frustaci A, Del BA, Fini M, Cesario A (2013) Multitarget drugs of plants origin acting on Alzheimer's disease. Curr Med Chem 20(13):1686–1693
Sato F, Kumagai H (2013) Microbial production of isoquinoline alkaloids as plant secondary metabolites based on metabolic engineering research. Proc Jpn Acad Ser B Phys Biol Sci 89(5):165–182
Sihasar B, Rahimi M, Tavassoli A, Raissi A (2011) Application of biotechnology in production of medicinal plants. American-Eurasian J Agric Environ Sci 11(3):439–444
Smita S, Ashok KS (2013) Biotechnology and genetic engineering for alkaloid production. In: Natural products. Springer, Berlin
Verpoorte R, Contin A, Memelink J (2002) Biotechnology for the production of plant secondary metabolites. Phytochem Rev:13–25
Yang YT, Bennet GN, San KY (1998) Genetic and metabolic engineering. Electron J Biotechnol 1(3)

Agricultural, Pharmaceutical, and Therapeutic Interior of *Catharanthus roseus* (L.) G. Don

Md. Abul Barkat, Harshita Abul, and Md. Akhlaquer Rahman

Abstract *Catharanthus roseus* (*C. roseus*) (L.) G. Don known as Madagascar periwinkle (MP) is a popular ornamental plant found in gardens and homes across the world. *C. roseus* (L.) G. Don has been widely distributed for long enough and gained popularity as diverse application in medicinal uses for a variety of purposes such as antimicrobial, antioxidant, anthelmintic, antifeedant, antisterility, antidiarrheal, and antidiabetic effect and have been used in the treatment of leukemia, Hodgkin's disease, malignant lymphomas, neuroblastoma, Wilms tumor, Kaposi's sarcoma, and mycosis fungoides to improve cerebral blood flow and treat high blood pressure. The pharmacology of the plant was found to be associated mostly especially with the alkaloids that occupies almost all parts of the plant. *C. roseus* (L.) G. Don is a legendary medicinal plant mostly because of possessing two invaluable antitumor terpenoid indole alkaloids (TIAs), vincristine and vinblastine. The ethnobotanical significance of *C. roseus* (L.) G. Don is exemplified by its international usage as a traditional remedy for abundant ailments and not only for cancer. TIAs are present only in micro quantities in the plant and are highly poisonous per se rendering a challenge for researchers to increase yield and reduce toxicity. Good agronomic practices ensure generous propagation of healthy plants that serve as a source of bioactive compounds and multitudinous horticultural applications. In this chapter, an attempt has been made to summarize the agricultural, pharmaceutical,

Md.A. Barkat, M.Pharm, Ph.D (✉)
Department of Pharmaceutics, Nanomedicine Research Lab, Faculty of Pharmacy, Jamia Hamdard, New Delhi 110062, India

Department of Pharmaceutics, Faculty of Pharmacy, Integral University, Kursi Road, Lucknow, India
e-mail: abulbarkat05@gmail.com

H. Abul, M.Pharm
Department of Pharmaceutics, College of Pharmacy, SGIT, Ghaziabad, U.P., India
e-mail: harshiab001@gmail.com

Md.A. Rahman
Department of Pharmaceutics, Faculty of Pharmacy, Integral University, Kursi Road, Lucknow, India
e-mail: rahmanpharma@gmail.com

M. Naeem et al. (eds.), *Catharanthus roseus*,
DOI 10.1007/978-3-319-51620-2_5

and pharmacological applications in a precise way to help the scientists and learners to understand the basic values of the plant.

Keywords *Catharanthus roseus* • Madagascar periwinkle • Hodgkin's disease • Wilms tumor • Kaposi's sarcoma • Terpenoid indole alkaloids

List of Abbreviations and Symbols

2,4-D	2,4-Dichlorophenoxyacetic acid
CCC	(2-chloroethyl) trimethylammonium chloride
GA	Gibberellic acid
GTP	Guanosine triphosphate
IAA	Indole acetic acid
ITIS	Integrated Taxonomic Information System
mAb or moAb	Monoclonal antibodies
MDR	Multiple drug resistance
P-gp	P-glycoprotein
SGD	Strictosidine β-D-glucosidase
TAA	Tumor-associated antigens
TIAs	Terpenoid indole alkaloids
TSN	Taxonomic Serial Number
VBL	Vinblastine
VCR	Vincristine
VDS	Vindesine
VRLB	Vinorelbine

1 Introduction

The therapeutic uses of natural products is considered primordial as human civilization and, for an extended period of time, therapeutic agents from mineral, plant, and animal were the foremost sources (De Pasquale 1984). The industrial revolt and the advancement of organic chemistry resulted in a first choice for synthetic products with therapeutic value meant for the treatment of ailments. The causes for this were that pure compounds were easily synthesized, structural changes to bring into being pharmacologically more active and safer drugs possibly will be straightforwardly performed, and the economic power of the pharmaceutical companies was escalating. Besides, during the development of human culture, the use of naturally found drugs has had magical-religious importance and different perspective concerning the concepts of health and disease existed within every culture. Undoubtedly, this approach was beside the latest modus vivendi of the industrialized western societies, in which medicinal agents from natural resources were well thought-out either an alternative for poorly educated or low earnings natives or just as religious superstition of no pharmacological importance. Near about 25% of the medicinal agents prescribed globally obtained from plant sources, 121 such phytoconstituents being used currently. World Health Organization (WHO) considered 252 drugs of natural origin as basic and necessary, from which 11% are entirely of plant derive and a considerable amount are synthetic drugs come from natural precursors.

More than 130 terpenoid indole alkaloids were obtained from *C. roseus* (L.) G. Don plant, its exception no other single plant species is revealed to have such a wide array of complex nitrogen containing basic compound (alkaloids). It is as well one of the most comprehensively explored medicinal herbs. It derives its importance mostly from its expensive, high demand, and limited shoot-specific anticancerous bisindole alkaloids, vinblastine and vincristine. At present, the plant remains the most imperative resource of these bisindole alkaloids (formed in plant by the condensation of monomeric TIAs, vindoline, and catharanthine) even though a few reports of their overall synthesis.

C. roseus (L.) G. Don sparked the attention of scientists and researchers of western laboratories, when the natives of Madagascar had started to use the plant for medical benefits. Nowadays, the plant is in jeopardy of destruction in its natural surroundings, beyond all due to the devastation of its habitat, whereas at the same, being well known globally. In its natural surroundings, the plant found in rosy and white varieties. One of the noteworthy peculiarities of *C. roseus* (L.) G. Don is its capability to stay alive under dry environment. Excluding its beauty and its resistance, it was in the foremost line its use as a medicinal plant, in which scientists were mostly paying attention (Rates 2001; Ashutosh and Suman 2013). In this chapter, we aimed to discuss the recent advancement with respect to botanical, agrotechnological, geographical, pharmaceutical, and pharmacological aspects (mainly anticancerous bisindole alkaloids, namely,vinblastine and vincristine) of *C. roseus* (L.) G. Don and also draw attention to future prospects.

2 Botanical Description of *C. roseus* (L.) G. Don Plant

The Madagascar periwinkle (*C. roseus* (L.) G. Don) is one of the most comprehensively investigated medicinal plants. The *Catharanthus* (genus) consists of eight species; all derived from Madagascar with the exception of *Catharanthus pusillus* (Murr.) G. Don, which is limited to India and Sri Lanka. *Catharanthus* is very closely correlated to Vinca. No consistent qualitative distinctions have been found in terms of alkaloids content in relation to the divergent colors of the corolla of *C. roseus* (L.) G. Don (Gurib-Fakim and Brendler 2004). *C. roseus* (L.) G. Don is considered as an imperative floral variety in horticulture and is one of the few medicinally important plants that has an extensive record. *C. roseus* (L.) G. Don is easily growing, erect, perennial herb or annual evergreen, small shrub, or herbaceous plant with woody base that is found at the height of up to 100 cm and produces milky latex. The branching is initiated from its base. The diameter of stem ranges from 1.1 to 8.9 cm and the distance between internode ranges from 0.5 to 6.8 cm. The shape of leaves is elliptical to oblong, 2.5–9 cm long, 1–3.5 cm broad.

The upper surface of the leaves is glossy green while lower surface pale green amid a pale midrib and differing in orientation. The base of leaf is acute whereas the apex is rounded to mucronate and the margin is entire. Leaf shows the presence of prominent veins on the lower surface. Green or red color petioles are present with the size range of 1 to 1.8 cm long. The leaves of the plant that contain stomata are 7–30 μm long and 7–30 μm wide. The inflorescence is racemose. Flowers are violet colored (var. *roseus*), white (var. *alba*) or white with a red eye (var. *ocellatus*) and pentamerous, actinomorphic with the size ranging from 2.0 to 5.1 cm in diameter. The lobes of calyx are linear to subulate and pubescent. The calyx is five parted and the size of sepal ranges from 0.09 to 0.70 cm long. The corolla tube is cylindrical in shape, 2–3 cm long along with five petal-like lobes. Stamens are placed 0.4–0.6 cm underneath the corolla mouth that comprises very small white colored filaments and filiform, subsessile anthers. The anthers are attached to the filament and sagittate to narrow lanceolate in shape. The pollen is elliptical in all sections through the long-axis or subglobose, smooth, and the size ranges from 10 to 60 μm in length. The anthers are devoid of stigma. The style is attenuated in shape with the size range of 0.5–3.0 cm long. There are two divergent carpels, in each carpel 10–30 ovules in two series are present. The fruit comprises of two 0.6–3.5 cm long elongated, cylindrical, piercing follicles (mericarps) and diverging or parallel. Each follicle contains numerous (3–35) blackish seeds. The shape of seeds is oblong or cylindrical amid of size ranges from 0.05 to 0.70 cm. The hilum is situated in a longitudinal depression on one side. The surface of seed is minutely forming a reticulum (reticulate). The roots pull out to 70 cm depth with the diameter of 0.5–6.0 cm (Mishra and Kumar 2000; Sreevalli et al. 2002; Plaizier 1981; Chaudhary et al. 2011). Janaki Ammal and Bezbaruah (1963) stated that the tetraploid plants develop more robustly with larger flowers. The floral morphology of *C. roseus* is advantageous for self and insect or cross-pollination (Janaki Ammal and Bezbaruah 1963).

In *C. roseus* (L.) G. Don, intraflower self-pollination is widespread, for the reason that the stigma may move toward into contact with the anthers, yet subsequent to anthesis. The rate of outcrossing may possibly differ with environmental circumstances and the occurrence of seasonal pollinating agent (butterflies). *C. roseus* (L.) G. Don will flower and fruit the entire year round in warm climate condition. The seeds generally fall nearby to the parent plant, however occasionally transported by ants. The first flowers will come into view after 6–8 weeks subsequent to germination. Some branches or even the complete plant will die when the temperature goes beyond 5 °C. While the temperature increases, the plant will regrow from basal axillary buds, particularly following hard pruning of shoots and roots. With no pruning, the plant redevelops mostly from the top (Ross 2003; van der Heijden et al. 2004).

2.1 Adulterations and Substitutes

Anticancerous bisindole alkaloids (vincristine, vinblastine) and interrelated compounds put off mitosis in a diverse way from colchicine (obtained from *Colchicum autumnale* L.), one more potent antineoplastic agent. Other plant species of

Apocynaceae family like Rauvolfia spp also shows the presence of ajmalicine (Raubasine) and its derivatives (Neuwinger 2000; Snoeijer 2001; Ross 2003; van der Heijden et al. 2004).

2.2 Scientific Classification

In the beginning, there was an assortment of uncertainty in the scientific literature concerning the accurate classification of the genus of the plant. Despite various names, such as *Ammocallis rosea* and *Lochnera rosea*, *Vinca rosea* is the most commonly used name (Mishra and Kumar 2000; Stearn 1975). In 1753, Linnaeus recognized the genus *Vinca* into two distinguished species *V. major* and *V. minor* in his species Plantarum (1:209). Further in 1759, Linnaeus added the tropical *Vinca rosea* (now *C. roseus* (L.) G. Don) to these two originally European species, even though it did not fit his generic explanation with reference to the stamens. Ludwig Reichenbach was the first to distinguish *V. rosea* as being generically different from the genus *Vinca* and further in 1828 he suggested generic name *Lochnera* for it. Nevertheless, the name did not obtain nomenclatural legitimacy as he was unable to endow with any generic explanation for the genus Lochnera. George Don in 1835 allocated the name *Catharanthus* to the genus represented by *V. rosea* in his general system of gardening and botany (4: 95). The word *Catharanthus* is derivative of the Greek terms katharos (pure) and anthos (flower), mentioning to the neatness, beauty, and elegance of the flower. The explanatory variances among the two genera *Vinca* and *Catharanthus* have currently been well accepted (Stearn 1966). However, the name periwinkle is recognized for both genera, they vary distinctly in their habitation, floral make-up, allocation, biochemistry as well as cytology. Vincas have funnel-shaped corolla tube (increasingly widens from below upward), filaments bend onward and then directed towards the back like a knee (Linnaeus termed them as inflexa, retroflexa), and the anthers that are each topped with a small hairy flap-like addendum, arch onto the stigma. In contrast, the corolla tube of *Catharanthus* is cylindrical in shape and the filaments are very small in size and anthers directly emerge on the corolla tube, while *C. lanceus* has long filaments. Furthermore, in *Catharanthus* species anthers deficit terminal appendages and merge conically over the stigma (Virmani et al. 1978). *C. roseus* (L.) G. Don obtained several vernacular names throughout the course of its extent and naturalization over the tropic and subtropic regions, as listed in Table 1 (Plaizier 1981; Virmani et al. 1978; Joy et al. 1998; Mishra and Kumar 2000).

The taxonomic classification of genus *Catharanthus* represented as Family: Apocynaceae; Subfamily: Plumeroideae; Tribe: Plumerieae; Subtribe: Alstoniiae; Genus: *Catharanthus*. In Table 2, information regarding the designation, origin, and distribution of the eight species are given (Stearn 1975; Mishra and Kumar 2000; Whiting et al. 2011).

Table 1 Vernacular names of *Catharanthus roseus* (L.) G. Don (Mishra and Kumar 2000; Virmani et al. 1978; Plaizier 1981; Joy et al. 1998)

Country	Vernacular names
India	Sadabahar, Sadaphul (Sadaphuli), Nayantara, Rattanjot, Billaganneru, Gul Feringhi, Ainskati, Sudukadu mallikai, Cape-periwinkle, Nityakalyani, Baramasi, Church-yard blossom, Dead-man's flower, Sudukattu mallikai, Ushamanjairi
West Indies	Old maid, Ramgoat rose, Cayenne jasmine, Magdalena, Vicaria
Ethiopia	Phlox
China	Chang Chun Hua
Germany	Zimmerimmergrun
Philippines	Chichirica
Japan	Nichinchi
England	Bright-eyes, Cape periwinkle, graveyard plant, Madagascar periwinkle, old-maid, old-maid-flower, rose periwinkle, rosy periwinkle
Indonesia	Indische maagdepalm, Soldatenbloem, Kembang sari tijna, Kembang tembaga, Tapak dara

Table 2 The species of the genus *Catharanthus*

Species name	Place of origin
C. roseus (L.) G. Don	Madagascar, now naturalized throughout the tropics
C. ovalis Markgraf	Madagascar
C. trichophyllus (Baker) Pichon	Madagascar
C. longifolius (Pichon) Pichon	Madagascar
C. coriaceus Markgraf	Madagascar
C. lanceus (Bojer ex A. DC.) Pichon	Madagascar
C. scitulus (Pichon) Pichon	Madagascar
C. pusillus (Murray) G. Don	India, Sri Lanka

As per the Integrated Taxonomic Information System (ITIS), the Taxonomic Serial Number (TSN) designed for *C. roseus (L.) G. Don* (L.) G. Don is 30, 168 and at present established taxonomic hierarchy is given as follows:

Kingdom: Plantae, plants
Subkingdom: Tracheobionta, vascular plants
Superdivision: Spermatophyta, seed plants
Division: Magnoliophyta, flowering plants
Class: Magnoliopsida, dicotyledons
Subclass: Asteridae
Order: Gentianales
Family: Apocynaceae, dogbane family
Genus: *Catharanthus* G. Don, periwinkle
Species: *C. roseus (L.) G. Don* (L.) G. Don, Madagascar periwinkle

2.3 Historical and Geographic Distribution

The *C. roseus* (L.) G. Don was recognized for its medical principles even in 50 BC (Virmani et al. 1978). Though, it is prevalently known as Madagascar periwinkle for the reason that it is supposed that it is native to Western Indian Ocean's large island of Madagascar adjacent to Africa. *C. roseus* (L.) G. Don has been suggested as a well-liked ornamental plant in various tropical and subtropical regions globally. In the early eighteenth century in Paris, the plant was cultivated from seeds brought from Madagascar and was afterward spread from European botanical gardens to the tropical regions. In India, *C. roseus* (L.) G. Don is thought to have been spread by Portuguese missionaries in the mid of the eighteenth century via Goa (Mishra and Kumar 2000). In India, the plant was at the outset developed in cemeteries on account of its continually flowering character. Afterward it grows all over the tropical as well as subtropical regions of the country and possibly will be seen growing uncultivated onto the plains, lower foothills, and terai areas in Northern India and the hilly areas of Southern India. The plant, *C. roseus* (L.) G. Don is extensively grown commercially in the United States, Africa, Australia, China, India, Spain, and Southern Europe for medicinal purposes. The antineoplastic bisindole alkaloids obtained from this plant discover major markets in the United States, Hungary, West Germany, Italy, the Netherlands, and the UK (Whiting et al. 2011; PROTA 2011; Łata 2007).

The anticancer principles of alkaloids obtained from *C. roseus* (L.) G. Don was revealed by coincidence in the late 1950s during searches for hypoglycemic agents. *C. roseus* (L.) G. Don frequently found in sandy locations beside the coast, however also inshore on river banks, in savanna vegetation and in dry waste spaces and roadsides, occasionally in open forest or scrub and on rocky soils, typically on sandy soils. It is extremely salt tolerant and is typically found close to sea-level, however infrequently up to 1500 m altitude. *C. roseus* (L.) G. Don is preferably grown in a soil with pH range of 5.5–6.5. It flourishes well in dry, frost-free, and humid surroundings with favorable moisture setting, in complete sunny or partial shade, and in well-drained soil. *C. roseus* (L.) G. Don cannot confront so much water, damp soils, or a cool spring. *C. roseus* (L.) G. Don is a fast-growing plant which is easily cultivated in hot climate conditions. It is valued for its durability in dry and nutrient poor situation. Under unfavorable conditions or in inadequately drained soils, *C. roseus* (L.) G. Don changes into yellowish green while overwatering possibly will cause the bacterial and fungal rot diseases of stem and root (Plaizier 1981; Thomas and Latimer 1996; van Bergen 1996). Twofold increment in alkaloidal content of full-grown leaves was found under harsh water stress, but no variation in stems and immature leaves and it decreased in roots.

3 Agrotechnology

The technical aspects regarding agriculture of *C. roseus* (L.) G. Don have been well worked out and explanatory preview are well documented here.

3.1 Soil and Climatic Conditions

Although a lot of research and examination have been made over the last few decades on the phytochemical and medicinal value of *C. roseus* (L.) G. Don, only some studies have been carried out on the agronomic and genetic characteristic of this medicinal herb. The plant grows well in a large variety of soils and typical weather conditions. Even though the plant is extremely tolerant to environmental factor like heat and drought (Gupta 1977), it preferably grows in tropical and subtropical regions in typical climatic conditions (Mishra and Kumar 2000). In India, *C. roseus* (L.) G. Don propagates easily in Tamil Nadu, Karnataka, Gujarat, Madhya Pradesh, and Assam (Virmani et al. 1978). The plant also grows in North India especially subtropical regions, where climate changes during winter season is deleterious and affects its growth, flowering, and seed formation. The plant propagates on almost all types of soil although highly alkaline and drenched soils are inappropriate. For the commercial cultivation of *C. roseus* (L.) G. Don preferred light, sandy loam soil rich in humus or treated with raw liquid cow/buffalo dung supplies an aerated bedrock to the crop (Virmani et al. 1978; Mishra and Kumar 2000). In such type of soils, roots are easily harvested. In the region of Southern India, the plant is extensively grown on red laterite soils and rarely on black cotton soils (Gupta 1977). The plant is easily able to tolerate mild salty soils. A well-distributed rainfall of 100 cm and higher is greatly advantageous for such type of rainfed medicinal crop.

3.2 Manures and Fertilizers

The use of farmyard manures at the rate of about 15 ton/ha is appropriate for the development of roots and aerial parts of *C. roseus* (L.) G. Don. In place of application of farmyard manures, green manuring the field can also be greatly advantageous. In case of nonavailability of organic manure, a basal dose of 30 kg K_2O, 30 kg P_2O_5, and 20 kg urea per hectare is advisable to use when the availability of organic manure is not sure. During the growing period, top covering of 20 kg nitrogen in two equivalent divided doses with the time duration of 30–45 days is exceedingly favorable to the crop (Virmani et al. 1978).

3.3 Weeding and Irrigation

In the course of agriculturing process, crop possibly will have need of at least two weeding. The first weeding possibly done after 2 months of seed sowing/transplanting, and the second may possibly be done 2 months afterward. The application of fluchoraline at 0.75 kg/ha or alachor at 1 kg/ha gives an efficient control in excess of a wide variety of weeds in the plant (Mishra and Kumar 2000). In the locality where rainfall is circulating consistently all through the year, the plant doesn't have the need of any irrigation. On the other hand, in areas wherever the rainfall is limited for a particular period of time, 4–5 irrigations are required to acquire best possible yield. Some extent of drought can be tolerated by the plant. Although under critical water stress, the water potential is lower than −1.9 Mpa at leaf, which causes slowing down of growth related to a little enhancement in alkaloidal content of the plant (Saenz et al. 1993). Talha et al. (1975) reported that 25–50% soil moisture deficit (SMD) level was the optimum for the enhancement of dry weight of the leaf, stem, and root (Talha et al. 1975). On the other hand, a further increment in SMD from 50% to 75% caused a decrease in herb yield, even though with considerable increase in alkaloidal content.

3.4 Propagation

The freshly obtained seeds are the most acceptable propagules for *C. roseus* (L.) G. Don because the seeds are viable for a short period of time (<8 months). Propagation was started either by seeds directly sown in the field or germinated seeds (seedlings) transplanted into the field from nurseries. Before sowing the seeds, fields are well prepared by pulverization at the beginning of the rainy season. In direct sowing, seeds are sown in rows 45 cm apart, and afterward the seedlings acquired are thinned out to keep a distance of a minimum of 30 cm among plants. One hectare of land requires about 2500 g of seeds for sowing. On the other hand, seedlings are grown in a nursery to cut down on seed consumption. In a nursery, about 0.5 kg of seeds is sufficient to give seedlings for 1 ha of land. In nursery beds, seeds are sown about 2 months in advance of the projected transplanting of the new germinated seeds, which ought to bring into line through the commencement of the rainy time of year. Germination of seed takes place within 10 days of time period, and 60 days old seedlings are appropriate for transplanting. About 74,000 new germinated seeds per hectare are accommodates in an area of 45 × 30 cm. As researchers reported that the uneven plant spacings of 45 × 15 cm, 45 × 20 cm, and 45 × 30 cm have no effect on root yield although more leaf yield is found at a plant spacing of 45 × 20 cm (Mishra and Kumar 2000). A plant spacing of 30 × 20 cm has also been established to be most favorable. Kulkarni et al. (1987) concluded that the impenetrable planting of tetraploids leads to better yields. Vegetative reproduction of the plant is also feasible by means of cuttings planted in the field by using similar approach as used for seedlings (Kulkarni et al. 1987).

3.5 Growth Regulators

Application of weekly doses of 100 mg gibberellic acid (GA) to *C. roseus* (L.) G. Don plants for a period of 7 weeks and reaped twice a week exhibited a remarkable improvement in the height of plant and dry weight of stems. The change in leaf shape has seen from obovate to lanceolate although no effect was seen on flowering (Masoud et al. 1968). At the end of final harvesting process, the presence of total alkaloids in the leaves and roots decreased. Although in GA-treated plants, bisindole alkaloid mainly vinblastine content is enhanced in the leaves. The application of nitrogen fertilizer as well as long day circumstances may lead to enhancement of alkaloidal content of the roots in typical situation but unable to counterbalance the negative consequence of GA on overall alkaloidal content of plants in the roots. Plant hormone 2,4-dichlorophenoxyacetic acid (2,4-D) works as a defoliator, while the use of indole acetic acid (IAA) and GA causes enhancement of leaf number and height of plant (Virmani et al. 1978). After fruiting period, the above effect was more prominent. Exogenous treatment of boron in autotetraploids caused fall of leaf width along with follicle lengths (Mishra and Kumar 2000). As the researcher reported that the inclusion of growth retardants, for example, (2-chloroethyl) trimethylammonium chloride (CCC) resulted in the production of lateral branches as well as alkaloid content of *Catharanthus pusillus* enhanced (Basu 1992). Improvement in the extent of the photosynthetic period enhances the production of root alkaloids along with the weight of aerial parts (Virmani et al. 1978).

3.6 Harvesting and Yield

Harvesting must be completed 12 months subsequent to sowing. The harvest is obtained through cutting on 7.5 cm exceeding the ground and dried out for stems, leaves, and seeds. The preferred moisture level was achieved by further irrigation of the field, stalked by ploughing and roots collection. The collected roots are clean properly followed by shade dry. The alkaloid contents in the root is maximum while the plants in flowering state. Prior to the final harvest leaves can be harvested twice. The initial disrobing is made 6 months subsequent to planting and the next 3 months later than the first cutting. Lastly, the entire plant is harvested subsequent to a further 2–3 months. The collected leaves are also dehydrated in shadow for the period of 15 days until it turns out to be crispy, trailing about 75% of their initial weight. The leaves preserved for drying should be spread evenly and turned from time to time to avoid any fungal contamination and fermentation. The obtained dry plant leaves are kept in a cool and dry place. In irrigated situations, a harvest planted in 1 ha land, the total yields obtained is about 3.6 ton of fully dried leaves, 1.5 ton of stems as well as 1.5 ton of roots. The total yield is affected in rainfed situations; the obtained yield per hectare is approximately 2 ton of leaves, 1 ton of stems along with 0.75 ton of roots (Virmani et al. 1978). In normal field surroundings, plants exhibit natural difference in the biosynthesis and buildup of secondary metabolites owing to genotypic variations in addition to ecological and seasonal changes (Choudhury and Gupta 2002).

The harvested plant seeds require being dry prior to storing them ideally in a cool, dark, and previously dried container at temperatures below 20 °C. The collected aerial plant parts and roots of *C. roseus* (L) G. Don are properly cleaned, subsequent to which they are dried at optimal temperatures, afterward packed it for further delivery. Preserved plants to utilize as ornamentals are generally traded in potted parcels. At this situation, it is being marketable for the period of 18 days, and has no need of watering throughout this period (Łata 2007; Staszewski et al. 2007).

3.7 Pests, Diseases, and Control Measures

In Argentina (Torres et al. 2004), Egypt (Omar et al. 2008), India (Chaturvedi et al. 2009), Malaysia (Khew et al. 1991; Nejat et al. 2010), and Myanmar (Win and Jung 2012), *C. roseus* (L.) G. Don plants are identified to be vulnerable to the aster yellows (16SrI) group phytoplasma and it is also susceptible to spirea stunt (16SrIII-E), peach yellow leaf roll (16SrIII-A), clover proliferation (16SrVI), potato witches' broom (16SrVIA) (Lee et al. 1998), Mexican periwinkle virescence (16SrXIII) (Gundersen et al. 1994), as well as Malaysian periwinkle virescence (16SrXXXII) (Nejat et al. 2009, 2013). In United States, *C. roseus* (L.) G. Don plants were the foremost nonrutaceous plant naturally infected with *Spiroplasma citri* (Allen 1975). This pathogen was consequently revealed in Mediterranean countries like Morocco, Cyprus, Syria, and Turkey (Bove 1986), as well as in Malaysia, Southeast Asia (Nejat et al. 2011). *C. roseus* (L.) G. Don plants are also at risk of cucumber mosaic virus (CMV), and infection occurrence has been proclaimed in Australia (Shukla et al. 1980), India (Samad et al. 2008), and Malaysia (Mazidah et al. 2012). In *C. roseus* (L.) G. Don plants, leaf mosaic disease is caused by a viral pathogen, consequences in irregular yellow patches and malformation, accompanied by necrosis and wilting.

Bedding plants, for example *C. roseus* (L.) G. Don, are prone to loads of damping-off diseases. *C. roseus* (L.) G. Don is susceptible to blight/top-rot or die back, leaf spot, and root rot. Disease carrying fungi, which are responsible for *Alternaria* for leaf spot, *Rhizoctonia solani* for stem, crown, and root rot, as well as *Phytophthora parasitica* Dast for foliars and stems. According to reports from India and the United States, in *C. roseus* (L.) G. Don plants, *P. parasitica,* a soil-borne fungus is present which causes significant losses and death (Dastur 1916; Chase 1999; Keim 1977; McMillan and Garofalo 2004). *Fusarium* root rot disease has been revealed from Taiwan (Chung et al. 1998). In Florida, *P. nicotianae* is responsible for blight, which is considered as one of the most devastating diseases of *C. roseus* (L.) G. Don (McGovern et al. 2003). *Colletotrichum dematium* causes other notorious blights like twig blight (McMillan and Graves 1996), in Virginia foliar blight carried out by *P. tropicalis* (Hao et al. 2010), and in the United States, Italy, and Taiwan gray mold blight exaggerated by *Botrytis cinerea* (Daughtrey et al. 1995; Garibaldi et al. 2009; Ou-Yang and Wu 1998). Black root rot by *Thielaviopsis* infection is the most devastating root disease in *C. roseus* (L.) G. Don as it is very complicated and not easy to manage. Periwinkle rust, another disease noted to occur due to the attack of *Puccinia vincae* (http://donsgarden.co.uk/). Humid or wet conditions that arise due to overwatering are the major causes of fungal infection so care has to be taken to combat such type of problems. Insect pests are not considered as major problem; however, spider mites, aphids, mealy bugs, thrips, and scale insects can infiltrate this plant (Thomas et al. 2009). However, *C. roseus* (L.) G. Don plants are enduring and quite resistant to pests and diseases, a number of disease-causing agents of viral, fungal, bacterial, and mycoplasmal origins have been encountered in this plant. Thrips and aphids are the general pests of this plant. *C. roseus* (L.) G. Don plants are vulnerable to a disease analogous to the spike disease of *Santalum album*. It is distinguished by a hairy exterior of the plant; in that condition leaves of the plant turn into smaller and shorter internodes. In the next stages, immoderate branching is developed. The disease-containing plants show evidence of hyperplasia or enhanced vegetative function. In entire cycle of the disease, the colors of flowers are green and produce phyllody. Green rosette disease has infrequently been noticed on these plants. In this disease, size of the leaf and flower is shortened and no fruiting occurs but there is no change in color and alteration of any floral part observed (Virmani et al. 1978; Garga 1958).

Diseased plants are ameliorated through regular treatment with the sprays of oxytetracycline, GA, tetracycline, as well as nicotine sulfate (0.5% aqueous solution). A few of the fungal diseases are able to be managed by the use of fungicide-containing foliar sprays including Carbendazim Bavistin 50 WP, Captafol Foltaf 80 WP, and Benomyl 50 WP (Kalra et al. 1991).

Environmentally, a full-sun planting place with optimal humidity levels, and better air exchange in the region of the plants should be selected. Potting mixes should be free from any kind of pathogen and fresh containers constructed for every planting. In relation to sanitation, loads of weeds should be kept minimum, upon revealing of any kind of contamination or infection in soils and plants should be removed instantaneously. All the equipments used for this purpose should be kept clean, and the planting region is cleared of previously used potting media and plant waste. Overhead irrigation and extended periods of leaf dampness must be avoided along with watering done only in the day time. Soil fertility has to be appropriately calibrated to pH maintained at 5.5. Transplants have to be distanced at least 10–12 in. apart. Biological control over soil-borne disease carrying fungi may possibly be tried with *Trichoderma virens*, binucleate *Rhizoctonia, or Burkholderia cepacia* (Thomas et al. 2009; McGovern et al. 2003; Benson 1995; Yandoc et al. 2007).

4 Natural and Induced Variation

The naturally grown populace of *C. roseus* (L.) G. Don has been observed to harbor substantial genetic variability. The marketable harvest shows an extensive morphological and chemical variation, with diverse ranges of heritability owing to a reasonable total of outcrossing existing among the species (Kulkarni et al. 1984). However, the plant is predominantly self-pollinated; Krishnan et al. (1979) reported recurrent outcrossing (13.6%), while Kulkarni (1999) mentioned in a research that outcrossing betwixt white × white-flowered plants varied from 28.3% to 65.3%, among white × pink-flowered plants it varied from 12.2% to 15.1%, and 11.4% was reported in the case of *die back* disease-resistant plant variety Nirmal (Krishnan et al. 1979; Kulkarni 1999). The natural inconsistency has been utilized to evolve horticultural and drug types. Variance occurring due to polyploidy, cross breeding, and induced mutagenesis has been varyingly engaged in *C. roseus* (L.) G. Don. Autotetraploids of dissimilar hereditary backgrounds, while differentiated with analogous diploids, were found to be comparatively more tolerable to die back and the collar- and root-rot disorder (Kulkarni et al. 1987). The autotetraploids were seen in certain strains to have larger leaves with shorter length/width ratio, decreased pollen fertility, bigger pollens, smaller follicles size, bulky/weighty seeds, lesser dry leaf matter, lesser vinblastine content, and increased entire alkaloidal content (Mishra and Kumar 2000). An elevated ploidy-level derivative was found to be typically well sterile with lesser seed germinability (Janaki Ammal and Bezbaruah 1963). Kulkarni et al. (1987) reported about the potential of diploids and induced autotetraploids of *C. roseus* (L.) G. Don in divergent measures of nitrogen and plant spacing (Kulkarni et al. 1987). Investigational hybrids have been explained among *C. roseus* (L.) G. Donand *C. trichophyllus* (Sevestre-Rigouzzo et al. 1993). Levy et al. (1983) reported that the considerable heterosis for yields of leaf and root in crosses required three actual lines although heterosis was not revealed for quantity of ajmalicine alkaloid in the roots (Levy et al. 1983).

Kulkarni et al. (1999) have revealed major differences in the morphological character of plant along with the presence of alkaloids in leaf and root of three induced mutants of *C. roseus* (Kulkarni et al. 1999). Nirmal is a superior variety of *C. roseus* developed and launched by CIMAP show high level of tolerance to die back disease. Another variety of *C. roseus* Prabal is developed and launched by CIMAP (Dwivedi et al. 2001). Kulkarni et al. (2003) have developed another distinct variety of *C. roseus* Dhawal (U.S. Patent No. 6,548,746) by means of mutation breeding of superior quality with improved alkaloid-producing potential and show high level of tolerance to die back disease (Kulkarni et al. 2003). Dhawal may possibly differentiate morphologically through its markedly undulating/wavy leaf periphery. At present, it is one of the finest varieties of *C. roseus* accessible for advanced molecular level research, and in addition it is utilized for commercial cultivation.

5 Occurrence and Distribution of Alkaloids

The *C. roseus* alkaloids at this time consist of a group of almost 130 TIAs, present mostly in Apocynaceae, Loganiaceae, and Rubiaceae families. It represents one of the leading groups of alkaloids along with more than 3000 members showing magnificent structural assortment (van der Heijden et al. 2004). There are a lot of differences seen in the constitutions of the alkaloids obtained from the aerial parts (leaves/shoots) as well as from underground parts of plant (root/rhizomes/true tubers/corms). The presence of alkaloid in different plant tissues varies significantly

Table 3 Variation in alkaloid content of various parts/tissues of *C. roseus* (L.) G. Don

Tissue/parts of plant	% Alkaloid content (dry weight basis)
Flower	0.005–0.84
Leaf	0.32–2.56
Fruit	~0.40
Pericarp	~1.14
Seed	~0.18
Stem	0.07–0.46
Root	0.125–2.60
Root bark	2.50–9.00

as the researcher reported (Table 3) (Mishra and Kumar 2000; Virmani et al. 1978). The variations found may be due to diverse agro-climatic conditions, hereditarily different genotypes, and the processes employed for alkaloid extraction.

6 Role of Alkaloids in *C. roseus* (L.) G. Don Plant

TIAs, similar to nearly all other secondary metabolites, are responsible for defensive/protective roles to the plant. The biosynthesis of plant secondary metabolite is not regulated separately as of the role of these yields for the plant. Though, it is not easy to identify the precise *in planta* role of secondary metabolites similar to the TIAs. In relation to *C. roseus* (L.) G. Don, an antifeedant potential in contrast to *Spodoptera* larvae has been indicated for vinblastine and catharanthine (Meisner et al. 1981). The antifeedant action in contrast to *Spodoptera* caterpillars in case of *C. roseus* (L.) G. Don leaf extracts has been explained also (Chockalingam et al. 1989; Meisner et al. 1981). The nematocidal potential of serpentine has been also described (Verpoorte et al. 1997). Luijendijk et al. (1996a) examined and analyzed the participation of strictosidine in the antimicrobial and antifeedent activities of leaves of *C. roseus* (L.) G. Don (Luijendijk et al. 1996a). Strictosidine and its deglucosylation product particularly produced by strictosidine β-D-glucosidase (SGD) were responsible for the action against several microorganisms. In contrast, neither the entire glucoside nor the aglycone commodity was observed to reveal antifeedent potential against *Spodoptera exigua* larvae, the same was found for whole *C. roseus* leaves as well as leaf extracts. In addition to alkaloids, other chemical compounds are also accountable for the antifeedent potential of *C. roseus* (L.) G. Don leaves, as validated by Singh et al. (2003), who observed that the n-hexane fraction of *C. roseus* (L.) G. Don leaves extract shows the presence of insect growth regulators, namely, α-amyrin acetate and oleanolic acid, which act against tobacco caterpillar (*S. litura* F.) and gram pod borer (*Helicoverpa armigera* Hub.) (Singh et al. 2003). The combination regimen of strictosidine and SGD exhibits strong antifungal potential (Luijendijk et al. 1996a; Verpoorte et al. 1997). Strictosidine and SGD contents are predominantly high in young leaves of *C. roseus* (L.) G. Don (260 μg/g fresh weights). According to the fact, which reveals the presence of strictosidine in the vacuole, while the vastly precise SGD is found somewhere else, it is supposed that this amalgamation of substrate and enzyme plays a vital role in the plant's defense system in relation to wounding (Luijendijk et al. 1996a; Verpoorte et al. 1997), and so as to strictosidine is a phytoanticipin. Therefore, it is vastly feasible that the phytoalkaloids have imperative ecochemical functions in the defense system of the plant in opposition to pathogens and herbivores.

7 Pharmaceutical and Pharmacological Applications

C. roseus (L.) G. Don has a long history of traditional usage for medicinal purposes around the world. The traditional and folkloric practice of various parts of the plant for the management of a variety of disorder has been examined in detail by earlier researchers (Mishra and Kumar 2000; Virmani et al. 1978). The folklore use of *C. roseus* (L.) G. Don decoctions is suggested for the cure of diabetes, diarrhea dysentery, dengue fever, malaria, insect bites, skin infection, cancer, dyspepsia, toothache, eye irritation, sore throat, and respiratory congestions (Sukumar and Osmani 1981; Duke 1985; Virmani et al. 1978). The plant roots are indicated for the management of hypertension, and it is considered as a tonic for general debility as well as shows sedative and tranquillizing properties (Narayana and Dimri 1990). In the Ayurvedic system of medicine, it is preferred for the treatment of diabetes. Also in modern system of medicine, the hypoglycemic potential of aqueous extracts of *C. roseus* (L.) G. Don has been proved (Vega-Avila et al. 2012). In the region of Madagascar, the bitter and astringent leaves of *C. roseus* (L.) G. Don has been traditionally used as an emetic and roots as a purgative, depurative, vermifuge, hemostatic agent and for toothache. In Philippines, leaf decoction is used as an herbal remedy for the management of diabetes; immature plant leaves are used for stomach cramps and root decoction for intestinal parasitism. In Mauritius, leaves infusion traditionally used for indigestion and dyspepsia relief. In Indian states mainly Orissa and Assam, leaves juice are employed to cure wasp stings whereas roots and leaves are used as an anticancer agents. *C. roseus* (L.) G. Don floral extracts offers an extensive application in many countries as a medicine for several diseases for example: Indo-China - dysmenorrhea; Cuba and Jamaica - eyewash for infants; Bahamas - respiratory problem (asthma); Bermuda - hypertension; Surinam and all over the Caribbean - eye irritation/infections, malaria, menstrual pains, and diaphoresis. The people of Uganda have belief in leaf infusions to cure digestive ulcers whereas in Batswana powdered leaves in milk for full-grown abscesses. A root decoction in Togo is used to cure dysmenorrhea (PROTA 2011; Neuwinger 2000). Still in European countries, *C. roseus* (L.) G. Don is employed as a folk remedy to treat diabetes/hyperglycemia for centuries. In China, it is a widely used versatile herb in the management of blood cancer, high blood pressure, lymphoma, and Brill-Symmers disease and it is also used as an astringent, diuretic, and cough remedy. Correspondingly, in central and south America, it is accepted as a household remedy for cold, lung congestion, inflammation, and sore throat. In Hong Kong and

Korea, *C. roseus* (L.) G. Don is employed as an alternative medical remedy for the treatment of blood cancer. In Japan, *C. roseus* (L.) G. Don has been used for the management of diabetes and Malignant Lymphoma (Guo et al. 2001). Besides that, modern research has discovered a wide range of therapeutic applications for the *C. roseus* (L.) G. Don (Table 4).

Besides, the significance of *C. roseus* (L.) G. Don plant in modern system of medicine has been recognized only after the coincidence discovery of anticancer alkaloids present in its leaves. Currently, there are four main *Catharanthus* alkaloids that have clinical acceptance such as vinblastine (VBL), vincristine (VCR), vinorelbine (VRLB), and vindesine (VDS); however, only VCR, VBL, and VRL are permitted for clinical use in the United States. Since 2008, in Europe, vinflunine, a latest synthetic *Catharanthus* alkaloid is used for the treatment of cancer (Moudi et al. 2013). The other notable alkaloid 'ajmalicine' is extracted from stems and roots of *C. roseus* (L.) G. Don, because of their clinical importance in the management of high blood pressure and obstructive circulatory problems as well as enhanced cerebral blood circulation (Verpoorte et al. 1991). *C. roseus* (L.) G. Don roots also contain serpentine, which is used in the management of hypertension (Mishra and Kumar 2000).

The discovery of vincristine and vinblastine (bisindole alkaloid) isolated from *C. roseus* (L.) G. Don symbolizes one of the most imperative introductions of plant-derived products into the cancer chemotherapy. These plant alkaloids are used to treat malignant and nonmalignant cancer. Vincristine and vinblastine are also employed in the treatment of thrombocytopenic disorders like immune thrombocytopenia (idiopathic thrombocytopenic purpura), Moschcowitz syndrome (thrombotic thrombocytopenic purpura), and microangiopathic hemolytic anemia arises due to chemotherapy. Even though vincristine and vinblastine are functional in platelet and platelet-associated ailments, it is an obligatory part of the pharmacopoeia that is used to treat malignancy (Neuss and Neuss 1990). Vincristine and vinblastine show a broad spectrum of biochemical action within the cells and tissues; the exact mechanisms of cytotoxic action is stated as the interactions with tubulin protein and disruption of microtubule (spindle fibers) function, particularly of microtubules comprising the mitotic spindle apparatus, directly inducing metaphase arrest. They connect speedily and reversibly to obligatory sites on tubulin that are different from those of the taxanes (paclitaxel and docetaxel), podophyllotoxin, colchicines, and guanosine triphosphate (GTP). They do not show cross-resistance with drugs which alkylate DNA and act through different mechanisms of action. The prime application of vinblastine is to treat Hodgkin's disease, whereas the principal use of vincristine is indicated to treat acute lymphocytic leukemia in children. Leukopenia and bone marrow depression are the dose-limiting toxicity of vinblastine, while neurotoxicity in the case of vincristine (Weiss et al. 1974). Vinblastine is also employed to treat testicular cancer effectively. Other types of blood cancer including Hodgkin lymphomas and the non-Hodgkin lymphomas may possibly be treated with combination regimens that consist of vinblastine. Vinblastine is also considered as an effective therapy to treat Kaposi's sarcoma, Alibert-Bazin syndrome (granuloma fungoides), and breast cancer. Vincristine is

Table 4 List of the most important studies on the medicinal effects of *Catharanthus roseus*

Application	Modern researches		References
	In vitro	In vivo	
Antibacterial (antiseptic) activity	Bacteria		(Virmani et al. 1978; Goyal et al. 2008; Patil and Ghosh 2010; Ramya et al. 2008; Verma and Singh 2010)
Acetyl cholinesterase and cholinergic antagonism inhibition	Microplate assay	Male Wistar rats (ex vivo)	(Pereira et al. 2009, 2010)
Antiangiogenesis activity	Chicken eggs		(Wang et al. 2004)
Alzheimer's syndrome		Human (clinical trial)	(Singh et al. 2001)
Antidysenteric activity	Wistar rats		(Hassan et al. 2011)
Antihypercholesterolemic activity (antihyperlipidemic)	Rabbit, rat		(Chauhan et al. 2011; Chattopadhyay et al. 1992)
Anthelmintic activity	*Pheretima posthuma*		(Agarwal et al. 2011)
Antineoplastic activity	Mice, rat	Clinical use	(Nobili et al. 2009; Cragg and Newman 2005; Dong et al. 1995; El-Merzabani et al. 1979; El-Sayed and Cordell 1981; El-Sayed et al. 1983; Johnson et al. 1960; Mukherjee et al. 2001; Noble 1990)
Larvicidal activity	*Anopheles stephensi* (malaria vector); *Aedes aegypti*		(Kuppusamy et al. 2009; Remia and Logaswamy 2010)
Regression of accessory reproductive organs	Male Wistar rats		(Akbarsha et al. 1995)
Antifertility effect	Male rat		(Sherines and Howard 1978; Mathur and Chaudan 1985; Prajapati et al. 1998)
Antiandrogenic activity	Mice		(Murugavel and Akbarsha 1991)
Blood cleanser			(Moerman 2009)
Antifungal activity	*Trichophyton rubrum* *Hendersonula toruloidea*		(Chile and Vyas 1984; Barde and Singh 1983)
Regression of entire reproductive system	Male rat		(Stanley et al. 1993)
Stomachic	—		(Narayana and Dimri 1990; Kurian and Sankar 2007)
Tranquilizing and sedative action	—		(Daniel 2006; Narayana and Dimri 1990; Kurian and Sankar 2007)

(continued)

Table 4 (continued)

Application	Modern researches		References
	In vitro	In vivo	
Tonic	—		(Narayana and Dimri 1990; Kurian and Sankar 2007)
Cytochrome P450 inhibition	CYP2D6		(Usia et al. 2005)
Antioxidant activity	*Rat*		(Jaleel et al. 2006; Chauhan et al. 2011; Zheng and Wang 2001)
Anti-inflammatory activity	Rat		(Chattopadhyay et al. 1992)
Wound healing	Rat		(Nayak and Pinto Pereira 2006)
Antiplasmodial activity	Human erythrocytes		(Gathirwa et al. 2007; Ponarulselvam et al. 2012)
Enhances kidney and liver functions	Wistar rat		(Iweala and Okeke 2005; Adekomi 2010)
Epididymal dysfunction	Rat		(Averal et al. 1996)
Generate giant spermatogonial cells	Albino rat		(Stanley and Akbarsha 1992)
Hypotensive activity	Rat		(Narayana and Dimri 1990)
Cytotoxic activity	Human cell line		(Hostettmann et al. 2000; Siddiqui et al. 2010)
Antihyperglycemic activity (antidiabetic)	Mice, rat, rabbit	Wistar albino rats	(Vega-Avila et al. 2012; Benjamin et al. 1994; Bnouham et al. 2006; Chattopadhyay 1999; Chauhan et al. 2011; Iweala and Okeke 2005; Jarald et al. 2008; Nammi et al. 2003; Singh et al. 2001)
Antiproliferative activity	Human cells		(Mans et al. 2000; Ueda et al. 2002)
Antispermatogenic	Male rat, mice		(Gupta and Sharma 2006; Joshi and Ambaye 1968)
Antimutagenic activity	Micronucleated erythrocytes		(Lim-Sylianco and Blanco 1981)

also an effective regimen of cancer therapy with recognized importance for the treatment of Hodgkin's lymphoma and other lymphomas, as well as pediatric tumors, for example, nephroblastoma (Wilms tumor) and embryonal rhabdomyosarcoma (McCormack 1990). Initially available in the 1960s, the *C. roseus* (L.) G. Don alkaloids are currently incorporated in each efficient combination chemotherapy curriculum, due to their uniqueness in relation to therapeutic action as well as toxicities (Neuss and Neuss 1990). Vincristine and vinblastine are remarkably imperative in both regimens (curative and palliative) (Table 5).

Table 5 Curative and palliative regimens containing *C. roseus* alkaloids (Verpoorte et al. 1991)

	Drugs	Disease
Curative regimens for cancer	Nitrogen mustard, vincristine, procarbazine, prednisone	Hodgkin's lymphoma
	Cyclophosphamide, daunomycin, vincristine, prednisone	Non-Hodgkin's lymphoma
	Daunomycin, bleomycin, vinblastine, dacarbazine	Hodgkin's lymphoma
	Cisplatinum, vinblastine, bleomycin	Testicular cancer
	Methotrexate, daunomycin, cyclophosphamide, vincristine, prednisone, bleomycin	Lymphoma
	Methotrexate, vinblastine, daunomycin, cyclophosphamide	Bladder cancer
Palliative regimens for cancer	Cyclophosphamide, vincristine, prednisone	Lymphoma
	Procarbazine, vincristine, cyclophosphamide	Melanoma, small-cell lung cancer
	Vinblastine, daunomycin, Thio-TEPA, Halotestin	Breast cancer
	Vinblastine/vindesine, cisplatinum	Non-small-cell lung cancer

7.1 Toxicity and Side Effects

If *C. roseus* (L.) G. Don consumed orally, it may possibly be unsafe for health. It may possibly be hallucinogenic and is reported as such (under its synonym *Vinca rosea*) in the Louisiana State Act 159. The combined cancer therapy regimens are planned based on the confidence that more cumulative actions may possibly be seen by employing drugs with varying sites (inside the cell) and period (inside the cell cycle) of action. Cumulative toxicities are possibly less as compared to single toxicities due to varying toxicities of different agents. The way of administration and the dosage range employed for the vincristine and vinblastine, similar to several other anticancer drugs, are narrowly attached to their medical toxicity. Watchful titration/selection of the dose to accomplish negligible toxicity consistent with therapeutic advantage is essential for successful clinical application. To lessen the toxicity of these anticancer agents on "innocent bystander" tissues, novel scientific and clinical approaches for selective drug targeting have been applied. Monoclonal antibodies (mAb or moAb), which are reactive with tumor-associated antigens (TAA), have been assessed as carriers for targeted delivery of cytotoxic *catharanthus* alkaloids to specific antigens containing tissues (Pearce 1990). The therapeutic index of vincristine may possibly improve considerably by the application of novel drug delivery system such as liposomal delivery system and also dose-related drug toxicity decreased (Waterhouse et al. 2005).

The tumor cells acquired resistance against a variety of antineoplastic agents; this phenomenon of resistance is known as multiple drug resistance (MDR) that shows an additional obstacle to triumph over in the management of cancer. MDR is expressed through reduction of intracellular drug accumulation consequential from augmented drug efflux by P-glycoprotein (P-gp) or P-170 that is an ATP-dependent efflux pump encrypted through *mdr-1* gene. The over-expression of *mdr-1* gene transcript P-gp is accountable for MDR and is persuaded by bisindole alkaloids such as vinblastine and vincristine. Due to the possible function of P-gp in medical drug resistance, a lot of investigations are paying attention on approaches to restrict the role or expression of this protein. A glucobrassicin indole-3-carbinol is a metabolite of cruciferous plants, resulted in <80% reversal of the bisindole alkaloid-induced P-gp expression in contrast with 65–70% reversal by a distinguished MDR-reversing agent, verapamil (Arora and Shukla 2003). Furthermore, several undesirable effects have been proclaimed for these drugs such as myelotoxicity, alopecia, abdominal cramps, constipation, nausea/vomiting, paralytic ileus, ulcerations of the mouth, kidney impairment, urinary retention, hepatocellular damage, pulmonary fibrosis, amenorrhea, azoospermia, orthostatic hypotension, and hypertension (Nejat et al. 2015). The drug dosage and route of administration should be carefully managed to lessen undesirable toxic effects (Sherines and Howard 1978; Leveque and Jehl 2007; James et al. 2007; Nobili et al. 2009).

7.2 *Pharmaceutical Production and International Trade*

The world trade in the fields of plant-derived raw materials intended for the drugs, pharmaceuticals, perfumeries, and cosmetics is rapid and escalating as manifested through trade statistics. Even though, India exported and imported huge quantity of medicinal herbs/plants, phytopharmaceuticals and plant-derived products, its contribution in the world trade has been somewhat negligible. *C. roseus* (L.) G. Don bedding plants and pot plants have huge market prospective in the United States. Several cultivars are being produced, and there is a new focus on cultivars with resistance to fungal attacks, in order to be able to propagate them under more humid situations. In southern Europe and Japan, limited figures of cultivars existed. In the beginning of 1990s, the consumption of vincristine and vinblastine in the international market is about 5–10 kg, with an aggregate value of US$ 25–50 million. In 2005, the consumption of vincristine and vinblastine in the world market was estimated with total value of US$ 150–300 million. In 1991, the consumption of ajmalicine in the international market is about 3–5 ton, with an aggregate value of US$ 4.5–7.5 million. Worldwide, there are two pharmaceutical formulations for the treatment of cancers, namely, Oncovin® (Vincristine sulfate) and Velban® (Vinblastine sulfate), originally obtained from *C. roseus* (L.) G. Don, being sold for a sum of US$ 100 million annually. *C. roseus* (L.) G. Don is extensively propagated in the United States, Spain, and China for its pharmaceutically accepted phytoconstituents. The two important pharmaceutically acquired bisindole alkaloidal phytoconstituents, VBL and VCR, are present mostly in the aerial parts of the plant in enormously low concentrations, the quantity of latter much less than the former (Tyler 1988; Verpoorte et al. 1993; Schmelzer and Gurib-Fakim 2008;

Pezzuto 1997). Vincristine sulfate, initially recognized as leurocristine, which is used to treat leukemia (blood cancer) effectively decreases white blood cell count significantly; since the 1950s, it has improved the survival rate of children who suffered with blood cancer from 20% to 80%. It is consider as one of the most expensive plant-derived compounds in the international market with significant undesirable effects. Vinblastine correspondingly reduces the number of white blood cells in the blood (Leveque and Jehl 2007). In the world trade, vinblastine sulfate has now been sold for over 40 years as an antineoplastic agent. It is being effectively used to treat Hodgkin's disease (Schmelzer 2007; Schmelzer and Gurib-Fakim 2008). Preparing food material using fresh leaves of *C. roseus* (L.) G. Don gives rise to the financial significance of the herb whereas such yields hold both pharmaceutical and nutritional properties, concurrently (Salim et al. 2013). These injectable anticancer medicines and their semisynthetic analogues, for example, VRLB and VDS impede with the cell division (Debnath et al. 2006; Foster 2010; Khan et al. 2009; Mujib et al. 2002; Sottomayor and Barcelo 2005; van der Heijden et al. 2004). 5-Noranhydrovinblastine (Navelbine) is a semisynthetic *Catharanthus* alkaloid which shows complete microtubule depolymerization, broader antitumor action, and exhibits low neurotoxicity as compared to VBL and VCR because it selectively impedes with tubulin assembly (Binet et al. 1990). The synthetic vincristine is less effective (only 20% efficiency) in contrast to the naturally derived vincristine from *C. roseus* (L.) G. Don, and therefore the significance of the species and its bioactive phytoconstituents are unchallenged due to their multifaceted structures. The other pharmaceutically and therapeutically important alkaloid "ajmalicine" is a component of hypotensive drugs used to treat high blood pressure. Near about 200–300 ton of *C. roseus* (L.) G. Don roots are needed for 3600 kg per annum world production of ajmalicine (Verpoorte et al. 1993).

8 Present and Future Prospects

In the last few decades, extensive research has been envisaged on *C. roseus* (L.) G. Don. Researchers find numerous issues and has been taken as a challenge, were experienced over the span of research and relating arrangements were likewise contrived. It was a monumental problem to supply the required quantity of raw materials from hundreds to thousands of kilogram per year, and this adequate supply of natural plant materials had to be maintained to fulfill the demand. Hence, a balance between demand and supply has to be maintained. Plantation with the determination that the desired alkaloids found in the leaves allowed for stripping of the plant, thereby affording a healthier and more profuse regrowth. Collection of plant from the wild eventually progressed to farm plantation or cultivation, allowing for greater control over growth.

Despite the political interference that would become a hurdle in reliable crude drug supply, planting and cultivation were started in the United States. The introduction of new crops is a great risk as well as there is a major concern related to economics involved in it. Hand collection of plant and "native" wages are big problems and out of the question. Despite a lot of risk, plantations and cultivation were started in many countries like Texas; India and Madagascar had lost a viable cash crop. Substitution of hand works and, a solitary planting as new procedures by rummage harvesters was turned out to be advantageous, and could give a few harvests/season, surviving harvester-cutting with moderately fast regrowth. *C. roseus* (L.) G. Don cultivation represents a renewable resource.

Interests in herbal products enter the boundary of American pharmaceutical industry, started working in the field of herbal products, and sustained by natural products from the higher order plants. It is an enigma that little work in this realm is being pursued by US industry, particularly since approximately 25% of new and refilled prescriptions from community pharmacies contain plant products (Farnsworth and Morris 1976). There is no dearth of plants to be collected and screened for specific or general biological activity. Of the Earth's 500,000–750,000 higher order plants, less than 10% have been investigated phytochemically. The investigator has an almost unlimited choice for selection from different regions of the world. Success potential should be extremely high. However, it is very important that selection of plant should be judiciously and must have appropriate facilities of biological test systems and relevant instrumental facilities for isolation and purification of herbal extracts.

The selection of plant is a very important factor for success potential, and it can be based on reported folkloric usage, but said usage must be scientifically rational in both an investigative and practical sense. Previous report has been already cited regarding the use of *C. roseus* (L.) G. Don as an oral insulin substitute. The use of *C. roseus* (L.) G. Don has also been reported against hemorrhage, scurvy, as a mouthwash for toothache and for healing and cleaning chronic wounds did not stimulate either scientific or practical adrenalin (SVOBODA et al. 1959). And yet, these uses could well prove valuable and a boon for the patient. Consideration of both botanical and chemotaxonomic relationships as well as reported chemical constituents in literature is important for the selection of plants and can be useful and beneficial. It is the fact that majority of US pharmaceutical houses are no longer working in the field of pharmacognostical/phytochemical research, particularly as it relates to palliative/curative measures against human cancer, may well stem from the erroneous concept that any product would be a "not-for-profit" item. It may also stem from ignorance on the part of research administrators who conceive, and most certainly maintain, that antibiotics will treat or cure all ills, mainly those designated as "profits."

Some of the pharmaceutical companies have been extremely kind to the periwinkle alkaloids, particularly leurocristine (vincristine, ONCOVIN), took a big space in global market and has become the highest percent profit item in the product line of Eli Lilly & Co., bearing the almost insignificant cost-of-sales of 12%. Revenue generation of nearly tens of millions of dollars per year indicate the absolute success of this natural product. Public perception that the item was to be sold "at a price calculated to yield no profit to the company" misled other companies that were considering of entering the field. The plant bears active phytoconstituents and exhibits various pharmacological activities like antidiabetic, antioxidant, antihypertensive, antimicrobial, cytotoxic, etc. *Catharanthus roseus* produces a spectrum of TIAs vinblastine and vincristine, the anticancer lead molecules. Being a source of these important secondary metabolites, an extensive study has to be carried out on *C. roseus.* The present need is to search the secondary metabolites derived from this plant, investigate its pharmacological activities, the biotechnological approaches should be introduced to enhance the production of TIAs and the prospects of potential endophytes residing inside the host tissue.

9 Conclusion

Catharanthus roseus (L.) G. Don is an important medicinal plant with a wide range of uses. The dried plant extracts contain many alkaloids of medicinal use. These alkaloids are produced in very small quantities inside the plant although attempts have been made over the past years to increase their production through various biotechnological applications. The plant has been proven useful not only in the field of medicine but has also been recently put into use for the phytoremediation of radiocesium from low level nuclear waste. The leaves of this plant have been found to be of immense medicinal use as most of the pharmacological activities of this plant are attributed to its leaves. So the cultivation as well as the conservation of this plant must be promoted on a large scale. Besides this alternative means, that are less time consuming, sustainable and more economical, must be developed and adopted for the production of these active constituents. In the present times, when the emphasis is being placed on the use of natural materials in the control and treatment of various diseases and infections because of the undesirable side effects of synthetic drugs, there is a need for further research especially on bioactive compounds, their production from alternative sources, methods for increasing their production, herbal remedies, effectiveness of plants for various uses, and bioprospecting new sources of natural bioactive products which can provide unlimited scope for the development of new drug leads. Endophytes could thus be exploited as the sources of the valuable secondary metabolites of medicinal, agricultural, and industrial importance.

References

Adekomi DA (2010) Madagascar periwinkle (*Catharanthus roseus*) enhances kidney and liver functions in Wistar rats. Eur J Anat 14(3):111–119

Agarwal S, Jacob S, Chettri S et al (2011) Evaluation of in vitro anthelminthic activity of *Catharanthus roseus* extract. Int J Pharm Sci Drug Res 3(3):211–213

Akbarsha MA, Stanley A, Averal HI (1995) Effect of vincristine on Leydig cell and accessory reproductive organs. Curr Sci 68(10):1053–1057

Allen RM (1975) Spiroplasma organism found in naturally infected periwinkle. Citrograph 60:428–446

Arora A, Shukla Y (2003) Modulation of vinca-alkaloid induced P-glycoprotein expression by indole-3-carbinol. Cancer Lett 189:167–173

Ashutosh KS, Suman PSK (2013) OMICS applications in crop science. In: Barh D (ed) *Catharanthus roseus*: the metabolome that represents a unique reservoir of medicinally important alkaloids under precise genomic regulation. CRC Press, New York

Averal HI, Stanley A, Murugaian P, Palanisamy M, Akbarsha MA (1996) Specific effect of vincristine on epididymis. Indian J Exp Biol 34(1):53–56

Baligh M, Delgado MA, Conway KE (1999) Evaluation of *Burkholderia cepacia* strains: root colonization of *Catharanthus roseus* and in-vitro inhibition of selected soil-borne fungal pathogens. Proc Oklahoma Acad Sci 79:19–27

Barde AK, Singh SM (1983) Activity of plant extracts against *Scytalidium anamorph* of *Hendersonula toruloidea* causing skin and nail diseases in man. Indian Drugs 20:362–364

Basu D (1992) Effects of growth retardant on growth, flowering and alkaloid production in *Catharanthus pusillus*. Ind For 118:659–661

Benjamin BD, Kelkar SM, Pote MS, Kaklij GS, Sipahimalani AT, Heble MR (1994) *Catharanthus roseus* cell cultures: growth, alkaloid synthesis and antidiabetic activity. Phytother Res 8(3):185–186

Benson DM (1995) Aluminum amendment of potting mixes for control of phytophthora damping-off in bedding plants. HortScience 30(7):1413–1416

Binet S, Chaineau E, Fellous A et al (1990) Immunofluorescence study of the action of navelbine, vincristine and vinblastine on mitotic and axonal microtubules. Int J Cancer 46(2):262–266

Bnouham M, Ziyyat A, Mekhfi H, Tahri A, Legssyer A (2006) Medicinal plants with potential antidiabetic activity—a review of ten years of herbal medicine research (1990–2000). Int J Diabetes Metab 14(1):1–25

Bove JM (1986) Stubborn and its natural transmission in the Mediterranean area and the Near East. FAO Plant Prot Bull 34:15–23

Burns JR, Benson DM (2000) Biocontrol of damping-off of *Catharanthus roseus* caused by *Pythium ultimum* with *Trichoderma virens* and binucleate Rhizoctonia fungi. Plant Dis 84(6):644–648

Chase AR (1999) Phytophthora aerial blight of vinca and other bedding plants. West Connect Turf Ornamentals 1:1–4

Chattopadhyay RR (1999) A comparative evaluation of some blood sugar lowering agents of plant origin. J Ethnopharmacol 67(3):367–372

Chattopadhyay RR, Banerjee RN, Sarkar SK, Ganguly S, Basu TK (1992) Antiinflammatory and acute toxicity studies with the leaves of *Vinca rosea* linn. in experimental animals. Indian J Physiol Pharmacol 36(4):291–292

Chaturvedi Y, Tewari AK, Upadhyaya PP, Prabhuji SK, Rao GP (2009) Association of Candidatus phytoplasma asteris with little leaf and phyllody disease of *Catharanthus roseus* in Eastern Uttar Pradesh, India. Int J Phytomed Relat Indust 1:103–108

Chaudhary S, Sharma V, Prasad M et al (2011) Characterization and genetic linkage mapping of the horticulturally important mutation leafless inflorescence (LLI) in periwinkle *Catharanthus roseus*. Sci Hortic 129(1):142–153

Chauhan K, Sharma S, Rohatgi K, Chauhan B (2011) Antihyperlipidemic and antioxidative efficacy of *Catharanthus roseus* Linn [Sadabahar] in streptozotocin induced diabetic rats. Asian J Pharm Health Sci 2(1):235–243

Chile SK, Vyas KM (1984) Efficacy of *Vinca rosea* extracts against protease from human pathogenic strains of *Trichophyton rubrum* Sab. Hindustan Antibiot Bull 26(3–4):114–116

Chockalingam S, Nalina Sundari MS, Thenmozhi S (1989) Impact of the extract of *Catharanthus roseus* on feeding and enzymatic digestive activities of *Spodoptera litura*. J Environ Biol 10:303–307

Choudhury S, Gupta K (2002) Influence of seasons on biomass and alkaloid productivity in *Catharanthus roseus*. J Med Arom Plant Sci 24:664–668

Chung WC, Huang JW, Sheu JC (1998) *Fusarium* root rot of periwinkle in Taiwan. Plant Prot Bull 40:177–183
Cragg GM, Newman DJ (2005) Plants as a source of anticancer agents. J Ethnopharmacol 100(1–2):72–79
Daniel M (2006) Medicinal plants: chemistry and properties. Science Publishers, Enfield
Dastur JF (1916) Phytophthora on *Vinca rosea*, in Memoirs of the Department of Agriculture in India. Bot Ser 8:233–242
Daughtrey ML, Wick RL, Peterson JL (1995) Compendium of flowering potted plant diseases. APS Press, St Paul
De Pasquale A (1984) Pharmacognosy: the oldest modern science. J Ethnopharmacol 11:1–16
Debnath M, Malik CP, Bisen PS (2006) Micropropagation: a tool for the production of high quality plant-based medicines. Curr Pharm Biotechnol 7(1):33–49
Dong JG, Bornmann W, Nakanishi K, Berova N (1995) Structural studies of vinblastine alkaloids by exciton coupled circular dichroism. Phytochemistry 40(6):1821–1824
Duke JA (1985) Handbook of medicinal herbs: *Catharanthus roseus*. CRC Press, Boca Raton
Dwivedi S, Singh M, Singh AP, Singh V, Uniyal GC, Khanuja SPS, Kumar S (2001) Registration of a new variety Prabal of *Catharanthus roseus*. J Med Arom Plant Sci 23:104–106
El-Merzabani MM, El-Aaser AA, Attia MA, El-Duweini AK, Ghazal AM (1979) Screening system for Egyptian plants with potential anti-tumour activity. Planta Med 36(2):150–155
El-Sayed A, Cordell GA (1981) Catharanthus alkaloids. XXXIV. Catharanthamine, a new antitumor bisindole alkaloid from *Catharanthus roseus*. J Nat Prod 44(3):289–293
El-Sayed A, Handy GA, Cordell GA (1983) *Catharanthus* alkaloids, XXXVIII. Confirming structural evidence and antineoplastic activity of the bisindole alkaloids Leurosine-N′b-oxide (pleurosine), roseadine and vindolicine from *Catharanthus roseus*. J Nat Prod 46(4):517–527
Farnsworth NR, Morris RW (1976) Higher plants-the sleeping giant of drug development. Am J Pharm 148:46–52
Foster S (2010) From herbs to medicines: the Madagascar periwinkle's impact on childhood leukemia: a serendipitous discovery for treatment. In: Simon J, Cooper R, Hughes K (eds) Alternative & complementary therapies: a new bimonthly publication for health care practitioners. Mary Ann Liebert, Inc., New York
Garga RP (1958) Studies on virus diseases of plants in Madhya Pradesh. I. Green rosette of *Vinca*. Curr Sci 27:493–494
Garibaldi A, Bertetti D, Gullino ML (2009) First report of botrytis blight caused by *Botrytis cinerea* on periwinkle (*Catharanthus roseus*) in Italy. Plant Dis 93(5):554
Gathirwa JW, Rukunga GM, Njagi ENM et al (2007) In vitro anti-plasmodial and in vivo antimalarial activity of some plants traditionally used for the treatment of malaria by the Meru community in Kenya. J Nat Med 61(3):261–268
Goyal P, Khanna A, Chauhan A, Chauhan G, Kaushik P (2008) In vitro evaluation of crude extracts of *Catharanthus roseus* for potential antibacterial activity. Int J Green Pharm 2:176–181
Gundersen DE, Lee IM, Rehner SA, Davis RE, Kingsbury DT (1994) Phylogeny of mycoplasma like organisms (phytoplasmas): a basis for their classification. J Bacteriol 176(17):5244–5254
Guo JX, Kimura T, But PPH, Sung CK, Han BH (2001) International collation of traditional and folk medicine, vol 4, World Scientific
Gupta R (1977) Periwinkle produces anti-cancer drug. Indian Farm 27(4):11–13
Gupta RS, Sharma R (2006) A review on medicinal plants exhibiting antifertility activity in males. Nat Prod Rad 5:389–410
Gurib-Fakim A, Brendler T (2004) Medicinal and aromatic plants of Indian Ocean Islands: Madagascar, Comoros. Seychelles and Mascarenes, Medpharm
Hao W, Richardson PA, Hong CX (2010) Foliar blight of annual vinca (*Catharanthus roseus*) caused by *Phytophthora tropicalis* in Virginia. Plant Dis 94(2):274
Hassan KA, Brenda AT, Patrick V, Patrick OE (2011) In vivo antidiarrheal activity of the ethanolic leaf extract of *Catharanthus roseus* linn. (Apocyanaceae) in wistar rats. Afr J Pharm Pharmacol 5(15):1797–1800

Hostettmann K, Marston A, Ndjoko K, Wolfender JL (2000) The potential of African plants as a source of drugs. Curr Org Chem 4(10):973–1010

Iweala EEJ, Okeke CU (2005) Comparative study of the hypoglycemic and biochemical effects of *Catharanthus roseus* (Linn) G. Apocynaceae (Madagascar periwinkle) and chlorpropamide (diabenese) on alloxan-induced diabetic rats. Biokemistri 17(2):149–156

Jaleel CA, Gopi R, Alagu Lakshmanan GM, Panneerselvam R (2006) Triadimefon induced changes in the antioxidant metabolism and ajmalicine production in *Catharanthus roseus* (L.) G. Don. Plant Sci 171(2):271–276

James SA, Bilbiss L, Muhammad BY (2007) The effects of *Catharanthus roseus* (L) G. Don 1838 aqueous leaf extract on some liver enzymes, serum proteins and vital organs. Sci World J 2:5–9

Janaki Ammal EK, Bezbaruah HP (1963) Induced tetraploidy in *Catharanthus roseus* (L.) G. Don. Proc Indian Acad Sci B 57:339–342

Jarald E, Joshi SB, Jain DC (2008) Diabetes and herbal medicines. Iran J Pharmacol Ther 7(1):97–106

Johnson IS, Wright HF, Svoboda GH, Vlantis J (1960) Antitumor principles derived from *Vinca rosea* Linn. I. Vincaleukoblastine and leurosine. Cancer Res 20:1016–1022

Joshi MS, Ambaye RY (1968) Effect of alkaloids from *Vinca rosea* L. on spermatogenesis in male rats. Indian J Exp Biol 6(4):256–257

Joy PP, Thomas J, Mathew S, Skaria BP (1998) Medicinal plants. Kerala Agricultural University, Aromatic and Medicinal Plants Research Station, Odakkali

Kalra A, Ravindra NS, Chandrashekhar RS (1991) Influence of foliar application of fungicides on dieback disease caused by *Pythium aphanidermatum* and alkaloid yield of periwinkle (*Catharanthus roseus*). Indian J Agric Sci 61:949–951

Keim R (1977) Foliage blight of perwinkle in southern California. Plant Dis Rep 61:182–184

Khan MY, Aliabbas S, Kumar V, Rajkumar S (2009) Recent advances in medicinal plant biotechnology. Indian J Biotechnol 8(1):9–22

Khew KL, Davis RE, Ong CA, Lee IM, Su HJ, Tsai MC (1991) Detection of a Malaysian mycoplasmalike organism (MLO) and its differentiation from other Asian, European, and North American MLOs by use of cloned chromosomal and extra chromosomal MLO DNA probes. J Plant Prot 8:167–180

Krishnan R, Naragund VR, Vasantha Kumar T (1979) Evidences for outbreeding in *Catharanthus roseus*. Curr Sci 48:80–82

Kulkarni RN (1999) Evidence for phenotypic assortative mating for flower colour in periwinkle. Plant Breed 118:561–564

Kulkarni RN, Dimri BP, Rajagopal K, Suresh N, Chandrashekar RS (1984) Variability for quantitative characters in periwinkle (*Catharanthus roseus* (L) G. Don). Indian Drugs 22:61–64

Kulkarni RN, Rajagopal K, Chandrashekar RS, Dimri BP, Suresh N, Rajeshwar Rao BR (1987) Performance of diploids and induced autotetraploids of *Catharanthus roseus* under different levels of nitrogen and plant spacings. Plant Breed 98:136–140

Kulkarni RN, Baskaran K, Chandrashekara RS, Kumar S (1999) Inheritance of morphological traits of periwinkle mutants with modified contents and yields of leaf and root alkaloids. Plant Breed 118:71–74

Kulkarni RN, Baskaran K, Chandrashekara RS, Khanuja SPS, Darokar MP, Shasany AK, Uniyal GC, Gupta MM, Kumar S (2003) 'Dhawal', a high alkaloid producing periwinkle plant. US Patent 6,548,746

Kuppusamy C, Murugan K, Arul N, Yasodha P (2009) Larvicidal and insect growth regulator effect of α-amyrin acetate from *Catharanthus roseus* Linn against the malaria vector *Anopheles stephensi* Liston (Diptera: Culicidae). Entomol Res 39(1):78–83

Kurian A, Sankar A (2007) Medicinal plants: horticulture science series—2. New India Publishing Agency, New Delhi

Łata B (2007) Cultivation, mineral nutrition and seed production of *Catharanthus roseus* (L.) G. Don in the temperate climate zone. Phytochem Rev 6(2–3):403–411

Lee IM, Gundersen-Rindal DE, Davis RE, Bartoszyk IM (1998) Revised classification scheme of phytoplasmas based on RFLP analyses of 16S rRNA and ribosomal protein gene sequences. Int J Syst Bacteriol 48(4):1153–1169

Leveque D, Jehl F (2007) Molecular pharmacokinetics of *Catharanthus* (*Vinca*) alkaloids. J Clin Pharmacol 47(5):579–588

Levy A, Milo J, Ashri A, Palevitch D (1983) Heterosis and correlation analysis of the vegetative components and ajmalicine content in the roots of the medicinal plant—*Catharanthus roseus* (L.) G. Don. Euphytica 32:557–564

Lim-Sylianco CY, Blanco F (1981) Antimutagenic effects of some anti-cancer agents. Bull Philipp Soc Biochem Mol Biol 4:1–7

Luijendijk TJC, Meijden EVD, Verpoorte R (1996a) Involvement of strictosidine as a defensive chemical in *Catharanthus roseus*. J Chem Ecol 22:1355–1366

Mans DRA, da Rocha AB, Schwartsmann G (2000) Anti-cancer drug discovery and development in Brazil: targeted plant collection as a rational strategy to acquire candidate anti-cancer compounds. Oncologist 5(3):185–198

Masoud AN, Sciuchetti LA, Farnsworth NR, Blomster RN, Meer WA (1968) Effect of gibberellic acid on the growth, alkaloid production, and VLB content of *Catharanthus roseus*. J Pharm Sci 57:589–593

Mathur R, Chaudan S (1985) Antifertility efficacy of *Catharanthus roseus* Linn: a biochemical and histological study. Acta Eur Fertil 16(3):203–205

Mazidah M, Yusoff K, Habibuddin H, Tan YH, Lau WH (2012) Characterization of cucumber mosaic virus (CMV) causing mosaic symptom on *Catharanthus roseus* (L.) G. Don in Malaysia. Pertanika J Trop Agric Sci 35(1):41–53

McCormack JJ (1990) Pharmacology of antitumor bisindole alkaloids from *Catharanthus*. In: Brossi A, Suffness M (eds) The alkaloids, vol 37. Academic, San Diego, pp 205–228

McGovern RJ, Seijo TE, Davis TA (2003) Evaluation of fungicides and a fertilizer for control of Phytophthora blight in Madagascar periwinkle. Fungic Nematicide Tests 58:48

McMillan RT Jr, Garofalo JF (2004) Phytophthora parasitica wilt of new cultivars of *Catharanthus roseus*. Proc Fla State Hortic Soc 117:316–317

McMillan RT Jr, Graves WR (1996) Periwinkle twig blight caused by *Colletotricum dematium* on *Catharanthus roseus* L. Proc Fla State Hortic Soc 9(109):19–20

Meisner J, Weissenberg M, Palevitch D, Aharonson N (1981) Phagodeterrency induced by leaves and leaf extracts of *Catharanthus roseus* in the larva of *Spodoptera littoralis*. J Econ Entomol 74:131–135

Mishra P, Kumar S (2000) Emergence of Periwinkle *Catharanthus roseus* as a model system for molecular biology of alkaloids: phytochemistry, pharmacology, plant biology and in vivo and in vitro cultivation. J Med Arom Plant Sci 22:306–337

Moerman DE (2009) Native American medicinal plants: an ethnobotanical dictionary. Timber Press, Portland

Moudi M, Go R, Yien CYS, Nazre M (2013) Vinca alkaloids. Int J Prev Med 4(11):1231–1235

Mujib A, Ilah A, Gandotra N, Abdin MZ (2002) In vitro application to improve alkaloid yield in *Catharanthus roseus*. In: Govil JN, Kumar PA, Singh VK (eds) Recent progress in medicinal plants, Biotechnology and genetic engineering, vol 4. Sci Tech Publication, Houston, pp 415–440

Mukherjee AK, Basu S, Sarkar N, Ghosh AC (2001) Advances in cancer therapy with plant based natural products. Curr Med Chem 8(12):1467–1486

Murugavel T, Akbarsha MA (1991) Anti-spermatogenic effect of *Vinca rosea* Linn. Indian J Exp Biol 29(9):810–812

Nammi S, Boini MK, Lodagala SD, Behara RBS (2003) The juice of fresh leaves of *Catharanthus roseus* Linn. reduces blood glucose in normal and alloxan diabetic rabbits. BMC Complement Altern Med 3(1):4

Narayana MR, Dimri BP (1990) Periwinkle and its cultivation in India. CIMAP, Lucknow

Nayak BS, Pinto Pereira LM (2006) *Catharanthus roseus* flower extract has wound-healing activity in Sprague Dawley rats. BMC Complement Altern Med 6(1):41

Nejat N, Sijam K, Abdullah SNA, Vadamalai G, Dickinson M (2009) Phytoplasmas associated with disease of coconut in Malaysia: phylogenetic groups and host plant species. Plant Pathol 58(6):1152–1160

Nejat N, Sijam K, Abdullah SNA, Vadamalai G, Dickinson M (2010) Molecular characterization of an aster yellows phytoplasma associated with proliferation of periwinkle in Malaysia. Afr J Biotechnol 9(15):2305–2315

Nejat N, Sijam K, Vadamalai G, Dickinson M (2011) First report of *Spiroplasma citri* (-induced) associated with periwinkle lethal yellows in Southeast Asia. Plant Dis 95(10):1312

Nejat N, Vadamalai G, Davis RE (2013) *Candidatus* Phytoplasma malaysianum', a novel taxon associated with virescence and phyllody of Madagascar periwinkle (*Catharanthus roseus*). Int J Syst Evol Microbiol 63(2):540–548

Nejat N, Valdiani A, Cahill D, Tan Y-H, Maziah M, Abiri R (2015) Ornamental exterior versus therapeutic interior of madagascar periwinkle (*Catharanthus roseus*): the two faces of a versatile herb. Sci World J 2015:19

Neuss N, Neuss MN (1990) Therapeutic use of bisindole alkaloids from *Catharanthus*. In: Brossi A, Suffness M (eds) The alkaloids, vol 37. Academic, San Diego, pp 229–240

Neuwinger HD (2000) African traditional medicine: a dictionary of plant use and applications. Medpharm Scientific, Stuttgart

Nobili S, Lippi D, Witort E et al (2009) Natural compounds for cancer treatment and prevention. Pharmacol Res 59(6):365–378

Noble RL (1990) The discovery of the *Vinca* alkaloids-chemotherapeutic agents against cancer. Biochem Cell Biol 68(12):1344–1351

Omar OF, Emeran AA, Abass JM (2008) Detection of phytoplasma associated with Periwinkle virescence in Egypt. Plant Pathol J 7(1):92–97

Ou-Yang W, Wu WS (1998) Survey of periwinkle diseases in Taiwan. Plant Pathol Bull 7:147–149

Patil PJ, Ghosh JS (2010) Antimicrobial activity of *Catharanthus roseus*—a detailed study. Br J Pharmacol Toxicol 1:40–44

Pearce HL (1990) Medicinal chemistry of bisindole alkaloids from *Catharanthus*. In: Brossi A, Suffness M (eds) The alkaloids, vol 37. Academic, San Diego, pp 145–204

Pereira DM, Ferreres F, Oliveira J, Valentao P, Andrade PB, Sottomayor M (2009) Targeted metabolite analysis of *Catharanthus roseus* and its biological potential. Food Chem Toxicol 47(6):1349–1354

Pereira DM, Ferreres F, Oliveira JMA et al (2010) Pharmacological effects of *Catharanthus roseus* root alkaloids in acetylcholinesterase inhibition and cholinergic neurotransmission. Phytomedicine 17(8–9):646–652

Pezzuto JM (1997) Plant-derived anticancer agents. Biochem Pharmacol 53(2):121–133

Plaizier AC (1981) A revision of *Catharanthus roseus* (L.) G. Don (Apocynaceae). Meded Landbouwhogesch Wagening 81(9):1–12

Ponarulselvam S, Panneerselvam C, Murugan K, Aarthi N, Kalimuthu K, Thangamani S (2012) Synthesis of silver nanoparticles using leaves of *Catharanthus roseus* Linn G Don and their antiplasmodial activities. Asian Pac J Trop Biomed 2(7):574–580

Prajapati V, Tripathi AK, Jain DC, Sharma S, Khanuja SPS (1998) Sensitivity of Spilarctia obliqua to root extracts of *Catharanthus roseus*. Phytother Res 12(4):270–274

PROTA (2011) Plant resources of Tropical Africa. African ornamentals. Proposals and examples. PROTA Foundation, Wageningen

Ramya S, Govindaraji V, Navaneetha K, Jayakumararaj R (2008) *In vitro* evaluation of antibacterial activity using crude extracts of *Catharanthus roseus* L. (G.) Don. Ethnobot Leaflets 12:1067–1072

Rates SM (2001) Plants as source of drugs. Toxicon 39(5):603–613

Remia KM, Logaswamy S (2010) Larvicidal efficacy of leaf extract of two botanicals against the mosquito vector *Aedes aegypti* (Diptera: Culicidae). Indian J Nat Prod Resour 1(2):208–212

Ross IA (2003) Medicinal plants of the world. In: Chemical constituents, traditional and modern uses, vol 1, 2nd edn. Humana Press, Totowa

Saenz L, Santamaria JM, Villanueva MA, Loyola-Vargas VM, Oropeza C (1993) Changes in the alkaloid content of plants of *Catharanthus roseus* (L) G. Don as a result of water stress and treatment with abscisic acid. J Plant Physiol 142:244–247

Salim V, Yu F, Altarejos J, de Luca V (2013) Virus-induced gene silencing identifies *Catharanthus roseus* 7-deoxyloganic acid-7-hydroxylase, a step in iridoid and monoterpene indole alkaloid biosynthesis. Plant J 76(5):754–765

Samad A, Ajayakumar PV, Gupta MK et al (2008) Natural infection of periwinkle (*Catharanthus roseus*) with Cucumber mosaic virus, subgroup IB. Australas Plant Dis Notes 3:30–34

Schmelzer GH (2007) *Catharanthus roseus* (L.). In: Schmelzer GH, Gurib-Fakim A (eds) Medicinal plants/Plantes medicinales 1. PROTA, Wageningen

Schmelzer GH, Gurib-Fakim A (2008) Plant resources of Tropical Africa: medicinal plants. PROTA Foundation, Backhuys

Sevestre-Rigouzzo M, Nef-Campa C, Ghesquière A, Chrestin H (1993) Genetic diversity and alkaloid production in *Catharanthus roseus*, *C. trichophyllus* and their hybrids. Euphytica 66:151–159

Sherines RJ, Howard SS (1978) Male infertility. In: Harrison JH, Gittes RF, Perlmutter AD, Stamey TA, Walsh PC (eds) Campbells urology, 4th edn. WB Saunders, Philadelphia

Shukla DD, Teakle DS, Gough KH (1980) Periwinkle, a latent host for broad bean wilt and cucumber mosaic viruses in Australia. Plant Dis 64:802–803

Siddiqui MJ, Ismail Z, Aisha AFA, Abdul Majid AMS (2010) Cytotoxic activity of *Catharanthus roseus* (Apocynaceae) crude extracts and pure compounds against human colorectal carcinoma cell line. Int J Pharmacol 6(1):43–47

Singh SN, Vats P, Suri S et al (2001) Effect of an antidiabetic extract of *Catharanthus roseus* on enzymic activities in streptozotocin induced diabetic rats. J Ethnopharmacol 76(3):269–277

Singh D, Mehta SS, Neoliya NK, Shukla YN, Mishra M (2003) New possible insect growth regulators from *Catharanthus roseus*. Curr Sci 84:1184–1186

Singh N, Pandey BR, Verma P (2011) An overview of phytotherapeutic approach in prevention and treatment of Alzheimer's syndrome & dementia. Int J Pharm Sci Drug Res 3(3):162–172

Snoeijer W (2001) International register of *Catharanthus* cultivars. First supplement. Division of Pharmacognosy, Leiden & Amsterdam Center for Drug Research, Leiden University, Leiden

Sottomayor M, Barcelo AR (2005) The Vinca alkaloids: from biosynthesis and accumulation in plant cells, to uptake, activity and metabolism in animal cells. Stud Nat Prod Chem 33: 813–857

Sreevalli Y, Kulkarni RN, Baskaran K (2002) Inheritance of flower color in periwinkle: orange-red corolla and white eye. J Hered 93(1):55–57

Stanley A, Akbarsha MA (1992) Giant spermatogonial cells generated by vincristine and their uses. Curr Sci 63(3):144–147

Stanley A, Averal HA, Akbarsha MA (1993) Reproductive toxicity of vincristine in male rats. Indian J Exp Biol 31(4):380–382

Staszewski Z, Bodzon Z, Lata B (2007) Possibilities and limitations of seed production of *Catharanthus roseus* (L.) G. Don under Polish climatic conditions. Phytochem Rev 6(2–3): 397–402

Stearn WT (1966) *Catharanthus roseus*, the correct name for the Madagascar periwinkle. Lloydia 29:196–200

Stearn WT (1975) A synopsis of the genus *Catharanthus* (Apocynaceae). In: Taylor WI, Farnsworth NR (eds) The *Catharanthus* alkaloids: botany, chemistry, pharmacology, and clinical uses. Marcel Dekker, New York, pp 9–44

Sukumar K, Osmani Z (1981) Insect sterilants from *Catharanthus roseus*. Curr Sci 50(12): 552–553

Svoboda GH, Neuss N, Gorman M (1959) Alkaloids of *Vinca rosea* Linn. (*Catharanthus roseus* G. Don.). V. Preparation and characterization of alkaloids. J Am Pharm Assoc Am Pharm Assoc 48:659–666

Talha M, Radwan AS, Negm S (1975) The effect of soil moisture deficit on growth and alkaloidal content of *Catharanthus roseus* G. Don. Curr Sci 44:614–616

Thomas PA, Latimer JG (1996) Growth of vinca as affected by form of nitrogen, presence of bark, and type of micronutrients. J Plant Nutr 18:2127–2134

Thomas PA, Woodward J, Stegelin FE, Pennisi SV (2009) A guide for commercial production of *Vinca*. Bulletin 1219

Torres L, Galdeano E, Docampo D, Conci L (2004) Characterization of aster yellows phytoplasma associated with *Catharanthus* little leaf in Argentina. J Plant Pathol 86(3):209–214

Tyler VE (1988) Medicinal plant research: 1953–1987. Planta Med 54(2):95–100

Ueda JY, Tezuka Y, Banskota AH et al (2002) Antiproliferative activity of Vietnamese medicinal plants. Biol Pharm Bull 25(6):753–760

Usia T, Watabe T, Kadota S, Tezuka Y (2005) Cytochrome P_{450} 2D6 (CYP2D6) inhibitory constituents of *Catharanthus roseus*. Biol Pharm Bull 28(6):1021–1024

van Bergen MA (1996) Revision of *Catharanthus roseus* G. Don series of revisions of Apocynaceae XLI. Agric Univ Wageningen Pap 96(3):9–46

van der Heijden R, Jacobs DI, Snoeijer W, Hallard D, Verpoorte R (2004) The *Catharanthus* alkaloids: pharmacognosy and biotechnology. Curr Med Chem 11(5):607–628

Vega-Avila E, Cano-Velasco JL, Alarcon-Aguilar FJ, Fajardo Ortız MDC, Almanza-Perez JC, Roman-Ramos R (2012) Hypoglycemic activity of aqueous extracts from *Catharanthus roseus*. Evid Based Complement Alternat Med 5:35–40

Verma AK, Singh RR (2010) Induced dwarf mutant in *Catharanthus roseus* with enhanced antibacterial activity. Indian J Pharm Sci 72(5):655–657

Verpoorte R, Van der Heijden R, Van Gulik WM, Ten Hoopen HJG (1991) Plant biotechnology for the production of alkaloids: present status and prospects. In: Brossi A (ed) The alkaloids, vol 40. Academic, San Diego, pp 1–187

Verpoorte R, van der Heijden R, Schripsema J, Hoge JHC, Ten Hoopen HJG (1993) Plant cell biotechnology for the production of alkaloids: present status and prospects. J Nat Prod 56(2):186–207

Verpoorte R, Van der Heijden R, Moreno PRH (1997) Biosynthesis of terpenoid indole alkaloids in *Catharanthus roseus* cells. In: Cordell GA (ed) The alkaloids, vol 49. Academic, San Diego, pp 221–299

Virmani OP, Srivastava GN, Singh P (1978) *Catharanthus roseus*: the tropical periwinkle. Indian Drugs 15:231–252

Wang S, Zheng Z, Weng Y et al (2004) Angiogenesis and antiangiogenesis activity of Chinese medicinal herbal extracts. Life Sci 74(20):2467–2478

Waterhouse DN, Madden TD, Cullis PR, Bally MB, Mayer LD, Webb MS (2005) Preparation, characterization, and biological analysis of liposomal formulations of vincristine. Methods Enzymol 391:40–57

Weiss HD, Walker MD, Wiernik PH (1974) Neurotoxicity of commonly used antineoplastic agents (second of two parts). N Engl J Med 291:127–133

Whiting D, O'Connor A, Jones J, McMulkin L, Potts L (2011) Taxonomic classification. CMG Gard Notes 122:1–8

Win NKK, Jung HY (2012) The distribution of phytoplasmas in myanmar. J Phytopathol 160(3):139–145

Yandoc CB, Rosskopf EN, Shah DA, Albano JP (2007) Effect of fertilization and biopesticides on the infection of *Catharanthus roseus* by *Phytophthora nicotianae*. Plant Dis 91(11): 1477–1483

Zheng W, Wang SY (2001) Antioxidant activity and phenolic compounds in selected herbs. J Agric Food Chem 49(11):5165–5170

Elicitors Enhance Alkaloid Yield in *Catharanthus roseus*

Dipti Tonk, A. Mujib, Muzamil Ali, and Nadia Zafar

Abstract *Catharanthus roseus* (L.) G. Don is an important plant of medicinal importance. A number of active compounds of this plant are anti-cancerous in nature. Two such compounds are vinblastine and vincristine; unfortunately, the level of these phytocompounds is very low. In this chapter, the influence of elicitation, one of the important biotechnological techniques, has been discussed for improving yield. Elicitation, the elicitor types, the role of biotic and abiotic elicitors, and molecular mechanism of elicitation have also been described by presenting elicitation model in *C. roseus*. Besides, the importance of in vitro culture, the role PGRs, precursor feeding, and other factors that have some role in enriching yield have also been highlighted.

Keywords Alkaloid yield • Elicitors • Abiotic stress • In vitro culture • Metabolic engineering • Phytocompounds

1 Elicitation and Elicitors

An elicitor is defined as a substance that induces the synthesis of compounds, used in defense responses (Koga et al. 2006). Elicitor used for this purpose can be either biotic or abiotic, viz. jasmonic acid, glucan polymers, glycoproteins, fungal cell materials, UV irradiation, salts of heavy metals, and many other chemicals (Zhong 2002). Elicitation is based on the concept that secondary metabolites are produced by plants as part of their defense against pathogen attack and is noted to be very effective in certain compound producing cultures (Rao and Ravishankar 2002). In *Rauvofja canescens* and *Eschscholzia californica*, addition of yeast extract (elicitor) increased alkaloids yield. Methyl jasmonate (MEJA), a secondary messenger, is also widely used and observed to have similar effects on a number of plant species (Rijhwani and Shanks 1998). Endogenous elicitors can also be released by digesting plant cell walls with fungal enzyme preparations as macerozyme. Macerozyme was

D. Tonk (✉) • A. Mujib • M. Ali • N. Zafar
Cellular Differentiation and Molecular Genetics Section, Department of Botany, Hamdard University, New Delhi 110062, India
e-mail: dipti10785@gmail.com

M. Naeem et al. (eds.), *Catharanthus roseus*,
DOI 10.1007/978-3-319-51620-2_6

also added to *C. roseus* hairy root cultures and after treatment with 1% macerozyme, TDC and phenylalanine ammonialyase were induced (Moreno-Valenzuela et al. 1999). Addition of an alginate monomer to a *C. roseus* cell culture promoted ajmalicine and 5′-phosphodiesterase (an antibiotic protein) production. After studying the effects of chitosan oligomer (exogenous elicitor) and oligogalacturonic acid (endogenous elicitor), it was realized that alginate acted as an endogenous elicitor. Esterification of carboxylic groups of alginate reduced the excretion of enzyme, indicating that the carboxyl groups have an important role in the initiation of elicitation reaction (Akimoto et al. 1999). The effect of mannitol as abiotic stress and *Aspergillus niger* as biotic stress was investigated by monitoring callus biomass growth and in the production of vinblastine and vincristine by Taha et al. (2009). Acetylsalicylic acid, a derivative of an important signaling compound, salicylic acid, also acted as a biotic elicitor (Pedras et al. 2002). Godoy-Hernandez et al. (2008) noted that the addition of various concentrations of acetylsalicylic acid to *A. tumefaciens* transformed cell lines grown on corn starch (as carbon source) increased alkaloid production five times, total phenolics 15 times, urano-coumarins six times, and anthocyanins 15 times. It is therefore suggested that the elicitation may modulate the expression of molecules of primary metabolism, involved in vacuolar transport and thereby regulates secondary metabolism (Vasconsuelo and Boland 2007).

2 Background

Most of the higher plants are sources of natural products, used as pharmaceuticals, agrochemicals, flavor, fragrance ingredients, food additives, and pesticides (Philipson 1990). In the search for alternatives to production of desirable medicinal compounds from plants, biotechnological approaches, specifically plant tissue cultures, are found to have enough potential in the production of bioactive plant metabolites (Rao and Ravishankar 2002). Plant cell suspension culture could be used for large-scale cultivation of plant cells from which secondary metabolites could be extracted. The advantage of this method is that it can provide a continuous, reliable source of natural products (Iantcheva et al. 2006). The possible use of plant cell cultures for the specific biotransformations of natural compounds has been demonstrated in several studies (Cheetham 1995; Scragg 1997; Krings and Berger 1998; Ravishankar and Rao 2000). The strong and growing demand for natural, renewable products compels attention on in vitro plant materials as potential factories for secondary products, and this paved the way for new research exploring secondary product synthesis in vitro (Karuppusamy 2009). Depending upon the culture medium composition and environmental conditions, the accumulation of secondary products in plant cell cultures varies. Due to several applications, tissue culture technology for production of plant chemicals has bloomed beyond expectations. One major application is it can enrich the level of alkaloids in culture and for this purposes various approaches including elicitation have been used.

3 Elicitor Types

An elicitor is defined as a compound that stimulates defense mechanism in synthesizing and accumulating secondary compounds including phytoalexins (Ebel and Scheel 1997). Elicitors are chemically complex biological compounds which include microbial cell wall (fungi or bacteria), cell wall derived polysaccharides (pectin or cellulose), proteins, G-protein phytoalexins, yeast extract, chitin, glucans, etc. Another groups of elicitors are substances of nonbiological origin such as inorganic salts ($CaCl_2$, $AlCl_3$, NaCl, Na_2SO_4), physical factors like UV radiation, heavy metal salts (Cu, Cd, Co, Ag ions, Ca^{2+}), high pH, salicylic acid, etc. Elicitors have been in use in promoting secondary metabolites synthesis in several investigated plant genera by utilizing cell/organ culture methods (van der Heijden et al. 2004; Wang 2008). An elicitor is also defined as a substance that induces the synthesis of compounds, used in defense responses (Koga et al. 2006). Elicitors used for this purpose are jasmonic acid, glucan polymers, glycoproteins, fungal cell materials, salts of heavy metals, and other chemicals (Zhong 2002). Elicitation is based on the concept that secondary metabolites are produced by plants as a part of defense response against pathogen attack and is noted to be very effective in producing certain compounds. In *Rauvofja canescens* and *Eschscholzia californica*, addition of yeast extract increased alkaloids yield; methyl jasmonate was also observed to have similar effects on a number of plant species (Rijhwani and Shanks 1998). Robins (1994) reported different strategies in order to improve the synthesis of secondary products in suspension cultures. Different media and employment of biotic and abiotic elicitors have been used as the elicitors have strong and rapid improving effects on indole alkaloid production (Zhang et al. 2000; Tung et al. 2002; Taha et al. 2009). There are many reports of cell culture in which production of secondary metabolites has been reported from various medicinal plants, such as tanshinone production in *Salvia miltiorrhiza* hairy roots on cocultivation with *Bacillus cereus* (Wu et al. 2007). In *Lepidium sativum*, lepidine content was noted to be dependent on source and type of explants (Pande et al. 2002). Since the biosynthetic efficiency of populations varies, a high yielding variety needs to be selected as a starting material (Tripathi and Tripathi 2003). Coste et al. (2011) reported that salicylic acid insignificantly influenced on biomass production in shoot culture of *Hypericum hirsutum* and *H. maculatum*. Jasmonic acid on the other at 250 M enhanced hypericin and pseudohypericin in *Hypericum hirsutum* and *H. maculatum* shoot cultures (Coste et al. 2011) (Tables 1 and 2).

Enhanced production of withanolides in shoot culture of *Withania somnifera* (L.) was influenced by cytokinin, the type, concentration, and exposure time of elicitor (Sivanandhan et al. 2013). The methyl jasmonate (MJ) and phenylalanine enhanced callus growth and flavonoids content in *Glycyrrhiza uralensis* (Guo et al. 2013). Abiotic elicitor, MeJ at 10l M, significantly improved the furanocoumarin production in shoot cultures of *Ruta graveolens* (Diwan and Malpathak 2010). In vitro cell culture offers an advantage for foreign protein synthesis in certain situations since the cell can be designed to produce therapeutic proteins, including monoclonal

Table 1 Abiotic elicitors and secondary metabolites (Source: Shilpa et al. 2010)

Abiotic elicitor	Plant species	Phytocompounds	Authors
Vanadium sulfate	*Catharanthus roseus*	Catharanthine	Smith et al. (1987)
Curdlane, Xanthan	*Capsicum frutescens*	Capsaicin	Johnson et al. (1991)
Arachidonic acid	*Capsicum annuum*	Capsidiol, Rishitin	Hoshino et al. (1994)
Salicylic acid	*Daucus carota*	Chitinase	Muller et al. (1994)
Diethyl amine ethyl dichloro phenyl ether	*Catharanthus roseus*	Indole alkaloids	Lee et al. (1998)
NaCl	*Catharanthus roseus*	Vincristine, vinblastine	Samar et al. (2015)
Cu^{2+}, Cd^{2+}	*Atropa belladonna*	Tropane alkaloids	Lee et al. (1998)
Oxidative stress	*Arabidopsis*	Camalexin	Zhao et al. (1998)
$CuSO_4$	*Hyoscyamus albus*	Phytoalexin	Mader (1999)
Dark	*Hydrangea macrophylla var. thunbergii*	Polyphenol	Yamamoto and Yamamoto (2000)
Copper chloride	*Matricaria chamomilla*	Hemiarin, Umbelliferone	Eliasova et al. (2004)
Electromagnetic treatment	*Ammi majus L.*	Umbelliferone	Krolicka et al. (2006)
$CaCl_2$	*Catharanthus roseus*	Vincristine, vinblastine	Zahid and Mujib (2012)

Table 2 Biotic elicitors and secondary products (Source: Shilpa et al. 2010)

Biotic elicitor	Plant species	Phytocompounds	Authors
Aspergillus sp.	*Catharanthus roseus*	vincristine, vinblastine	Dipti et al. (2016)
Chitosan	*Ruta graveolens*	Rutacridone epoxide	Eilert et al. (1984)
Hemicellulose	*Brugmansia candida*	tryosumine	Sandra et al. (1988)
Arachidonic acid	*Taxus sp.*	Taxol	Ciddi et al. (1995)
Cellulose	*Capsicum annuum*	Capsidol	Patrica et al. (1996)
Chitosan	*Ocimum basilicum*	Rosmarinic acid and Eugenol	Kim et al. (2005)
Salicylic acid	*Daucus carota*	Chitinase	Muller et al. (1994)
Methyl Jasmonate	*Hyoscyamus albus*	Phytoalexins	Kuroyanagi et al. (1998)
Alginate oligomers	*Catharanthus roseus*	5′-Phosphodiesterase (Pdase)	Akimoto-Tomiyama et al. (2002)
Fungal elicitor	*Cupressus lusitanica*	Beta-thujaplicin	Zhao et al. (2001a, b)
Trichoderma viride	*Catharanthus roseus*	Ajmalicine	Namdeo et al. (2002)
Yeast elicitor	*Medicago truncatula*	Beta-amyrin	Broeckling et al. (2005)

antibodies, antigenic proteins that act as immunogenes, human serum albumin, interferon, immuno-contraceptive protein, ribosome unactivator trichosanthin, antihypersensitive drug angiotensin, leu-enkephalin neuropeptide, and human hemoglobin (Manson and Arntzen 1995; Wahl et al. 1995; Arntzen 1997; Hahn et al. 1997; La Count et al. 1997; Marden et al. 1997; Wongsamuth and Doran 1997; Doran 2000). The appeal of using natural products for medicinal purposes is increasing, and metabolic engineering can alter the production of pharmaceuticals and help to design new therapies. At present, researchers aim to produce substances with antitumor, antiviral, hypoglycemic, anti-inflammatory, antiparasite, antimicrobial, tranquilizer and immune modulating activities through tissue culture technology (Harbourne 1999). Research in plant tissue culture technology has resulted in production of a wide variety of pharmaceuticals like alkaloids, terpenoids, steroids, saponins, phenolics, flavonoids, and amino acids. The dimeric indole alkaloids vincristine and vinblastine from *C. roseus* have become valuable drugs in cancer chemotherapy due to their potent antitumor activity against various leukemias and solid tumors. The alkaloid vinblastine is composed of catharanthine and vindoline. Since vindoline is more abundant than catharanthine in intact plants, it is less expensive. Influence of various factors like stress, addition of bioregulators, elicitors, and synthetic precursors on indole alkaloids production was studied in *C. roseus* (Zhao et al. 2001a, b). An improved synthesis of vinblastine and vincristine by using NaCl as an elicitor was reported recently (Fatima et al. 2015). The influence of fungus elicitor *Aspergillus flavus* on alkaloid yield was also investigated in *C. roseus* (Dipti et al. 2016). The study revealed increased yield of vinblastine and vincristine in cultivated tissues. Metabolic rate-limitations methodologies, precursor feeding, and effect of elicitor dosage on biosynthesis of indole alkaloids in *C. roseus* hairy root cultures were also reported earlier (Rijhwani and Shanks 1998; Morgan and Shanks 2000). Namdeo et al. (2002) reported higher accumulation of ajmalicine in *C. roseus* when treated with different concentrations of elicitor of *T. viride*, *A. niger*, and *F. moniliforme*.

Plant cells/tissues are exposed to stresses and it is a remarkable adaptive plasticity of plant genome that it deciphers and responds to novel in vitro stresses (Nuernberger 1999). In recent time, techniques are used to enhance secondary metabolites yield by triggering stress response by using precursors, biotransformation, changing environment conditions, altering medium constituents, etc. (Angelova et al. 2006). These are compounds stimulating plant defense response by synthesizing metabolites. The definition of elicitors includes both substances of pathogen origin (exogenous elicitors) and compounds released from plants (endogenous elicitors) by the action of the pathogen (Patel and Krishnamurthy 2013).

These groups of elements are used in enhancing plant secondary metabolite synthesis. The biotic elicitors are biological origin, derived from pathogens or from plant itself, while abiotic elicitors are other than biological origin and are grouped in physical factor and chemical compounds (Radman et al. 2003). On the basis of plant–elicitor interaction, it may be classified as race specific and general elicitors (Vasconsuelo and Boland 2007). Elicitation of plant cells may be promising as it shows enhanced yield of antibiotics and many other fermented products. Elicitation enhances secondary metabolism in plants or plant cells in vitro and this provides an opportunity for exploitation of plant cells for manufacturing secondary metabolites. There has been a strong need for the identification of important elicitors to be able in synthesizing enhanced yield of alkaloids. A number of elicitors have been identified and used to induce phytoalexin accumulation in cultured plant cells. These elicitors will be used in other systems in synthesizing important secondary metabolites capable of giving protection against infectious diseases/agents (Savitha et al. 2006; Thangavel et al. 2008). Biotic elicitors comprise polysaccharides, proteins, glycoproteins, or cell wall fragments derived from fungi, bacteria, and even plants. Among those, fungal elicitors have been most widely used for enhancement of secondary metabolites (Taha et al. 2009). The concentration of elicitors and the incubation time differ with elicitor types and culture system. Another important area is the time of addition of an elicitor to the medium. Optimal induction occurs when the elicitors are at its late exponential or early stationary phase.

4 Biosynthesis, Pathways, and Metabolic Engineering of *C. roseus* Alkaloids

The biosynthesis of secondary metabolites is usually discussed by starting with the description of the metabolites involved. However these metabolites are in fact the end products of a complex process comprising the involvement of several enzymes, genes, regulatory genes and (transport through) intra- and intercellular compartments. The TIAs are condensation products of two biosynthetic routes, requiring intermediates supplied by both pathways. The process starts with amino acid tryptophan and the monoterpenoid geraniol; the biosynthesis of vinblastine requires the participation of at least 35 intermediates, 30 enzymes, 30 biosynthetic, two regulatory genes, and seven intra- and intercellular compartments. The produced (secondary) metabolites are stored or either directly available for their biological function or, as in case of phytoanticipins, require a final processing for displaying their action, e.g., hydrolysis by a glucosidase.

The first study on *Catharanthus* alkaloids biosynthesis was performed at the end of the 1950s (Noble 1990). Periwinkle plants were grown in controlled atmosphere with labelled $^{14}CO_2$ and after alkaloids extraction, column and paper chromatography, several labelled alkaloids were detected of which vinblastine was important. Since then studies on TIA biosynthesis have been performed and several alkaloids have been identified. The advantage of metabolic pathway is that these steps are easy to manipulate, both physiologically (rapid growth, ease of precursor feeding, etc.) and genetically. Major drawbacks of such system are that the cells normally do not accumulate vindoline (and thus no bisindole alkaloids are formed) and the limited differentiation of the cells (for example, no functional chloroplasts are developed). TIA production in *C. roseus* cells can be induced by abiotic and biotic elicitors (Zhao and Verpoorte 2007; Siddiqui et al. 2010). Abiotic elicitor includes heavy metal ions and UV light and second category comprises yeast extract, culture filtrate of phytopathogenic fungus *Pythium aphanidermatum*, etc.

These biotic elicitors are complex mixtures, and often specific constituents like oligosaccharides or (glyco-) peptides, which are responsible for elicitation effect (Shirsau et al. 1997). Yeast extract induced transcription of the biosynthetic gene encoding strictosidine synthase (STR) in cultured *C. roseus* cells and alkalinization of the culture medium. The active compound from yeast extract was partially purified and found to be of a proteinaceous nature (Menke et al. 1999).

As the dosage is crucial, the preparation of biotic elicitor needs to be optimized carefully. The culture age of the elicitor is another important aspect that needs to be considered wisely (Moreno et al. 1995; Ramos-Valdivia et al. 1997), and often added a few days after inoculation of the culture, when the microbial cells are rapidly dividing. The elicitor rapidly dispersed homogeneously throughout the culture medium and the elicitor molecules come into contact with their specific receptors in plant cell wall but in *C. roseus*, no specific receptors have so far been identified. Downstream of the receptor, a signal transduction pathway is involved (Fig. 1, Memelink et al. 2001). Perception of an elicitor induces alkalinization of the medium (caused by proton uptake of the cells) and an influx of extracellular Ca^{2+}.

This requires the reaction of a protein kinase. The increased concentration of Ca^{2+} results in an activation of the octadecanoid pathway. This pathway starts from linolenic acid and results in the biosynthesis of jasmonic acid (JA). The secondary messenger JA is perceived by a (hypothetical) receptor and a downstream protein

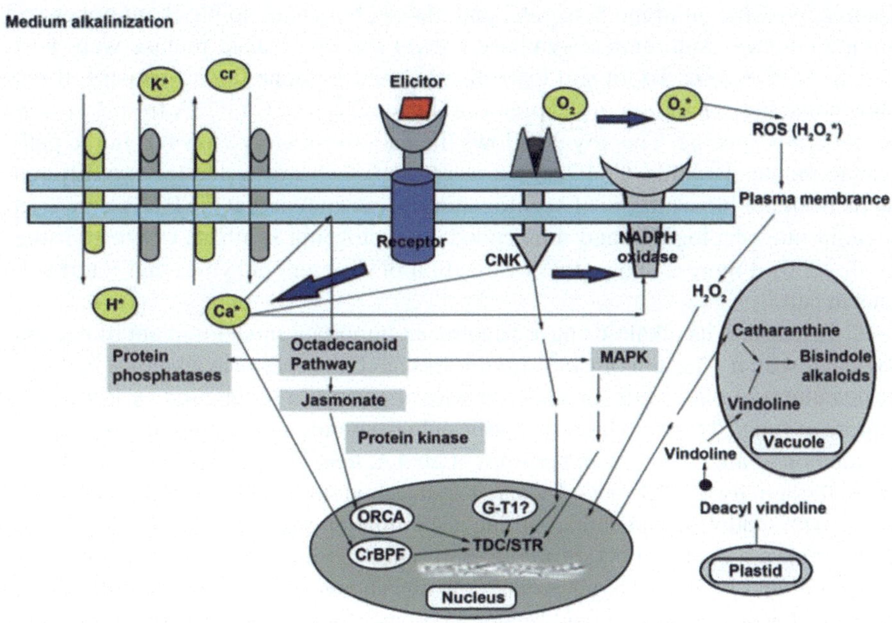

Fig. 1 Elicitation model (Source and courtesy: Memelink et al. 2001)

phosphorylation event leads to the de novo synthesis of nuclear proteins (ORCA2 and ORCA3). These proteins interact with the jasmonate- and elicitor-responsive elements in the promotors of several biosynthetic genes to activate gene expression (Rijhwani and Shanks 1998). The overall process is very quick, and induced mRNA levels may be observed shortly. The evidence for the signal transduction pathway was noted by using specific inhibitors for various steps (Menke et al. 1999). The pathway may be shortened by using JA or methyl jasmonate (MeJA) as elicitor, as there may have been JA receptors on cell wall, which perceive signal quickly (Rijhwani and Shanks 1998).

The post-strictosidine biosynthetic steps have been studied largely in *C. roseus* seedlings. When applied exogenously, MeJA is able to enhance alkaloid accumulation in these seedlings (Rijhwani and Shanks 1998). In contrast to monomeric alkaloids, the vinblastine content was not increased by MeJA (Aerts et al. (1996). MeJA promoted vindoline biosynthesis by induction of TDC and D17H activity in developing seedlings. Salicylic acid, another signaling compound in plants, was not active (Vázquez-Flota et al. 2009). The necessity of a functional octadecanoid pathway for alkaloid production in auxin-starved *C. roseus* cells was demonstrated by various feeding experiments. When auxin-starved cells were treated with octadecanoid pathway inhibitors, the alkaloid production was strongly reduced. This could be restored by addition of MeJA (Gantet et al. (1998).

In *Catharanthus*, the early steps leading to the biosynthesis of tryptamine and secologanin have been extensively studied. Chorismate forms an important branching point in the shikimate pathway, with branches leading to tryptophan, phenylalanine, tyrosine, *p*-amino benzoate, and via isochorismate to phylloquinones and anthraquinones. Anthranilate synthase (ASA) and chorismate mutase were purified from *C. roseus*. By in situ hybridization and immune histochemistry, it was shown that the expression of the previously characterized CYP72A from *C. roseus* is epidermis specific. Thereby it follows the pattern for early enzymes in the pathway to indole alkaloids. CYP72A expressed in *E. coli* was tested for two different steps from the biosynthesis of secologanin. It was known that CYP72A converted loganin into secologanin and thus encodes secologanin synthase enzyme (Irmler et al. 2000). Figure 2 shows the localization of different enzymes and intermediates in cell.

C. roseus and its alkaloids have become an important model in plant biotechnological research. The aim of *Catharanthus* research is to produce alkaloids by large-scale culture of plant cells, in a similar way of penicillin production by large-scale culture of fungi. There are large scientific achievements; still there is no commercial production of alkaloids yet. In undifferentiated *A. tumefaciens* transformed cell cultures, the activity of DAT (acetyl-CoA: 4-*O*-deacetylvindoline 4-Oacetyltransferase) along with vindoline and catharanthine accumulation was reported (O'Keefe et al. 1997). Palazon et al. (1998) showed the presence of vindoline and catharanthine in hairy root and higher alkaloid production was shown to be related with thin morphology, lower growth rate, and higher amounts of the rol C gene product (rolC is one of the genes on the T-DNA of the *A. rhizogenes* and is involved in the induction and development of the roots). The subculture cycle affected both growth and

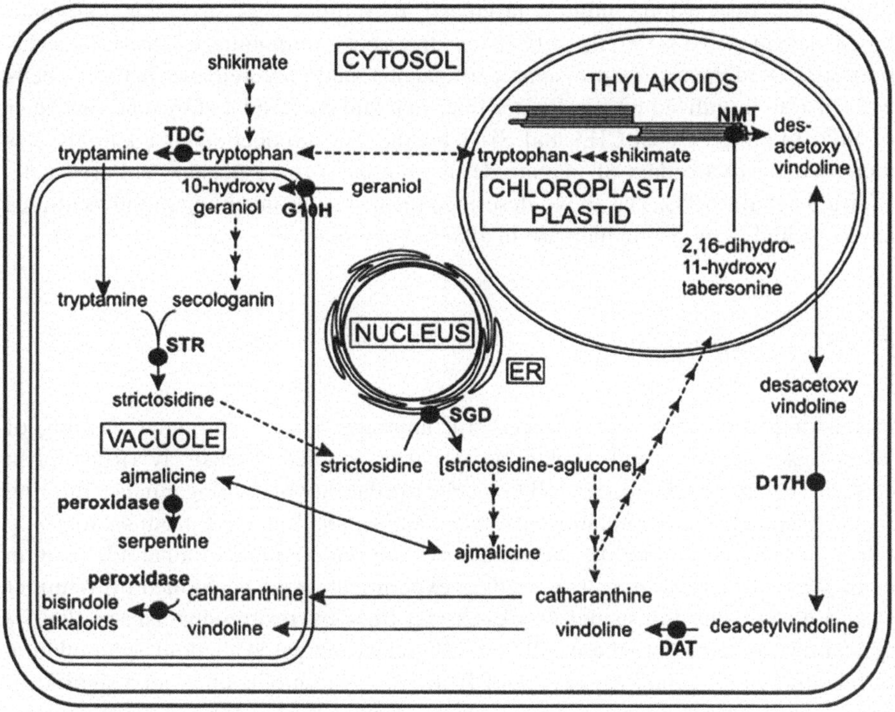

Fig. 2 Participation of enzymes and intermediates (Source: van der Heijden et al. 2004)

alkaloid production of hairy root cultures in *C. roseus*. A 2-week subculture cycle improved growth, while a 4-week cycle lowered growth of culture. Lochnericine yield was reported to be high in faster growing cultures while the highest serpentine concentration was noted in slow-growing cultures (Rijhwani and Shanks 1998). Hairy root cultures were elicited in late exponential phase with pectinase and JA, both at different concentrations and with different exposure times. Pectinase treatment resulted in a 150% increase in tabersonine after 48 h and immediately after elicitation, catabolism of serpentine, tabersonine, and lochnericine was observed. Addition of JA increased the yields of ajmalicine (80%), serpentine (60%), lochnericine (150%), and horhammericine (500%).

The effect on tabersonine was dose dependent, a decrease was observed at low levels, and an increase at high concentrations of JA was found. Aminobenzotriazole, a P450-dependent monooxygenase inhibitor, inhibited the accumulation of horhammericine in hairy root cultures, while clotrimazole, another oxygenase inhibitor, inhibited the synthesis of lochnericine, suggesting that two different P450 monooxygenases are involved in the formation of lochnericine and horhammericine (Morgan and Shanks 1999).

By manipulation of the culture medium, e.g., increasing the concentrations of micronutrients, hairy root cultures could be changed into cell suspension cultures.

These cell suspension cultures produced five times less alkaloids (Moreno-Valenzuela et al. 1998). Cells of *C. roseus* have been immobilized into various carriers such as alginate, agarose, agar, carrageenan, and polyurethane foam; the cells were also immobilized on a polyester fiber mat and encapsulated by a porous layer of SiO_2 modified by Si-CH3 and Si-H bonds. The production of alkaloids was increased by two orders of magnitude as compared to suspension-cultured cells (Carturan et al. 1998). The above-described technologies involving in vitro cultures have resulted in moderate increase in alkaloid content.

5 Precursor Feeding

Precursor feeding has been a successful approach and in particular, feeding of loganin, secologanin, and tryptamine has been studied extensively (Runguphan et al. 2010). In most *C. roseus* cell lines, the availability of secologanin is a limiting factor for alkaloid accumulation. An increase of inoculums size from 40 to 160 g FW/L in medium favored the accumulation of secologanin and alkaloids (Contin et al. 1998). Extensive precursor feeding experiments were performed by Whitmer et al. (2002) using transgenic *C. roseus* cell lines overexpressing *Tdc* and/or *Str* gene. Loganin feeding to these cell lines efficiently increased alkaloid accumulation as feeding with loganin, geraniol, and 10-hydroxygeraniol resulted in a significant increase of tabersonine accumulation (Morgan and Shanks 2000). A synthetic precursor derived from tryptamine, MIA (*N*-(methoxycarbonylethyl)-*N*-(2-(1H-indol-3yl)-ethyl)-β-methylalaninate, was fed to *C. roseus* plants and this resulted enhanced synthesis of ajmalicine in roots (90%), while the other alkaloid serpentine levels remained unchanged. In the aerial parts, however, not much effect was observed and no new alkaloids were identified (Bonzom et al. 1997).

6 Alkaloid Yield and Elicitation: Some Past Observations

In general, two common strategies are followed for enhancement of alkaloid yield: Modification or metabolic engineering of metabolic pathways, which includes (1) overexpression of biosynthetic rate limiting enzymes (Verpoorte et al. 1999, 2000; Verpoorte and Memelink 2002; Hughes and Shanks 2002) and (2) Genetic manipulation of in vitro cultures, and then by screening and selection for high producing cell lines and optimization of culture conditions (Verpoorte et al. 1991; Moreno et al. 1995; Junaid et al. 2008). Elicitors are applied in culture for the enhancement of alkaloid yield. MeJA, a secondary messenger, is widely used. Endogenous elicitors can also be released by digesting plant cell walls with fungal enzyme preparations as macerozyme. Macerozyme was also added to *C. roseus* hairy root cultures and after treatment with 1% macerozyme, TDC and phenylalanine ammonialyase were induced (Moreno-Valenzuela et al. 1999). Addition of an alginate monomer to a *C. roseus* cell culture promoted ajmalicine and 5′-phosphodiesterase (an antibiotic protein) production. After studying the effects of chitosan oligomer (exogenous elicitor) and oligogalacturonic acid (endogenous elicitor), it was realized that alginate acted as an endogenous elicitor. Esterification of carboxylic groups of alginate reduced the excretion of enzyme, indicating that the carboxyl groups have an important role in the initiation of elicitation reaction (Akimoto et al. 1999). The effect of mannitol as abiotic stress at the concentrations 0.0, 2000, 4000, or 8000 ppm or *Aspergillus niger* as biotic stress at the concentrations 0.0, 0.05, 0.15, and 0.25% on calli growth parameters and production of vinblastine and vincristine was investigated by Taha et al. (2009).

Acetylsalicylic acid, a derivative of an important signaling compound, salicylic acid, also acted as a biotic elicitor (Pedras et al. 2002). Addition of various concentrations of acetylsalicylic acid to a *A. tumefaciens* transformed cell line growing on corn starch as carbon source increased the alkaloid production five times, total phenolics 15 times, urano-coumarins six times, and anthocyanins 15 times (Godoy-Hernandez et al. 2008). Taha et al. (2009) reported *Aspergillus niger* mediated improved callus growth in *C. roseus*. In *Rosmarinus officinalis* L., treatment with biotic elicitors (*Pseudomonas aeruginosa*, *F. oxysporum*) and abiotic elicitors ($CaCl_2$) increased the productivity of callus tissue (Rashid et al. 2011). *C. roseus* treated with 150 and 200 kg P_2O_5/ha along with Arbuscular Mycorrhizal Fungi (AMF), i.e., *Glomus mosseae*, showed maximum plant height, leaves, root biomass, phosphorus content, root colonization, spore count, and ajmalicine content (120 days after planting) when compared with the control plants. The inoculation of AMF and other beneficial soil microorganisms significantly increased the biomass of different medicinal plants (Sena and Das 1998; Kothari et al. 1999). High dosage of elicitor has been reported to induce hypersensitive response leading to cell death, whereas an optimum level facilitates alkaloid accumulation (Collinge and Susarenka 1987; Mukandan and Hjorosto 1990; Roewer et al. 1992). Tung et al. (2002) reported vincristine accumulation by using *Fusarium oxysporum*, an endophyte of *C. roseus*, while Guo and Kunming (1998) obtained and analyzed vinblastine level through TLC and HPLC method by utilizing *Alternaria* sp.

Zhao et al. (2001a, b) noted reduced Ca^{2+} influx with EGTA and verapamil application, which lowered TIA production (ajmalicine+serpentine+catharanthine) in fungal-elicitated *C. roseus* suspensions. The concentration plays a very important role in elicitation process (Namdeo et al. 2002) and higher accumulation of ajmalicine was noted when treated with different concentrations of *T. viride*, *A. niger*, and *F. moniliforme* extract. Zhao et al. (2000) reported enhanced production of catharanthine in *C. roseus* cell culture by combined elicitor treatment of *Aspergillus niger* and tetramethyl ammonium bromide in shake flasks and bioreactors. Several other studies also revealed fungal elicitors profoundly affect the regulation of indole alkaloid biosynthesis (Sayed and Verpoorte 2007). On application of biotic elicitor, the vincristine and vinblastine yield can also be enhanced in culture.

The study is important as the plant is the only source of expensive and important agents with activity against several kinds of cancer. Kumar et al. (2013) isolated endophytic fungi, i.e., *Fusarium oxysporum*, from *C. roseus* plant and found a fungus AA-CRL-6 which produced vinblastine and vincristine in appreciable amounts. *F. oxysporum* extract significantly improved the production of KBBA (11-keto-β-boswellic acid); AKBBA (acetyl-11-keto-β-boswellic acid); BBA (β-boswellic acid); and of ABBA (acetyl-β-boswellic acid), respectively, in the callus biomass (Ghorpade et al. 2011). Catharanthine production in *C. roseus* was reported to be high in optimized cultivation conditions. The use of immobilization technique (Facchini and DiCosmo 1991) and application of different elicitors like homogenates of fungal mycelium or non-biotic elicitors such as vanadium improved the yield of alkaloids. Moreno et al. (1996) studied the effect of elicitation on different metabolic pathways involved in secondary metabolism of *C. roseus* cell suspension.

Namdeo (2007) reported successful protocol development involving precursor feeding by identifying compounds, which may be converted to desire secondary metabolites by select plant cell line. Andrade et al. (2013) noted mycorrhizal application increased ajmalicine and serpentine contents in *C. roseus* roots, suggesting that mycorrhization has a good influence on alkaloid accumulation. A few other reports indicated vincristine and/or vinblastine enhancement in leaves and ajmalicine in roots with arbuscular mycorrhizal fungi (AMF) association (Ratti et al. 2010; Dela Rosa-Mera et al. 2011). Paclobutrazol (PBZ), gibberellic acid, and *Pseudomonas fluorescens* treatments had profound effect on antioxidant metabolism, caused an enhancement in nonenzymatic antioxidant potentials in *C. roseus* (Jaleel et al. 2009).

7 Abiotic Stress and Alkaloid Synthesis

Beside biotic elicitation, the content of secondary metabolites in plants is greatly affected by abiotic stresses like salinity, wounding- and water stress (Jaleel et al. 2009). The addition of acetic anhydride influenced the biosynthesis of serpentine, catharanthine, tabersonine, and vindoline in *C. roseus* cell suspension but the content significantly declined with increased acetic anhydride concentrations (Guo et al. 2013). Similarly, NaCl treatment increased root alkaloid ajmalicine (Jaleel 2009). Catharanthine and vindoline increased threefold and 12-fold, respectively, on treatment with a 5-min UV-B irradiation (Ramani and Jayabaskaran 2007). The addition of methyl jasmonate significantly promoted the synthesis of tabersonine, catharanthine, and serpentine (Guo et al. 2013). The addition of dithiothreitol (DTT) to *C. roseus* influenced the biosynthesis of serpentine, catharanthine, tabersonine, and vindoline and the alkaloid level declined sharply as the concentration of DTT increased (Guo et al. 2013). How biotic elicitors elicit fast cultural growth and later stimulate enriched level of alkaloids is not well elucidated. van der Heijden et al. (2004) reported that the biotic elicitors often contain compounds like oligosaccharides and glycopeptides which evoke elicitation effect.

8 Elicitation Mechanism

The elicitation mechanism is based on "elicitor-receptor" interaction (Radman et al. 2003), which activates signal transduction pathways and stimulates transcriptional control of several defense genes participating in alkaloid synthesis pathways (Memelink et al. 2001; Mujib et al. 2012). Thus, the experimentation on elicitation is important and valuable as it promotes biomass and improves alkaloid biosynthesis. Although, our current knowledge of elicitors' mode of action is related exclusively to secondary metabolism, in recent years, it has been noted that the primary metabolism is also affected by elicitation (Yu and Facchini 2000; Dixon 2001; Broeckling et al. 2005). It is therefore suggested that the elicitation may modulate the expression of molecules of primary metabolism, involved in vacuolar transport and thereby regulates secondary metabolism (Vasconsuelo and Boland 2007). Based on available information, the following role/involvement of elicitor in secondary metabolism is hypothesized: binding of elicitor to plasma membrane receptor, changes of Ca^{2+} influx in cytoplasm from extracellular and intracellular sources; decrease of pH in cytoplasm; activation of NADPH oxidases, protein phosphorylation patterns and protein kinase activation; changes in cell wall structure (lignification) and in generating reactive oxygen species (ROS); synthesis of jasmonic acid and salicylic acid as secondary messengers; and activation of genes, which produce defense-related proteins, plant defense molecules like phytoalexins and secondary compounds/alkaloids.

The fundamental requirement is good yield of compound, and reduced cost compared to natural synthesis. Elicitation may be used as one of the important strategies in order to improve the productivity of bioactive secondary metabolites (Roberts and Shuler 1997). In order to obtain high yields, efforts now have been focused on screening and selecting of high- producing cells/strains, employing precursor feeding, transformation methods, and immobilization techniques (Dicosmo and Misawa 1995). Transgenic hairy root cultures have revolutionized the role of plant tissue culture in secondary metabolite production. The cultures are unique in their genetic and biosynthetic stability, faster in growth, and more easily maintained. Using this methodology, a wide range of chemical compounds have been synthesized (Shanks and Morgan 1999; Giri and Narasu 2000).

Genome manipulation is another important area in which relatively large amounts of desired compounds can be produced by plants, infected with an engineered virus, whereas transgenic plants can maintain constant levels of production of proteins without additional intervention (Sajc et al. 2000). Kieran et al. (1997) reported the impact of specific engineering-related factors on cell suspension cultures. Current developments indicate that transcription factors are efficient new molecular tools for plant metabolic engineering to increase the production of valuable compounds (Gantet and Memelink 2002).

9 Conclusion

A number of studies suggest that the alkaloid enrichment in *C roseus* is quite complex. Several strategies involving chemical, semi-chemical, and biotechnology have recently been tried to improve alkaloid yield. Elicitor-based synthesis provides an excellent opportunity for exploitation of in vitro cultivated plant cells for improving secondary metabolites. The identification of signaling molecule/moiety of elicitors and their molecular action on synthesis are still not elucidated clearly. This missing information is essential in knowing the regulatory mechanism for enhanced synthesis of anti-cancerous compounds in *C. roseus*.

References

Aerts RJ, Schafer A, Hesse M, Baumann TW, Slusarenko A (1996) Signalling molecules and the synthesis of alkaloids in *Catharanthus roseus* seedlings. Phytochemistry 42(2):417–422

Akimoto C, Aoyagi H, Tanaka H (1999) Endogenous elicitor-like effects of alginate on physiological activities of plant cells. Appl Microbiol Biotechnol 52:429–436

Akimoto-Tomiyama C, Aoyagi H, Ozawa T, Tanaka H (2002) Production of 5′-phosphodiesterase by *Catharanthus roseus* cells promoted by heat-degraded products generated from uronic acid. J Biosci Bioeng 94:154–159

Andrade SAL, Malik S, Sawaya ACHF, Bottcher A, Mazzafera P (2013) Association with arbuscular mycorrhizal fungi influences alkaloid synthesis and accumulation in *Catharanthus roseus* and *Nicotiana tabacum* plants. Acta Physiol Plant 35:867–880

Angelova Z, Georgiev S, Roos W (2006) Elicitation of plants. Biotechnol Biotechnol Eq 20: 72–83

Arntzen CJ (1997) High tech herbal medicine: plant based vaccines. Nat Biotechnol 15(3): 221–222

Bonzom MP, Andary C, Gueiffier A, Roussel JL, Gargadennec A (1997) Effect of tertiary amine on alkaloid accumulation in *Catharanthus roseus*. Phytochemistry 46:235–239

Broeckling CD, Huhman DV, Farag MA, Smith JT, May GD, Mendes P, Dixon RA, Sumner LW (2005) Metabolic profiling of Medicago truncatula cell cultures reveals the effects of biotic and abiotic elicitors on metabolism. J Exp Bot 56:323–336

Carturan G, Dal Monte R, Pressi G, Secondin S, Verza P (1998) Production of valuable drugs from plant cells immobilized by hybrid sol-gel SiO_2. J Sol Gel Sci Technol 13:273–276

Cheetham PSJ (1995) Biotransformations: new routes to food ingredients. Chem Ind 7:265–268

Ciddi V, Srinivasan V, Shuler ML (1995) Elicitation of Taxus sp. cell cultures for production of taxol. Biotechnol Lett 17:1343–1346

Collinge DB, Susarenka AJ (1987) Plant gene expression in response to pathogens. Plant Mol Biol 9:389–410

Contin A, van der Heijden R, Ten Hoopen HJG, Verpoorte R (1998) The inoculum size triggers tryptamine or secologanin biosynthesis in a *Catharnthus roseus* cell culture. Plant Sci 139:205–211

Coste A, Vlase L, Halmagyi A, Deliu C, Coldea G (2011) Effects of plant growth regulators and elicitors on production of secondary metabolites in shoot cultures of *Hypericum hirsutum* and *Hypericum maculatum*. Plant Cell Tiss Org Cult 106:279–288

Dela Rosa-Mera CJ, Ferrera-Cerrato R, Alarcón A, Sánchez-Colin MJ, Mu-oz-Mu-iz OD (2011) Arbuscular mycorrhizal fungi and potassium bicarbonate enhance the foliar content of the vinblastine alkaloid in *Catharanthus roseus*. Plant Soil 349:367–376

Dicosmo F, Misawa M (1995) Plant cell and tissue culture: alternatives for metabolite production. Biotechnol Adv 13(3):425–453
Dipti T, Mujib A, Maqsood M, Ali M, Zafar N (2016) Aspergillus flavus fungus elicitation improves vincristine and vinblastine yield by augmenting callus biomass growth in *Catharanthus roseus*. Plant Cell Tiss Org Cult. doi:10.1007/s11240-016-0998-1
Diwan R, Malpathak N (2010) Bioprocess optimization of furanocoumarin elicitation by medium renewal and re-elicitation: a perfusion-based approach. Appl Biochem Biotechnol 163:756–764
Dixon RA (2001) Natural products and plant disease resistance. Nature 411:843–847
Doran PM (2000) Foreign protein production in plant tissue cultures. Curr Opin Biotechnol 11:199–204
Ebel J, Scheel D (1997) Signals in host–parasite interactions. In: Carroll G, Tudzynski P (eds) The mycota, Plant relationships, vol 5. Springer, Berlin, pp 85–105
Eilert U, Ehmke A, Wolters B (1984) Elicitor-induced accumulations of acridone alkaloid epoxides in Ruta graveloens suspension cultures. Planta Med 50:508–512
Eliasova A, Repcak M, Pastirova A (2004) Quantitative changes of secondary metabolites of Matricaria chamomilla by abiotic stress. Z Naturforsch C 59:543–548
Facchini PJ, DiCosmo F (1991) Secondary metabolites biosynthesis in cultured cells of *Catharanthus roseus* (L.) Don immobilized by adhesion to glass fibers. Appl Microbiol Biotechnol 35:382–392
Fatima S, Mujib A, Tonk D (2015) NaCl amendment improves vinblastine and vincristine synthesis in *Catharanthus roseus*: a case of stress signalling as evidenced by antioxidant enzymes activities. Plant Cell Tiss Organ Cult 121:445–458
Gantet P, Imbault N, Thiersault M, Doireau P (1998) Necessity of a functional octadecanoic pathway for indole alkaloid synthesis by *Catharanthus roseus* cell suspensions cultured in an auxin-starved medium. Plant Cell Physiol 39:220–225
Gantet P, Memelink J (2002) Transcription factors: tools to engineer the production of pharmacologically active plant metabolites. Trends Pharmacol Sci 23:563–569
Ghorpade RP, Chopra A, Nikam TD (2011) Influence of biotic and abiotic elicitors on four major isomers of boswellic acid in callus culture of *Boswellia serrata Roxb*. Plant Omics J 4(4):169–176
Giri A, Narasu ML (2000) Transgenic hairy roots: recent trends and applications. Biotechnol Adv 18:1–22
Godoy-Hernandez G, Aviles-Berzunza E, Carillo-Pech M, Vazquez-Flota F (2008) Agrobacterium-mediated transient transformation of Mexican prickly poppy (*Argemone maxicana* L.). Electron J Biotechnol 11:1–5
Guo B, Kunming LH (1998) A middle vinblastine fungi isolated. J Yunnan Univ 20:214–215
Guo S, Man S, Gao W, Liu H, Zhang L, Xiao P (2013) Production of flavonoids and polysaccharide by adding elicitor in different cellular cultivation processes of *Glycyrrhiza uralensis*. Acta Physiol Plant 35:679–686
Hahn JJ, Eschenlauer CA, Narrol HM, Somers AD, Srienc F (1997) Growth kinetics, nutrient uptake, and expression of the Alcaligenes eutrophus poly (–hydroxybutarate) synthesis pathway in transgenic maize cell suspension cultures. Biotechnol Prog 13(4):347–354
Harbourne JB (1999) Classes and functions of secondary products. In: Walton NJ, Brown DE (eds) Chemicals from plants. Perspectives on secondary plant products. Imperial College Press, London, pp 1–25
Hoshino T, Chida M, Yamaura T, Yoshizawa Y, Mizuatani J (1994) Phytoalexin induction in green pepper cultures treated with arachidonic acid. Phytochemistry 36:1417–1419
Hughes EH, Shanks JV (2002) Metabolic engineering of plants for alkaloid production. Metab Eng 4(1):41–48
Iantcheva A, Vlahova M, Atanassov A (2006) Somatic embryogenesis in genera *Medicago*: an overview. Plant Cell Monogr 2:285–304
Irmler S, Schroder G, St-Pierre B, Crouch NP, Hotze M, Schmidt J, Strak D, Matern U, Schröder J (2000) Indole alkaloid biosynthesis in *Catharanthus roseus*: new enzyme activities and identification of cytochrome P450 CYP72A1 as secologanin synthase. Plant J 24:797–804

Jaleel CA, Gopi R, Panneerselvam R (2009) Alterations in nonenzymatic antioxidant components of *Catharanthus roseus* exposed to paclobutrazol, gibberellic acid and Pseudomonas fluorescens. Plant Omics J 2:30–40

Johnson TS, Ravishankar GA, Venkataraman LV (1991) Elicitation of capsaicin production in freely suspended cells and immobilized cell cultures of Capsicum frutescens mill. Food Biotechnol 5:197–205

Junaid A, Mujib A, Fatima S, Sharma MP (2008) Cultural conditions affect somatic embryogenesis in *Catharanthus roseus* L (G.) Don. Plant Biotechnol Rep 2:179–189

Karuppusamy S (2009) A review on trends in production of secondary metabolites from higher plants by in vitro tissue, organ and cell cultures. J Med Plants Res 3:1222–1239

Kieran PM, MacLoughlin PF, Malone DM (1997) Plant cell suspension cultures: some engineering considerations. J Biotechnol 59(1–2):39–52

Kim HJ, Chen F, Wang X, Rajapakse NC (2005) Effect of chitosan on the biological properties of sweet basil (Ocimum basilicum L.). J Agric Food Chem 53:3696–3701

Koga J, Kubota H, Gomi S, Umemura K, Ohnishi M, Toshiaki K (2006) Cholic acid, a bile acid elicitor of hypersensitive cell death, pathogenesis-related protein synthesis, and phytoalexin accumulation in rice. Plant Physiol 140:1475–1483

Kothari SK, Singh S, Singh VB, Kumar S (1999) Response of bergamot mint (*Mentha citrata*) to vesicular arbuscular mycorrhizal fungi and phosphorus supply. J Med Aroma Plant Sci 21:990–995

Krings U, Berger RG (1998) Biotechnological production of flavours and fragrances. Appl Microb Biotechnol 49:1–8

Krolicka A, Kartanowicz R, Wosinski SA, Szpitter A, Kaminski M, Lojkowska E (2006) Induction of secondary metabolite production in transformed callus of Ammi majus L. grown after electromagnetic treatment of the culture medium. Enzyme Microbial Technol 39:1386–1391

Kumar A, Patil D, Rajamohanan PR, Ahmad A (2013) Isolation, purification and characterization of vinblastine and vincristine from endophytic fungus *Fusarium oxysporum* isolated from *Catharanthus roseus*. PloS One 8(9):e71805

Kuroyanagi M, Arakawa T, Mikami Y, Yoshida K, Kawahara N, Hayashi T, Ishimaru H (1998) Phytoalexins from hairy roots of Hyascyamus albus treated with methyl jasmonate. J Nat Prod 61:1516–1519

La Count W, An G, Lee JM (1997) The effect of PVP on the heavy chain monoclonal antibody production from plant suspension cultures. Biotechnol Lett 19(1):93–96

Lee TK, Yamakawa T, Kodama T, Shimomura K (1998) Effects of chemicals on alkaloid production by transformed roots of Atropa belladona. Phytochemistry 49:2343–2347

Mader CJ (1999) Effects of jasmonic acid, silver nitrate and L-AOPP on the distribution of free and conjugated polyamines in roots and shoots of Solanum tuberosum, in vitro. J Plant Physiol 154:79–88

Manson HS, Arntzen CJ (1995) Transgenic plants as vaccine production system. Trends Biotechnol 3:388–392

Marden MC, Dieryck W, Pagnier J, Poyart C, Gruber V, Bournat P, Baudino S, Merot B (1997) Human haemoglobin from transgenic tobacco. Nature 342:29–30

Memelink J, Verpoorte R, Kijne JW (2001) ORCAnization of jasmonate-responsive gene expression in alkaloid metabolism. Trends Plant Sci 6:212–219

Menke FLH, Parchmann S, Mueller MJ, Kijne JW, Memelink J (1999) Involvement of the octadecanoid pathway and protein phosphorylation in fungal elicitor induced expression of terpenoid indole alkaloid biosynthetic genes in *Catharanthus roseus*. Plant Physiol 119:1289–1296

Moreno RH, van der Heijden R, Verpoorte R (1995) Cell and tissue cultures of *Catharanthus roseus*: a literature survey 2. Updating from 1988 to 1993. Plant Cell Tissue Org Cult 42:1–25

Moreno PRH, Poulsen C, Van der Heijden R, Verpoorte R (1996) Effects of elicitation on different secondary metabolic pathways in *Catharanthus roseus* cell suspension cultures. Enzyme Microb Technol 18:99–107

Moreno-Valenzuela OA, Galaz Avalos RM, Minero Garcia Y, Loyola Vargas VM (1998) Role of differentiation on the alkaloid production in *Catharanthus roseus* hairy roots. Plant Cell Rep 18:99–104

Moreno-Valenzuela OA, Monforte-Gonzalez M, Munoz-Sanchez J, Mendez-Zeel M (1999) Effect of macerozyme on secondary metabolism plant production and phospholipase C activity in *Catharanthus roseus* hairy roots. J Plant Physiol 155(s4-5):447–452

Morgan JA, Shanks JV (1999) Inhibitor studies of tabersonine metabolism in *C. roseus* hairy roots. Phytochemistry 51:61–68

Morgan J, Shanks J (2000) Determination of metabolic rate-limitation by precursor feeding cultures. J Biotechnol 79:137–145

Mujib A, Ilah A, Aslam J, Fatima S, Siddiqui ZH, Maqsood M (2012) *Catharanthus roseus* alkaloids: application of biotechnology for improving yield. Plant Growth Regul 68:111–127

Mukandan U, Hjorosto MA (1990) Effect of fungal elicitor on thiophene production in hairy root cultures of *Tagetes patula*. Appl Microb Biotechnol 33:145–147

Muller SS, Kurosaki F, Nishi A (1994) Role of salicylic acid and intrercelluar Ca^{2+} in the induction of chitinase activity in carrot suspension culture. Physiol Mol Plant Pathol 45:101–109

Namdeo AG (2007) Plant cell elicitation for the production of secondary metabolites: a review. Pharmacog Rev 1:69–79

Namdeo A, Patil S, Fulzele DP (2002) Influence of fungal elicitors on production of ajamalicine by cell cultures of *Catharanthus roseus*. Biotechnol Prog 18(1):159–162

Noble RL (1990) The discovery of the vinca alkaloids- chemotherapeutic agents against cancer. Biochem Cell Biol 68:1344–1351

Nuernberger T (1999) Signal perception in plant pathogen defence. Cell Mol Life Sci 55:167–182

O'Keefe BR, Mahady GB, Gills JJ, Beecher CWW (1997) Stable vindoline production in transformed cell cultures of *Catharanthus roseus*. J Nat Prod 60:261–264

Palazon J, Cusido RM, Gonzalo J, Bonfill M, Morales C, Pinol MT (1998) Inhibition of paclitaxel and baccatin III accumulation by mevinolin and fosmidomycin in suspension culture of *Taxus baccata*. J Biotechnol 101(2):157–163

Pande D, Malik S, Bora M, Srivastava PS (2002) Rapid protocol for in-vitro micropropagation of *Lepidium sativum* Linn and enhancement in the yield of lepidine. In Vitro Cell Dev Bio Plant 38:451–456

Patel P, Krishnamurthy R (2013) Feasibility studies on in vitro mass-scale propagation of Indian Ashwagandha (*Withania somnifera*) cultivars for commercial purpose. J Pharmacog Phytochem 2(2):168–174

Patrica M, Moctezuma L, Gloria LE (1996) Biosynthesis of sesqutiterpenic phytoalaxin captidiol in elicited root cultures of chilli peppers (C. annum). Plant Cell Rep 15:360–366

Pedras MSC, Nycholat CM, Montaut S, Xu Y, Khan AQ (2002) Chemical defenses of crucifers: elicitation and metabolism of phytoalexins and indole-3-acetonitrile in brown mustard and turnip. Phytochemistry 59:611–625

Philipson JD (1990) Plants as source of valuable products. In: Chalwood BV, Rhodes MJ (eds) Secondary products from plant tissue culture. Clarendon Press, Oxford, pp 1–21

Radman R, Saez T, Bucke C, Keshavarz T (2003) Elicitation of plants and microbial cell systems. Biotechnol Appl Biochem 37:91–102

Ramani S, Jayabaskaran C (2007) UV-B-induced signalling events leading to enhanced-production of catharanthine in *Catharanthus roseus* cell suspension cultures. BMC Plant Biol 7:61–78

Ramos-Valdivia AC, van der Heijden R, Verpoorte R (1997) Elicitor-mediated induction of anthraquinone biosynthesis and regulation of isopentyl diphosphate isomerase and farnesyl diphosphate synthase activities in cell suspension cultures of *Cinchona robusta* How. Planta 203:155–161

Rao RS, Ravishankar GA (2002) Plant cell cultures: chemical factories of secondary metabolites. Biotechnol Adv 20:101–153

Rashid KI, Ibrahim KM, Hamza SJ (2011) Effect of some biotic and abiotic elicitors on phenolic acids and diterpenes production from rosemary (*Rosmarinus officinalis* L.) leaf and callus analyzed by high performance liquid chromatography (HPLC). J Al-Nahrain Univ 14(3):104–109

Ratti N, Verma H, Gautam S (2010) Effect of Glomus species on physiology and biochemistry of *Catharanthus roseus*. Ind J Microbiol 50:355–360

Ravishankar GA, Rao S (2000) Biotechnological production of phyto-pharmaceuticals. J Biochem Mol Biol Biophys 4:73–102

Rijhwani SK, Shanks JV (1998) Effect of elicitor dosage and exposure time on biosynthesis of indole alkaloids by *Catharanthus roseus* hairy root culture. Biotechnol Prog 14:442–449

Roberts SC, Shuler ML (1997) Large-scale plant cell culture. Curr Opin Biotechnol 8:154–159

Robins RI (1994) Secondary products from cultured cells and organs: molecular and cellular approaches. In: Dixon RA, Gonzales RA (eds) Plant cell culture. IRI Press, New York

Roewer IA, Cloutier N, Van der Heijden R (1992) Transient induction of tryptophandecarboxylase (TDC) and strictosidine synthase, (SS) genes in cell suspension cultures of *Catharanthus roseus*. Plant Cell Rep 11(2):86–89

Runguphan W, Qu W, O'Connor SE (2010) Integrating carbon-halogen bond formation into medicinal plant metabolism. Nature 468:461–464

Sajc L, Grubisic D, Vunjak-Novakovic G (2000) Bioreactors for plant engineering: an outlook for further research. Biochem Eng J 4:89–99

Samar F, Mujib A, Dipti T (2015) NaCl amendment improves vinblastine and vincristine synthesis in *Catharanthus roseus*: a case of stress signalling as evidenced by antioxidant enzymes activities. Plant Cell Tissue Org Cult 121:445–458

Sandra I, Alvarez P, Giulietti AM (1988) Novel biotechnological approaches to obtain scopolamine and tryosumine: the influence of biotic elicitors and stress agents on cultures of transformed rats of *Bygumansia candida*. Phytother Res 12:518–520

Savitha BC, Timmaraju R, Bhagyalakshmi N, Ravishankar GA (2006) Different biotic and abiotic elicitors influence betalin production in hairy root cultures of *Beta vulgaris* in shake flask and bioreactor. Process Biochem 41:50–60

Sayed ME, Verpoorte R (2007) Catharanthus terpenoid indole alkaloids: biosynthesis and regulation. Phytochem Rev 6:277–305

Scragg AH (1997) The production of aromas by plant cell cultures. In: Schepier T (ed) Advanced biochemistry, engineering and. biotechnology, vol 55. Springer, Berlin, pp 239–263

Sena MK, Das PK (1998) Influence of microbial inoculants on quality of turmeric. Ind Cocoa Arecanut Species J 21:31–33

Shanks JV, Morgan J (1999) Plant hairy root culture. Curr Opin Biotechnol 10:151–155

Shilpa K, Varun K, Lakshmi BS (2010) An alternate method of natural drug production: Eliciting secondary metabolite production using plant cell culture. J Plant Sci 5(3):222–247

Shirsau K, Nakajima H, Krishnamachari RV, Dixon RA, Lamb C (1997) Salicylic acid potentiates an agonist- dependent gain control that amplifies pathogen signals in the activation of defense mechanisms. Plant Cell 9:261–270

Siddiqui ZH, Mujib A, Ahmad MA, Ali A (2010) Fungal elicitors: A potent approach for enhancing secondary metabolites in cultured cells. In: Gupta VK, Tuohy M, Gaur RK (eds) Fungal biochemistry and biotechnology. Lap Lambert Academic Publishing AG & CO. KG, Saarbrücken, pp 88–104

Sivanandhan G, Rajesh Arun MM, Jeyaraj M, Dev GK, Arjunan A, Manickavasagam M, Muthuselvam M, Selvaraj N, Ganapathi A (2013) Effect of culture conditions, cytokinins, methyl jasmonate and salicylic acid on the biomass accumulation and production of withanolides in multiple shoot culture of *Withania somnifera* (L.) Dunal using liquid culture. Acta Physiol Plant 35:715–728

Smith JI, Smart NJ, Misawa M, Kurg WG, Tallevi SG, Dicosmo F (1987) Increased accumulation of indole alkaloids by some lines of *Catharanthus roseus* in response to addition of vanadyl sulphate. Plant Cell Rep 6:142–145

Taha HS, El-Bahr MK, Seif-El-Nasr MM (2009) In vitro studies on Egyptian *Catharanthus roseus* (L.) G. Don.: manipulation of some amino acids as precursors for enhanced indole alkaloids production in suspension cultures. Aust J Basic Appl Sci 3:3137–3144

Thangavel K, Sasikala M, Maridass M (2008) In vitro antibacterial potential of *Aegle marmelos* (L.) callus extract. Pharmacol Online 3:190–196

Tripathi L, Tripathi JN (2003) Role of biotechnology in medicinal plants. Trop J Pharm Res 2(2):243–253

Tung CY, Yang DB, Gou M (2002) A preliminary study on the condition of then culture and isolate of endophytic fungus producing vincristine. J Chuxiong Normal Univ 6:39–41

Van Der Heijden R, Jacobs DI, Snoeijer W, Hallared D, Verpoorte R (2004) The Catharanthus alkaloids: pharmacognosy and biotechnology. Curr Med Chem 11:607–628

Vasconsuelo A, Boland R (2007) Molecular aspects of the early stages of elicitation of secondary metabolites in plants. Plant Sci 172:861–875

Vázquez-Flota F, Hernández-Domínguez E, De Lourdes M-HM, Monforte-González M (2009) A differential response to chemical elicitors in *Catharanthus roseus* in vitro cultures. Biotechnol Lett 31:591–595

Verpoorte R, Memelink J (2002) Engineering secondary metabolite production in plants. Curr Opin Biotechnol 13:181–187

Verpoorte R, Van der Heijden R, Van Gulik WM, Ten Hoopen HJG (1991) In: Brossi A (ed) The alkaloids, vol 40. Academic, San Diego, pp 1–187

Verpoorte R, Van der Heijden R, Ten Hoopen HJG, Memelink J (1999) Metabolic engineering of plant secondary metabolite pathways for the production of fine chemicals. Biotechnol Lett 21:467–479

Verpoorte R, Van der Heijden R, Memelink J (2000) Engineering the plant cell factory for secondary metabolite production. Transgenic Res 9:323–343

Wahl MF, An G, Lee JM (1995) Effects of DMSO on heavy chain monoclonal antibody production from plant cell culture. Biotechnol Lett 17(5):463–468

Wang Y (2008) Needs for new plant-derived pharmaceuticals in the post-genome era: an industrial view in drug research and development. Phytochem Rev 7:395–406

Whitmer S, Van der Heijen R, Verpoorte R (2002) Effect of precursor feeding on alkaloid accumulation by a strictosidine synthase over-expressing transgenic cell line S1 of *Catharanthus roseus*. Plant Cell Tiss Org Cult 69:85–93

Wongsamuth R, Doran MP (1997) Production of monoclonal antibodies by tobacco hairy roots. Biotechnol Bioeng 54(5):401–415

Wu JY, Janet NG, Ming S, Wu SJ (2007) Enhanced secondary metabolite (tanshinone) production of *Salvia miltirrhiza* hairy roots in a novel root-bacteria coculture process. Appl Microbiol Biotechnol 77:543–550

Yamamoto H, Yamamoto M (2000) Effects of various culture conditions on callus growth and polyphenol contents of *Hydrangea macrophylla* var. Thunbergii callus cultures. Bull Hokuriku Univ 24:25–33

Yu M, Facchini PJ (2000) Molecular cloning and characterization of a type III glutathione S-transferase from opium poppy cell suspension cultures treated with a fungal elicitor. Physiol Plantarum 108:101–109

Zahid SH, Mujib A (2012) Accumulation of vincristine in calcium chloride elicitated *Catharanthus roseus* cultures. Nat Prod J 2(9):307–315

Zhang LB, Gou LH, Zeng SV (2000) Preliminary study on the isolation of endophytic fungus of *Catharanthus roseus* and its fermentation to produce product of therapeutic value. Chin Trad Herb Drugs 11:805–807

Zhao J, Verpoorte R (2007) Manipulating indole alkaloid production by *Catharanthus roseus* cell cultures in bioreactors: from biochemical processing to metabolic engineering. Phytochem Rev 6:435–457

Zhao J, William CC, Last RL (1998) Induction of Arabidopsis tryptophan pathway enzymes and camalexin by amino acid starvation, oxidative stress, and an abiotic elicitor. Plant Cell 10:359–370

Zhao J, Zhu WH, Hu Q (2000) Promotion of indole alkaloid production in *Catharanthus roseus* cell cultures by rare earth elements. Biotechnol Lett 22:825–828

Zhao J, Hu Q, Zhu WH (2001a) Enhanced catharanthine production in *Catharanthus roseus* cell cultures by combined elicitor treatment in shake flasks and bioreactors. Enzyme Microb Technol 28:673–681

Zhao J, Hu Q, Guo Y-Q, Zhu W-H (2001b) Elicitor-induced indole alkaloid biosynthesis in *Catharanthus roseus* cell cultures is related to Ca^{2+} influx and the oxidative burst. Plant Sci 161:423–431

Zhong JJ (2002) Plant cell culture for production of paclitaxel and other taxnes. J Biosci Bioeng 94:591–599

Catharanthus roseus: The Cancer-Fighting Medicine

D. Sathya Prabhu and V. Devi Rajeswari

Abstract Cancer is a major disease around the world with high mortality rate. Cancer is caused due to internal and external factors. Currently, the available treatment methods are chemotherapy, radiotherapy and surgery. The current treatment methods cause severe side effects to the patient, and more over the treatment cost is highly expensive. Medicinal plants could be the better alternative to cut down these barriers. Here in this chapter, we discuss about the importance of the medicinal plant *Catharanthus roseus*. Vinblastine and vincristine are the two important bioactive alkaloids produced from this plant. These two are key compounds in treatment of various types of cancer. The potent anticancer compounds are synthesized in shikimate, mevalonate and methyl-erythritol phosphate (MEP) pathways. These two compounds have many biological properties, and it is discussed in detail in the following section. Recent years have seen nanotechnology-based synthesis, and these formulations have been considered to increase the efficacy of the bioactive compounds. This could be opening doors to a new era in the development of nanotechnology-based drugs. On the other hand, toxicity concerns of the prosperity of the vinca compounds were taken into consideration.

Keywords Anticancer activity • Antimicrobial activity • Anti-psoriasis activity • Anti-diabetic property • Anti-hypertension activity • Radioimmunoassay

1 Introduction

Cancer is a deadly group of disease which is highly prevalent around the globe. Cancer is the major cause for the increased mortality rate in the world. About 1,685,210 new cases were expected to be diagnosed in 2016, out of which half of it is expected to be in the United States. This is an alarming situation that demands immediate preventive measures against carcinogens. Cancer is caused due to various factors, and its treatment depends upon its types and stages. In initial stages of

D. Sathya Prabhu • V. Devi Rajeswari (✉)
Department of Biomedical Sciences, School of Biosciences and Technology, VIT University, Vellore 632014, Tamilnadu, India
e-mail: sdevirajeswari@gmail.com

M. Naeem et al. (eds.), *Catharanthus roseus*,
DOI 10.1007/978-3-319-51620-2_7

cancer, neoplasmic cells clump and grow into tumours which later advances to malignant tumour. Thereafter, it can develop into metastasis, an uncontrollable stage in which abnormal cancer cells spread throughout the human body and affect the normal healthy cells, thus leading to death (Siegel et al. 2015; Pan et al. 2016).

According to WHO, approximately 10–13% was the result of cancer. The major types of cancer are predominantly lung cancer, stomach cancer, liver cancer, colorectal cancer and breast cancer (de Gonzalez and Darby 2004). Cancer is prevalent among children, every year nearly about 1,50,000 children are diagnosed with cancer. The prevalence is high around the world despite of country, community, etc. The main reasons for this situation are due to the lack of patient care, particularly in developing countries. In turn, the survival rate of the patient can be increased with proper care. Global attention is needed to tackle the situation in terms of creating awareness, access to treatment, cutting the cost of medications, and moral support to sustain the patients who are surviving with cancer, particularly in low income countries. In order to raise the awareness of cancer, globally, International Childhood Cancer Day is celebrated on 15th of February, every year (Stewart and Wild 2014; Louis et al. 2016).

In India, a million new cases have been registered every year. Cancer is the leading cause for high mortality, and the numbers could be doubled in upcoming years. Cancer is caused to both sexes. In men, lung and oral cancers are the most common and among women cervix and breast cancer is found to be larger in number of cases. Moreover, prostate cancer is found to be common in elderly man. Even though, effective treatment is available in Indian subcontinent. However, the cost of the treatment is very high, and hence it is inaccessible to millions of cancer patients. Moreover, it has been reported that there is lack of doctors throughout the country. Currently, there is only one doctor per 2000 patients. Even though Indian government has taking many steps to decrease the mortality rate of cancer cases, the resources and expenditure on health by the Indian government is inadequate. It is suggestive that more steps should to be taken by the government of India in future (Krishnan et al. 2015; Gupta et al. 2016).

Nowadays, due to modernization and urbanization, the number of cancer patients is increasing globally. Although many steps have been taken to create awareness, it is unavoidable to change the situation, and the people are therefore in risk of cancer. Cancer is mainly influenced by two factors, i.e. internal and external factors. The internal factors are inherited genetic mutations, hormonal changes weak immune system, etc. The external factors are change in life style, smoking, microbes, unhealthy foods, expose to pollution, hazardous chemicals, etc. People who are highly exposed to air pollution are more vulnerable to lung cancer. Even then, taking preventive measures can help in not getting the disease. In turn, majority of cancers can be prevented by maintaining healthy life style and regular health check-ups. It was observed that proper physical activity, good nutritional foods, avoiding use of tobacco and alcohol consumption can deduct the number of cancer cases to half in the last decade. The achievement was made possible only as a result of creating awareness among people to quit smoking. Apart from this, it has been observed that an increase in physical activity has decreased the risk of almost 26 types of cancers. Certain cancers which have been caused by microbes can be prevented by vaccination. Skin cancer can be prevented by less exposure to the sunlight, also following simple protection steps like carrying an umbrella or applying sunscreen creams can be useful to avoid direct exposure of skin to the sunlight. Major cancers like breast, colorectal and cervical cancers can be checked regularly using screening tests by clinicians. Noticing the physical changes in the body could put a check for this deadly disease (Fajersztajn et al. 2013; Moore et al. 2016; Smith et al. 2016).

At present, the treatment methods for cancer include chemotherapy, radiotherapy and surgery (Winn et al. 2016; Board 2016). Although various treatment methods are available, the mortality rate of the patient depends upon the type and stage of the cancer (Early Breast Cancer Trialists' Collaborative Group (EBCTCG)). Chemotherapy is predominantly considered at primary stages of cancer; chemotherapy prevents the diffusion of cancer to the healthy normal cells effectively. However, the patient suffers with the side effects caused by chemotherapy. Moreover, the treatment is highly expensive (Nagao et al. 2011; Yunaiyama et al. 2014). Patients from middle income and poor countries cannot afford the treatment expenditure (Farmer et al. 2010).

In order to curb the burden of side effects, medicinal plants could be a better alternative. Medicinal plants are better aliments in treating various diseases. The secondary metabolites present in the medicinal plants fight against various microorganisms which include bacteria, fungi, protozoa and other parasites. These bioactive compounds present in medicinal plants like *C. roseus* are effective in killing the cancer cells by impairing mitosis of cancer cells and also causing apoptosis of cancer cells. The medicinal plants contain no side effects (Vickers 2004; Chong et al. 2009; Mohan et al. 2011). At present, majority of the anticancer drugs are obtained from the natural plant sources (Cragg et al. 2009). Bioactive compounds like alkaloids were present in most of the plants. Among alkaloids, monoterpene indole alkaloids (MIAs) are most important in the treatment of cancer. The compounds of MIAs are known for their medicinal values (Facchini and De Luca 2008).

The *Catharanthus roseus* (*C. roseus*) is a perennial tropical medicinal plant belonging to the family Apocynaceae (Table 1). This family produces more MIAs, which makes it attractive in the field of medicine (Gunatilaka 2006; Magnotta 2007). The plant *C. roseus* has got many names, and it was popularly known as

Table 1 Scientific classification of *Catharanthus roseus*

Kingdom	Plantae
Division/ phylum	Magnoliophyta
Class	Equisetopsida
Order	Genitianales
Family	Apocynaceae
Genus	Catharanthus
Species	*C. roseus*

Fig. 1 The *Catharanthus roseus* plant was taken from the Botanical Garden, VIT University, Vellore, Tamilnadu, India

"Madagascar periwinkle", "Cape periwinkle", "Rose periwinkle", "Rosy periwinkle" and "Old maid". In India, the *C. roseus* is known by different names depending upon the state and language as "Sadabahar" in Hindi and "Nayantara", "Sudukattu mallikai" in Tamil and "Billaganneru" in Telegu (Sain and Sharma 2013).

The *C. roseus* plant is widely grown as an ornamental plant. The perpetual *C. roseus* plant can grow up to 80 cm to 1 m. The leaves of the plant are dark green, silky in nature and oval in shape. The colour of the flower ranges from pale pink to pink and the flower blossoms throughout the summer (Fig. 1). The fruit is broad and long in size, and it ranges from 2 to 4 cm (Taylor and Farnsworth 1975; Jaleel et al. 2006, 2007). The *C. roseus* can grow under any climatic conditions, and its seeds can be stored for years. At room temperature, the germination rate of seeds is stable up to 7 years. The stability can be extended up to 15 years when the seeds are stored at 5 °C. But the germination rate is dropped and is limited to 4–5 years if the seeds are stored at −10 °C (Buchwald et al. 2007).

The cultivation of *C. roseus* plant is a very feasible process as the plant grows under most of the climatic conditions. The favourable growth is observed at high temperatures under sunlight. The best temperature for the plant growth is believed to be 25 °C, and the plant growth is retarded if there is a slight drop or rise in the temperature. However, amount of alkaloids present in the plant is not affected at room temperature. In turn, the concentration of alkaloids can be doubled if the plant is grown under greenhouse conditions with the supplement of macro- and micronutrients that are essential for the growth of the plant (Buchwald et al. 2007). High temperatures favour the increase in concentration of vinca alkaloids in *C. roseus*. The alkaloid contents increase as temperature level rises. The highest level of alkaloids was found at 45 °C. Whereas, long-term experiments found VBR and VCR contents in *C. roseus* plant were at higher rate at 35 °C (Guo et al. 2007). So the temperature should be taken into consideration in *C. roseus* cultivation for better yield.

Alternatively, there are several approaches made by the researchers to increase the yield quantity of vinca alkaloids. The alkaloid concentration can be increased by inoculating the plant with arbuscular mycorrhizal fungi (AMF). And it should be noted that the alkaloid concentration was declined when the plant is stressed with potassium bicarbonate ($KHCO_3$) and sodium chloride both separately or in combination (De la Rosa-Mera et al. 2011). The growth of the plant is highly favoured in regions where high carbon dioxide is present and notably the alkaloid concentration is also elevated. Whereas, few studies found nitrogen essential for the growth of the plant. But it is important to note, under high nitrogen concentration the plant grows well but the alkaloid concentration remained the same (Singh and Agrawal 2015).

In Ayurveda, *C. roseus* plant was used to treat many diseases, such as malaria, diabetes and Hodgkin's lymphoma (Fabricant and Farnsworth 2001; Elujoba et al. 2005; Okigbo and Mmeka 2006). Medicinally, the plant *C. roseus* plays a vital role as the plant is solely responsible for synthesizing monoterpenoid indole alkaloids. Among those alkaloids, two are very important alkaloids, namely, vinblastine (VBR) and vincristine (VCR) (Fig. 2). These two alkaloids are strong anticancer agents. Apart from anticancer activity, the two bioactive compounds were widely studied for their various medical applications. The USFDA approved vinca alkaloids for the treatment of leukaemia in children since 1963 also it is widely used to treat many types of cancers (Waterhouse et al. 2005; Sisodiya 2013). The important mechanism by which this anticancer action takes place is via *vinca* alkaloids which suppresses the action of the tumour at the mitosis level by binding to tubulin and depolymerizes the microtubules (Fig. 3) (Okouneva et al. 2003). The VBR and VCR are derived from the precursor vindesine and vinorelbine, and its synthesis is discussed below. The powerful anticancer compounds, VBR and VCR, are isolated from the leaves and other plant parts of *C. roseus* using various techniques (Moreno et al. 1995; Jacobs et al. 2004). VBR was first isolated by Robert Noble and Charles Thomas Beer at the University of Western Ontario from the Madagascar periwinkle plant (Wright 2002; Kumar et al. 2013). Usually, VBR is synthesized from callus, crown gall, shoot, and somatic embryo and vincristine is synthesized from shoot and somatic embryo (Potier 1980; Ishikawa et al. 2009). The production of VBR depends upon the age and type of tissue of the plant. The germinating leaf callus is found to be containing high amount of VBR (Aslam et al. 2010). Initially, out to be anticancer drug as it reduced a number of white blood cells (Noble et al. 2009; Kalidass et al. 2009).

Earlier around 1950s and 1960s worldwide cultivation of the plant was seen only in a few countries. Hence, very less quantity of compounds were extracted and purified from the available sources. Even today, synthesis of these plant compounds is not viable in bulk through the industries, thus limiting its synthesis to laboratory. The production involves many stages, and *C. roseus* is the only source for the production of these anticancer compounds till date. Therefore, this situation raises the demand for the compound thus making these bioactive substances very expensive (Svoboda 1961; Facchini and De Luca 2008).

The genome size of *C. roseus* is 1500 Mbp located on the chromosome number 16 (Balamani and Rao 1981; Mendioro et al. 2005). By using complete plastid sequence genome analysis, it was found that almost 41 *C. roseus*-specific simple sequence repeats were present. This information shows the high divergence rate of

Fig. 2 Structure of vinblastine and vincristine (Ball and stick model)

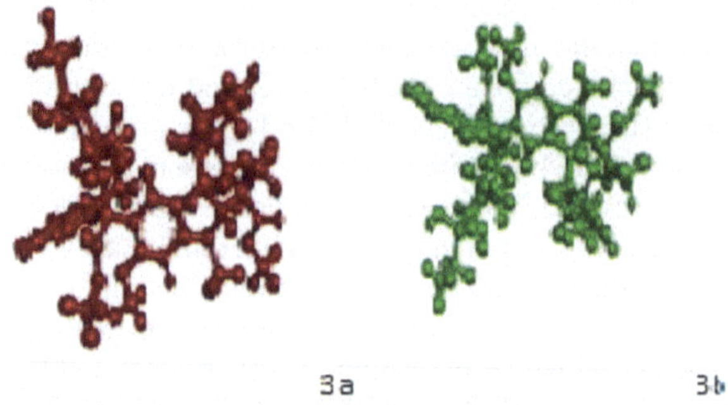

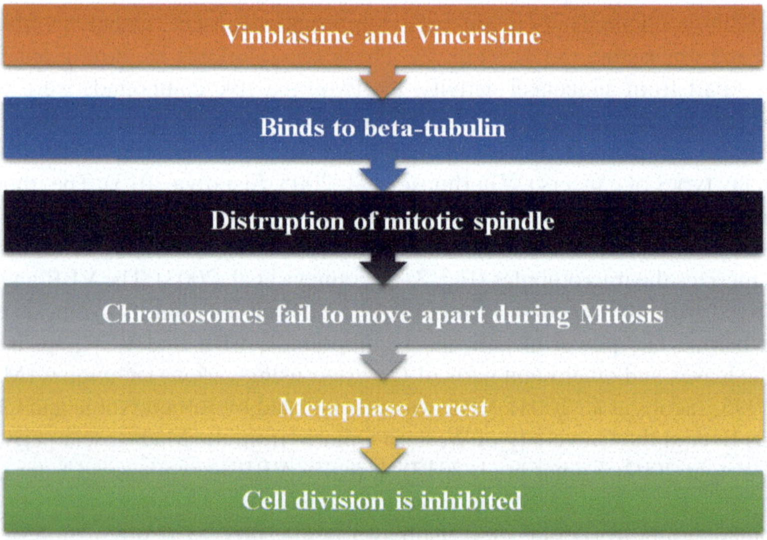

Fig. 3 Vinblastine and vincristine drug mode of action against cancer cell metabolism

C. roseus (Ku et al. 2013). Van Moerkercke et al. (2015) carried out transcriptome analysis in suspension cell lines MP183L and found out a transcription factor called jasmonate-regulated basic helix-loop-helix (bHLH). This transcriptional factor is the key factor responsible for the activation of the genes, those involving in high production of monoterpenoid indole alkaloid which means it is possible to produce high amounts of anticancer compounds like VCR and VBR by activating these genes (Van Moerkercke et al. 2015).

Many researchers have worked on alternative ways of inexpensive amplification of the production VBR and VCR. Pure form of VBR and VCR compounds were isolated from the fungus *Fusarium oxysporum* from Indian *C. roseus*. This method of synthesis can help to produce the anticancer compounds (VBR and VCR) in bulk amount and can prove to be cost-effective. Moreover, the production can be increased by using transformed cultures (Begum et al. 2009).

2 Synthesis of Alkaloids (VBR and VCR)

Alkaloids are basic compounds which contain nitrogen. The vinca alkaloids were hence named as they are composed of carbon, hydrogen, nitrogen and oxygen (Dalton 2000; Hesse 2002). They are synthesized from common amino acid precursors such as phenylalanine, tyrosine, tryptophan, lysine, ornithine and anthranilic acid. Some of the common reactions involved in synthesis of alkaloids are transamination and decarboxylation. These reactions lead to the formation of Schiff's base and further it reacts with the carbanion. In general, alkaloid formation requires single or many amino acids of same kind (Svoboda et al. 1959).

It has been challenging to find the metabolites present in the *C. roseus*. Till now, it is well known that *C. roseus* produces more than hundreds of bioactive metabolites. The major alkaloids produced are VCR, VBR, serpentine, vinorelbine, etc. And it is important to notice that *C. roseus* was the only source for producing the major anticancer compounds like VBR (Fig. 2) and VCR (Fig. 2) (Campone et al. 2012; Kellner et al. 2015). The vinca alkaloids from *C. roseus* were synthesized from the precursor amino acid tryptophan. In the first step, tryptophan is converted to tryptamine. Simultaneously, secologanin is synthesized from geraniol. The tryptamine and secologanin combines to form strictosidine. Strictosidine undergoes series of reaction and forms tabersonine and catharanthine. Tabersonine forms vindoline. Catharanthine and vindoline forms the 3, 4 AnhydroVBR, which ultimately undergoes series of reactions and finally forms VBR and VCR (Neuss et al. 1964; Kalaus et al. 1993). VCR can be obtained from VBR, when it is incubated with cell suspension cultures of *C. roseus* for 2 days (Hamada and Nakazawa 1991). The synthesis of alkaloids is given in detail below.

For identifying the vinca compounds present in the *C. roseus* plant, different extractions are prepared using various solvents like ethanol, ether, chloroform, hexane, etc. Further, the presence of alkaloids was detected using analytical techniques like thin-layer chromatography (TLC), ultraviolet-visible (UV-VIS) spectroscopy and capillary electrophoresis. The high pressure liquid chromatography (HPLC) method is the most reliable method to identify the large number of known and unknown compounds of *C. roseus* (Hisiger and Jolicoeur 2007).

Synthesis of vinca compounds, VCR and VBR, starts with the formation of common central precursor isopentenyl diphosphate (IPP). IPP is formed from the two pathways, namely, the mevalonate pathway and methyl-erythritol phosphate (MEP) pathway (Fig. 4) (Nejat et al. 2015). The vinca alkaloids are generally synthesized in three pathways that are given below:

1. Shikimate pathway
2. Mevalonate pathway
3. Methyl-erythritol phosphate (MEP) pathway

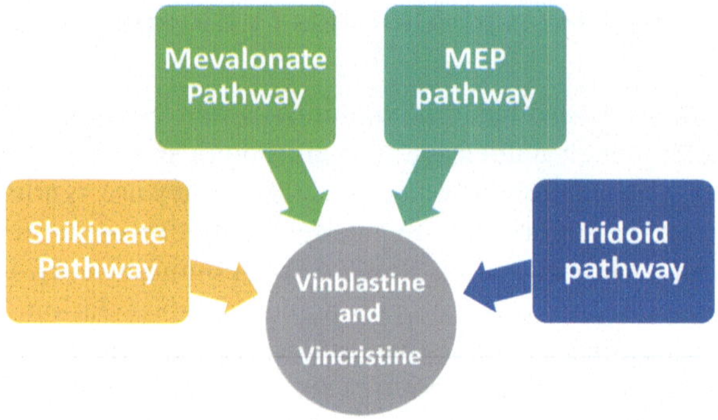

Fig. 4 Overview of the pathways to vinblastine and vincristine

2.1 Shikimate Pathway

The formation of tryptophan from chorismate takes place in this pathway. In the first step, chorismate forms anthranilate. This reaction is catalyzed by anthranilate synthase. In the second step, anthranilate is converted to N-(5-Phosphoribosyl) anthranilate in the presence of enzyme phosphoribosyl diphosphate (PR) anthranilate transferase. In the third step, N-(5-Phosphoribosyl) anthranilate is converted to 1-(o-carboxyphenylamino)-1-deoxyribulose phosphate in the presence of PR-anthranilate isomerase. In the fourth step, 1-(o-carboxyphenylamino)-1-deoxyribulose phosphate is converted to indole-3-glycerol phosphate in the presence of indole-3-glycerol phosphate synthase. In the fifth step, indole-3-glycerol phosphate is converted to indole in the presence of tryptophan synthase a. In the sixth step, indole is finally converted to tryptophan, and this reaction is catalyzed by tryptophan synthase b. The tryptophan is the precursor for the synthesis of vinca alkaloids (Fig. 5) (Whitmer et al. 1998; Tzin and Galili 2010; Verma et al. 2012).

2.2 Mevalonate Pathway

Mevalonate pathway takes place in mitochondria and cytoplasm (Fig. 6), particularly in higher plants. Initially, the two molecules of acetyl CoA react to form acetoacetyl CoA, this reaction is catalyzed by enzyme acetoacetyl CoA thiolase. The acetoacetyl CoA condenses with acetyl CoA and forms 3-hydroxy-3-methylglutaryl CoA (HMG CoA) in the presence of HMG CoA synthase. The enzyme HMG CoA reductase converts HMG CoA to mevalonate. The mevalonate is phosphorylated by mevalonate kinase to 5-diphosphomevalonate. The enzyme diphosphomevalonate de carboxylase converts 5-diphosphomevalonate to isopentenyl diphosphate. A key step in synthesizing isoprenoids is as follows: the isopentenyl diphosphate isomerized to dimethylallyl diphosphate by isopentyl isomerase. Further, dimethylallyl diphosphate is condensed with isopentenyl diphosphate and forms geranyl diphosphate. The geranyl diphosphate is the precursor for synthesis

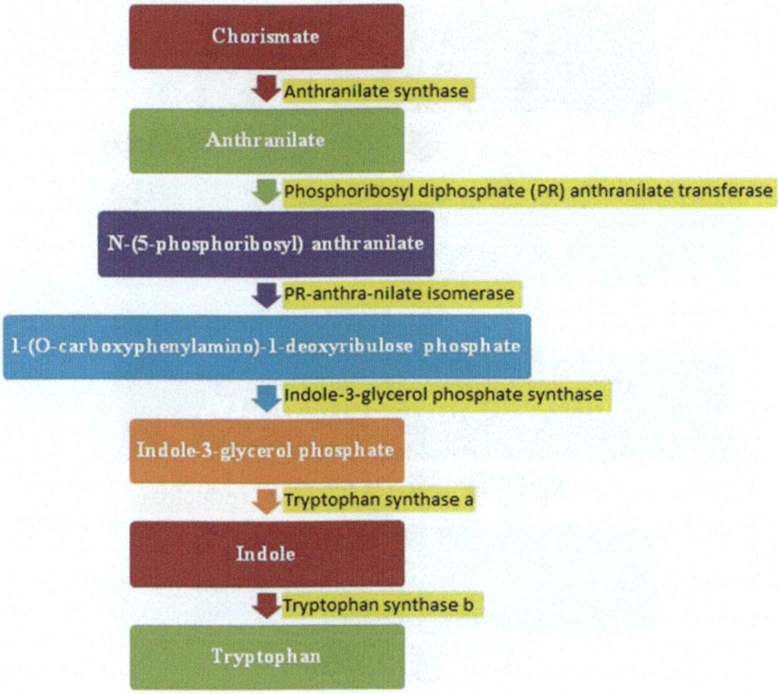

Fig. 5 Schematic representation of Shikimate pathway

of monoterpenoids. The monoterpenoid synthesis is catalyzed by prenyl transferase, monoterpene synthase and geranyl diphosphate synthase (Veau et al. 2000; Dubey et al. 2003).

2.3 *Methyl-Erythritol Phosphate (MEP) Pathway*

Methyl-erythritol phosphate (MEP) pathway mainly takes place in plasmids (Fig. 7). In MEP pathway, mono, di and tri terpenoids are formed. Initially, pyruvate is condensed with glyceraldehyde-3-phosphate and yields 1-deoxy-D-xylulose-5-phosphate in the presence of 1-deoxy-D-xylulose-5-phosphate synthase. 1-deoxy-D-xylulose-5-phosphate is converted to 2-C-methyl-D-erythritol 4-phosphate by redecto isomerase. The 2-C-methyl-D-erythritol 4-phosphate is converted to 4-cytidyl diphospho-2C-methyl-D-erythritol, in the presence of synthase. The 4-cytidyl diphospho-2C-methyl-D-erythritol is converted to 2-C- methyl-D-erythritol 2,4 cyclodiphosphate in the presence of synthase. The 2-C-methyl-D-erythritol 2,4 cyclodiphosphate converted to 1-hydroxy-2-methyl 2(E) butenyl 4 phosphate in the presence of kinase, and finally forms isopentenyl diphosphate (Cordoba et al. 2009; Salim and De Luca 2013).

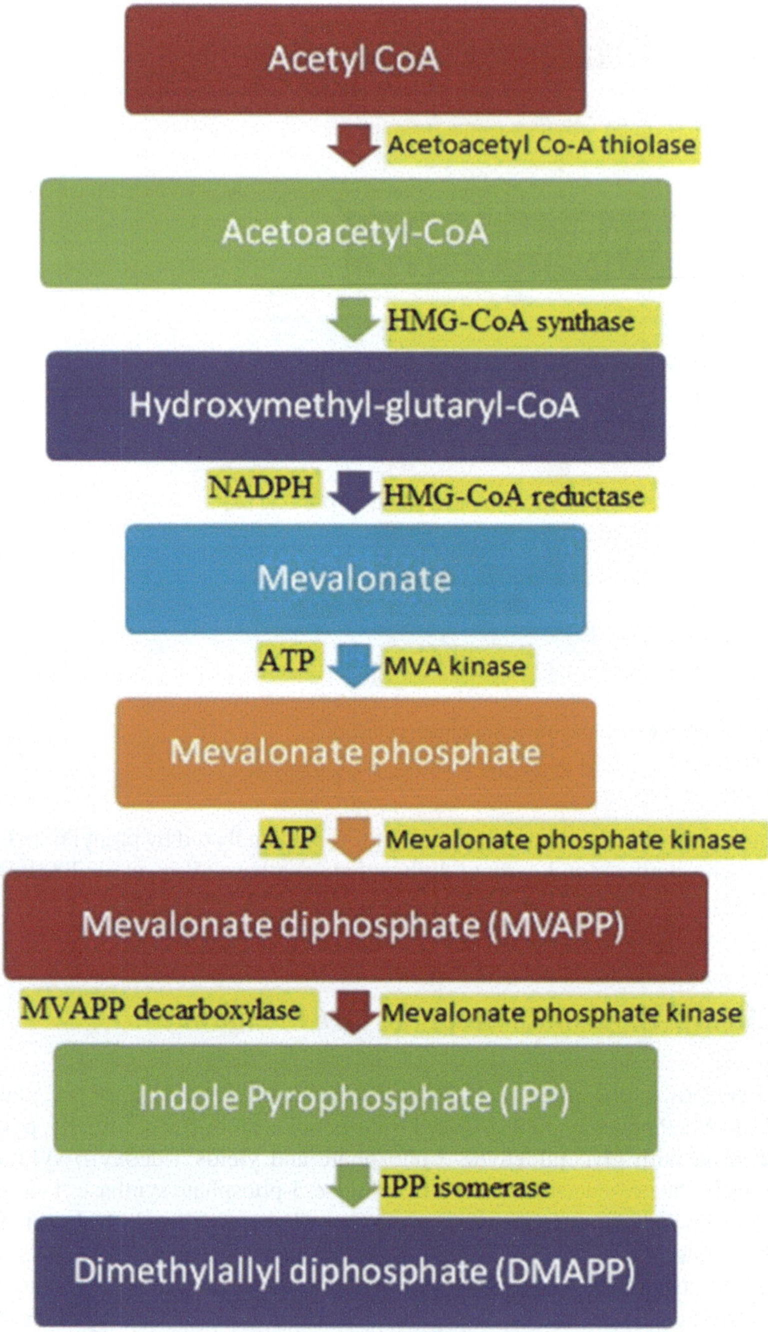

Fig. 6 Schematic representation of Mevalonate pathway

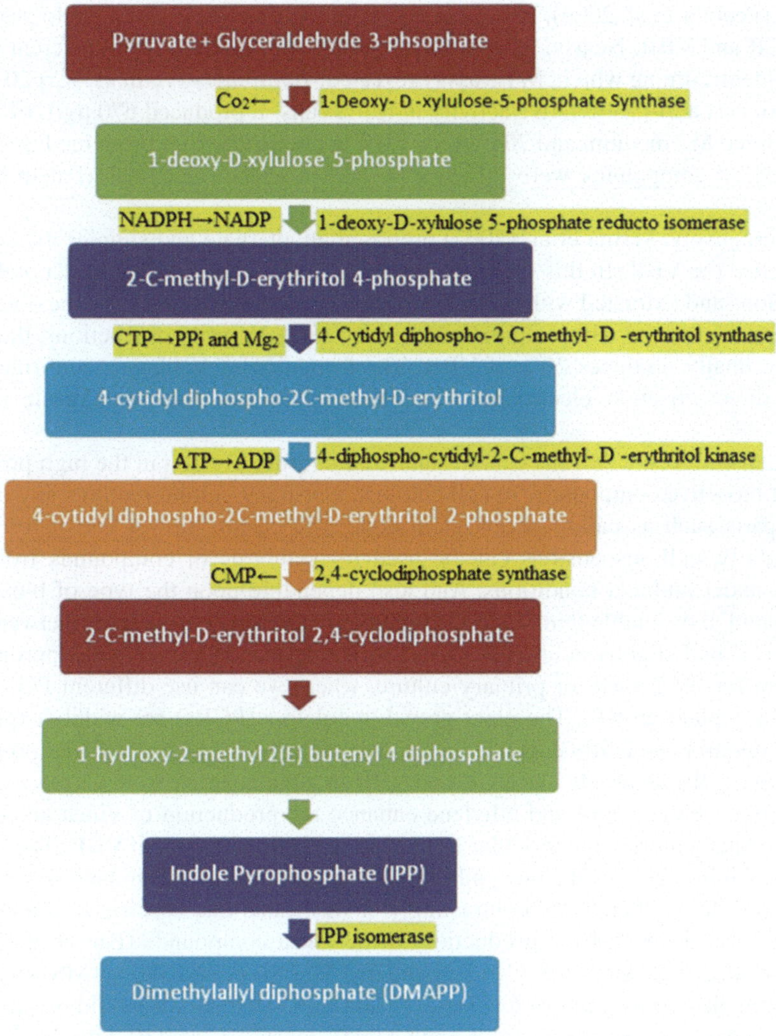

Fig. 7 Schematic representation of MEP pathway

Recently, researchers are focused in increasing the production of vinca compounds. Some of them are discussed here. The yield of bioactive compounds from *C. roseus* can be increased by using different plant transformation techniques such as insertion via genetic vectors (*Agrobacterium tumefaciens*, *A. rhizogenes*), gene insertion, protoplast fusion, electroporation and microinjection. Though these techniques are aimed to yield high quantity of compounds, they are set to certain limitations (Zárate and Verpoorte 2007). The terpenoids are the precursors, which are vital in production of VBR and VCR. Interestingly, induction of respective enzymes of these precursors will increase the production of the active compounds to a large extent (Peebles et al. 2006). Fungi derived from the *C. roseus* are shown to produce the VCR and VBR. Nearly, 22 endophytic fungi were isolated from different parts of the plant. Among which, MTT assay revealed, *Talaromyces radicus* –Crp20 have the strongest anti-proliferative activity in HeLa cells. It produced 670 μg/L of VCR in modified M2 medium and 70 μg/L of VBR in potato dextrose broth medium. The synthesized compounds were effective in human cancer cell lines (Palem et al. 2015).

Alternatively, Verma et al. (2007) proposed an alternate and simpler method to synthesize the VBR. In this method, leaves of *C. roseus* were dried under ambient conditions and extracted with 0.1 M hydro chloric acid (HCl). Further, the obtained compound is treated with sodium hydroxide (NaoH). In series of reactions, the precipitate finally produces 20% of VBR. The formation of VBR was confirmed by using direct-injection electrospray ionization mass spectrometry (Verma et al. 2007).

In cell cultures, *C. roseus* can be stabilized in order to find out the high production of bioactive compounds. In cell cultures, a primary culture requires any of the plant parts such as callus, shoot, stem node, suspension, flower and hairy root. Particularly, cell suspensions can be used for synthesis of compounds from *C. roseus* under ambient conditions. And also depending upon the type of bioactive compound, the suitable temperature for maximum production varies in between 25 and 30 °C and slightly acidic pH (5.8–6.0) The agar medium was the appropriate medium for the growth of primary culture, where we can use different PGRs for promoting plant growth. The plant growth regulators (PGRs) are widely explored by the researchers. PGRs enhance the growth of the plant. And also PGRs promote synthesis of the alkaloids like VCR and VBR in large quantity from *C. roseus*. The PGRs like salicylic acid and ethylene enhance the production of vinca alkaloids. Chlormequat chloride in particular enhances the production rate of VBR. The auxin and cytokinin regulates the biosynthesis of alkaloids. Methyl jasmonate is found to be neutral. In another study, contrastingly, it was found that abscisic acid and gibberellic acid decreased the production of the vinca compounds (Pan et al. 2010; Mujib et al. 2012). Ajay et al. (2013) found that Murashige and Skoog (MS) medium were suitable to grow buds of *C. roseus*. In addition, the medium should be supplied with 6-benzylaminopurine (1 mg/L) and naphthaleneacetic acid (0.2 mg/L) (Verma et al. 2012).

3 Medicinal Properties

C. roseus is a popular medicinal plant. It's been reported since 1910, the leaves of *C. roseus* had been used as mouthwash and for curing ailments like scurvy, wounds, stomach ache, ulcer, to control the hemorrhage, and also as antioxidant, antimicrobial, anti-helminthic, anti-diabetic, anti-obesity and majorly anticancer activity (Gajalakshmi et al. 2013; Balaji 2014). The important medicinal properties are discussed below.

3.1 Antioxidant Activity

Medicinal plants are rich in antioxidants, which act as soldiers and engulf the free radicals and also overcome the stress conditions (Prochazkova et al. 2001). The antioxidant enzymes are increased when the seeds of this plant is treated with diazotrophs (*Azospirillum* and *Azotobacter*). Almost all parts of this plant *C. roseus* were naturally filled with high concentrations of antioxidants (Kar et al. 2003). Ohadoma and Michael (2011) has reported the presence of phytochemicals such as phenols, tannins, flavanoids, terpenoids, glycoside and alkaloids. The antioxidant present in VCR is an excellent protector against cardiac necrosis in animal models. The dosage used for this experiment is about 25 mg/kg body weight in isoproterenol-oxidative stress-induced albino rats. However, the exact mechanism for this role is yet to be determined (Panda et al. 2014).

3.2 Elemental Composition

Nutrients are the important factors responsible for functioning of normal metabolism. Plants supply essential nutrients required to our body. The atomic absorption spectrophotometer analysis reveals that the medicinal plant *C. roseus* is composed of nutrients such as Chromium (Cr), Aluminum (Al), Copper (Cu), Nickel (Ni), Magnesium (Mg), Cadmium (Cd), Iron (Fe), Calcium (Ca), Sodium (Na), Potassium (K), Lead (PB) and Zinc (Zn). The above essential macro- and micro-nutrients are found to be present in both leaves and flowers (Aziz et al. 2016).

3.3 Antimicrobial Activity

The *C. roseus* plant has a strong antimicrobial activity against different microorganism such as *Candida albicans, Aspergillus fumigatus, Baccilus fusiformis* and *Escherichia coli*. Particularly, leaves are very active than other plant parts. The Minimum Inhibitory Concentration is proven to be 20–100 mg/mL (Kumari and Gupta 2013). Antimicrobial activity of *C. roseus* flower extract (200 μg/mL) was carried out against *Pseudomonas aeruginosa, Beta-hemolytic streptococci, Enterobacter agglomerans* and *Staphylococcus aureus*. The extract is a strong antimicrobial agent against *Pseudomonas aeruginosa* and *Staphylococcus aureus,* but the other two strains are resistant to *C. roseus* flower extract (Nayak and Pereira 2006).

The Silver nanoparticles (AgNPs) were synthesized from dried leaves of *C. roseus*. The characterization studies revealed 27–30 nm in size, and the synthesized AgNPs was proven to have a strong antimicrobial activity against Gram-positive bacteria. Such as *Escherichia coli and Pseudomonas fluorescens* (Kotakadi et al.

2013). Ravindra et al. (2012) synthesized AgNPs from the aqueous extract of leaf, callus and root of *C. roseus* using biological methods. Further the AgNPs were proven to be strong antibacterial against all different microbes such as *Bacillus subtilus, Staphylococcus aureus, Escherichia coli, Klebsiella pneumonia* and *Candida albicans,* which are taken from patient samples (Malabadi et al. 2012).

3.4 Wound Healing Activity

Using incision and dead space model, the ethanolic extract of *C. roseus* flower healed the wounds in rat models. The flowers were shade dried and the powder was suspended in 100 mL ethanol for 20 h, and finally the extract is given about 100 mg/kg/day for each Sprague Dawley rat. The flower extract actively healed the wounds faster when compared with control groups. Hence, it can be used as a medicine for wound healing (Nayak and Pereira 2006).

3.5 Anti-Psoriasis Activity

Psoriasis is an incurable skin disease, where keratinocytes differentiate abnormally. Even though it is controllable with chemotherapy, it is not permanently curable. This makes patients' life miserable. Nattaporn et al. (2014) studied the effectiveness of *C. roseus* stem and *Gloriosa superba* leaves against psoriasis at molecular level. In this study, ethanolic extracts of both the plants were tested for expression of psoriatic marker, Keratin 17 (K17) in human keratinocytes. Both the plant extracts down regulated the expression of K17. Hence, it is well understood that both plants were effective individually and showed better activity when combined (Pattarachotanant et al. 2014).

3.6 Anti-Diabetic Property

It is well known that diabetes mellitus is a deadly disorder with serious complications. The medicinal plant *C. roseus* is proven to have anti-diabetic property. Almost all parts of the plant possess anti-diabetic activity (Kar et al. 2003). Particularly, the ethanolic extract of *C. roseus* possesses more anti-diabetic activities compared to aqueous extract. It has also got significant anti-diabetic activity with standard anti-diabetic drug metformin. The recommended ethanolic extract concentration is 100 and 200 mg/kg body weight in animal models (Al-Shaqha et al. 2015).

Since many years in folk medicine, *C. roseus* leaf juice or decoction had been used to treat diabetes mellitus. This anti-diabetic activity of *C. roseus* leaf juice was scientifically proved in alloxan-induced diabetic rats. The different dose-dependent

C. roseus fresh leaf juice had effective anti-diabetic activity when compared with the standard anti-diabetic drug glibenclamide. This juice can be used to treat the patients with diabetes; as an adjuvant, it may prevent the diabetic complications. And also it is believed to treat urinary problems (Nammi et al. 2003; Lans 2006).

In a study, anti-diabetic activity was evaluated for the ethanolic leaf extract of *C. roseus* plant. The plant extract was given to alloxan rats in combination with the standard anti-diabetic drugs, metformin and glibenclamide. The dosage of different concentrations was given to animals which were divided into six groups. Each group has five rats, among those first group was control, group two was treated with plant extract (250 mg/kg body weight), three treated with metformin (100 mg/kg body weight) and four with glibenclamide (1 mg/kg body weight), group five treated with plant extract along with metformin (250+100 mg/kg body weight) and group six with plant extract and glibenclamide (250+1 mg/kg body weight) for a period of 7 days. At last, it was found that all the treatment doses either separately or in combinations had got activity. It was also found that plant extract and metformin combination had highest activity when compared with other groups (Ohadoma and Michael 2011).

Whereas, leaves of methanolic extract along with dichloromethane in equal ratio (500 mg/kg body weight) was tested in streptozotocin-induced rats for 30 days. A complete decline in blood sugar level to normal value in rats was observed. Moreover, the metabolic enzymes such as glycogen synthase, glucose-6-phosphate dehydrogenase were found to be declined. Also methanolic extract in combination with dichloromethane enhanced the cure rate. It was concluded that ethanolic extract of *C. roseus* can be formulated to modern medicine and readily used for treating diabetes mellitus (Singh et al. 2001; Gajalakshmi et al. 2013).

Plant parts of *C. roseus* proved to be anti-diabetic. The aqueous and organic extracts of stem, root, flower and leaf were tested and studied for anti-diabetic activity in alloxan-induced diabetic mice. Almost all parts of *C. roseus* plant were potent anti-diabetic. In that, aqueous extract proved to be better than organic extract (Vega-Ávila et al. 2012). In 2010, Mohammed et al. found that methanolic whole plant extract had effective anti-diabetic activity and it also regenerated the β-cells of pancreas, when the plant extract at the dose of 500 mg/kg body weight was given to alloxan-induced diabetic rats for 14 days. It also had additional benefits like body weight gain and well-maintained lipid profile (Sivalingam and Sriram 2013).

3.7 Hodgkin's Disease

Hodgkin's disease (HD) is one the common lymphomas (Küppers et al. 2006). VBR is used effectively in the treatment of Hodgkin's disease. In patients with advanced stages of Hodgkin's disease, the VBR sulfate is reportedly found to be as effective as nitrogen mustard. And VBR could be in the first preference for treatment of patients with Hodgkin's disease (Alison and Whitelaw 1970).

3.8 Anticancer Activity

Cancer can be defined as unregulated cell growth which leads to formation of malignant tumours. The cancer cells spread and convert normal cells into cancer cells (Anand et al. 2008). Historically, the miracle plant *C. roseus* is used in the treatment of cancer since many years in Ayurveda, traditional Chinese medicine. The most medicinally valuable plant kills the cancer cells without any side effects (Gurib-Fakim 2006). Vinca alkaloids are synthetically produced and used as drugs in cancer therapy and it also acts as an immunosuppressive drug. These compounds include VBR, VCR, vindesine, and vinorelbine. VCR, VBR and vinorelbine are approved for its use in the United States. The VBR is commonly traded as VBR sulfate injection (Rowinsky 2003).

The VBR was the first of vinca to be used in the treatment of cancer. VBR is a chemotherapy drug used to treat different cancers, including lymphomas, bladder, breast and testicular cancer, and Kaposi's sarcoma (Cragg and Newman 2005; Epstein et al. 1989). VBR is the salt of a naturally occurring vinca alkaloid obtained from the flowering herb periwinkle (Pharma 2003). Vinca alkaloids act by preventing the polymerization of tubulin to form microtubules, as well as inducing depolymerization of formed tubules. VBR may also interfere with nucleic acid and protein synthesis by blocking glutamic acid utilization. Vinca alkaloids are cell cycle phase specific for M phase and S phase. VBR exerts some immunosuppressive activity. It is believed that glutamic acid has a role in reducing the neurotoxicity; however, it is not found to be effective in children, when given to children who are taking treatment in children oncology (Bradfield et al. 2015).

VCR also known as leurocristine is used to treat various types of cancer. It is a cancer chemotherapy drug that is usually used with other chemotherapy drugs to slow or stop cancer cell growth. VCR is used in combination with other chemotherapy drugs to treat certain types of leukaemia (cancer of the white blood cells), including acute myeloid leukaemia and acute lymphoblastic leukaemia (Kumar et al. 2013). VBR is used in treating a rare malignant tumour, angiosarcoma, in combination with β-blockers. Angiosarcoma is rare but it is aggressive if caused. Unlike the treatment with radiation and surgery, this regimen (β-blockers+VBR) will effectively cure the disease. Moreover, this treatment is economically feasible (Pasquier et al. 2016).

In cancer cells, the two anticancer compounds of vinca bind to the β-tubulin and disrupt mitotic spindle. As a result, the chromosomes fail to move apart during mitosis, and metaphase gets arrested thus inhibiting cell division (Schläger and Dräger 2016). Primarily, the vinca drugs interfere with the polymerization of tubulin to prevent microtubule formulation thus arresting cell division. The drugs are large molecules and are lipophilic compounds that are formed of two structural multi-ringed units. Other compounds like vindoline and catharanthine, linked by a carbon–carbon chain, differ only in a single substitution on the catharanthine group. The vindoline and catharanthine works by preventing mitosis in metaphase. These alkaloids bind to tubulin, thus, preventing the cell from making the spindles (Kumar 2016).

The in silico docking analysis of vinca alkaloids such as VBR, VCR, catharanthine and vinorelbine with α/β-tubulin was performed. All the vinca compounds were docked with the target tubulin. Finally, the study showed that these vinca compounds have binding affinity towards the target. Among which VBR, VCR and vinorelbine have strong affinity, while vindoline and catharanthine showed a weak affinity. Based on this it is evident that, particularly, VCR and VBR have strong anticancer activity (Sertel et al. 2011). In cell line study, VBR is believed to be inhibiting the proteasomes which could lead to block the cell progression and induce apoptosis. This could be the possible mechanism of target for VCR (Piccinini et al. 2001).

Callegaro-Filho et al. (2014) performed a retrospective study of all patients evaluated at MD Anderson Cancer Center (MDACC) between May 2004 and April 2014 with the diagnosis of small cell carcinoma of the ovary—Hypercalcemic type (SCCOHT) and identified eight patients and they were treated with VBR and cisplatin on Day 1 and cyclophosphamide and bleomycin on Day 2 and with doxorubicin and etoposide on Day 3 (VPCBAE) followed by pegfilgrastim as primary treatment followed by surgery. The median age at diagnosis was 28 years (range, 21–41). The authors' results suggested that the combination of VPCBAE was found to be effective in patients with SCCOHT. Further, they also noted no treatment-related deaths as the treatment very effectively cured the disease (Callegaro-Filho et al. 2014).

Dichloromethane: methanol extract of *C. roseus* is the only plant which has been proved to be effective on both MCF-7 and HeLa cancer cell lines compared with ten plants of the same Apocynaceae family, respectively (Siu et al. 2011). MTS [3-(4,5-dimethylthiazol-2-yl)-5(3-carboxymethoxyphenyl)-2-(4-sulfophenyl)-2H-tetrazolium] assay shows that the aqueous extract of *C. roseus* was effective against the Jurket cancer T-cells (Nor et al. 2010). Recently, two new compounds similar to VBR have been discovered in *C. roseus* leaves, namely, 17-desacetoxyVBR N'b-oxide and 20′-deoxyVBR N'b-oxide. Using MTT assay, it was found that the two alkaloids were effective against human hepato cellular carcinoma (HepG2) cell line, human colorectal carcinoma (Lovo) cell line and human breast carcinoma (MCF-7) cell line (Zhang et al. 2013).

Breast cancer is one of the frequent malignant cancers. VBR in combination with mitomycin C and cisplatin (MVC) was given as treatment to 87 patients, who were suffering from metastatic breast cancer (MBC). The patients had already taken chemotherapy as primary treatment. After 21 days of treatment, the MVC regimen was effective against MBC; however, mild toxicity was observed (Urruticoechea et al. 2005).

In animal model, cancer was induced by an agent Diethylnitrosamine (DEN). The DEN altered the elemental composition, increased the elemental composition levels (Ni,Zn, Fe,Br and Cr) and decreased some elements concentration levels (Ca, Fe) indicating that alteration in elemental composition was the cause for cancer. In order to check the effect of *V. rosea* against the DEN's effect, the animals were treated with *V. rosea* extract for a period of 4 weeks. Surprisingly, the treatment reverted the changes made by DEN. However, the changes were limited to tissue level (Mohanta et al. 2007).

In a clinical study, 30 patients with metastatic condition were treated with chemotherapy (Gemcitabine and cisplatin). The patients under chemotherapy were found to be relapsed, and in some patients the disease condition progressed. Those patients were treated with the regimen methotrexate, VBR, doxorubicin and cisplatin (M-VAC). Treatment with M-VAC regimen was found to be significant for the improvement of patient's health and also it prolonged the life of the patient. However, the need to be carried out in large sample size is essential for further validation (Han et al. 2008).

Vinca alkaloids such as VCR in combination with cytostatic drugs like anthracylines, irradiation and dexamethasone in tumour cell lines diminished the cell death up to 86% and in preclinical trails the cytostatic drug anthracyclines inhibited the action of vinca alkaloids. However, caffeine reverses the effect caused by the cytostatic drug against vinca alkaloids (Ehrhardt et al. 2013).

In a case study, the rhabdomyosarcoma (RMS) a soft-tissue sarcoma usually caused cancer in childhood and adolescence affected a 17-year-old patient. The patient was treated with multidrug therapy (MDT) along with chemotherapy. The regimen is composed of cyclophosphamide (CPA), doxorubicin (ADM), dacarbazine (DTIC) and VCR (VCR) (CYVADIC). This MDT regimen treatment which is for 17 cycles along with chemotherapy had extended the patient's life for 4 years. This gives evidence that VCR is a life-saver drug even in advanced stages of cancer (Isono et al. 2016). But it leads to damage in the nervous system and causes neuropathy, widely known as VCR-induced neuropathy (Shimoda et al. 2007). In animal models, the fruit extract of *Momordica charantia* L. prevents the VCR-induced neuropathy (Jain et al. 2015).

Apart from the rise in treatment for different types of cancers using VCR, its resistance has been observed in colon cancer cells. Next-generation sequencing analysis revealed about 121 long non-coding RNAs (IncRNAs) and showed different expression in HCT-8 colon cancer cells. This data further could be used as a marker in finding the patients who are VCR resistant (Sun et al. 2015). In another study, VCR in combination with the cyclophosphamide, doxorubicin, prednisolone (CHOP) along with monoclonal antibody rituximab is widely used to treat diffuse large B-cell lymphoma (DLBCL) and cardio toxicity (Pfreundschuh et al. 2011; Yung-Cheng et al. 2011). Patients with thymic carcinoma were successfully treated with weekly dose of cisplatin, VCR, doxorubicin and etoposide (CODE) followed by a surgery. This regimen is a promising therapy for expanding longevity of patient's life (Kawasaki et al. 2014).

In the United States, VCR sulfate liposome injection (VSLI) has been approved for relapsed and refractory acute lymphoblastic leukaemia (O'Brien et al. 2013). VCR is also used to treat neuroblastoma. Neuroblastoma is a common cancer that occurs during childhood. Chemotherapy with VCR eventually cures the disease by arresting the cell cycle and inducing apoptosis. This mechanism is found in SH-SY5Y human neuroblastoma cells in which VCR regulates the cyclin B and C, thus leading to mitotic arrest. And the activation of caspase-3 and -9 leads to apoptosis (Biedler et al. 1978; Tu et al. 2013).

3.9 Anti-Hypertension Activity

Resistant hypertension can be cured by VCR by a novel method called chemical sympathetic denervation. In this method, VCR was successfully delivered in Landrace swine animal models using catheter. After 28 days of treatment, histopathological sample observation revealed a significant result that the intact nerve is lower in VCR-treated group when compared with the placebo group. However, further confirmation in other animal models is required (Stefanadis et al. 2013). Apart from this, in animal models, heme oxygenase 1 (HO-1), which has many properties degrades heme to carbon monoxide, free iron and biliverdin. The antioxidant-rich HO-1 mediated the pain induced by the vincristine (Ham et al. 2012; Shen et al. 2015).

4 Larvicidal Effect

The vector-borne diseases are very dangerous to mankind, particularly malaria, which causes high mortality rate around the world. Even though, there are a lot of programmes to control vectors, new alternatives are essential. The plant extract of *C. roseus* showed larvicidal activity, particularly the petroleum ether extract showed maximum activity against the *Anopheles stephensi* malarial larvae. The bioactive of this compound can be used against the malarial vector. For better results, the extract of *C. roseus* can be used in combination with standard drugs in vector control programmes (Panneerselvam et al. 2013).

5 Nanotechnology

5.1 Silver Nanoparticles

Biologically synthesized nanoparticles are very effective, eco-friendly and economical. Silver nanoparticles were synthesized from the leaves of *C. roseus*. The biologically synthesized nanoparticles were characterized and it was found that the silver nanoparticles were 35–55 nm in size. Further, the biologically synthesized nanoparticles were tested for antiplasmodial activity and found to be potentially active against the malaria parasite *(Plasmodium falciparum)*. This is a positive sign, and these nanoparticles could be used in killing deadly malaria parasite

(Ponarulselvam et al. 2012). Kannan et al. (2011) observed silver nanoparticles which were synthesized biologically from *C. roseus* were 45–70 nm in size and stable for 4 months (Kannan et al. 2011).

5.2 Titanium Dioxide Nanoparticles

The titanium dioxide nanoparticles (TiO_2NPs) were synthesized through green approach. The characterization showed the nanoparticle size ranges between 25 and 110 nm. By using Bragg's law and Scherer's constant, it was confirmed to be 65 nm. The TiO_2NPs from *C. roseus* leaf extract were proven to be effective against parasites such as *Hematophagous fly*, *Hippobosca maculata*, leach, sheep-biting louse and *Bovicola ovis* Schrank. However, the maximum activity has been observed in Hippobosca maculata, leach and *Bovicola ovis* Schrank (Velayutham et al. 2012).

Titanium dioxide nanoparticles (TiO_2 NPs) are synthesized biologically from the plant extract of *C. roseus*. The synthesized nanoparticles are characterized using standard techniques at the size of the TiO_2 NPs ranging from 25 to 110 nm. The TiO_2 NPs is active against the parasites hematophagous fly and sheep-biting louse. This biologically synthesized TiO_2 NPs are nontoxic and feasible in cost and can be used to control the flies and parasites which target the livestock (Velayutham et al. 2012). In the same manner, biologically synthesized zinc oxide nanoparticles (ZnO-NPs) from the leaves of *C. roseus* were studied and it was proved to be a strong antibacterial agent against *Pseudomonas aeruginosa* (Bhumi and Savithramma 2014).

5.3 Liposomal Formulations

VCR is used in treating various human cancers. Even though, some of the reports claim its neurotoxic and tissue damage effect, thus limiting its application. In order to curb the side effects and to enhance the activity of VCR, it is encapsulated with liposomes using various loading methods such as passive loading, pH gradient loading and ionophore-assisted loading. The encapsulated VCR with liposomes is characterized and standardized using various methods. The encapsulated VCR is better than the non-encapsulated VCR in in vivo experiments. It has also been observed that it has a moderate decrease in the drug toxicity (Waterhouse et al. 2005). The enhanced liposomal drug formulations will reduce the drug toxicity and are used to deliver the VCR which reaches the target tissues and doesn't affect the healthy cells. Liposomal drug delivery also increases the dosage concentration to the target sites (Gabizon 1992; Gabizon et al. 1994; Maeda 2001; Kato et al. 2012).

Liposomal formulation of VBR was prepared against non-small cell lung cancer (NSCLC), a common lung cancer. The formulation was at the size of 100 nm in size. Cholesterol was used as cationic material and peanut agglutinin was modified. The formulation was very effective both in vivo and in vitro methods against NSCLC, the drugs work in a mechanism by which it initiates apoptosis by circulating in blood for a long period. This sustainably cures the NSCLC (Li et al. 2015). The half-life of VCR sulfate is increased many folds by formulating it as dextran microspheres (Dectran-MSs) and in fused form with chitosan-β-glycerophosphate thermosensitive gel (VCS-Dextran-MSs-Gel) (Thakur et al. 2016).

VCR and VBR alkaloids were formulated to liposomal and immuno-liposomal formulation. The in vivo drug release test in rats concluded that VCR would get easily cleared than VBR. Both the formulations were tested in human mammary carcinoma cell lines (SKBR-3 and BT474-M2) and found to be effective in killing the cancer cells. These formulations can be studied further for treating solid tumours (Noble et al. 2009).

The VCR is effective in treating various types of cancer. But it has many disadvantages like dosing, pharmacokinetic limitations. A cholesterol-based nanoparticles formulation was developed. It is known as Marqibo, a VCR sulfate liposome injection (VSLI). The FDA approved VSLI to overcome the limitations of VCR. The VSLI is effective in delivering the required doses at target sites in a sustainable rate and time without toxicity (Silverman and Deitcher 2013).

Nanoparticles play an important role in medicine and are widely attracted for its applications. Green syntheses of nanoparticles are preferably safe and feasible. Size-controlled chitosan nanoparticles with the size of 45 nm at pH 3 are biologically synthesized from *C. roseus* leaf extract (3:1 ratio). Further, the nanoparticles are entrapped with ketoconazole and chloramphenicol drugs. The in vitro drug release studies showed a sustainable release of the two drugs up to 12 h. The synthesized chitosan nanoparticles are less toxic and are cost-effective. These nanoparticles could be a promising approach in future for increasing the stability and to increase drug efficacy (Nagaonkar et al. 2015).

The above formulation was confined mostly to animal studies, and it will be effectively available for the treatment of humans. Also, exploring new methods and adjuvant was need of the hour. It is believed that Omega 3 fatty acids increase the chemo-sensitivity to VCR and few other cancer drugs (Fahrmann and Hardman 2013). The discoveries of new advances in the field of medicine are most welcomed for treating the patients effectively.

5.4 Radioimmunoassay

In 1977, Teale and co-workers developed radioimmunoassay for VBR and VCR. In this assay, at first using Mannich reaction, the VBR and VCR were conjugated to albumin. Secondly, these conjugate where injected to rabbits to develop antisera. Finally, the collected serum was used to determine the antibodies, and radioimmunoassay was carried out as per the developed protocol. This method is used to determine the amount of drug present in plasma. But it is sensitive only after 24 h of injection with 15 mg of drug. This a highly sensitive method for detection of VBR (2.1 ng/mL) and VCR (3.8 ng/mL) (Teale et al. 1977).

Using radioimmunoassay, John et al. in 1979 proposed less than 1 pmol sample of vinca alkaloids is enough to measure the level of alkaloids present in plasma. The hapten-antibody reaction assay is used to determine the compound concentration of vinca plant (Langone et al. 1979). The amount of VCR dose present in tumours can be determined using ultra-high performance liquid chromatography and positive electrospray ionization-high resolution mass spectrometric (Tina et al. 2013). Achanta et al. (2013) identified plasma and urine samples from canine which precisely determines the amount of VBR and its metabolite desacetyl VBR using liquid chromatography-atmospheric pressure chemical ionization-mass spectrometry (Achanta et al. 2013).

5.5 *Toxicity*

Vinca alkaloids lead to peripheral toxicity (Moudi et al. 2013). VBR can cause severe side effects like leukopenia, gastrointestinal disturbances, cellulitis and phlebitis. And toxicity of VCR can cause hepatic mpairment, respiratory distress, neuromuscular neurotoxicity, pulmonary, hypersensitivity, haematologic disorder, gastrointestinal disorder, urogenital disorder, dermatological disorder, endocrine disorder and cardiovascular disorder (Muna et al. 2014).

Clinically, leukopenia is caused by the side effects of VBR sulfate. In general, the larger the dose administered the more profound and longer lasting the leukopenia will be. In a study, therapy with VBR sulfate leads to a decrease in the white blood cell count and the condition lasted longer. Recovery of the white blood count is fairly rapid thereafter and is usually complete within another 7–14 days. With the smaller doses employed for maintenance therapy, leukopenia may not be a problem (Warwick et al. 1961; Beier et al. 1963).

Peripheral neurotoxicity is dose limiting and disables side effect of several important chemotherapeutic agents. In particular, VCR, cisplatin, oxaliplatin, paclitaxel and docetaxel are frequently used antineoplastic agents, which are known causes of a peripheral neuropathy. VCR causes an axonal sensorimotor neuropathy often early during treatment, heralded by paresthesias and followed by severe motor weakness if treatment is continued. Pain is not a prominent feature. Neurotoxicity is the dose-limiting toxicity of VCR, in particular related to dose per cycle. Other vinca alkaloids are far less neurotoxic (Casey et al. 1973).

Toxicity of vinca compounds concerns widely. Apart from this potential anticancer activity, the compounds are extensively neurotoxin (Himes 1991). It severely affects the central nervous system leading to mental confusions, hallucinations, depression, agitation, insomnia and also coma. It also leads to imbalance of the antidiuretic hormone. Even though antinodes like B-complex are considered but they are not up to the mark. Gastrointestinal disturbances like bloating, constipation, abdominal pain and in severe cases vomiting, nausea and diarrhea were observed most commonly by both VBR and VCR (McGuire et al. 1989; Kufe et al. 2003).

Due to severe side effects, it's been advised to pregnant and lactating women, and those who are under vaccination that not to take VCR or VBR as treatment. Because in pregnant women the toxic effect of vinca compounds may lead to birth defects of the child and also can cause serious consequences (Johnson et al. 1963). Advice from the clinicians must be considered before taking the treatment, and the clinicians should thoroughly examine the history of the patient before starting the treatment with these compounds.

Lee et al. (2016) investigated the effect of curcumin on VBR in HeLa human cervical cells. Curcumin is widely used in diet and studied for its potent anticancer activity. In this experiment, they pretreated the HeLa cells with curcumin and further treated with VBR. In pretreatment, microtubule filaments were disordered. Cell death and DNA condensation were decreased in pretreated cells when compared with the control. So, it has been suggested that curcumin inhibits the anticancer activity of VBR and human consumption of curcumin can be avoided while on medication with VBR (Lee et al. 2016). Also in the same manner, antifungal compounds along with VCR causes severe neurotoxicity. The azole compounds such as itraconazole, posaconazole, voriconazole and ketoconazole had ceased the metabolism of VCR leading to toxicity. So it is advisable for the patients who are on medication with VCR to avoid these antifungal agents (Moriyama et al. 2012).

Recently, Simvastatin, a cardiovascular drug used to treat cardiovascular disease was found to reverse the VCR-induced neuropathic pain in patient with nerve damage and neuropathic pain caused by the effect of VBR (Bhalla et al. 2015). Overall, vinca alkaloids are the second most-used class of cancer drugs and will stay among the original cancer therapies. In future, more studies have to be carried out on vinca alkaloids and its medical applications.

References

Achanta S, Ngo M, Veitenheimer A, Maxwell LK, Wagner JR (2013) Simultaneous quantification of vinblastine and desacetylvinblastine concentrations in canine plasma and urine samples using LC–APCI–MS/MS. J Chromatogr B 913:147–154

Ajay K, Krishna P, Rajesh K, Nitish K (2013) In vitro plant propagation of *Catharanthus roseus* and assessment of genetic fidelity of micropropagated plants by RAPD marker assay. Appl Biochem Biotechnol 169(3):894–900

Alison R, Whitelaw D (1970) A comparison of nitrogen mustard and vinblastine sulfate in the treatment of patients with Hodgkin's disease. Can Med Assoc J 102(3):278

Al-Shaqha WM, Khan M, Salam N, Azzi A, Chaudhary AA (2015) Anti-diabetic potential of *Catharanthus roseus* Linn. and its effect on the glucose transport gene (GLUT-2 and GLUT-4) in streptozotocin induced diabetic wistar rats. BMC Complement Altern Med 15(1):1–8

Anand P, Kunnumakara AB, Sundaram C, Harikumar KB, Tharakan ST, Lai OS, Sung B, Aggarwal BB (2008) Cancer is a preventable disease that requires major lifestyle changes. Pharmaceut Res 25(9):2097–2116

Aslam J, Mujib A, Fatima Z, Sharma MP (2010) Variations in vinblastine production at different stages of somatic embryogenesis, embryo, and field-grown plantlets of *Catharanthus roseus* L.(G) Don, as revealed by HPLC. In Vitro Cell Dev Biol Plant 46(4):348–353

Aziz S, Saha K, Sultana N, Nur HP, Ahsan MA, Ahmed S, Hossain MK (2016) Comparative studies of elemental composition in leaves and flowers of *Catharanthus roseus* growing in Bangladesh. Asian Pacific J Trop Biomed 6(1):50–54

Balaji DH (2014) Versatile therapeutic effects of *Vinca rosea* Linn. Int J Pharmaceut Sci Health Care 1(4):59–76

Balamani G, Rao R (1981) Chromosome number reports LXXIII. Taxon 30:855–856

Begum F, Nageswara Rao S, Rao K, Prameela Devi Y, Giri A, Giri C (2009) Increased vincristine production from *Agrobacterium tumefaciens* C58 induced shooty teratomas of *Catharanthus roseus* G. Don. Nat Prod Res 23(11):973–981

Beier FR, Thatcher LG, Lahey ME (1963) Treatment of reticuloendotheliosis with vinblastine sulfate: preliminary report. J Pediatr 63(6):1087–1092

Bhalla S, Singh N, Jaggi AS (2015) Dose-related neuropathic and anti-neuropathic effects of simvastatin in vincristine-induced neuropathic pain in rats. Food Chem Toxicol 80:32–40

Bhumi G, Savithramma N (2014) Biological synthesis of zinc oxide nanoparticles from *Catharanthus roseus* (l.) G. Don. Leaf extract and validation for antibacterial activity. Int J Drug Dev Res 6:208–214

Biedler JL, Roffler-Tarlov S, Schachner M, Freedman LS (1978) Multiple neurotransmitter synthesis by human neuroblastoma cell lines and clones. Cancer Res 38(11 Part 1):3751–3757

Bradfield SM, Sandler E, Geller T, Tamura RN, Krischer JP (2015) Glutamic acid not beneficial for the prevention of vincristine neurotoxicity in children with cancer. Pediatr Blood Cancer 62(6):1004–1010

Buchwald W, Dedio I, Kozłowski J, Łata B (2007) Hydroponic culture of *Catharanthus roseus* (L.) G. Don and studies on seed production. Phytochem Rev 6(2–3):413–417

Callegaro-Filho D, Ramirez P, Nick A, Kavanagh J, Euscher E, Dickens A, Patel D, Schmeler K (2014) 903p effectiveness and toxicity of vinblastine, cisplatin, cyclophosphamide, bleomycin, doxorubicin and etoposide (VPCBAE) in the management of patients with small cell carcinoma of the ovary-hypercalcemic type (SCCOHT). Ann Oncol 25(suppl 4):iv314

Campone M, Isambert N, Bourbouloux E, Roché H, Bonneterre J, Milano G, Fumoleau P (2012) A phase I study of vinflunine in combination with capecitabine in patients with metastatic breast cancer previously treated with anthracyclines and taxanes. Cancer Chemother Pharmacol 69(4):871–879

Casey E, Jellife A, Quesne P, Millett Y (1973) Vincristine neuropathy. JACC Clin Electrophysiol 96:69–86

Chong F, Chakravarthi S, Nagaraja H, Thanikachalam P, Lee N (2009) Expression of transforming growth factor-and determination of apoptotic index in histopathological sections for assessment of the effects of Apigenin (4′, 5′, 7′-Trihydroxyflavone) on Cyclosporine A induced renal damage. Malays J Pathol 31(1):35–43

Cordoba E, Salmi M, León P (2009) Unravelling the regulatory mechanisms that modulate the MEP pathway in higher plants. J Exp Bot 60(10):2933–2943

Cragg GM, Newman DJ (2005) Plants as a source of anti-cancer agents. J Ethnopharmacol 100(1):72–79

Cragg GM, Grothaus PG, Newman DJ (2009) Impact of natural products on developing new anti-cancer agents†. Chem Rev 109(7):3012–3043

Dalton DR (2000) Alkaloids. In: Kirk-Othmer encyclopedia of chemical technology. Wiley, New York. http://dx.doi.org/10.1002/0471238961.0112110104011220.a01; doi: 10.1002/0471238961.0112110104011220.a01

de Gonzalez AB, Darby S (2004) Risk of cancer from diagnostic X-rays: estimates for the UK and 14 other countries. Lancet 363(9406):345–351

De la Rosa-Mera CJ, Ferrera-Cerrato R, Alarcón A, de Jesús S-CM, Muñoz-Muñiz OD (2011) Arbuscular mycorrhizal fungi and potassium bicarbonate enhance the foliar content of the vinblastine alkaloid in *Catharanthus roseus*. Plant soil 349(1-2):367–376

Dubey VS, Bhalla R, Luthra R (2003) An overview of the non-mevalonate pathway for terpenoid biosynthesis in plants. J Biosci 28(5):637–646

Ehrhardt H, Pannert L, Pfeiffer S, Wachter F, Amtmann E, Jeremias I (2013) Enhanced anti-tumour effects of Vinca alkaloids given separately from cytostatic therapies. Br J Pharmacol 168(7):1558–1569

Elujoba AA, Odeleye O, Ogunyemi C (2005) Traditional medicine development for medical and dental primary health care delivery system in Africa. Afr J Trad Complement Altern Med 2:46–61

Epstein JB, Lozada-Nur F, McLeod WA, Spinelli J (1989) Oral Kaposi's sarcoma in acquired immunodeficiency syndrome. Review of management and report of the efficacy of intralesional vinblastine. Cancer 64(12):2424–2430

Fabricant DS, Farnsworth NR (2001) The value of plants used in traditional medicine for drug discovery. Environ Health Perspect 109(Suppl 1):69

Facchini PJ, De Luca V (2008) Opium poppy and Madagascar periwinkle: model non-model systems to investigate alkaloid biosynthesis in plants. Plant J 54(4):763–784

Fahrmann JF, Hardman WE (2013) Omega 3 fatty acids increase the chemo-sensitivity of B-CLL-derived cell lines EHEB and MEC-2 and of B-PLL-derived cell line JVM-2 to anti-cancer drugs doxorubicin, vincristine and fludarabine. Lipids Health Dis 12(1):1

Fajersztajn L, Veras M, Barrozo LV, Saldiva P (2013) Air pollution: a potentially modifiable risk factor for lung cancer. Nat Rev Cancer 13(9):674–678

Farmer P, Frenk J, Knaul FM, Shulman LN, Alleyne G, Armstrong L, Atun R, Blayney D, Chen L, Feachem R (2010) Expansion of cancer care and control in countries of low and middle income: a call to action. Lancet 376(9747):1186–1193

Gabizon AA (1992) Selective tumor localization and improved therapeutic index of anthracyclines encapsulated in long-circulating liposomes. Cancer Res 52(4):891–896

Gabizon A, Catane R, Uziely B, Kaufman B, Safra T, Cohen R, Martin F, Huang A, Barenholz Y (1994) Prolonged circulation time and enhanced accumulation in malignant exudates of doxorubicin encapsulated in polyethylene-glycol coated liposomes. Cancer Res 54(4):987–992

Gajalakshmi S, Vijayalakshmi S, Devi RV (2013) Pharmacological activities of *Catharanthus roseus*: a perspective review. Int J Pharma Biosci 4(2):431–439

Gunatilaka AL (2006) Natural products from plant-associated microorganisms: distribution, structural diversity, bioactivity, and implications of their occurrence. J Nat Prod 69(3):509–526

Guo X-R, Yang L, Yu J-H, Tang Z-H, Zu Y-G (2007) Alkaloid variations in *Catharanthus roseus* seedlings treated by different temperatures in short term and long term. J Forest Res 18(4):313–315

Gupta S, Morris SK, Suraweera W, Aleksandrowicz L, Dikshit R, Jha P (2016) Childhood cancer mortality in india: direct estimates from a nationally representative survey of childhood deaths. J Global Oncol 2(6):403–411. doi:10.1200/JGO.2015.000935

Gurib-Fakim A (2006) Medicinal plants: traditions of yesterday and drugs of tomorrow. Mol Asp Med 27(1):1–93

Ham M, Akiba Y, Takeuchi K, Montrose MH, Kaunitz JD (2012) Gastroduodenal mucosal defense. Physiol Gastrointest Tract 1:1169–1208

Hamada H, Nakazawa K (1991) Biotransformation of vinblastine to vincristine by cell suspension cultures of *Catharanthus roseus*. Biotechnol Lett 13(11):805–806

Han K, Joung J, Kim T, Jeong I, Seo H, Chung J, Lee K (2008) Methotrexate, vinblastine, doxorubicin and cisplatin combination regimen as salvage chemotherapy for patients with advanced or metastatic transitional cell carcinoma after failure of gemcitabine and cisplatin chemotherapy. Br J Cancer 98(1):86–90

Hesse M (2002) Alkaloids: nature's curse or blessing? Wiley, New York

Himes RH (1991) Interactions of the catharanthus (Vinca) alkaloids with tubulin and microtubules. Pharmacol Ther 51(2):257–267

Hisiger S, Jolicoeur M (2007) Analysis of *Catharanthus roseus* alkaloids by HPLC. Phytochem Rev 6(2-3):207–234

Ishikawa H, Colby DA, Seto S, Va P, Tam A, Kakei H, Rayl TJ, Hwang I, Boger DL (2009) Total synthesis of vinblastine, vincristine, related natural products, and key structural analogues. J Am Chem Soc 131(13):4904–4916

Isono M, Sato A, Asano T (2016) A case of long-term survival of advanced paratesticular rhabdomyosarcoma treated with a multimodal therapy including a combination of cyclophosphamide, vincristine, doxorubicin and dacarbazine. Urol Case Rep 7:3–6

Jacobs DI, Snoeijer W, Hallard D, Verpoorte R (2004) The Catharanthus alkaloids: pharmacognosy and biotechnology. Curr Med Chem 11(5):607–628

Jain V, Pareek A, Ratan Y, Singh N (2015) Standardized fruit extract of *Momordica charantia* L. protect against vincristine induced neuropathic pain in rats by modulating GABAergic action, antimitotoxic, NOS inhibition, anti-inflammatory and antioxidative activity. South Afr J Bot 97:123–132

Jaleel CA, Gopi R, Lakshmanan GA, Panneerselvam R (2006) Triadimefon induced changes in the antioxidant metabolism and ajmalicine production in *Catharanthus roseus* (L.) G. Don. Plant Sci 171(2):271–276

Jaleel CA, Gopi R, Sankar B, Manivannan P, Kishorekumar A, Sridharan R, Panneerselvam R (2007) Studies on germination, seedling vigour, lipid peroxidation and proline metabolism in *Catharanthus roseus* seedlings under salt stress. South Afr J Bot 73(2):190–195

Johnson IS, Armstrong JG, Gorman M, Burnett JP (1963) The vinca alkaloids: a new class of oncolytic agents. Cancer Res 23(8 Part 1):1390–1427

Kalaus G, Greiner I, Kajtar-Peredy M, Brlik J, Szabo L, Szantay C (1993) Synthesis of vinca alkaloids and related compounds. 63. A new synthetic pathway for preparing alkaloids and related compounds with the aspidosperma skeleton. Total syntheses of (.+−.)-vincadifformine,(.+−.)-tabersonine, and (.+−.)-oxotabersonine. J Org Chem 58(6):1434–1442

Kalidass C, Mohan VR, Daniel A (2009) Effect of auxin and cytokinin on vincristine production by callus cultures of *Catharanthus roseus* L. (apocynaceae). Trop Subtrop Agroecosyst 12(2):283–288

Kannan N, Mukunthan K, Balaji S (2011) A comparative study of morphology, reactivity and stability of synthesized silver nanoparticles using *Bacillus subtilis* and *Catharanthus roseus* (L.) G. Don. Colloid Surf B Biointerfaces 86(2):378–383

Kar A, Choudhary B, Bandyopadhyay N (2003) Comparative evaluation of hypoglycaemic activity of some Indian medicinal plants in alloxan diabetic rats. J Ethnopharmacol 84(1):105–108

Kato J, Li Y, Xiao K, Lee JS, Luo J, Tuscano JM, O'Donnell RT, Lam KS (2012) Disulfide cross-linked micelles for the targeted delivery of vincristine to B-cell lymphoma. Mol Pharmaceut 9(6):1727–1735

Kawasaki H, Taira N, Ichi T, Yohena T, Kawabata T, Ishikawa K (2014) Weekly chemotherapy with cisplatin, vincristine, doxorubicin, and etoposide followed by surgery for thymic carcinoma. Eur J Surg Oncol 40(9):1151–1155

Kellner F, Kim J, Clavijo BJ, Hamilton JP, Childs KL, Vaillancourt B, Cepela J, Habermann M, Steuernagel B, Clissold L (2015) Genome-guided investigation of plant natural product biosynthesis. Plant J 82(4):680–692

Kotakadi VS, Rao YS, Gaddam SA, Prasad T, Reddy AV, Gopal DS (2013) Simple and rapid biosynthesis of stable silver nanoparticles using dried leaves of *Catharanthus roseus*. Linn. G. Donn and its anti microbial activity. Colloid Surf B Biointerfaces 105:194–198

Krishnan S, Dhillon PK, Bhadelia A, Schurmann A, Basu P, Bhatla N, Birur P, Colaco R, Dey S, Grover S (2015) Report from a symposium on catalyzing primary and secondary prevention of cancer in India. Cancer Cause Contr 26(11):1671–1684

Ku C, Chung W-C, Chen L-L, Kuo C-H (2013) The complete plastid genome sequence of Madagascar periwinkle *Catharanthus roseus* (L.) G. Don: plastid genome evolution, molecular marker identification, and phylogenetic implications in asterids. PLoS One 8(6):e68518

Kufe DW, Pollock RE, Weichselbaum RR, Bast RC, Gansler TS, Holland JF, Frei E (2003) Holland-Frei cancer medicine, 6th edn. BC Decker, Hamilton, ON

Kumar A (2016) Vincristine and vinblastine: a review. IJMPS 6:23–30

Kumar A, Patil D, Rajamohanan PR, Ahmad A (2013) Isolation, purification and characterization of vinblastine and vincristine from endophytic fungus Fusarium oxysporum isolated from *Catharanthus roseus*. PLoS One 8(9):e71805

Kumari K, Gupta S (2013) Phytopotential of *Catharanthus roseus* L.(G.) Don. var."Rosea" and "Alba" against various pathogenic microbes in vitro. Int J Res Pure Appl Microbiol 3(3):77–82

Küppers R, Yahalom J, Josting A (2006) Advances in biology, diagnostics, and treatment of Hodgkin's disease. Biol Blood Marrow Transpl 12(1):66–76

Langone JJ, D'Onofrio MR, Van Vunakis H (1979) Radioimmunoassays for the vinca alkaloids, vinblastine and vincristine. Anal Biochem 95(1):214–221

Lans CA (2006) Ethnomedicines used in Trinidad and Tobago for urinary problems and diabetes mellitus. J Ethnobiol Ethnomed 2(1):1

Lee JW, Park S, Kim SY, Um SH, Moon EY (2016) Curcumin hampers the antitumor effect of vinblastine via the inhibition of microtubule dynamics and mitochondrial membrane potential in HeLa cervical cancer cells. Phytomedicine 23(7):705–713

Li X-T, He M-l, Z-y Z, Jiang Y, Cheng L (2015) The antitumor activity of PNA modified vinblastine cationic liposomes on Lewis lung tumor cells: in vitro and in vivo evaluation. Int J Pharmaceut 487(1):223–233

Louis DN, Perry A, Reifenberger G, von Deimling A, Figarella-Branger D, Cavenee WK, Ohgaki H, Wiestler OD, Kleihues P, Ellison DW (2016) The 2016 World Health Organization classification of tumors of the central nervous system: a summary. Acta Neuropathol 131(6): 803–820

Maeda H (2001) The enhanced permeability and retention (EPR) effect in tumor vasculature: the key role of tumor-selective macromolecular drug targeting. Adv Enzyme Regul 41(1): 189–207

Magnotta M (2007) Metabolic, biochemical and molecular profiling of *Catharanthus roseus* flower cultivars and transformed hairy roots for monoterpenoid indole alkaloid accumulation. MSc thesis

Malabadi RB, Chalannavar RK, Meti NT, Mulgund GS, Nataraja K, Kumar SV (2012) Synthesis of antimicrobial silver nanoparticles by callus cultures and *in vitro* derived plants of *Catharanthus roseus*. Res Pharm 2:18–31

McGuire SA, Gospe SM, Dahl G (1989) Acute vincristine neurotoxicity in the presence of hereditary motor and sensory neuropathy type I. Med Pediatr Oncol 17(5-6):520–523

Mendioro MS, Diaz MGQ, Alcantara MTB, Hilario OJ, Mateo P, Maghirang R (2005) Cytological Studies of Selected Medicinal Plants: *Euphorbia pulcherrima* Willd. ex Klotz., *Moringa oliefera* Lam., *Catharanthus roseus* (L.) Don. and *Chrysanthemum indicum* Linn. Philippine J Sci 134(1):31–37

Mohan S, Bustamam A, Ibrahim S, Al-Zubairi AS, Aspollah M, Abdullah R, Elhassan MM (2011) In vitro ultramorphological assessment of apoptosis on CEMss induced by linoleic acid-rich fraction from Typhonium flagelliforme Tuber. Evid Based Complement Altern Med 2011:421894

Mohanta B, Sudarshan M, Boruah M, Chakraborty A (2007) Potential of Vinca rosea extracts in modulating trace element profile. Biol Trace Element Res 117(1-3):139–151

Moore SC, Lee I-M, Weiderpass E, Campbell PT, Sampson JN, Kitahara CM, Keadle SK, Arem H, de Gonzalez AB, Hartge P (2016) Association of leisure-time physical activity with risk of 26 types of cancer in 1.44 million adults. JAMA Intern Med 176(6):816–825

Moreno PR, van der Heijden R, Verpoorte R (1995) Cell and tissue cultures of *Catharanthus roseus*: a literature survey. Plant Cell Tiss Org Cult 42(1):1–25

Moriyama B, Henning SA, Leung J, Falade-Nwulia O, Jarosinski P, Penzak SR, Walsh TJ (2012) Adverse interactions between antifungal azoles and vincristine: review and analysis of cases. Mycoses 55(4):290–297

Moudi M, Go R, Yien CYS, Nazre M (2013) Vinca alkaloids. Int J Prev Med 4(11):1231–1235

Mujib A, Ilah A, Aslam J, Fatima S, Siddiqui ZH, Maqsood M (2012) *Catharanthus roseus* alkaloids: application of biotechnology for improving yield. Plant Growth Regul 68(2):111–127

Muna S, Jeboori KHA, Ban A (2014) Immunopatholgical study on the effects of vincristine sulfate in white mice and rats. Glob J Biosci Biotechnol 3:56–62

Nagao T, Chaturvedi P, Shaha A, Sankaranarayanan R (2011) Prevention and early detection of head and neck squamous cell cancers. J Oncol 2011:318145

Nagaonkar D, Gaikwad S, Rai M (2015) *Catharanthus roseus* leaf extract-synthesized chitosan nanoparticles for controlled in vitro release of chloramphenicol and ketoconazole. Colloid Polym Sci 293(5):1465–1473

Nammi S, Boini MK, Lodagala SD, Behara RBS (2003) The juice of fresh leaves of *Catharanthus roseus* Linn. reduces blood glucose in normal and alloxan diabetic rabbits. BMC Complement Altern Med 3(1):1–4

Nattaporn P, Varaporn R, Tewin T (2014) Effect of Gloriosa superba and *Catharanthus roseus* extracts on IFN-γ-Induced Keratin 17 expression in HaCaT human keratinocytes. J Evid Based Complement Alternat Med 1(2014):1–11

Nayak B, Pereira LMP (2006) *Catharanthus roseus* flower extract has wound-healing activity in Sprague Dawley rats. BMC Complement Altern Med 6(1):41–47

Nejat N, Valdiani A, Cahill D, Tan Y-H, Maziah M, Abiri R (2015) Ornamental exterior versus therapeutic interior of Madagascar periwinkle (*Catharanthus roseus*): the two faces of a versatile herb. Sci World J 2015:982412

Neuss N, Gorman M, Hargrove W, Cone NJ, Biemann K, Buchi G, Manning R (1964) Vinca Alkaloids. XXI. 1 The structures of the oncolytic alkaloids vinblastine (VLB) and vincristine (VCR) 2. J Am Chem Soc 86(7):1440–1442

Noble CO, Guo Z, Hayes ME, Marks JD, Park JW, Benz CC, Kirpotin DB, Drummond DC (2009) Characterization of highly stable liposomal and immunoliposomal formulations of vincristine and vinblastine. Cancer Chemother Pharmacol 64(4):741–751

Nor HA, Rohanizah AR, Ishak M (2010) *Catharanthus roseus* aqueous extract is cytotoxic to Jurkat Leukaemic T-cells but induces the proliferation of normal peripheral blood mononuclear cells. Trop Life Sci Res 21(2):101–113

O'Brien S, Schiller G, Lister J, Damon L, Goldberg S, Aulitzky W, Ben-Yehuda D, Stock W, Coutre S, Douer D (2013) High-dose vincristine sulfate liposome injection for advanced, relapsed, and refractory adult Philadelphia chromosome–negative acute lymphoblastic leukemia. J Clin Oncol 31(6):676–683

Ohadoma S, Michael H (2011) Effects of co-administration of methanol leaf extract of *Catharanthus roseus* on the hypoglycemic activity of metformin and glibenclamide in rats. Asian Pacific J Trop Med 4(6):475–477

Okigbo R, Mmeka E (2006) An appraisal of phytomedicine in Africa. KMITL Sci Technol J 6(2):83–94

Okouneva T, Hill BT, Wilson L, Jordan MA (2003) The effects of vinflunine, vinorelbine, and vinblastine on centromere dynamics 1. Mol Cancer Ther 2(5):427–436

Palem PP, Kuriakose GC, Jayabaskaran C (2015) An endophytic fungus, *Talaromyces radicus*, isolated from *Catharanthus roseus*, produces vincristine and vinblastine, which induce apoptotic cell death. PLoS One 10(12):e0144476

Pan Q, Chen Y, Wang Q, Yuan F, Xing S, Tian Y, Zhao J, Sun X, Tang K (2010) Effect of plant growth regulators on the biosynthesis of vinblastine, vindoline and catharanthine in *Catharanthus roseus*. Plant Growth Regul 60(2):133–141

Pan ST, Li ZL, He ZX, Qiu JX, Zhou SF (2016) Molecular mechanisms for tumor resistance to chemotherapy. Clinic Exp Pharmacol Physiol 43:723–737

Panda S, Kar A, Ramamurthy V (2014) Cardioprotective effect of vincristine on isoproterenol-induced myocardial necrosis in rats. Eur J Pharmacol 723:451–458

Panneerselvam C, Murugan K, Kovendan K, Kumar PM, Ponarulselvam S, Amerasan D, Subramaniam J, Hwang JS (2013) Larvicidal efficacy of *Catharanthus roseus* Linn. (family: Apocynaceae) leaf extract and bacterial insecticide Bacillus thuringiensis against Anopheles stephensi Liston. Asian Pacific J Trop Med 6(11):847–853

Pasquier E, André N, Street J, Chougule A, Rekhi B, Ghosh J, Philip DS, Meurer M, MacKenzie KL, Kavallaris M (2016) Effective management of advanced angiosarcoma by the synergistic combination of propranolol and vinblastine-based metronomic chemotherapy: a bench to bedside study. EBioMedicine 6:87–95

Pattarachotanant N, Rakkhitawatthana V, Tencomnao T (2014) Effect of *Gloriosa superba* and *Catharanthus roseus* extracts on IFN-γ-induced keratin 17 expression in HaCaT human keratinocytes. Evid Based Complement Altern Med 2014:249367

PDQ Adult Treatment Editorial Board (2002) Gastric Cancer Treatment (PDQ®): Health Professional Version. 2016 Jun 30. PDQ Cancer Information Summaries [Internet]. National Cancer Institute, Bethesda. http://www.ncbi.nlm.nih.gov/books/NBK65766/ PubMed PMID: 26389209.

Peebles CA, Hong SB, Gibson SI, Shanks JV, San KY (2006) Effects of terpenoid precursor feeding on *Catharanthus roseus* hairy roots over-expressing the alpha or the alpha and beta subunits of anthranilate synthase. Biotechnol Bioeng 93(3):534–540

Pfreundschuh M, Kuhnt E, Trümper L, Österborg A, Trneny M, Shepherd L, Gill DS, Walewski J, Pettengell R, Jaeger U (2011) CHOP-like chemotherapy with or without rituximab in young patients with good-prognosis diffuse large-B-cell lymphoma: 6-year results of an open-label randomised study of the MabThera International Trial (MInT) Group. Lancet Oncol 12(11):1013–1022

Pharma E (2003) IV Busulfex (busulfan) injection product monograph. ESP Pharma, Edison, NJ

Piccinini M, Tazartes O, Mezzatesta C, Ricotti E, Bedino S, Grosso F, Dianzani U, Pier-angelo T, Mostert M, Musso A (2001) Proteasomes are a target of the anti-tumour drug vinblastine. Biochem J 356(3):835–841

Ponarulselvam S, Panneerselvam C, Murugan K, Aarthi N, Kalimuthu K, Thangamani S (2012) Synthesis of silver nanoparticles using leaves of *Catharanthus roseus* Linn. G. Don and their antiplasmodial activities. Asian Pacific J Trop Biomed 2(7):574–580

Potier P (1980) Synthesis of the antitumor dimeric indole alkaloids from Catharanthus species (vinblastine group). J Nat Prod 43(1):72–86

Prochazkova D, Sairam R, Srivastava G, Singh D (2001) Oxidative stress and antioxidant activity as the basis of senescence in maize leaves. Plant Sci 161(4):765–771

Ravindra B, Neelambika T, Meti GS, Mulgund K, Nataraja S, Vijaya K (2012) Synthesis of antimicrobial silver nanoparticles by callus cultures and in vitro derived plants of *Catharanthus roseus*. Res Pharm 2(6):18–31

Rowinsky E (2003) The vinca alkaloids, 6th edn. BC Decker, Hamilton, ON

Sain M, Sharma V (2013) *Catharanthus roseus* (an anti-cancerous drug yielding plant). A review of potential therapeutic properties. Int J Pure Appl Biosci 1(6):139–142

Salim V, De Luca V (2013) Towards complete elucidation of monoterpene indole alkaloid biosynthesis pathway: *Catharanthus roseus* as a pioneer system. Adv Bot Res 68:1–37

Schläger S, Dräger B (2016) Exploiting plant alkaloids. Curr Opin Biotechnol 37:155–164

Sertel S, Fu Y, Zu Y, Rebacz B, Konkimalla B, Plinkert PK, Krämer A, Gertsch J, Efferth T (2011) Molecular docking and pharmacogenomics of vinca alkaloids and their monomeric precursors, vindoline and catharanthine. Biochem Pharmacol 81(6):723–735

Shen Y, Zhang Z-J, Zhu M-D, Jiang B-C, Yang T, Gao Y-J (2015) Exogenous induction of HO-1 alleviates vincristine-induced neuropathic pain by reducing spinal glial activation in mice. Neurobiol Dis 79:100–110

Shimoda K, Kwon S, Utsuki A, Ohiwa S, Katsuragi H, Yonemoto N, Hamada H, Hamada H (2007) Glycosylation of capsaicin and 8-nordihydrocapsaicin by cultured cells of *Catharanthus roseus*. Phytochemistry 68(10):1391–1396

Siegel RL, Miller KD, Jemal A (2015) Cancer statistics, 2016. CA Cancer J Clin 66:7–30

Silverman JA, Deitcher SR (2013) Marqibo® (vincristine sulfate liposome injection) improves the pharmacokinetics and pharmacodynamics of vincristine. Cancer Chemother Pharmacol 71(3):555–564

Singh A, Agrawal M (2015) Effects of ambient and elevated CO_2 on growth, chlorophyll fluorescence, photosynthetic pigments, antioxidants, and secondary metabolites of *Catharanthus roseus* (L.) G Don. grown under three different soil N levels. Environ Sci Pollut Res 22(5):3936–3946

Singh SN, Vats P, Suri S, Shyam R, Kumria M, Ranganathan S, Sridharan K (2001) Effect of an antidiabetic extract of *Catharanthus roseus* on enzymic activities in streptozotocin induced diabetic rats. J Ethnopharmacol 76(3):269–277

Sisodiya PS (2013) Plant derived anticancer agents: a review. Int J Res Dev Pharm Life Sci 2:293–308
Siu K, Yau Y, Noor R, Fariza J (2011) Antiproliferative and phytochemical analyses of leaf extracts of ten Apocynaceae species. Pharmacognosy Res 3(2):100–106
Sivalingam G, Sriram N (2013) Anti-diabetic activity of Ribes nigrum fruit extract in alloxan induced diabetic rats. Int J Pharmaceut Sci Res 4(3):1196
Smith RA, Andrews K, Brooks D, DeSantis CE, Fedewa SA, Lortet-Tieulent J, Manassaram-Baptiste D, Brawley OW, Wender RC (2016) Cancer screening in the United States, 2016: a review of current American Cancer Society guidelines and current issues in cancer screening. CA Cancer J Clin 66:95–114
Stefanadis C, Toutouzas K, Synetos A, Tsioufis C, Karanasos A, Agrogiannis G, Stefanis L, Patsouris E, Tousoulis D (2013) Chemical denervation of the renal artery by vincristine in swine. A new catheter based technique. Int J Cardiol 167(2):421–425
Stewart BW, Wild CP (2014). World cancer report 2014. World Health Organization (WHO). http://site.ebrary.com/id/11014806
Sun Q-L, Zhao C-P, Wang T-Y, Hao X-B, Wang X-Y, Zhang X, Li Y-C (2015) Expression profile analysis of long non-coding RNA associated with vincristine resistance in colon cancer cells by next-generation sequencing. Gene 572(1):79–86
Svoboda GH (1961) Alkaloids of Vinca rosea (*Catharanthus roseus*). IX. Extraction and characterization of leurosidine and leurocristine. Lloydia 24:173–178
Svoboda GH, Neuss N, Gorman M (1959) Alkaloids of Vinca rosea Linn. (*Catharanthus roseus* G. Don.) V. Preparation and characterization of alkaloids. J Am Pharmaceut Assoc 48(11):659–666
Taylor WI, Farnsworth NR (1975) The Catharanthus alkaloids. Botany, chemistry, pharmacology, and clinical use. Marcel Dekker, New York Includes: Stearn WT. A synopsis of the genus *Catharanthus* (Apocynaceae)
Teale J, Clough JM, Marks V (1977) Radioimmunoassay of vinblastine and vincristine. Br J Clin Pharmacol 4(2):169–172
Tina K, Tanja D, Darja G, Ester H, Primož S, Gregor S, Maja Č (2013) Determination of vinblastine in tumour tissue with liquid chromatography-high resolution mass spectrometry. Talanta 116(11):887–893
Thakur V, Kush P, Pandey RS, Jain UK, Chandra R, Madan J (2016) Vincristine sulfate loaded dextran microspheres amalgamated with thermosensitive gel offered sustained release and enhanced cytotoxicity in THP-1, human leukemia cells: in vitro and in vivo study. Mater Sci Eng C 61:113–122
Tu Y, Cheng S, Zhang S, Sun H, Xu Z (2013) Vincristine induces cell cycle arrest and apoptosis in SH-SY5Y human neuroblastoma cells. Int J Mol Med 31(1):113–119
Tzin V, Galili G (2010) New insights into the shikimate and aromatic amino acids biosynthesis pathways in plants. Mol Plant 3(6):956–972
Urruticoechea A, Archer C, Assersohn L, Gregory R, Verrill M, Mendes R, Walsh G, Smith I, Johnston S (2005) Mitomycin C, vinblastine and cisplatin (MVP): an active and well-tolerated salvage regimen for advanced breast cancer. Br J Cancer 92(3):475–479
Van Moerkercke A, Steensma P, Schweizer F, Pollier J, Gariboldi I, Payne R, Bossche RV, Miettinen K, Espoz J, Purnama PC (2015) The bHLH transcription factor BIS1 controls the iridoid branch of the monoterpenoid indole alkaloid pathway in *Catharanthus roseus*. PNAS 112(26):8130–8135
Veau B, Courtois M, Oudin A, Chénieux J-C, Rideau M, Clastre M (2000) Cloning and expression of cDNAs encoding two enzymes of the MEP pathway in *Catharanthus roseus*. Biochim Biophys Acta 1517(1):159–163
Vega-Ávila E, Cano-Velasco JL, Alarcón-Aguilar FJ, Fajardo Ortíz MC, Almanza-Pérez JC, Román-Ramos R (2012) Hypoglycemic activity of aqueous extracts from *Catharanthus roseus*. Evid Based Complement Altern Med 2012:934258

Velayutham K, Rahuman AA, Rajakumar G, Santhoshkumar T, Marimuthu S, Jayaseelan C, Bagavan A, Kirthi AV, Kamaraj C, Zahir AA (2012) Evaluation of *Catharanthus roseus* leaf extract-mediated biosynthesis of titanium dioxide nanoparticles against Hippobosca maculata and Bovicola ovis. Parasitol Res 111(6):2329–2337

Verma A, Laakso I, Seppänen-Laakso T, Huhtikangas A, Riekkola M-L (2007) A simplified procedure for indole alkaloid extraction from *Catharanthus roseus* combined with a semi-synthetic production process for vinblastine. Molecules 12(7):1307–1315

Verma P, Mathur AK, Srivastava A, Mathur A (2012) Emerging trends in research on spatial and temporal organization of terpenoid indole alkaloid pathway in *Catharanthus roseus*: a literature update. Protoplasma 249(2):255–268

Vickers A (2004) Alternative cancer cures:"unproven" or "disproven"? CA Cancer J Clin 54(2):110–118

Warwick O, Alison R, Darte J (1961) Clinical experience with vinblastine sulfate. Can Med Assoc J 85(10):579

Waterhouse DN, Madden TD, Cullis PR, Bally MB, Mayer LD, Webb MS (2005) Preparation, characterization, and biological analysis of liposomal formulations of vincristine. Method Enzymol 391:40–57

Whitmer S, Canel C, Hallard D, Gonçalves C, Verpoorte R (1998) Influence of precursor availability on alkaloid accumulation by transgenic cell line of *Catharanthus roseus*. Plant Physiol 116(2):853–857

Winn AN, Ekwueme DU, Guy GP, Neumann PJ (2016) Cost-utility analysis of cancer prevention, treatment, and control: a systematic review. Am J Prev Med 50(2):241–248

Wright JR (2002) Almost famous: E. Clark Noble, the common thread in the discovery of insulin and vinblastine. Can Med Assoc J 167(12):1391–1396

Yunaiyama D, Saito K, Funatsu T, Nakayama H, Shimizu A, Ito H, Suzuki M, Akata S, Tokuuye K (2014) Effectiveness of CT assists for intraarterial chemotherapy: therapeutic outcome of chemoradiation for advanced head and neck cancer extending across the anatomical midline. Jpn J Radiol 32(12):708–715

Yung-Cheng S, Lee-Yung S, Chung-Chi H (2011) Rituximab-induced acute eosinophilic pneumonia with diffuse alveolar damage: a case report. J Exp Clin Med 3(6):314–317

Zárate R, Verpoorte R (2007) Strategies for the genetic modification of the medicinal plant *Catharanthus roseus* (L.) G. Don. Phytochem Rev 6(2–3):475–491

Zhang W-K, Xu J-K, Tian H-Y, Wang L, Zhang X-Q, Xiao X-Z, Li P, Ye W-C (2013) Two new vinblastine-type N-oxide alkaloids from *Catharanthus roseus*. Nat Prod Res 27(20): 1911–1916

In Vitro Studies, Biosynthesis of Secondary Metabolites and Pharmacological Utility of *Catharanthus roseus* (L.) G. Don.: A Review

P. Senbagalakshmi, M.V. Rao, and T. Senthil Kumar

Abstract *Catharanthus roseus* (L.) G. Don (formerly *Vinca rosea* L.) belongs to the family *Apocynaceae*, it has been used to control cancer, diabetes, malaria etc., by folklore and traditional medicinal herbalists of India over two millennia. It is one of the most studied legendary medicinal plants due to the presence of monoterpene indole alkaloids (MIAs) or terpenoid indole alkaloids (TIAs). The active constituents from above ground parts extract contain most well recognized invaluable anticancer drugs vinblastine and vincristine, some antifibrillic and hypertensive agents, whereas the root contains several bioactive drugs such as, ruabasine (ajmalicine), serpentine, vinceine, vincamine and reserpine. In this review, the botanical information, ethnobotanical significance, update in tissue culture, secondary metabolites biosynthesis, cellular compartmentation and their pharmacological properties discovered and proved in the past decades and their potential in further exploitation of *C. roseus* are discussed.

Keywords Anticancer • Antidiabetic • Botanical description • *Catharanthus roseus* • Cellular compartmentation • In vitro studies, Medicinal uses • Pharmacological activity • Secondary metabolites production

1 Introduction

Catharanthus roseus (L.) G. Don is an important medicinal plant, traditionally used to treat various diseases. This plant has a long history of use as medicine in ayurveda, unani, traditional and folk medicine in India and in several other countries (Table 1).

P. Senbagalakshmi • M.V. Rao (✉)
Department of Plant Science, Bharathidasan University, Tiruchirappalli, Tamil Nadu 620 024, India
e-mail: mvrao_456@yahoo.co.in

T. Senthil Kumar
Department of Plant Science, Bharathidasan University,
Tiruchirappalli, Tamil Nadu 620 024, India

Department of Industry University Collaboration, Bharathidasan University,
Tiruchirappalli, Tamil Nadu 620 024, India

M. Naeem et al. (eds.), *Catharanthus roseus*,
DOI 10.1007/978-3-319-51620-2_8

Table 1 Vernacular names and ethnobotanical significance of *Catharanthus roseus*

Country	Common names/language name	Traditional uses	Reference
India	Ainskati, Billaganneru, Nayantara, Nithiyakalyani, Periwinkle, Rattanjot, Sadaphul, Sadabahar,Sadapulli, Ushamanjairi, Sudhukattu Arali or Mallikai, Savakkottappacha, SavamNaari, Ushamalari, Baramassi, Ainskati, Ushamanjairi, Sadapushpa	The whole plant has been used for relieving muscle pain, depression of the central nervous system, and wasps stings	Don (1999)
		It is used in the cases of nose bleed, bleeding gums, mouth ulcers, and sore throats	Virmani et al. (1978)
		The root is considered tonic and stomachic	Anonymous (1985)
		The infusion of the leaves used in the treatment of menorrhagia, Toothache, memory loss, and blood circulation	Tiong et al. (2013)
		The hot water extract of dried entire plant is taken orally by human for cancer. Hot water extract of dried leaves is taken orally to Hodgkin's disease	El-Sayed and Cordell (1981)
		The root extract is taken orally for menorrhagia	Sain and Sharma (2013)
Hawaii	Kihapai	The plant was boiled to make a poultice to stop bleeding	Fransworth (1961)
Philippines	Atay-biya, Chichirica, Kantotan, Periwinkle, Tsitsirika	Hot water extract of dried root is taken orally as an emmenagogue	Virmani et al. (1978)
		Hot water extract of leaves is taken orally for diabetes mellitus, amenorrhea, and hot water extract of root is taken orally by pregnant women to produce abortion, as an effective emmenagogue, dysmenorrheal with scanty flow abortion	Zaguirre (1944) and Anonymous (1903)

(continued)

Table 1 (continued)

Country	Common names/language name	Traditional uses	Reference
West Indies	Brown man's fancy, Consumption brush, Old maid, Periwinkle, Pink flower, Ram goat rose, Red rose, Sailor's flower, white tulip	Hot water extract of leafy stems is taken orally for diabetes	Nguywen (1977)
Sri Lanka	Mini –Mal, Patti-Poo	Regulating blood sugar, reducing toothaches and high blood pressure, soothing insect bites, improving memory and circulation, healing wounds	Samara (2015)
Kew (London)	Rosy periwinkle	Folk treatment for diabetes, cancer	Hussey (1964)
Brazil	Boa-noite, Congorca	Decoction of dried root is taken orally for fevers and malaria. Hot water extract of root bark, dried entire plant is taken orally by human for diabetes mellitus	Brandao et al. (1985) and De Mello (1980)
Thailand	Phaengphoifarang, Phang-Puai-Fa-Rang	Hot water extract of dried entire plant is taken orally for diabetes	Yang et al. (1987)
Rodriguez Islands	Saponaire	An extract of the flower was commonly administered as eyewash for the eyes of infants	Duke (1985)
Cook Islands	Tiare- tupapu-kimo	Decoction of dried leaves to treat diabetes, hypertension, and cancer	Holdsworth (1990)
Marshall Islands	Raan nan raan	Medicinally it is used to treat kidney problems, diabetes, and arthritis	Taafari et al. (2006)
Japan	Nichinichi –So, SunnichiTahome	To cure stomach ulcer, sedative, and anticancer	Perry (1980)
Mexico	Ninfa	Root infusion used for diabetes vincaria	Ross (2003)
		Infusion of whole extract of the aerial parts is taken orally for diabetes	Andrade-Cetto and Heinrich (2005)
Kenya	Maua	Decoction of dried roots is taken orally for stomach problem	Morrison and West (1982)
		Hot water extract of dried leaves is taken orally for diabetes	Ross (2003)

(continued)

Table 1 (continued)

Country	Common names/language name	Traditional uses	Reference
England	Bright-eyes, Cape periwinkle, graveyard plant, Madagascar periwinkle, Old-maid, Old-maid-flower, Rosy periwinkle	Hot water extract of dried entire plant is taken orally as an antigalactagogue	Don (1999)
		Hot water extract of the leaf is taken orally by human adults for diabetes	Thompson (1976)
Dutch	Roze maagdenpalm	Take the plant for high blood pressure	Posthouwer et al. (2016)
Vietnam	Dua can	Hot water extract of arial parts is taken orally as a menstrual regulator; hot water extract of entire plant is taken orally by human adults as an antigalactogogue	Fransworth (1961) and Virmani et al. (1978)
France	Madagascar, Rose amère, Sorcerer's violet	The bitter and astringent leaves used as vomitive; roots used as purgative, vermifuge, depurative, hemostatic; and toothache remedy	Fransworth (1961)
	Pervenchede, Kihapai	Hot water extract of entire plant is taken as an antigalactagogue	Brun et al. (2001)
Ethiopia	Phlox	The plant is considered poisonous especially for children	Getahun (1976)
Guatemala	Chatilla	Hot water extract of dried leaves is taken orally for diabetes	Duke (1985)
Dominica	Caca poule, Cangrejera	Hot water extract of the leaf is taken orally by pregnant women to combat primary inertia in childbirth; the tea is used to treat diabetes	Hodge and Taylor (1956)
Cuba	Vicaria	Used for diabetes	Seaforth (2006)
Peru	Chavetilla	Hot water extract of dried entire plant is taken orally by human adults for cancer, heart disease, and leishmaniasis	Ramirez et al. (1988)

(continued)

Table 1 (continued)

Country	Common names/language name	Traditional uses	Reference
Venda	Liluvha	Take the plant for high blood pressure	Siegel (1976)
		Water extract of dried root is taken orally for venereal disease	Ross (2003)
Jamaica	Periwinkle	Hot water extract of dried leaves is taken orally for diabetes	Fransworth (1961)
		The flower extract was commonly administered as an eye wash for the eyes of infants	Duke (1985)
USA	Periwinkle, Bright eyes	Gargling with an infusion of the plant is considered to relieve pain from a sore throat, laryngitis, and chest complaints	Morton (1976)
	Bright eyes	In central and South America, it was used as a homemade cold remedy to ease luncogestion and inflammation	Aslam et al. (2013)
Caribbean	Churchyard blosson, Doctor dyette, Every day flower, Old maid, periwinkle, Pervanche de, Madagascar, Twelve o'clock, Guajaca	An extract from the flowers was used to make a solution to treat eye irritation and infections	Seaforth (2006)
	Periwinkle, Twelve o'clock, Madagascar	Folk cure or control of diabetes when taken as a tea used as sedative	Nejat et al. (2015)
Europe	Periwinkle	Decoction of dried leaves is taken orally for diabetes mellitus	Flatt et al. (1989); Swanston- Flatt et al. (1989)
Pakistan	Sadabahar	Hot water extract of dried ovules is taken orally for diabetes	Rahman (1982)
Indonesia	Indischemaagdepalm, Soldatenblom, Kembang Sari cina, Tapakdara, Tijna, KembangTembaga	Use the stalks and leaves for dysmenorrheal	Taafari et al. (2006)
Australia		Hot water extract of dried leaves is taken orally for menorrhagia, human adults for diabetes. Hot water extract of root bark is taken orally as a febrifuge	Bhandari and Mukerji (1959) and Webb (1984)

(continued)

Table 1 (continued)

Country	Common names/language name	Traditional uses	Reference
Myanmar	Thin-Baw-MA-Hnyo	Decoction of whole plant is orally taken for diabetes in a dosage of 300–400 mL. Expressed juice together with the same amount of honey is orally taken in a dosage of 20–40 mL for anemia. Crush fresh leaves are externally used on various ulcers, inflammation	Monograph (2016)
Spain	Chatas, Chula, Pervinca de Madagascar, Vincapervinca, Hierbadoncella	Traditionally used for diabetes	Seaforth (2006)
China	Chang Chun Hua	Hot water extract of the arial parts is taken orally as a menstrual regulator, malaria	Fransworth (1961)
		It was used as an astringent; diuretic and cough remedy	Aslam et al. (2013)
Mozam-bique	Madagascar periwinkle, Rosy periwinkle	Hot water extract of leaves is taken orally for diabetes and rheumatism, hot water extract of root is taken orally as a hypertensive and febrifuge	Amico (1977)
Africa	Kanniedood	Hot water extract of dried leaves is taken orally for menorrhagia and diabetes	Duke (1985)
Malaysia	Kemuntingcina	Plant decoction used for diabetes	Letchuman et al. (2010) and Kevin et al. (2012)
Taiwan		Decoction of dried entire plant is taken orally by human adults to treat diabetes mellitus and liver diseases	Anonymous (1903), Hsu and Cheng (1992) and Yang et al. (1987)

There are 3000 plant species used or recommended for cancer treatment in different parts of the world. Among these, *C. roseus* is the only species produces dimeric alkaloids Vincristine (VLC) and Vinblastine (VLB). These drugs are first time clinically proven for the cancer therapy and have high commercial value. Vinblastine or vinleucoplastin (vincaleukoblastine, VelbeR) was introduced in 1960 and is used in the treatment of Hodgkin's disease, non-Hodgkin lymphomas, testis carcinomas, breast cancer and chorio carcinomas. Vincristine or leucocristine also known as "miracle drug" is an oxidized form of vinblastine was introduced in 1963 and marketed as ONCOVIN. It is used against acute leukemia, Hodgkin's disease, non-Hodgkin lymphomas, rhabdomyosarcomas, wilm's tumors in children and breast cancer (Gueritte and Fahy 2005). Although vincristine and vinblastine are chemically related, minor differences in molecular level observed by Washington (1983). Some other alkaloid such as ajmalicine marketed as HydroserpanR, LamuranR was introduced in 1957 for the treatment of hypertension. Anhydrovinblastine has also been used as an antineoplastic agent in the treatment of cervical and lung cancer (Schmidt et al. 1998).

Several scientists during the period of 1949–2016 (67 years) have attempted and succeeded in the pharmacology, phytochemistry (isolation and structure elucidation) and plant tissue culture studies of *C. roseus* (Fig. 1). However, there are many references about the traditional uses of this plant, often they are contradictory. In the late 1940s *C. roseus* was used to develop cell culture technique using crown-gall tumors or *Agrobacterium* transformation (De Ropp 1947; Hildebrandt and Riker 1949). The antidiabetic use of this plant was published by Garcia (1954). This report invited American researchers to use *C. roseus* in their screens (Svoboda 1966). *C. roseus* was also studied separately by the Canadian group. In the interest of anti-cancer, alkaloids were discovered (reviewed by Noble 1990). The earlier reviews related to pharmacology and clinical applications (Kumar et al. 2015; Sutrisna 2015; Barik et al. 2016), phamacognosy, phytochemistry, biosynthetic pathway and its regulation (Mustafa and Verpoorte 2007; Pan et al. 2016) and biotechnology of *C. roseus* were made (Heijden et al. 2004; Pietrosiuk et al. 2007; Aslam et al. 2010; Nejat et al. 2015). Particularly, they focused one or two areas only and also they deals only with minimum literature and not discussed about the contradictory views,

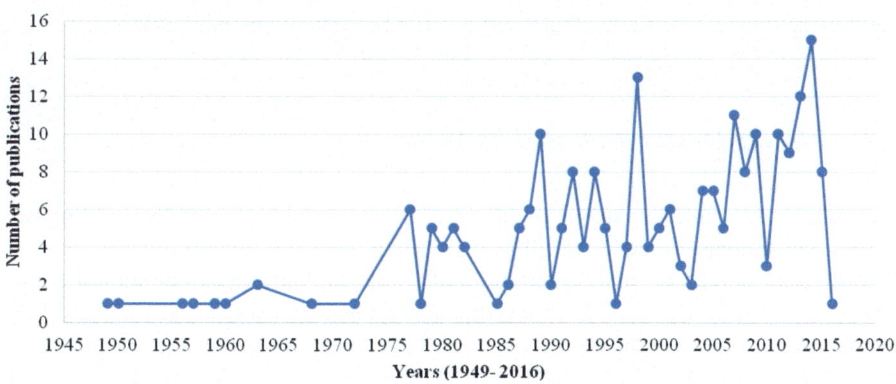

Fig. 1 Number of publications in *Catharanthus roseus* during the year 1949–2016

on biotechnological aspects and future prospects. The present review was compiled to give maximum information for future researchers.

Alternative in vitro technology would be beneficial in accelerating mass propagation, scale-up production of secondary metabolites and in conservation of plant population. Additionally, it require extensive studies to encounter numerous challenges including the selection and implementation of appropriate high-throughput screening bioassays, scale-up of active compounds are used by elicitors or precursors, biosynthesis of alkaloids and cellular compartmentation. At present, much is known about the biosynthesis of the alkaloids, despite the length and the complexity of the pathway. Very few gene codes for the enzymes in alkaloid biosynthesis pathway have identified for *C. roseus* (Fig. 2). The characterization of all the genes will be helpful to perform the synthesis of more alkaloids through metabolomic engineering.

2 Description of the Plant

Catharanthus roseus is fast growing evergreen tropical species in the family Apocynaceae. The species of the genus *Catharanthus* are brushy, erect, ever blooming perennial herb or subshrub, latex with flexible long branches with flower. The name *Catharanthus* comes from the Greek word *Katharos* means "pure flower" and *roseus* means red, rose, rosy, and their vernacular names are in Table 1. It is a wonderful garden plant and has thick glossy oppositely arranged leaves. The leaves were chewed to suppress the sensations of hunger and fatigue. Pretty five-petalled fragrant flowers, are usually solitary in mauve, red or white. Over the years, plenty of colourful cultivars have been developed. The genus *Catharanthus* comprises eight known species, in which seven species are endemic to Madagascar (*C. roseus, C. ovalis, C. tricophyllus, C. longifolius, C. coriaceus, C. lanceus, C. scitulus* and *C. pusillus*), three variants: (i) Rose purple flowered, (ii) White- flowered and (iii) White- flowered with rose purple spot in the centre, but horticulturists have developed more than 100 varieties that yield prolific blooms of splendid colors.

2.1 Flower Description and its Value

It also has a reputation as magic plant, the Germans called it the "flower of immortality", the Italians called it "the flower of death" and the French referred to it as "violet of the sorcerers" and "an emblem of friendship". European thought it could ward off evil spirits, it was also used to garland those awaiting execution, and laid on the biers of dead children. It was one of the plants believed to have power to expel evil spirits. Apuleius (1480) in his herbarium writes: "this wort is of good advantage for many purposes, first against devil sinks and demoniacal possessions

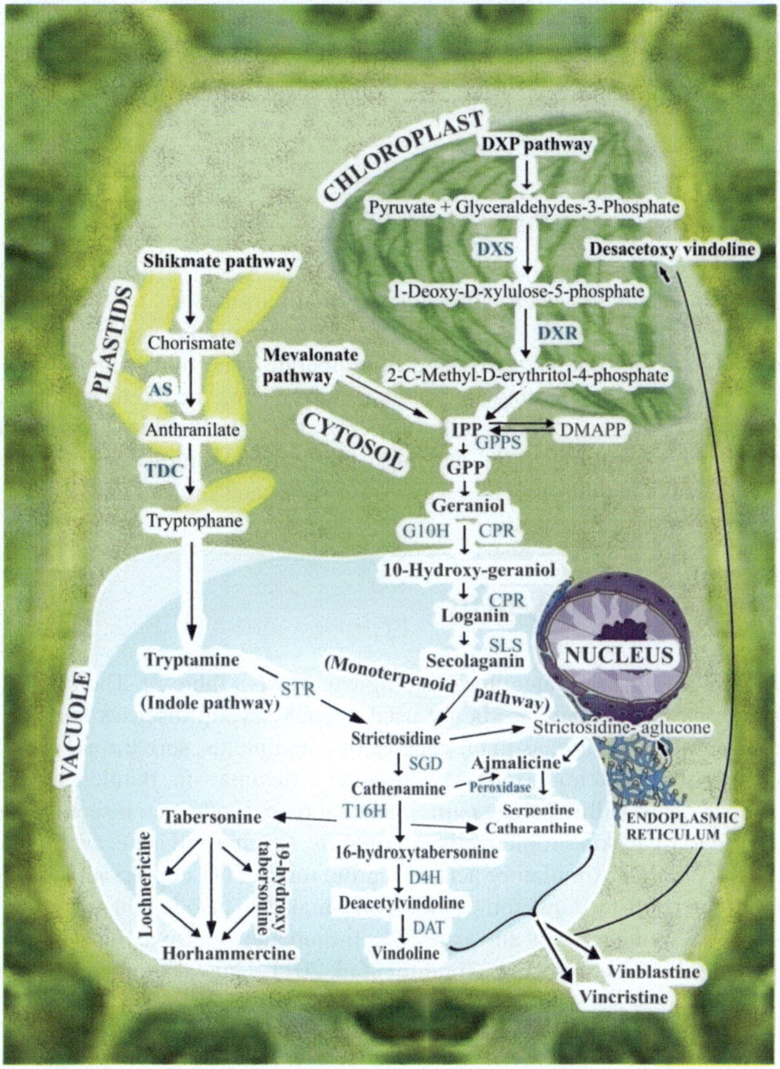

Fig. 2 Cellular compartmentation of monoterpenoid indole alkaloid (MIA) pathway. *G10H* geraniol 10-hydroxylase, *TDC* tryptophan decarboxylase, *DXS* 1-deoxy-0-xylulose-5-phosphate, *STR* strictosidine synthase, *SLS* secologanin synthase, *AS* Anthranilate synthase, *SGD* Strictosidine-β-D-glucosidase, *Tl6H* tabersonine 16-hydroxylase, *NMT* 2,3-dihydro-3-hydroxytabersonine-*N*-methyl transferase enzymes involved in alkaloid biosynthesis are associated with diverse subcellular compartments including the cytosol, vacuole, tonoplast membrane, endoplasmic reticulum, chloroplast stroma, thylakoid membranes, and perhaps unique "biosynthetic" or transport vesicles [with modification from Memelink et al. (2001), El-sayed and Verpoorte (2007) and Mahroug et al. (2007)]

against snakes, wild beast and against poisons and for various wishes and for envy and for terror". "The periwinkle is a great binder" said an old herbalist, and both Discorides and Galen commented it against flux. It was considered a good remedy for cramp (Kintzios and Barberaki 2003).

2.2 Cultivation and Propagation

Catharanthus roseus has been cultivated for dual purposes, as an ornamental for decorative and herbal medicine. It is self-fertile plant, propagated by seeds under the tropical and subtropical zones, and in temperate zone by cuttings (Sadowska et al. 1989). Cuttings containing 1–3 knots are taken from the top or middle part of the shoot soaked 24 h to remove milk juice, covered with the thin layer of sand or perlite under temperature 20 °C and high humidity conditionsfor rooting (Lata 2007). Plants obtained through this way flower faster and set fruit and seeds.

3 Pharmacological Activity

Catharantus roseus has many pharmacological effects (Table 2). The plant (roots, shoots and leaves) extracts are being used against several diseases such as diarrhoea, alzheimer's disease, asthma, coughs, throat ailments, sore throat, prevention of dementia, water retention (edema), dysentery, rheumatism, flatulence, tuberculosis, dyspepsia, tonsillitis, chest pain, intestinal pain, toothache, insect's sting, for external use to treat skin problems like dermatitis, eczema and acne, swelling (anti-inflammatory), brain stimulatory actions, cardio tonic, CNS depressant, anti-angiogenesis effects, anti-feedant, anti-sterility, anti-malarial, also potent anti-microbial, anti-oxidant activity, possess anti-cancer, anti-diabetic, cytotoxic, high blood pressure, hypolipidemic activity etc. (Ibrahim et al. 2011; Gajalakshmi et al. 2013).

3.1 Pharmacognosy and Medicinal Uses

The aerial parts of the plant contain 0.2–1.0% alkaloids (Bruneton 1993). About 130 alkaloids have been isolated from *C. roseus*, of particular interest is a group of 20 dimeric alkaloids (Evans 1996). Svoboda (1966) has reported isolation of 74 alkaloids from mature plants and an additional 21 chemical compounds from immature plants. This research and the results thereof constitute a classic in the annals of pharmacognosy and phytochemistry as well as tissue culture (Table 3). Definitely it is assumed that research will be able to find out more suitable and specific drug plant having particular activity in specific season.

Table 2 Pharmacological activity of *Catharanthus roseus*

Source	Bioactive compounds	Activity	Reference
Ethanol extract of leaf			Khalil (2012)
All parts		Antimicrobial	Kumari and Gupta (2013)
Leaves		Antimicrobial	Patel and Ghosh (2010)
All parts (different solvents)		Antibacterial	Govindasamy and Srinivasan (2012)
Leaves, stems, and flowers (different solvents)		"	Chaman et al. (2013)
Whole plant		Antimicrobial	Wagay et al. (2013)
Entire plant	Phytocystatins (thiol protease inhibitors)	Antibacterial	Sharma et al. (2011)
Benzene extract of flower		Antibacterial	Rajas and Cuellar (1981)
Ethanol extract of leaves		"	Ross et al. (1980)
Root	Total alkaloids	"	Chopra et al. (1959)
Water extract of entire plant		"	Negi and Bhatia (1956)
Water extract of callus		Antiviral activity	Misawa (1977)
Leaves	Selected alkaloids	Antiviral	Farmsworth et al. (1968)
Acetone and water extract of aerial parts		Antifungal	Kubas (1972)
Ethanol extract of leaves		"	Balaabirami and Patharajan (2012)
Hot water extract of dried leaves		"	Rai and Upadhyay (1988)
Hot water extract of dried stem		"	Chile et al. (1981)
Crude extract of whole plant		Antioxidant and antibacterial	Ethalsha and Retna (2014)
Crude extract of whole plant		Antibacterial	Ramya et al. (2008)
Leaves and root			Kulkarni and Ravindra (1988)
Chloroform extract of root		Antimalarial	Spencer et al. (1949)
Entire plant	Alkaloid fraction	Antidiuretic	Negi and Bhatia (1956)
Ethanol extract of leaf		Antidiarrheal	Hassan et al. (2011)
Whole plant extract		Anthelminthic activity	Hoskeri et al. (2011)

(continued)

Table 2 (continued)

Source	Bioactive compounds	Activity	Reference
Whole plant extract		Anthelminthic	Archana et al. (2016)
Methanol extract of leaf		Antioxidant	Kumar et al. (2012)
Ethanol extract of leaf		Antioxidant	Bhutkar and Bhise (2011)
Shoot	Alkaloid fraction	"	Rasool et al. (2011)
Leaves and flower		Larvicidal efficacy	Prasad et al. (2014)
Aqueous leaf extract		"	Alaguchamy and Jayakumararaj (2015)
N Hexane fraction		Larvicidal	Kuppusamy et al. (2009)
Entire plant	Total alkaloids	Anti-spermatogenetic	Murugavel and Akbarsha (1991)
		Spermatogenesis	Joshi and Ambaye (1968)
Methanol extract of leaf		Antidiabetic, wound Healing	Singh et al. (2014)
Chloromethane and methanol extract of whole plant		Antimicrobial, antidiabetic	Ibrahim et al. (2011)
Aqueous leaf extract		Reduce blood glucose level	Jafri et al. (2014)
Aqueous flower extract		Alloxan-induced diabetes	Natarajan et al. (2012)
Leaves (different extracts)		Antidiabetic and hypolipidemic	Islam et al. (2009)
Methanol leaf extract			Kevin et al. (2012)
Dichloromethane extract of leaves	Major alkaloids	Antidiabetic and antioxidant	Tiong et al. (2013)
Flower extract of methanol		Antidiabetic	Khan (2016)
Leaf powder		Antihyperglycemic	Rasineni et al. (2010)
The juice of fresh leaves		Reduces blood glucose	Nammi et al. (2003)
Aqueous extract of whole plant		Antidiabetic	Mostofa et al. (2007)
Organic and aqueous extracts of all plant parts		Hypoglycemic activity	Vega-Avila et al. (2012)
Dichloromethane and methanol extracts of leaf		Antihypoglycemic effect	Jayanthi et al. (2010)

(continued)

Table 2 (continued)

Source	Bioactive compounds	Activity	Reference
Aqueous extract of leaves		Hyperglycemia and hyperlipidemia (Insulin resistance)	Bisla et al. (2013)
Ethanolic extract of leaves		Hypotensive and hypolipidemic	Ara et al. (2009)
Ethanolic extract of whole Plant		Antidiabetic activity	Navitha et al. (2012)
Dichloromethane and methanol extracts of leaves, twig, flower		Antidiabetic	Singh et al. (2001)
Fresh juice of whole plant		Hypolipidemic	Patel et al. (2011)
Fresh juice of whole plant		Cholesterol, triglyceride, and lipoproteins levels in normal rats	Anita and Okokin (2005)
Ethanol extract of leaf		Hepatoprotective	Ahmed and Rao (2013)
Ethanol extract of leaf		Hypoglycemic	Mohan et al. (2015)
Aqueous extract of flower and leaves		Blood glucose lowering effect	Abbasi et al. (2014)
Leaf powder		Antidiabetic	Banakar et al. (2007)
Methanol extract of whole plant		Antidiabetic	Ahmed et al. (2010)
Aqueous extract of leaf		Hypoglycemic and biochemical remedies	Muralidharan (2015)
Ethanolic extract of leaf		Antidiabetic	Shaqha et al. (2015)
Cold methanol extract of leaf		"	Ohadoma and Michael (2011)
Dried leaves		Antihyperglycemic	Flatt et al. (1989)
Hot water extract of dried arial parts		"	Shorti et al. (1963)
Water extract of whole plant		"	Benjamin et al. (1994)
Ethanolic extract of leaf and flower		Free radical scavenging activity in type 2 diabetes mellitus	Asha et al. (2015)
Hot water extract of dried leaves		Hypoglycemic agent	Asthana and Misra (1979)
Hot water extract of dried leaves	Ajmalcine	Circulatory disorder	Misra et al. (1996)
Hot water extract of dried leaves	Ajmalcine, serpentine	"	Jaleel et al. (2006)

(continued)

Table 2 (continued)

Source	Bioactive compounds	Activity	Reference
Aqueous extract of leaf		Anticancer and free radical scavenging potency	Widowati et al. (2013)
Aqueous extract of leaf		Cytotoxicity	Ahmad et al. (2010)
Stem and leaves		Leukemia in children	Sain and Sharma (2013)
Leaves		Hodgkin's diseases	Widowati et al. (2013)
Leaves		Lymphocytic leukemia	Mukhapadhyay and Cordell (1981)
Leaves		Antitumor	Ruskin and Aruna (2014)
Methanol extract of leaf		Antitumor	Sayeed et al. (2014)
Methanol extract of leaf		Cytotoxic	Siddiqui et al. (2010)
Ethanolic extract leaves		Antimitotic	El-Merzabani et al. (1979)
Ethanolic extract leaves		Anti-inflammatory	Chattopadhyay et al. (1992)
Methanol and water dried leaves		Anti-fertility	Anonymous (1979)
Dried leaves	Alkaloid fraction	Cytotoxic activity	El-Sayed and Cordell (1981)
Aerial parts		Antitumor	Ruskin and Aruna (2014)
Leaves		Antioxidant and anticancer properties	Thingujam et al. (2015)
Aqueous leaf extracts		Antiangiogenesis	Wang et al. (2004)
Vinblastine		Hodgkin's disease	Schmeller and Wink (1998)
Vincristine		Leukemia	Schmeller and Wink (1998)
Catharanthamine		Antitumor	El-Sayed and Cordell (1981)
Flower		Wound healing	Nayak and Pereria (2006)

4 Alkaloid Production through In Vitro Culture

Plant biotechnology is an attractive and alternative approach to sustainable production of pharmaceutical compounds and it offers an opportunity to exploit cells, tissues, organs or entire organism by growing them in vitro and to genetically manipulate to get desired valuable compounds (Rao and Ravishankar 2002). Plant

Table 3 Bioactive compounds and pharmacological activity of various plant parts of *Catharanthus roseus*

Plant part	Bioactive compounds	Medicinal property	Reference
Roots	Ajmaline	Cardiovascular disease, high blood pressure	Farnsworth et al. (1967)
	Catharanthine	Hypertensive,	Rijhwani and Shanks (1998)
	Raubasin	Antidiabetic, pain relieving	Elisabetsky and Costa-Campos (2006)
	Reserpine	Tranquilizers	Elisabetsky and Costa-Campos 2006
	Serpentine	Antipsychotic agent	Elisabetsky and Costa-Campos (2006)
	Lochnerinine	Cancer chemotherapy	Kai et al. (2012)
	Vindoline	Stimulates insulin secretion	Yao et al. (2013)
	Vincaleukoblastine	Antitumor	Aslam et al. (2013)
	Alstonine	Antipsychotic agents	Elisabetsky and Costa-Compos (2006)
	Leurosine	Antitumor	Aslam et al. (2013)
	Tabersonine	Anticancer	Farnsworth et al. 1967 and Laflamme et al. (2001)
	Coronaridine	–	Nejat et al. (2015)
	Horhammericine	–	–
	Echitovenine	Anticancer	Laflamme et al. (2001)
Leaves	Vinblastine	Antitumor, antiviral, antidiabetic	Farnsworth et al. (1967) and Cordell et al. (2001)
	Vincristine	Hypertensive, antiviral	Abraham and Farnsworth (1969)
	Vincamin	α-adrenergic blocker, anticancer	Cordell et al. (2001)
	Camptothecin	Anticancer	Cordell et al. (2001)
	Vindoline	Hypoglycemic activity, anticancer	Kutney et al. (1980) and Laflamme et al. (2001)
	Yohimbine	Antiviral, α-adrenergic blocker	Schmelzer and Fakim (2008)
	Leurosine	Antitumor	Aslam et al. (2013)
	Catharanthine	antineuroinflammatory agents	Manigandan et al. (2014)
	Vindolidine	Hypoglycemic activity	Tiong et al. (2013)
	Vindolicine	Hypoglycemic activity	Tiong et al. (2013)
	Vindolinine	Hypoglycemic activity	Tiong et al. (2013) and Schmelzer and Fakim (2008)
	Lochnerine	–	Nejat et al. (2015)
	Raubasine	Antidiabetic, pain relieving	Elisabetsky and Costa-Campos (2006)

(continued)

Table 3 (continued)

Plant part	Bioactive compounds	Medicinal property	Reference
Flower	Apparicine	Antimicrobial, Antiprotozoal, Antiviral activity (Active against Polio Virus III)	Ferrari et al. (1971)
	Catharanthine	Antiviral, antineuroinflammatory agents	Manigandan et al. (2014) and Svoboda and Blake (1975)
	Perivine	Hypotensive activity	Schmelzer and Fakim (2008)
	Leurocristine	Hypoglycemic activity	Schmelzer and Fakim (2008)
	Coronaridine		Nejat et al. (2015)
	Tetrahydroalstonine	Hypotensive activity	Schmelzer and Fakim (2008)
	Tabersonine	Anticancer	Farnsworth et al. (1967) and Laflamme et al. (2001)
	11-methoxy tabersonine		Nejat et al. (2015)
	Vindolidine	Hypoglycemic activity	Kutney et al. (1980)
	Vindoline	Hypoglycemic activity	Kutney et al. (1980)
	Vinblastine	Anticancer	Laflamme et al. (2001) and Cordell et al. (2001)
	Catharicine	–	–
	Carosine	Hypotensive activity	Schmelzer and Fakim (2008)
	Leurosine	Hypoglycemic activity, antitumor	Kutney et al. (1980) and Aslam et al. (2013)
	Lochnerine	Hypotensive activity	Schmelzer and Fakim (2008)
	Tricin (Flavones)	–	–
Seed	Tabersonine	Anticancer	Farnsworth et al. (1967) and Laflamme et al. (2001)
	Strictosidine	Antiviral activity, antidiabetic	Jossang et al. (1998)
	Vinsedicine	Hypertensive	
	Vinsedine	–	–
	Vingramine	–	–
	Vingramine	–	–
	Methylvingramine	–	–
Seedling	Tabersonine	Anticancer	Farnsworth et al. (1967) and Laflamme et al. (2001)
	Serpentine	Antiviral and antitumor activity	Svoboda and Blake (1975)
	Preakuammicine	–	–
	Periformyline	Hypoglycemic activity	Schmelzer and Fakim (2008)
	Vindoline	Hypoglycemic activity	Kutney et al. (1980)
	Deacetoxy-Vindoline	Anticancer	Laflamme et al. (2001)
	Vinblastine	Anticancer	Laflamme et al. (2001) and Cordell et al. (2001)

cell and tissue culture hold great promise for controlled production of myriad of useful secondary metabolites (Filova 2014). During the last 50 years, vinblastine (VLB) and vincristine (VCR) belonging to the group of terpenoid-indole alkaloids (TIAs) are significantly well recognized therapeutic agents have been used for the treatment and cure all types of cancers and antidiabetic, because of their unique mode of action and their effectiveness. Nevertheless, their occurrence in wild plants are very low (it takes about 900 kg of dried leaves to isolate 1 g of VLB) and they are present only as minor constituents of the complex mixture of about 130 alkaloids produced by this plant (Noble 1990; Julsing et al. 2006). From the early work in the late 1940s it was already shown that *C. roseus* (gall) tissue was relatively amenable to culture however, isolation and purification of this antitumor agents are not economically feasible. Hence, the valuable leading commercial secondary product (TIAs) having difficulties in continuous supply. These alkaloids having great demand and are excellent targets for production by biotechnological means. Therefore, the cell culture of this plant has been carried out extensively in many laboratories for the production of valuable secondary metabolites (Heijden et al. 1989; Hirata et al. 1994). India is the third largest producer of vinblastine and vincristine.

There are only a few reports available on in vitro propagation of *Catharanthus* species using different explants (Table 4). In the early 1980s it was shown that plant cells, as a suspension of single cells in a liquid medium, could quite readily be grown in bioreactors. Due to the inability of the cell suspension cultures to accumulate vindoline, various types of differentiated tissue cultures techniques including, shoot culture, foliar (organ) culture and hairy root culture have been applied to produce alkaloids (Miura et al. 1988; Doran 1997). The relation between vinblastine production and tissue differentiation was studied and the accumulation of VLB increased as seedlings matured, reaching a steady concentration when the plants became more than 3 months old. On an average, whole seedlings, young plants and mature plants contained 7, 11.5 and 12 μg/g dry weight VLB, respectively. In callus derived from various explants, VLB was detected. After induction of shoot formation, the VLB contents increased rapidly to similar levels of in vitro seedlings of the same age (Datta and Srivastava 1997). The production of leaf-specific bisindole alkaloids depends on the different developmental and physiological stages of the plant and these are obtained mostly from wild plants. Alternatively, there are several fundamental factors viz. media optimization, control of pH, temperature, light, aeration and plant growth regulators influence the accumulation of alkaloids production through in vitro techniques. In addition, other stress factors including both elicitors and precursors (abiogenic and/or biogenic) and, other techniques *viz.* bioreactors and immobilization are being employed to enhance the alkaloids production.

Table 4 In vitro studies on *Catharanthus roseus*

Explants	Growth response	Culture medium	Reference
Shoot tip and node	Micropropagation	NAA 1.0 mg/L BAP(A) 1.0 mg/L KN(B) 2.0 mg/L NAA(C) 0.5 mg/L BAP+1.0 mg/L NAA(D)	Alen et al. (1995)
Node		WPM containing 5 μm BA and 5 μm NAA	Swanberg and Dai (2008)
Axillary bud and shoot tip		MS medium with BAP (4.0 mg/L) + NAA (4.0 mg/L)	Bakrudeen et al. (2011)
Shoot tip and node		MS medium with BAP (0.5 mg/L) + NAA (1.0 mg/L)	Faheem et al. (2011)
Node		BAP and NAA, (1.0 mg/L)	Haq et al. (2013)
		MS medium BAP (1.0 mg/L)	Pandey et al. (2014)
Hypocotyls and leaves	Organogenesis	MS medium with BAP alone or along with NAA	Dhruva et al. (1977), Rohtas (2001), Ramavat et al. (1978), Abou-Mandour et al. (1979) and Yuan and Hu (1994)
Hypocotyl		MS medium with BAP (1.0 mg/L) + NAA (1.0 mg/L)	Singh et al. (2011)
Pre-plasmolyzed leaves		Half-strength MS medium with 7.0 mg/L 6-benzyladenine(BA) and 3.0 mg/L a-naphthalene acetic acid (NAA)	Verma and Mathur (2011)
Leaf and internode		WPM containing 5 mM BA and 5 mM NAA	Swanberg and Dai (2008)
Stem nodes and Meristem tips	Somatic embryogenesis	MS medium with either 2.5 μM TDZ or 5.3 μM NAA and 2.2 μM BA	Dhandapani et al. (2008)
Anthers		MS medium with 1.0 mg/L NAA and 0.1 mg/L KN	Kim et al. (2004)
Shoot tip explants		MS medium with 2.0 μM 2, 4-D and 5.0 μM triacontanol (TRIA)	Malabadi et al. (2009)
Hypocotyls		MS medium containing 2,4-D 1.0 mg/L	Aslam et al.(2014)
		MS medium with 150 mg/L- casein hydrolysate, 250 mg/L- L-proline, 30 g/L sucrose, and 3 g/L gelrite	Yuag et al. (2011)
		MS with 2, 4,-D 4.52 μM and BAP 2.22 μM added liquid medium	Mujib et al. (2014)
Leaf explants		½ MS + 6% sucrose 2, 4-D 0.5 mg/L and BA 1.0 mg/L	Verma et al. (2012a, b)

4.1 Factors Influencing TIAs Production

4.1.1 Media Compositions

The effect of carbon (C), inorganic nutrients phosphate (P), nitrogen (N) on TIAs production from *C. roseus* cell cultures have been extensively studied. Medium optimization by altering nutrients (organic and inorganic) and plant growth regulators at different stages of cultures helps in increasing alkaloid production (Knobloch and Berlin 1980; Zenk et al. 1977). In the first stage, rapid callus biomass is produced and in the second stage accumulation of alkaloids content is induced. Morris (1986a) and Zhao et al. (2001c, d) have devloped a protocol for large callus biomass and synthesis of alkaloids together. The type of carbon source is also an important factor for TIAs production. An increased concentration of sucrose (6%) improved the biomass accumulation and enhanced ajmalicine, serpentine and catharanthine production (Scragg et al. 1990; Verma et al. 2012a, b), in addition a low level of glucose or fructose were also observed to enhance ajmalicine or catharanthine production, respectively (Jung et al. 1992; Schlatmann et al. 1995). Both nitrogen and phosphate promote the alkaloid production (Knobloch and Berlin 1980; Kubota et al. 1989; Gulik et al. 1993). The effect of nitrogen is always dependent on the carbon availability of the cells which makes the carbon-to-nitrogen (C/N) ratio, it influences the alkaloid production. By the determination of cellular C/N ratio, Rho and Andre (1991) identified three distinct growth phases namely an active phase, accumulation phase and biomass decline phase (endogenous metabolism). Low concentration of nitrate, ammonia and high concentration of phosphate increased the ajmalicine production (Schlatmann et al. 1992). Cellular and tissue level differentiation might be responsible for the important sources of C, N, P concentration in the media and can facilitate the different alkaloid biosynthesis. Vitamins and organic components had a marginal effect on the production of TIAs (Mujib et al. 2012).

4.1.2 Light

Light is thought to influence enzyme induction and activation. Light significantly influenced the biosynthesis of vindoline and other alkaloids, as well as acidic and basic peroxidase activities. Light stimulates plastid development, peroxidase activity and promotes the concentration of vindoline, ajmalicine and serpentine production in cultured cells, leaves, seedlings and plants (Zhao et al. 2001b). Higher serpentine accumulation was reported in light induced chloroplast (Loyola-Vargas et al. 1992). 12 h light improved embryo maturation and germination process (Junaid et al. 2008), while16 h light proved to be very effective for faster embryo proliferation, multiple shoot culture and vindoline synthesis of *C. roseus* (Campos-Tamayo et al. 2008). Hirata et al. (1992) observed the dark photoperiods longer than 12 h stimulate shoot elongation and catharanthine production. It was also noticed that vindoline and catharanthine were enhanced under intense light (20–30 w/m^2).

The results of gene expression investigation have demonstrated that upregulation of TDC, D4H, and DAT has been observed in *C. roseus* cultures after light expression (De Luca et al. 1988; Ouwerkerk et al. 1999; Schroder et al. 1999; Vazquez-Flota et al. 2009; Zhao et al. 2001a; He et al. 2011; Liu et al. 2011).

4.1.3 Temperature and pH

The temperature and pH are important elements that influence in vitro culture growth and alkaloid metabolism in *C. roseus*. The biomass growth was found maximum at 35°C and highest production of serpentine at 25 °C (Morris 1986b). The influence of buffered media upon the growth and alkaloid productivity was studied in *C. roseus* hairy root culture as the buffers minimized the shifts of media pH. It was found that the specific yield of lochnericine was significantly lower, tabersonine yield was higher, while the specific yields of ajmalicine, serpentine, and horhammericine remained unchanged (Morgan and Shanks 2000). It was known that the cell-culture itself acidified the medium rapidly and caused differential response (Sakano et al. 1997). Low and higher values of pH were used to release intracellular alkaloids into the culture medium (Asada and Shuler 1989). Medium adjusted with pH 5.5 was very effective for serpentine biosynthesis (Doller et al. 1976; Moreno et al. 1995) and pH 8.3 for vinblastine synthesis (Verma et al. 2007).

4.1.4 Aeration

The gaseous atmosphere plays an important role in plant cell growth and the production of TIAs. High rates of aeration are often required for high biomass densities which can remove volatiles that are apparently important for some plants grown in culture. The availability of the O_2, CO_2 (Maurel and Parellieux 1985) and ethylene greatly enhanced ajmalicine accumulation (Yahia et al. 1998).

4.1.5 Effect of Plant Growth Regulators (PGRs)

Different explants have been tested on several basal media with or without plant growth regulators (PGRs) for obtaining callus of *C. roseus* (Akcam and Yurekli 1995; Moreno et al. 1995; Heijden et al. 2004; Mujib et al. 2012). The auxins were found the best regulators for callus proliferations and growth. Low auxin and higher cytokinin concentrations to be better for callus proliferation from leaf explants (Kodja et al. 1989; Verma et al. 2012a, b). Roots induction from the callus tissue of *C. roseus* was first reported by Dhruva et al. (1977). An externally applied auxins induced both abnormal development and tryptophan decarboxylase (TDC) activity in the radicles of *Catharanthus* seedlings and although auxins slightly delayed the light-mediated induction of the cotyledon-specific vindoline biosynthesis (Aerts et al. 1992). IAA application as foliar spray has recommended to obtain better biomass production and vincristine content (Muthulakshmi and Pandiyarajan 2013), while, 2,4-D was failed to increase the alkaloid content in cell cultures (Gantet et al. 1998). Cytokinins have a positive role on indole alkaloid accumulation in some lines of *C. roseus*. Addition of zeatin to a medium containing 2,4-D decreased tryptamine levels and increased the bioconversion of secologanin to ajmalicine.

Zeatin also enhanced the geraniol -10 hydroxylase activities and modified the TIAs pattern (Decendit et al. 1993). However, abscisic acid decreased the alkaloid accumulation (El-Sayed and Verpoorte 2002). Gibberellin treatment enhanced the total alkaloid accumulation in leaves, stems and roots (Srivastava and Srivastava 2007). Gibberellin application by soil drenching also resulted in profound increase in ajmalicine content (Jaleel et al. 2007). Gibberellin and cytokinin have antagonistic effect on TIA biosynthesis (Amini et al. 2009). Enhancement of catharanthine in the hairy roots and accumulation of vindoline in shoot cultures are observed after ethylene treatment. The various aspects of plant tissue culture of *Cathranthus roseus* are presented in Table 4.

4.1.6 Effect of Additives and Polyamines (PAs)

Nitrogenous compounds can increase both cell viability and as well productivity in cell cultures. The combined treatment of cyclodextrins and methyl jasmonate (MJ) induced alkaloid synthesis which resulted in enhancement of ajmalicine and catharanthine (Almago et al. 2014) and salicylic acid significantly increased in phenylalanine ammonialyase (PAL) activity, antioxidant enzymes and also enhanced malondialdehyde content as well as alkaloid accumulation in salt stressed condition (Misra et al. 2014). High salinity and drought conditions favor the production of vincristine since, high salinity and drought resulted in increased amount of amino acids serine, methionine, arginine and these are precursors of various polyamines (Mohamed et al. 2007). Polyamines regulate DNA synthesis and cell division in suspension cultures of *C. roseus*. Levels of PAs were found to increase markedly prior to the synthesis of DNA in the S phase and prior to the cytokinesis (Minocha et al. 1991). Putrescine and tryptophan enhanced the cell growth as well as total alkaloid content and foliar application of putrescine increased the endogenous growth hormones like GA_3, IAA, cytokinins and ABA (Iman et al. 2005). Putresine plays an important role in cell division (Minocha et al. 1990) and also suppressed *Catharanthus* cell proliferation by prevention of the progression of the cell cycle in synchronized culture (Maki et al. 1991).

4.2 Factors Influencing Enhanced TIAs Production: Elicitors and Precursors

Using of specific elicitors and precursors in accurate dosage, or their mixtures in different proportions, time of exposure, more precise manipulations could be used to probe the complex secondary metabolite pathways and their regulation to stimulate a targeted branch for enhancement in production of desired compound. There are different types of biogenic and abiogenic elicitors (signal molecules) or biosynthetic precursors in cell cultures frequently increased the yield. Biogenic elicitors include microbial filtrates while abiogenic elicitors comprise of simple inorganic and organic molecules.

4.2.1 Abiogenic (Exogenous and endogenous) Elicitors

Organic and inorganic chemical compounds increased the accumulation of catharanthine, serpentine and tryptamine in culture. Potassium chloride, sorbitol and sodium chloride treatment increased the content of ajmalicine production (Jaleel 2009). The combined elicitors of malate and sodium alginate resulted in the yield of 26 mg/L catharanthine and ajmalicine 41 mg/L in the cell cultures (Zhao et al. 2001e). Calcium flux is an essential component of several intracellular signaling processes. Accumulation of vincristine was observed after 20 days of culture in embryogenic suspended cells in response to calcium ($CaCl_2$) elicitation (Siddiqui and Mujib 2016). Wang et al. (2008) has reported that salt stress increases the alkaloid content in *C. roseus* seedlings where higher NaCl concentration (2.9 g/L) was found to increase the content of vindoline, catharanthine, vinblastine and vincristine. UV-B light induced the stimulation of threefold vindoline and twelve fold catharanthine was observed by Ramani and Jayabaskaran (2008). Vanadylsulphate stimulated intracellular accumulation of catharanthine and ajmalicine (Smith et al. 1987). Addition of vanadylsulphateand cariophyllene oxide increased the production of ajmalicine (500 μg/g), catharanthine (131.0 μg/g), and tryptamine levels (Tallevi and Dicosmo 1988; Saifullah 2011). Treatment with 250 mM mannitol and 4 g/L KCl yielded 42.3 and 33.6 mg/L of ajmalicine content respectively, which were about fourfold increase than the control. Higher yields of ajmalicine and serpentine have been reported in cell cultures of *C. roseus* upon addition of Betaine, malic acid, tetramethyl ammonium bromide and rare elements (Zhao et al. 2000a, b). The relationship between changes in cell morphology (more long cells) with the increased content of catharanthine was observed after tryptophan treatment. 50.96 μg/g amount dry weight of catharanthine content occurred on day 14 after treatment with 150 mg/L of tryptophan induced on 75% long cells (Pandiangan et al. 2013).

Liu et al. (2014) demonstrated that artemisinic acid as a novel elicitor to enhance the yield of TIAs by up-regulating the transcriptions enzymes (TDC, G10H, T16H, D4H, DAT) involved in the biosynthetic pathway of vinblastine in the suspension-cultured cells of *C. roseus.* H_2O_2 treatment alone could mimic the oxidative burst and NADPH pretreatment promoted elicitor- induced alkaloid production (Zhao et al. 2001f). Alginate acted as an endogenous elicitor and promotes ajmalicine production, due to the presence of carboxyl groups, oligomers and trans-4, 5-dihydroxy-2-cyclopenten-1-one (Akimoto et al. 1999; Aoyagi et al. 2006). The use of trans-cinnamic acid, an inhibitor of PAL activity also results in significant increase in the alkaloid production of *C. roseus* (Valluri 2009). Yeast extracts (YE) induced transcription of the biosynthetic genes STR and TDC encoding TIAs production in *Catharanthus* cells and alkalinization of the culture medium (Pauw et al. 2004; Moreno et al. 1995). An exogenous supplementation of jasmonate, α-linolenic acid or MJ precursors induced jasmonate biosynthetic pathway, is an integral part of the elicitor-triggered signal transduction pathway that results in the coordinate expression of the STR and TDC genes and that protein kinases act both upstream and downstream level (Aerts et al. 1994; Vazquez-Flota and De Luca 1998; Gantet et al. 1998; Rijhwani and Shanks 1998; Menke et al. 1999). Cell suspensions of *C. roseus* elicitated with MJ alters the precursor availability for TIA biosynthesis. MJ increased ajmalicine production by threefold, induced cultures were limited by terpenoid precursors (Lee-Parsons and Royce 2006). Lee-Parsons and Erturk (2005) investigated the interaction between Ca^{2+} and MJ in modulating defense responses by monitoring ajmalicine production in suspension cultures. They found that ajmalicine production in MJ-induced cultures depended on the intracellular Ca^{2+} concentration and a low extracellular Ca^{2+} concentration (3 mM) enhanced MJ-induced ajmalicine production.

4.2.2 Biogenic Elicitors

Induction of alkaloid biosynthesis has been reported in response to fungal elicitors. Mollers and Sarkar (1989) induced calluses from phytoplasma-infected stem tissues. These callus tissues differentiated into plants on the MS medium with 0.25 mg/L 6-benzyladeine (BA), 1.0 mg/L naphthalene acetic acid (NAA), and 10.0mg/L gentamicin. Different fungal cell wall fragments of *Aspergillus niger, Fusarium moniliforme* and *Trichoderma viride* were used to study the effects of elicitor dosage, exposure time, and age of subculture on ajmalicine accumulation. Maximum yield 166 μg g/L DW of ajmalicine was synthesized in 20 days old suspension culture treated with *T. viride*. A longer period of incubation of cell cultures with elicitors (*A. niger, F. moniliforme* and *T. viride*) adversely affected the ajmalicine synthesis (Namedo et al. 2002). Different culture filtrates of *Pythium aphanidermatum* and other species of *Chrysosporium palmorum, Eurotium rubrum, Micromucor isabellina* were added to cell suspensions of *C. roseus* to increase rapidly the production of indole alkaloids and phenolics (Dicosmo and Towers 1984; Asada and Shuler 1989).

4.2.3 Combined Effect of Elicitors (Biogenic and Abiogenic)

A synergistic effect on alkaloid accumulation was observed in *C. roseus* cell cultures when treated with some combined elicitors of fungal preparations and chemicals. The combination of tetramethylammonium bromide and *A. niger* mycelial homogenate induced ajmalicine and catharanthine production (Zhao et al. 2001e). Mannitol and *A. niger* enhanced the biosynthesis of indole alkaloids of *Catharanthus roseus* in leaf, stem and root derived cell culture (Taha et al. 2009). He et al. (2011) suggested that the combined effect of light and MJ induced on the transcription of biosynthetic genes as well as promote the accumulation of vindoline in cell suspension culture. Elicitors also induce many intracellular events, including cytoplasmic calcium concentration, ion transport, production of reactive oxygen species and protein phosphorylation. The induction of JA biosynthesis by the combination of the above-mentioned events is still largely unknown (Zhao et al. 2005).

4.2.4 Precursors

Feeding with specific biosynthetic precursors or metabolites either individual or combined effect has proved to be a successful strategy to increase the levels of TIAs using cultures. Specifically, tryptophane, tryptamine, geraniol 10-hydroxygeraniol, loganin, loganic acid, succinic acid, and secologanin have been can act as precursors most extensively added (Bordelius et al. 1979; Moreno et al. 1993; Whitmer et al. 1998; Morgan and Shanks 2000; Zhao et al. 2001b; El-Sayed and Verpoorte 2002; El-Sayed et al. 2004; Kumar et al. 2015). *C. roseus* cell suspension cultures fed with stemmadenine resulted in the accumulation of catharanthine, tabersonine and condylocarpine (El-Sayed et al. 2004).

4.3 Techniques involved in TIAs Production

4.3.1 Hairy Root Culture

Hairy root culture is a unique system, often used for root specific TIAs production (Toivonen et al. 1989). In the late exponential phase, hairy root cultures of *C. roseus* were elicited with pectinase and JA, the selective effects on indole alkaloid yields were observed upon addition of both elicitors at different concentration and different time exposure. 150% of tabersonine specific yield was observed upon addition of 72 units of pectinase after 48 h but, the serpentine, tabersonine, and lochnericine levels was decreased when compared to control. Jasmonic acid addition caused an increase in the specific yields of ajmalicine (80%), serpentine (60%), lochnericine (150%), and horhammericine (500%) in dosage studies. Tabersonine, the likely precursor of lochnericine and horhammericine, decreased at lower levels of JA and then increased with increasing JA concentration. Transient studies showed that lochnericine and tabersonine levels go through a maximum, then decrease back to control levels and reduce below control levels, respectively. The yields of ajmalicine, serpentine and horhammericine increased continuously after the addition of JA (Rijhwani and Shanks 1998). In another case, the increased levels of tryptophane and tryptomine precursors, induced through TDC in *C. roseus* hairy roots (Whitmer et al. 2002; Hughes et al. 2004). In transgenic *Catharanthus* root, a significant increase of ajmalicine and catharanthine was noticed by Batra et al. (2005). Other groups used various types of bioreactors and fermenters to improve the growth of hairy roots for better production of secondary metabolites. Nearly 3600 kg per annum catharanthus roots need in the world market for ajmalicine production, to overcome the availability of these roots produced by this conventional method.

4.3.2 Bioreactor Application

Terpenoid indole Alkaloid production has mainly been conducted on cell suspension culture in which a rotary shaker is desirable but this alkaloid biosynthesis is extremely low, which prevents their industrial production. Shake flask studies with *C. roseus* demonstrated that alkaloid production commenced only after growth had slowed or ceased. Several researchers have looked for alternative sources and strategies to produce TIAs in high amounts. Paynee et al. (1988) have suggested the use of bioreactors in secondary metabolites production in plant cell culture of *Catharanthus*. For mass scale production larger size culture vessels, fermenters and bioreactors are essential (Schlatmann et al. 1994). To obtain high alkaloid productivities for extended period developed from shake-flasks to readily scale up process for both immobilized and suspended cell systems through bioreactor. Zenk et al. (1977), Wagner and Vogelmann (1977) first time used air lift bioreactors for alkaloid production in *Catharanthus*, while Doller et al. (1976) and Hegarty et al. (1986) have observed the inhibitions effect at high aeration rates on culture growth in airlift bioreactors. Zhao and Verpoorte (2007) proved that *C. roseus* cells can be cultivated in bioreactors. Various types and volume of bioreactors and complex fermenters were tried to improve the better biomass production and accumulation of secondary metabolites from hairy-roots (Ten Hoopen et al. 1994; Nuutila 1994). Various design and types of fermentors have been used for providing enhanced oxygen level. Based on the oxygen requirements of the cells and the oxygen transfer capabilities, agitated bubble column, established conditions for growth and production dynamics, Carboy system, roller flask, V-shaped reactors, stirred-jar fermentors, airlift bioreactor were commonly employed (Moreno et al. 1995; Heijden et al. 2004). Another group, cultured hairy roots on a large scale in bioreactors of up to 14L under these conditions, ajmalicine and catharanthine were produced (Moreno et al. 1999). The biomass growth could easily be improved by using a variety of bioreactors (Schlatmann et al. 1994). Moreno et al. (1995) reported the simple fermentors were well suitable for maintenance of cell culture growth and alkaloid formation. Taha et al. (2014) studied the physical conditions such as pH and different aeration (0.5; 0.8 and 1.0 L/min) state affecting cell suspension biomass [615.35 g/run F/W and 50.76 g/run D/W] and relative percentage of vincristine 5.61 g/run D/W and vinblastine 14.52 (fold) were produced in stirred tank bioreactor. However, a bench-top bioreactor allowing continuous extraction of secondary metabolites was also designed for *C. roseus* cell suspensions (Valluri 2009). The bioreactors with low shear stress and an impeller speed of 100 rpm were most appropriate for the accumulation of alkaloids; however, higher impeller speed increased callus/suspension growth (Paynee et al. 1988). The effect of scaling-up from shake flask to bioreactor on biomass and ajmalicine production was increased by Ten Hoopen et al. (1994). Production of phenolic compound from 5-methyl tryptophan over producing cell suspensions of *Catharanthus*, the cell biomass yield was enhanced to 30-folds through bioreactor (Verma et al. 2013). The exploitation of medium to large culture vessel or bioreactor may make the process more efficient in getting large number of *Catharanthus* plant propagated from somatic embryos and obtained novel alkaloids from co culturing of *C. roseus* and *Rauwolfia serpentina* in shake flask and stirred tank bioreactor (Verma et al. 2012a, b; Mujib et al. 2014). The eco-friendly indigenous bioreactor technology is useful for the scale-up production of bioactive compounds and to protect the natural source.

4.3.3 Immobilization

The immobilization of plant cell has been suggested for better accumulation of secondary metabolites in *Catharanthus* sp. An immobilization process can maintain cells for longer period of time and generates extracellular accumulation of secondary metabolites that could cut down cost considerably through increased production (Moreno et al. 1995). Considering other potential applications, various immobilized systems such as alginate, agarose, agar, carrageen, polyurethane foam etc. were developed (Heijden et al. 1989). Agar and agarose were found to be effective for long term maintenance of cell in *C. roseus.* The alginate mediated immobilized cells retained respiratory and biosynthetic activity for extended period of time resulting in enhanced tryptamide, ajmalicine and serpentine accumulation. During the last few years, surface immobilization has been proposed using different types of matrices for large scale production of alkaloids (Facchini and Dicosmo 1990; Moreno et al. 1995). Cells immobilized with fiberglass accumulated reduced level of tryptamide, ajmalicine and catharanthine compared to suspended cells was observed (Facchini et al. 1988). Gel, matrices entrapment in polysaccharide beads, polyacrylamide sheet, polyester fiber mat with porous layer of SiO_2 were used in *Catharanthus* in which better accumulation of alkaloids was induced (Carturan et al. 1998). Protoplast immobilized with agarose gel produced cell wall accumulative enzymes (Mera et al. 2003).

Further research is in need of use coirpith, luffa sponge, cellulose fiber and cotton matrix entrapment by membrane barrier which are cost effective. Cell growth in matrix is very slow as compared to free suspension which leads to a prolongation of the exponential phase and hence an enhanced production period. Accumulation of products within the cell, low yield of product and genetic instability of the cell line could be reduced or overcome by immobilization. The increased use of genetic tools and more in depth understanding of the molecular mechanisms for secondary metabolism in plants will facilitate researchers to get benefit by further improvement in immobilization technology.

4.4 *Metabolites or Biosynthetic Derivatives from Cell and Tissue Culture of* C. roseus

Metabolites are in fact the end products of a complex process comprising the involvement of several enzymes, genes, regulatory genes and (transport through) intra-and inter-cellular compartments. Beside alkaloids, various biosynthetic metabolites have been isolated from *C. roseus*, which include Tryptamine, N, N-dimethyltryptamine, N-acetyltryptamine, derivatives of tryptophan (Kurtz et al. 1980; Seitz et al. 1989), monoterpenoids, glucosides (loganin, secologanin, deoxyloganin, dehydrologanin), dehydroconiferyl glucosides, 2,3-dihydroxybenzoic acid act as an intermediates of phenolics (Knobloch et al. 1982; Lynn et al. 1987; Verpoorte et al. 1993) and anthocyanins -petunidine, malvidin, hirsutidin (Knobloch et al. 1982). Steroidal metabolites such as catasteron, brassinolides, campesterol, sitosterol, stigmasterol, cholesterol, isofucosterol, 24-methylene-cholesterol, brassinolidecatasterone, α –amyrin, β- amyrin, ursolic acid were screened by Park et al. (1989) and Duperon et al. (1992). Fatty acids like palmitic acid, oleic acid and linoleic acid were reported by Seitz et al. (1989).

The above described in vitro technique results have sufficient plotform for manipulating the higher yields of alkaloids. The disadvantage of these approachesis that only few precursors like tryptophane, tryptamine, geraniol 10-hydroxygeraniol, loganin, loganic acid, secologanin, catharanthine, tabersonine, condylocarpine and stemmadenine were premeditatedfor enhancement of alkaloid content apart from these other biosynthetic intermediates are available such as strictosidine, cathenamine, etc., which are not yet studied. Then the procedure for selecting superior cell lines are quite elaborated so can need extensive study in future for enhancement of alkaloid content.

5 Biosynthesis of the Catharanthus Alkaloids

Alkaloids are low molecular weight nitrogen containing substances with a prominent pharmacological activity. Over 12,000 alkaloids have been produced from different plants. Especially, TIAs comprise a family of 3000 compounds that includes the antineoplastic agents vinblastine and camptothecin, the antimalarial drug quinine, and the rat poison strychnine (Facchini, 2001). Amino acids such as phenylalanine, tyrosine, lysine, ornithine and tryptophan serve as starting points or backbone for alkaloid biosynthesis. The first study on the biosynthesis of alkaloids was performed at the end of the 1950s for *Catharanthus*. The alkaloids biosynthesis is a very complex process in *Catharanthus*, there are more than 130 alkaloids have been identified that share many biosynthetic steps.

Alkaloid biosynthesis in *C. roseus* involved in three stages: (Rischer et al. 2006; Liu et al. 2007; El-Sayed and Verpoorte 2007). I) The formation of monomeric alkaloids: tryptamine and secologanin (iridoid and secoiridoid) derived from the terpenoid (isoprenoid) biosynthesis. II) Secologanin condensation with tryptamine by strictosidine synthase (STR) to form the central intermediate strictosidine which is further converted to monomeric alkaloids like vindoline and catharanthineand. III) Formation of bisindole alkaloids: Vinblastine and vincristine synthesize from the coupling of catharanthine and vindoline (Fig. 2).

5.1 Biosynthesis of Isoprenoids

In plants, isoprenoid biosynthesis is a highly multifarious metabolic network, generates a wide array of diverse molecular compounds via parallel but compartmentally two distinct pathways in plants. One is mevalonate pathway which takes place in cytosol region another one is non-mevalonate pathway in plant chloroplast. These two pathways interact with each other. The outputs of both the mevalonate pathway synthesize two five-carbon precursors: the isoprene building block of isopentenyl diphosphate (IPP) and its isomer, dimethylallyl diphosphate (DMAPP).

5.1.1 Non Mevalonate Pathway or MEP or DXP Pathway

Higher plants, bacteria and protozoa have the ability to produce isoprenoids using an alternative pathway called the 2C-methyl-D-erythritol-4-phosphate pathway (MEP), the non-mevalonate pathway, the mevalonate-independent pathway, the 1-deoxy-D-xylulose-5-phosphate (DXP or DOXP) pathway. DXP synthase (DXS-EC 2.2.1.7) catalyzes the condensation of pyruvate and D-glyceraldehyde 3-phosphate to form 1-Deoxy-D-xylulose 5-phosphate (DXP). DXP reductoisomerase (DXR-EC1.1.1.267), which simultaneously catalyzes the intramolecular rearrangement and reduction of DXP to form 2-C-methyl-D-erythritol 4-phosphate, constitutes a key enzyme of an alternative mevalonate-independent pathway for IPP biosynthesis (Veau et al. 2000). Three dehydrations and one phosphorylation process would be needed for the conversion of MEP into IPP (Lichtenthaler 1999). Akhila (2007) described the isoprene (C5) monoterpenes (C10), diterpenes (C20), and tetraterpenes (C40), synthesized via DXP pathway. The formation of IPP was occur in triple sub cellular regions e.g., mitochondrial, peroxisome and plastids in *Catharanthus* (Guirimand et al. 2012).

5.1.2 The mevalonate pathway (MVA Pathway)

The cytosolic mevalonate pathway consists of eight enzymatically controlled reactions (Hunter 2007). The pathway is trigged by Acetyl Coenzyme A where classical intermediate mevalonic acid which leads to the formation of IPP and DMAPP. These further combine to form sesquiterpenes (C_{15}) and triterpenes (C_{30}) reviewed by Akhila (2007).

5.2 Monoterpene indole alkaloids (MIAs) or Terpenoid Indole Alkaloids (TIAs) Biosynthesis:

Vinblastine and vincristine are high-valuable bioactive compounds with a wide spectrum of pharmacological importance. MIAs produced in extremely low levels, leading to high market prices and poor availability. Their biotechnological production is hampered by the incomplete knowledge of their biosynthesis.

5.2.1 Biosynthesis of Tryptophan

The enzymes of tryptophan synthesis organized in several ways in different organisms due to different patterns of gene function, it occurs in non-green plastids. The shikimate pathway links metabolism of carbohydrates consists of seven enzymatic reactions whose end product is chorismate. It is a branch point metabolite for aromatic amino acids of tryptophan. Anthranilate, the first intermediate catalysed by the nucleus localized enzyme anthranilate synthase (AS—EC 4.1.3.27) from chorismate. The conversion of anthranilate to tryptophane synthesis encoded by four genes and consists of five enzyme catalyzed reactions (Tzin and Galli 2010; Zhang et al. 2015).

5.2.2 Biosynthesis of Secologanin

Secologanin is formed through several steps starting from the DMAPP condensation with one IPP in a head-to-tail fashion generating geranyl diphosphate (GPP), the precursor for the monoterpenes including iridoids such as secologanin (Verpoorte et al. 1997; Contin 1999; Courdavault et al. 2013) derived from MEP pathways. GPP is then converted to geraniol by geraniol synthase (GES - EC 3.1.7.11). Geraniol 10-hydroxylase (G10H- EC 1.14.13.152), an enzyme of the cytochrome P450 family, catalyzes geraniol to 10 hydroxy geraniol, are responsible for the conversion of the dialdehyde into 7-deoxyloganin, which is converted into loganin by deoxyloganin 7-hydroxylase (DL7H - EC 1.14.13.74). A secologanin synthase (SLS - EC 1.3.3.9) is an enzyme that catalyzes loganin into secolaganin. G10H and SLS involved in secologanin biosynthesis are also expressed in epidermal and laticifercells, as well as in the vascular cells of leaves (Irmler et al. 2000; Burlat et al. 2004; Mahroug et al. 2006).

5.2.3 Biosynthesis of Strictosidine

The conversion of monoterpene geraniol into secologanin is thought to occur in the vacuole from MEP pathway and tryptophane which is synthesized via the shikmate pathway, has to be transported to the cytosol by the action of tryptophan decarboxylase (TDC - EC 4.1.1.27) to yield tryptamine (De Luca and Cutler 1987; Vazquez-Flota et al. (1997). Tryptamine is formed by the enzyme tryptophane carboxylase (TDC) while the strictosidine synthase (STR - EC 4.3.3.2) helps in coupling of trptamine and secologanin to form strictosidine (STR) was studied by Kutchan (1993). STR was reported to be localized either to the cytoplasm or vacuole (De Luca and Cutler 1987; McKnight et al. 1991; Stevens et al. 1993). The biosynthesis of secologanin to strictosidine is encoded by the expression of eight genes (Miettinen et al. 2014). TDC and STR produced strictocidine and ajmalicine. Strictosidine can be classified as a phytoanticipin, as it is stored in the vacuole and then translocated outside the vacuole by unknown transport systems. The high activity of strictosidine glucosidase (SGD - EC:3.2.1.105) and additional enzymes bound in the membrane of the endoplasmic reticulum that allow it be glycosylated by a multimerized complex of SGD present in the nucleus as well as chloroplastic and cytoplasmic enzymes, involved in the glucoalkaloid-strictosidine and or strictosidine-derived TIAs, including ajmalicine, catharanthine, serpentine, and tabersonine (Barleben et al. 2007).

5.2.4 Formation of Catharanthine

Catharanthine biosynthesis is quite limited and accumulates as waxy exudates of leaves. It is derived from cathenamine regulated via strictosidine through the intermediate of geissoschizine and stemmadenine (Zhu et al. 2015). Cathenamine catalysed by tabersonine 16-hydroxylase produced taberosine, 16-hydroxytabersonie and catharanthine.

5.2.5 Formation of Vindoline

Strictosidine-β-D-glucosidase (SGD) enzyme playing an important role in monoterpenoid indole alkaloid biosynthesis (El-Sayed and Verpoorte 2007). The removal of the glucose moiety of strictosidine by SGD leads to an unstable, highly reactive aglucon, which is thought to be converted to 4, 21-dehydrogeissoschizine (Geerlings et al. 2000). The latter is believed to be converted by cathenamine synthase to cathenamine (El-Sayed and Verpoorte 2007; Ruffer et al. 1979) subsequently, the cathenamine is converted into tabersonine through several steps, which are not clearly understood. Finally, tabersonine is transformed into vindoline by a sequence of six main steps namely aromatic hydroxylation, O-methylation, hydration of the 2,3-double bond, N(1)-methylation, hydroxylation at position 4, and 4-O-acetylation (De luca et al. 1986; El-Sayed and Verpoorte 2007; Liscombe and O'Connor 2011), five intermediates involved are 16-hydroxytabersonine, 16-methoxytabersonine, 16-methoxy-2,3-dihydro-3-hydroxy-tabersonine, desacetoxyvindoline, and deacetylvindoline and five enzymes, they are tabersonine 16-hydroxylase (T16H - EC 1.14.13.73), O-methyltransferase (OMT- EC 2.1.1.94), N-methyltransferase (NMT- EC 2.1.1.99), desacetoxyvindoline-4-hydroxylase (D4H - EC1.14.11.20), and deacetylvindoline-4-O-acetyltransferase (DAT- EC 2.3.1.107) in the biosynthetic pathway (Zhu et al. 2015). Among these tabersonine and T16H, together with 16-OH OMT, are found predominantly in leaf epidermis and NMT in non-epidermal cells inside the leaf. Vindoline biosynthesis takes place from the base part of young leaves, including epidermal cells, palisade mesophyll cells, palisade-assisted idioblast cells, cross-connected laticifer cells and vascular cells, where several genes associated with vindoline biosynthesis have been localized (Murata and De Luca 2005).

5.2.6 Formation of Bisindole Alkaloids

The bisindole alkaloids vinblastine and vincristine are of great interest. Starting from the amino acid tryptophan and the monoterpenoid geraniol, the biosynthesis of vinblastine requires the participation of at least 35 intermediates, 30 enzymes, 30 biosynthetic, two regulatory genes and seven intra and intercellular compartments (Heijden et al. 2004). They are synthesized from the coupling of vindoline and catharanthine, the process is catalyzed by the major class of vacuolar III peroxidase enzyme called anhydrovinblastine synthase (AVLBS - EC 1.11.1.7) in leaves (Sottomayor et al. 1998; Costa et al. 2008). These monomeric alkaloids produced anhydrovinblastine by a peroxidase which is a reduction product. This is the true precursor to the other bisindole alkaloids vinblastine, vincristine and leurosine. AVLBS the enzyme catalysis the formation of vinblastine from α-3′,4′-anhydrovinblastine and also influenced the conversion of vinblastine to vincristine (Zhu et al. 2015). Vincristine synthesized via N-desmethylvinblastine, and 4-desacetoxy-16-desmethoxyvinblastine both of which a series of key structural analogues bearing systematic modifications in the vindoline subunit (Ishikawa et al. 2009). Currently, the regulatory genes and the (sub) cellular compartments are of particular interest, at which these processes take place, which require transport of both intermediates and enzymes (targeting). The produced (secondary) metabolites are stored or either directly available for their biological function.

6 Conclusion

Catharanthus roseus considered as a famous chemical factory for biosynthesis of a huge array of alkaloids and many of these chemicals possesses different pharmacological studies which continuously being used in the treatments of numeral diseases. Biosynthesis of alkaloids by in vitro cell culture has the advantages to manipulate the physiological (rapid growth, ease of precursor feeding, etc) and genetical process. During the biosynthesis of alkaloids of *C. roseus* various types of proteinaceous compounds (Heijden et al. 2004), enzymes and metabolites (Koul et al. 2013) have been reported.

Deciphering of the biosynthetic routes for vital TIAs are an exciting research field. Research on the enzymology, chemistry, sites of alkaloid synthesisand translocationcan deliver the rational engineering of alkaloids to improve their drug. The steps involved in the conversion of geraniol to secologanin are unknown and formation of strictosidine generated by two secologanin and tryptamine precursors by STR and SLS mechanism as yet uncharacterized. Moreover, the enzyme that catalyzes the conversion of vinblastine to vincristine is also not isolated. However, the enzyme that catalyzes the conversion of 16-methoxytabersonine to 16-methoxy-2,3-dihydro-3-hydroxy- tabersonine and vinblastine from α-3′,4′-anhydrovinblastine still unknown at the molecular level. Investigation with plant biotechnological approaches has suggested that production of some metabolite is developmentally regulated and this may account for study in the future if commercial exploitation is to occur. Many aspects of alkaloid biosynthesis, such as the elaborate subcellular compartmetation of enzymes and the intercellular translocation of pathway intermediates, reveal intriguing new variations in the complexity of plant metabolism. TIAs are present only in micro quantities in all parts of the plant, are highly poisonous and a great task for researchers to increase yield and reduce toxicity.

Detail studies are required to know the proper enzyme functions at various levels, product membrane permeability and adsorption for improvements towards achieving a viable economic production methodology. Emerging knowledge of the biochemistry, cell biology, metabolomics and the expansion of molecular biology will promote efforts to identify regulators associated with the development of cell types that can accommodate alkaloid pathways will also lead to exciting opportunities to engineer alkaloid metabolism in transgenic plants. The inherent novelties of the pathways and the socio-economic importance of the products are sure to encourage greater interest in alkaloid biosynthesis. Diabetes and cancer are the most challenging and threatened diseases for human health, *Catharanthus roseus* is a rich source of wonderful bioactive compounds used to cure both diseases and reduce the death rate.

Acknowledgements The authors acknowledge the University Grants Commission (UGC), New Delhi, for providing meritorious fellowship to the first author, and emeritus fellowship to the corresponding author. The authors are also thankful to Mr. S. Parthibhan and Mr. R.G. Baradwaj, Department of Plant Science, Bharathidasan University for their valuable suggestions in constructing this manuscript.

References

Abbasi SP, Abbasi ST, Khoharo K, Talpur HK, Siddiqui SM (2014) Blood glucose lowering effect of *Catharanthus roseus* in alloxan induced diabetic rats. Eur J Mol Biol Biochem 1:63–66

Abou-Mandour AA, Fischer S, Czygan F (1979) Regeneration of intact plants from haploid and diploid callus cells of *Catharanthus*. Z Pflanzen Physiol Bot 91:83–88

Abraham DJ, Farnsworth NR (1969) Structure elucidation and chemistry of Cathranthus alkaloids III: structure of Leurosine, an active anticancer alkaloid. J Pharm Sci 58:694–698

Aerts RJ, Alarco AM, Luca VD (1992) Auxins induce tryptophan decarboxylase activity in radicles of *Catharanthus* seedlings. Plant Physiol 100:1014–1019

Aerts RJ, Gisi D, Carolis DE, Luca DV, Baumann TW (1994) Methyl jasmonate vapor increases the developmentally controlled synthesis of alkaloids in *Catharanthus* and Cinchona seedlings. Plant J 5:635–643

Ahmad NH, Rahim RA, Mat I (2010) *Catharanthus roseus* aqueous extract is cytotoxic to jurkat leukaemic t-cells but induces the proliferation of normal peripheral blood mononuclear cells. Trop Life Sci Res 21:101–113

Ahmed MF, Rao AS (2013) Comparative hepatoprotective activities of selected Indian medicinal plants. Glob J Med Res Pharma Drug Discov Toxicol Med 13(2):15–21

Ahmed MF, Kazim SM, Ghori SS, Mehjabeen SR, Ahmed SM, Ibrahim AM (2010) Antidiabetic activity of *Vinca rosea* extracts in alloxan-induced diabetic rats. Int J Endocrinol 2010:841090

Akcam E, Yurekli AK (1995) Effect of different nutrient media and explant sources on callus induction of *Catharanthus roseus* L. (G). Don plants. Turk J Bot 19:569–572

Akhila A (2007) Metabolic engineering of biosynthetic pathways leading to isoprenoids: mono- and sesquiterpenes in plastids and cytosol. J Plant Interact 2:1742–9153

Akimoto C, Aoyagi H, Tanaka H (1999) Endogenous elicitor-like effect of alginate on physiological activities of plant cells. Appl Microbiol Biotechnol 52:429–436

Alaguchamy N, Jayakumararaj R (2015) Larvicidal effect of *Catharanthus roseus* L (G) Don. Aqueous extracts on the larvae of *Heliocoverpa armigera* (Hubner). Int J Life Sci Educ Res 3:10–14

Alen K, Mohan Jain S, Huhtikangas A (1995) Micro-propagation of *Catharanthus roseus* for Vinblastine and vincristine production. European Research Conferences on Plant Cell Biology and Biotechnological Applications Medicinal Plants: Biology, Chemistry and Drug Development. Dourden, pp 14–19

Almago L, Gutierrez J, Pedreno AM, Sottomayar M (2014) Synergetic and additives influence of cyclodextrins and methyljasmonate on the expression of the terpenoid indole alkaloid pathway genes and metabolites in *Catharanthus roseus* cell cultures. Plant Cell Tiss Org Cult 119:543–551

Amico A (1977) Medicinal plants of Southern Zambesia. Fitoterapia 48:101–139

Amini A, Glevarco G, Andreu F, Rideau M, Creche J (2009) Low levels of gibberellic acid control the biosynthesis of ajmalicine in *Catharanthus roseus* cell suspension cultures. Planta Med 75:187–191

Andrade-Cetto A, Heinrich M (2005) Mexican plants with hypoglycaemic effect used in the treatment of diabetes. J Ethnopharmacol 99:325–348

Anita BS, Okokin JE (2005) Effect of leaf juice of *Catharanthus roseus* Linn on cholesterol, triglyceride and lipoproteins levels in normal rats. Ind J Pharmacol 37:401–402

Anonymous (1903) Description of the Philippines. Part I, Bureau of Public Printing, Manila 24:234–236

Anonymous (1979) Antifertility studies on plants. Annual Report of the Director General–Indian Council of Medical Research 47:40–43

Anonymous (1985) Ayurvedic drug to fight cancer. Probe 24:234–236

Aoyagi H, Akimoto TC, Tanaka H (2006) Preparation of mixed alginate elicitors with high activity for the efficient production of 5′-phosphodiesterase by *Catharanthus roseus* cells. Biotechnol Lett 28:1567–1571

Ara N, Rashid M, Amran MS (2009) Comparison of hypotensive and hypolipidemic effects of *Catharanthus roseus* leaves extract with atenolol on adrenaline induced hypertensive rats. Pak J Pharm Sci 22:267–271

Archana H, Vasist R, Sharma RB, Gupta A (2016) Pharmacological reputation of Vinca plant. World J Pharm Pharm Sci 5:1602–1610

Asada M, Shuler ML (1989) Stimulation of ajmalicine production and extraction from *Catharanthus roseus*: effect of adsorption in situ, elicitors and alginate immobilization. Int J Appl Microbiol Biotechnol 30:475–481

Asha A, Vinita P, Hetal J (2015) In vivo comparative study of different variants of *Catharanthus roseus* (L.) G. Don. for free radical scavenging activity in type 2 diabetes mellitus. J de Afrikana 2:74–86

Aslam J, Khan SH, Siddiqui ZH, Fatima Z, Maqsood M, Bhat MA, Nasim SA, Ilah A, Ahmad IF, Khan SA, Mujib A, Sharma MP (2010) *Catharanthus roseus* (L.) G. Don. An important drug: it's applications and production. Int J Comp Pharm 4:1–16

Aslam J, Ajaz S, Nadim MM (2013) Pharmacognosy, phytochemistry pharmacological and biotechnological approaches of *Catharanthus roseus* (L.) G. Don. In: Shahid M, Shahzad A, Malik A, Sahai A (eds) Recent trends in biotechnology and therapeutic application of medicinal plants. Springer, New York, p 189

Aslam J, Mujib A, Sharma MP (2014) Somatic embryos in *Catharanthus roseus*: a scanning electron microscopic study. Natulae Scientia Biologicae 6:167–172

Asthana RB, Misra MK (1979) Orally effective hypoglycemic agent from *Vinca rosea*. Ind J Biochem Biophys 16:30–35

Bakrudeen AAA, Suba Shanthi G, Gouthaman T, Kavitha MS, Rao MV (2011) In vitro micropropagation of *Catharathus roseus*- an anticancer medicinal plant. Acta Bot Hung 53:197–209

Balaabirami S, Patharajan S (2012) In vitro antimicrobial and antifungal activity of *Catharanthus roseus* leaves extract against important pathogenic organisms. Int J Pharm Pharm Sci 4:487–490

Banakar V, Malagi U, Naik R (2007) Impact of periwinkle leaves (*Catharanthus roseus*) on management of diabetes mellitus, Karnataka. J Agric Sci 20:115–119

Barik K, Sao S, Parihar DK (2016) Phytochemical and pharmaceutical panorama of *Catharanthus roseus*. Indo Am J Pharm Sci 3:288–293

Barleben L, Panjikar S, Ruppert M, Koepke J, Stockigta J (2007) Molecular architecture of strictosidine glucosidase: the gateway to the biosynthesis of the monoterpenoid indole alkaloid family. Plant Cell Rep 19:2886–2897

Batra J, Dutta A, Singh D (2005) Growth and terpenoids indole alkaloid production in *Catharanthus roseus* hairy root clones in relation to left-and right- termini-linked Ri T-DNA gene integration. Plant Cell Rep 23:148–154

Benjamin BD, Kelkar SM, Pote MS (1994) *Catharanthus roseus* cell culture: growth, alkaloid synthesis and antidiabetic activity. Phytother Res 8:185–186

Bhandari PR, Mukerji B (1959) *Lochnera rosea* Linn Reichb. Gauhati Ayurvedic Coll Mag 8:1–4

Bhutkar MA, Bhise SB (2011) Comparative studies on antioxidant properties of *Catharanthus roseus* and *Catharanthus alba*. Int J Pharm Technol Res 3:1551–1556

Bisla G, Choudhary S, Singh E, Chaudhary V (2013) Hyperglycemia and hyperlipidemia mitigating impact of *Catharanthus roseus* (sadabahar) leaves aquous extract on type 2 diabetes mellitus subjects. Asian J Plant Sci Res 3:170–174

Bordelius P, Deus K, Mosbach K, Zenk MH (1979) Immobilized plant cells for the production of natural product. FEBS Lett 103:93–97

Brandao M, Botelho M, Krettli E (1985) Antimalarial experimental chemotherapy using natural products. Cienc Cult 37(7):1152–1163

Brun G, Bessiere JM, Franca DMG, David B, Mariotte AM (2001) Valatile components of *Catharanthus roseus* (L) G. Don (Apocyanaceae). Flav Frag J 16:116–119

Bruneton J (1993) Phytochemistry medicinal plants, 2nd edn (trans: Hatton CK). Lavoisier, Paris, pp 1016–1018

Burlat V, Oudin A, Courtois M, Rideau M, St-Pierre B (2004) Co-expression of three MEP pathway genes and geraniol 10-hydroxylase in internal phloem parenchyma of *Catharanthus roseus* implicates multicellular translocation of intermediates during the biosynthesis of monoterpene indole alkaloids and isoprenoid-derived primary metabolites. Plant J 38:131–141

Campos-Tamayo F, Hernandez-Domınguez E, Vazquez-Flota F (2008) Vindoline formation in shoot cultures of *Catharanthus roseus* is synchronously activated with morphogenesis through the last biosynthetic step. Ann Bot 102:409–415

Carturan G, Dal Monte R, Pressi G, Secondin S, Verza P (1998) Production of valuable drugs from plant cells immobilized by hybrid Sol-Gel SiO_2. J Sol Gel Sci Technol 13:273–276

Chaman S, Sharma G, Anil S, Reshi K (2013) Study of antimicrobial properties of *Catharanthus roseus* by agar well diffusion method. Int Res J Pharm Appl Sci 5:65–68

Chattopadhyay RR, Banerjee RN, Sarkar SK (1992) Anti inflammatory and acute toxicity studies with the leaves of *Vinca rosea* Linn. in experimental animals. Ind J Physiol Pharmacol 36:291–292

Chile SK, Saraf M, Barde AK (1981) Efficacy of *Vinca rosea* extract against human pathogenic strains of *Trichophyton rubrum* Sab. Ind Drugs Pharm Ind 16:31–33

Chopra IC, Jamwal KS, Chopra CL (1959) Preliminary pharmacological investigations of total alkaloids of *Lochnera rosea* (Rattonjot). Ind J Med Res 47:40

Contin A (1999) The biosynthesis of secologanin in *Catharanthus roseus* cell suspension cultures. Ph.D thesis, Leiden University, The Netherlands

Cordell GA, Quinn-Beattie ML, Farnsworth NR (2001) The potential of alkaloids in drug discovery. Phytother Res 15:183–205

Costa MR, Hilliou F, Duarte P, Pereira LG, Almeida I, Leech M, Memelink J, Barceloo AR, Sottomayor M (2008) Molecular cloning and characterization of a vacuolar class III peroxidase involved in the metabolism of anticancer alkaloids in *Catharanthus roseus*. Plant Physiol 146:403–417

Courdavault V, Clastre M, Simkin AJ, Guivarch NG (2013) Prenylated proteins are required for methy-jasmonate induced monoterpenoid indole alkaloids biosynthesis in *Catharanthus roseus*. In: Bach TJ, Rohmer M (eds) Isoprenoid synthesis in plants and microorganism: new concepts and expremental approaches. Springer Science + Business Media, New York, p 285

Datta A, Srivastava PS (1997) Variation in vinblastine production by *Catharanthus roseus*, during in vivo and in vitro differentiation. Phytochemistry 46:135–137

De Luca V, Cutler AJ (1987) Subcellular localization of enzymes involved in indole alkaloid biosynthesis in *Catharanthus roseus*. Plant Physiol 85:1099–1102

De Luca V, Balsevich J, Tyler RT, Eilert U, Panchuk BD, Kurz WGW (1986) Biosynthesis of indole alkaloids: developmental regulation of the biosynthetic pathway from tabersonine to vindoline in *Catharanthus roseus*. J Plant Physiol 125:147–156

De Luca V, Fernandez JA, Campbell D, Kurz WGW (1988) Developmental regulation of enzymes of indole alkaloid biosynthesis in *Catharanthus roseus*. Plant Physiol 86:47–450

De Mello JF (1980) Plants in traditional medicine in Brazil. J Ethnopharmacol 2:49–55

De Ropp RS (1947) The response of normal plant tissues and of crown gall tumor tissue to synthetic groth hormones. Am J Bot 34:53–63

Decendit A, Petit G, Andreu F, Doireau P, Merillon J, Merillion M, Rideau M (1993) Putatives sites of cytokinins action during their enhancing effect on indole alkaloid accumulation in periwinkle cell suspensions. Plant Cell Rep 12:710–712

Dhandapani MD, Kim H, Hong SB (2008) Efficient plant regeneration via somatic embryogenesis from the explants of *Catharathus rosues*. In Vitro Cell Dev Biol 44:18–25

Dhruva B, Ramakrishnan T, Vaidyanthan C (1977) Studies in *Catharanthus roseus* callus cultures, callus initiation and differentiation. Curr Sci 46:364–365

Dicosmo F, Towers GHN (1984) Stress and secondary metabolism in cultured plant cells. In: Timmermann BN, Steelink C, Leowus FA (eds) Recent advances stress. Plenum Press, New York, pp 97–175

Doller G, et al. (1976) Production von indole alkaloid in callus kulturen von *Catharanthus roseus*. Planta Med 30:14–20

Don G (1999) *Catharanthus roseus,* medicinal plants of the world. Human press, Totowa, NJ, pp 109–118

Doran PM (1997) Hairy roots, culture and application. Harwood Academic, Newark, NJ

Duke JA (1985) Handbook of medicinal herbs. CRC, Boca Raton, FL, p 256

Duperon P, Allais JP, Dupaix AL (1992) Composition of the sterol content of tonoplast and microsomal fractions isolated from *Catharanthus roseus* suspension cultured cells. Plant Physiol Biochem 30:495–498

Elisabetsky E, Costa-Campos L (2006) The alkaloid alstonine: a review of its pharmacological properties. J Evid Based Complement Altern Med 3:39–48

El-Merzabani MM, Aaser EAA, Duweini EAK (1979) A bioassay of antimitotic alkaloids of *Catharanthus roseus*. Planta Med 36:87–90

El-Sayed A, Cordell GA (1981) *Catharanthus* alkaloids. XXXIV. Catharanthamine, a new antitumor bisindole alkaloid from *Catharanthus roseus*. J Nat Prod 44:289–293
El-Sayed M, Verpoorte R (2002) Effect of phytohormones on growth and alkaloid accumulation by a *Catharanthus roseus* cell suspension cultures fed with alkaloid precursors tryptamine and loganin. Plant Cell Tiss Org Cult 68:265–270
El-Sayed M, Verpoorte R (2007) *Catharanthus* terpenoid indole alkaloids: biosynthesis and regulation. Phytochem Rev 6:277–305
El-Sayed M, Choi YH, Frederich M, Roytrakul S, Verpoorte R (2004) Alkaloid accumulation in *Catharanthus roseus* cell suspension cultures fed with stemmadenine. Biotechnol Lett 26:793–798
Ethalsha P, Retna M (2014) Evaluation of antioxidant potential and antibacterial activity of crude extracts *Catharanthus roseus*. Int J Pharm Sci Res 5:3490–3495
Evans WC (1996) Trease and Evans. Pharmacognosy, 14th edn. pp 421
Facchini PJ (2001) Alkaloid biosynthesis in plants: biochemistry, cell biology, molecular regulation, and metabolic engineering applications. Annu Rev Plant Physiol Plant Mol Biol 52:29–66
Facchini PJ, Dicosmo F (1990) Immobilization of cultured *Catharanthus roseus* cells using a fiberglass substratum. Appl Microbiol Biotechnol 33:36–42
Facchini PJ, Dicosmo F, Radvanyi LG, Giguere Y (1988) Adhesion of *Catharanthus roseus* cells to surfaces: effect of substrate hydrophobocity. Biotechnol Bioeng 32:935–938
Faheem M, Singh S, Tanwer BS, Khan M, Shahzad A (2011) *In vitro* regeneration of multiplication shoots in *Catharanthus roseus*—an important medicinal plant. Adv Appl Sci Res 2:208–213
Farmsworth NR, Svoboda GH, Blomster RN (1968) Antiviral acvtivity of selected *Catharanthus* antiviral activity of selected *Catharanthus* alkaloids. J Pharmacol Sci 57:2174–2175
Farnsworth NR, Blomster RN, Buckley JP (1967) *Catharanthus* alkaloids XIII: antineoplastic and hypotensive activity of alkaloid fractions and certain alkaloids from *Catharanthus lanceus*. J Pharm Sci 56:23–27
Ferrari G, Fervidi O, Ferrari M (1971) Phytochemistry. In: Arnold B (ed) The alkaloid. Academic Press, London, pp 24–28
Filova A (2014) Production of secondary metabolites in plant tissue cultures. Res J Agric Sci 46:236–245
Flatt SK, Day C, Flatt PR (1989) Glycaemia effects of traditional European plant treatments for diabetes studies in normal and streptozotocin diabetic mice. J Diabetes Res 10:69–73
Fransworth NR (1961) The pharmacognosy of the Periwinkles: Vinca and *Catharanthus*. Lioydia 24:105–138
Gajalakshmi S, Vijayalakshmi S, Rajeswari DV (2013) Pharmacological activities of *Catharanthus roseus*: a perspective. Int J Pharm Biosci 4:431–439
Garcia F (1954) Proceeding. 8th Pacific Science, Congress National Council Research, Philippines, p 182
Gantet P, Imbaut N, Theiersaul M, Doireau P (1998) Necessity of a octadecanoic pathway for alkaloid synthesis by *Catharanthus roseus* cell suspension cultured in an Auxin starved medium. Plant Cell Physiol 39:220–225
Geerlings A, Ibanez MM, Memelink J, Heijden R, Verpoorte R (2000) Molecular cloning and analysis of strictosidine beta-D-glucosidase, an enzyme in terpenoid indole alkaloid biosynthesis in *Catharanthus roseus*. J Biol Chem 275:3051–3056
Getahun A (1976) Some common medicinal and poisonous plants used in ethiopian folk medicine. Addis Ababa University, Addis Ababa, Ethiopia, p 216
Govindasamy C, Srinivasan R (2012) In vitro antibacterial activity and phytochemical analysis of *Catharanthus roseus* (Linn.) G. Don. Asian Pac J Trop Biomed 2:155–158
Gueritte F, Fahy J (2005) The vinca alkaloids. In: Cragg GM, Kingston DGI, Newman DJ (eds) Anticancer agents from natural products. Runner-Routledge Psychology Press, Taylor & Francis Group, Boca Raton, FL, pp 123–136
Guirimand G, Guihu RA, Phillips MA, Oudin A, Glevarec G, Melin C, Papon N, Clastre M, Giglioli-Guivarch N, St-Pierre B, Rodriguez-Concepcion M, Burlat V, Courdavault V (2012)

Triple subcellular targeting of isopentenyl diphosphate isomerases encoded by a single gene. Plant Signal Behav 7:1495–1497

Gulik VWM, Ten Hoopen HJG, Heijnen JJ (1993) A structured model describing carbon and phosphate limited growth of *Catharanthus roseus* plant cell suspensions in batch and chemostat cultures. Biotechnol Bioeng 41:771–780

Haq R, Naz S, Aslam F, Manzoor F (2013) Comparison of in vitro response of micropropagation and callogenesis of medicinal plant, *Vinca rosea* L. J Agric Res 51:9

Hassan K, Brenda A, Patrick V, Patrick O (2011) In vivo antidiarrheal activity of the ethanolic leaf extract of *Catharanthus roseus* Linn. (Apocynaceae) in Wistar rats. Afr J Pharm Pharmacol 5(15):1797–1800

He L, Yang L, Tan R, Zhao S, Hu Z (2011) Enhancement of vindoline production in suspension culture of the *Catharanthus roseus* cell line C20hi by light and methyl jasmonate elicitation. Anal Sci 27(12):1243–1248

Hegarty PK, Smart NJ, Scargg AH, Fowler MW (1986) The aeration of *Catharanthus roseus* L. G. Don suspension cultures in airleft bioreactors; the inhibitions effect at high aeration rates on culture growth. J Exp Bot 37:1911–1920

Heijden R, Verpoorte R, Ten Hoopen HJG (1989) Cell and tissue cultures of *Catharanthus roseus* (L.) G. Don: a literature survey. Plant Cell Tiss Org Cult 18:231

Heijden VDR, Jacobs D, Snoeijer W, Hallard D, Verpoorte R (2004) The *Catharanthus* alkaloids: pharmacognosy and biotechnology. Curr Med Chem 11(5):607–628

Hildebrandt AC, Riker AJ (1949) The influence of various carbon components on the growth of marigold, Paris daisy, periwinkle, sunflower and tobacco tissue in vitro. Am J Bot 36:74–85

Hirata K, Horiuchi M, Asada M, Ando T, Miyamota K, Miura Y (1992) Stimulation of dimeric alkaloid production by near ultraviolet light in multiple shoot cultures of *Catharanthus roseus*. J Ferment Bioeng 74:222–225

Hirata K, Miyamoto K, Miura Y (1994) III *Catharanthus roseus* L. (Periwinkle): production of vindoline and catharanthine. In: Bajaj YPS (ed) Biotechnology in agriculture and forestry. Medicinal and aromatic plants. VI. Springer, Berlin, pp 46–54

Hodge WH, Taylor D (1956) The ethanobotany of the Island Caribes of Dominica. WEBBIA 12:513–644

Holdsworth DK (1990) Traditional medicinal plants of Rarotonga, Cooks Islands Part I. Int J Crude Drug Res 28:209–218

Hoskeri H, Agarwal S, Jacob S, Chettri N, Bisoyi S, Tazeen A, Vedamurthy A, Krishna V (2011) In-vitro anthelminthic activity of *Catharanthus roseus* extract. Int J Pharma Sci Drug Res 3:211–213

Hsu FL, Cheng JT (1992) Investigation in rats of the antihyperglycaemic effect of plant extracts used in Taiwan for the treatment of diabetes mellitus. Phytother Res 6:108–111

Hughes EH, Hong SB, Gibson SI, Shanks JV, San KV (2004) Metabolic engineering of the indole pathway in *Catharanthus roseus* hairy roots and increased accumulation of tryptamine and serpentine. Metab Eng 6:268–276

Hunter WN (2007) The non-mevalonate pathway of isoprenoid precursor biosynthesis. J Biol Biochem 282:21573–21577

Hussey RL (1964) Eli Lilly and Company, Personal Communication. pp 153

Ibrahim M, Mehjabeen SS, Narsu ML (2011) Pharmacological evaluation of *Catharanthus roseus*. Int J Pharm Appl 2:165–173

Iman M, Talaat MA, Bekheta MMH (2005) Physiological response of periwinkle plants (*Catharanthus roseus* L.) to tryptophan and putrescine. Int J Agric Biol 7:210–213

Irmler S, Schroder G, St-Pierre B, Crouch NP, Hotze M, Sch-midt J, Strack D, Matern U, Schroder J (2000) Indolealkaloid biosynthesis in *Catharanthus roseus*: new activities and identification of cytochrome P450 CYP72A1 as secologanin synthase. Plant J 24:797–804

Ishikawa H, Colby DA, Seto S, Van P, Tam A, Kakei H, Rayl TJ, Hwang I, Boger DL (2009) Total synthesis of vinblastine, vincristine, related natural products, and key structural analogues. J Am Chem Soc 131:4904–4916

Islam MA, Akhtar MA, Khan MRI, Hossain MS, Alam MK, Wahed MII, Rahman BMA, Anisuzzaman SM, Shaheen SM, Ahmed M (2009) Antidiabetic and hypolipidemic effects of different fractions of *Catharanthus roseus.* (Lin.) Don normal and streptozotocin-induced diabetic rats. J Sci Res 1:334–344

Jafri SA, Rehman K, Qasim M, Kalsoom (2014) Effect of *Murraya koenigii, Catharanthus roseus* and *Psidium guajava* leaves extract on blood glucose in alloxan induced diabetic rats. Am J Biol Life Sci 2:1-5

Jaleel CA (2009) Changes in non-enzymatic antioxidants and ajmalicine production in *Catharanthus roseus* with different soil salinity regimes. Bot Res Int 2:01–06

Jaleel CA, Gopi R, Lakshmana AGM, Panneerselvam R (2006) Triadimefon induced changes in the antioxidant metabolism and ajmalicine production in *Catharanthus roseus* (L.) G. Don. Plant Sci 171:271–276

Jaleel CA, Manivannen P, Sankar B, Kishorekumar A, Sankari S, Panneerselvam R (2007) Paclobutrazol enhances photosynthesis and ajmalicine production in *Catharanthus roseus.* Process Biochem 42:1566–1570

Jayanthi M, Sowbala N, Rajalakshmi G, Kanagavalli U, Sivakumar V (2010) Study of anti-glycemic effect of *Catharanthus roseus* in alloxan induced diabetic rats. Int J Pharm Pharm Sci 2(4):114–116

Joshi MS, Ambaye RY (1968) Effect of alkaloids from *Vinca rosea* on spermatogenesis in male rats. Ind J Exp Biol 6:256–257

Jossang A, Fodor P, Bodo B (1998) A new structural class of bisindole alkaloids from the seeds of *Catharanthus roseus*: vingramine and methylvingramine. J Org Chem 63:7162–7167

Julsing MK, Koulman A, Woerdenbag HJ, Quax WJ, Kayser O (2006) Combinatorial biosynthesis of medicinal plant secondary metabolites. Biomol Eng 23:265–279

Junaid A, Mujib A, Fatima S, Sharma M (2008) Cultural conditions affect somatic embryogenesis in *Catharanthus roseus* L. (G.) Don. Plant Biotechnol Rep 2:179–189

Jung KH, Kwak SS, Kim SW, Lee H, Choi CY, Liu JR (1992) Improvement of catharanthine productivity in hairy root cultures of *C. roseus* by using monosaccharides as carbon sources. Biotechnol Lett 14:695–700

Kai G, Zhang A, Guo Y, Li L, Cui L, Luo X, Liu C, Xiao J (2012) Enhancing the production of tropane alkaloids in transgenic *Anisodus acutangulus* hairy root cultures by over-expressing tropinone reductase I and hyoscyamine-6b-hydroxylase. Mol Biosyst 8:2883–2890

Kevin LYW, Hussin AH, Zhari I, Chin JH (2012) Sub-acute oral toxicity study of methanol leaves extract of *Catharanthus roseus* in rats. J Acute Dis 1:38–41

Khalil A (2012) Antimicrobial activity of ethanol leaf extracts of *Catharanthus roseus* from Saudi Arabia. Second International Conference on Environment Science and Biotechnology. IPCBEE, vol 48. IACSIT Press, Singapore

Khan A (2016) A comparitive study of antidiabetic activity of *Catharanthus roseus* and *Catharanthus alba* flower extracts on alloxan induced diabetic rats. World Int J Pharm Pharm Sci 5(2):527–543

Kim SW, Dong S, Choi P, Jang R (2004) Plant regeneration from immature zygotic embryo-derived embryogenic calluses and cell suspension cultures of *Catharanthus roseus*. Plant Cell Tiss Org Cult 76:131–135

Kintzios SE, Barberaki MG (2003) Plant that fight cancer. CRC Press, Washington, DC, pp 33–54

Knobloch KH, Berlin J (1980) Influence of medium composition on the formation of secondary compounds in cell suspension cultures of *Catharanthus roseus* (L.) G. Don. Z Naturforsch 35:551–556

Knobloch KH, Bast G, Berlin J (1982) Medium and light induced formation of serpentine and anthocyanins in cell suspension cultures *Catharanthus roseus*. Phytochemistry 21:591–594

Kodja HP, Liu D, Merillon JM, Andreau F, Raideau M, Cheniex JC (1989) Stimulation par les cytokinins de I, accumulation d' alcaloides indoliques dans des suspensions cellularies de *Catharanthus roseus.* (L.) G. Don. Z Naturforsch 35c:551–556

Koul M, Lakra SN, Chandra R, Chandra S (2013) *Catharanthus roseus* and prospects of its endophytes: a new avenue for production of bioactive metabolites. Int J Pharm Pharm Sci 4:2705–2716

Kubas J (1972) Investigation on known or potential antitumoral plants by means of microbiological test. Part III. Biological activity of some cultivated plant species in *Neurospora crassa* test. Acta Biol Cracov Ser Bot 15:87–100
Kubota K, Li XN, Ashihara H (1989) The short-term effects of inorganic phosphate on the levels of metabolites in suspension-cultured *Catharanthus roseus* cells. Z Naturforsch 44c:802–806
Kulkarni R, Ravindra N (1988) Resistance to *Pythium aphanidermatum* in diploid and induced autotetraploids of *Catharanthus roseus*. Planta Med 54:356–359
Kumar A, Singhal KC, Sharma RA, Govind K, Kumar VV (2012) Analysis of antioxidant activity of *Catharanthus roseus* L. and it's association with habitat temperature. Asian J Exp Biol Sci 3:706–713
Kumar KB, Rajeev Kumar SB, Dwivedia VB, Raia AB, Ashutosh K, Shuklab KS, Dinesh A, Nagegowdaa B (2015) Precursor feeding studies and molecular characterization of geraniol-synthase establish the limiting role of geraniol in monoterpene indolealkaloid biosynthesis in *Catharanthus roseus* leaves. Plant Sci 239:56–66
Kumari K, Gupta S (2013) Phytopotential of *Catharanthus roseus* L. (G.) Don. var. Rosea and alba against various pathogenic microbes In vitro. Int J Res Pure Appl Microbiol 3:77–82
Kuppusamy C, Murugan K, Arul N, Yasodh P (2009) Larvicidal and insect growth regulator effect of α-amyrin acetate from *Catharanthus roseus* Linn against the malaria vector *Anopheles stephensi* Liston (Diptera: Culicidae). Entomol Res 39:78–83
Kurtz WGW, Chatson KB, Constable F (1980) Alkaloid production in *Catharanthus roseus* cell cultures. Characterization of the 953 cell line. Helv Chim Acta 63:1891–1896
Kutchan TM (1993) Strictosidine-from alkaloid to enzyme to gene. Phytochemistry 32:493–505
Kutney JP, Honda T, Kazmaizer RM, Lewis NJ, Worth BR (1980) Helv Chem Acta 63:366
Laflamme P, St-Pierre B, De Luca V (2001) Molecular and biochemical analysis of a madagascar periwinkle root-specific minovincinine-19-Hydroxy-O-Acetyltransferase. Plant Physiol 125:189–198
Lata B (2007) Cultivation mineral nutrition and seed production of *Catharanthus roseus* (L.) G. Don in the temperate climate zone. Phytochem Rev 6:403–411
Lee-Parsons CWT, Erturk S (2005) Ajmalicine production in methyl jasmonate-induced *Catharanthus roseus* cell cultures depends on Ca^{2+} level. Plant Cell Rep 24:677–682
Lee-Parsons CWT, Royce AJ (2006) Precursor limitations in methyl jasmonate—induced *Catharanthus roseus* cell cultures. Plant Cell Rep 25:607–612
Letchuman GR, Nazaimoon WWM, Mohamad WWB, Chandran LR, Tee GH, Jamaiyah H, Isa MR, Zanariah H, Fatanah I, Faudzi AY (2010) Prevalence of diabetes in the Malaysian national health morbidity survey III 2006. Med J Malays 65:173–179
Lichtenthaler HK (1999) The 1-deoxy-d-xylulose-5-phosphate pathway of isoprenoid biosynthesis in plants. Annu Rev Plant Physiol Plant Mol Biol 50:47–65
Liscombe DK, O'Connor SE (2011) A virus-induced gene silencing approach to understanding alkaloid metabolism in *Catharanthus roseus*. Phytochemistry 72:1969–1977
Liu D, Jin H, Chen Y, Cui L, Ren W, Gong Y (2007) Terpenoid indole alkaloids biosynthesis and metabolic engineering in *Catharanthus roseus*. J Integr Plant Biol 49:961–974
Liu Y, Zhao D, Zu Y, Tang Z, Zhang Z, Jiang Y (2011) Effects of low light on terpenoid indole alkaloid accumulation and related biosynthetic pathway gene expression in leaves of *Catharanthus roseus* seedlings. Bot Stud 52:191–196
Liu J, Zhu J, Tang L, Wen W, Lv S, Yu R (2014) Enhancement of vindoline and vinblastine production in suspension-cultured cells of *Catharanthus roseus* by artemisinic acid elicitation. World J Microbiol Biotechnol 30:175–180
Loyola-Vargas VM, Mende-Zeel M, Monforte-Gonzales M, Miranda-Ham ML (1992) Serpentine accumulation during greening in normal and tumor tissues of *Catharanthus roseus*. J Plant Physiol 140:213–217
Lynn DG, Chen RH, Manning KS (1987) The structural characterization of endogenous factors from *Vinca rosea* crown gall tumors that promote cell division of tobacco cells. Proc Natl Acad Sci USA 84:615–619

Mahroug S, Courdavault V, Thiersault M, St-Pierre B, Burlat V (2006) Epidermis is a pivotal site of at least four secondary metabolic pathways in *Catharanthus roseus* aerial organs. Planta 223:1191–1200

Mahroug S, Burlat V, St-Pierre B (2007) Cellular and sub-cellular organisation of the monoterpenoid indole alkaloid pathway in *Catharanthus roseus*. Phytochem Rev 6:363–381

Maki H, Ando S, Kodama H, Komamine A (1991) Polyamines and the cell cycle of *Catharanthus roseus* cells in culture. Plant Physiol 96:1008–1013

Malabadi RB, Mulgund GS, Nataraja K (2009) Triacontanol induced somatic embryogenesis and plantlet regeneration in *Catharanthus* roseus. J Med Arom Plant Sci 31:147–151

Manigandan V, Gurudeeban S, Satyavani K, Ramanathan T (2014) Molecular docking studies of Rhizophora mucronata alkaloids against neuroinflammatory marker cyclooxygenase 2. Int J Biol Chem 8:91–99

Maurel B, Parellieux A (1985) Effect of carbon dioxide on the growth of cell suspensions of *Catharanthus roseus*. Biotechnol Lett 7:313–318

McKnight T, Bergey D, Burnett R, Nessler C (1991) Expression of enzymatically active and correctly targeted strictosidine synthase in transgenic Tobacco plants. Planta 185:148–115

Memelink J, Verpoorte R, Kijne W (2001) ORCAnization of jasmonate-responsive gene expression in alkaloid metabolism. Trend Plant Sci 6:212–219

Menke FLH, Parchmann S, Mueller MJ, Kijne JW, Memelink J (1999) Involvement of the octadecanoid pathway and protein phosphorylation in fungal elicitor-induced expression of terpenoid indole alkaloid biosynthetic genes in *Catharanthus roseus*. Plant Physiol 1:1289–1296

Mera N, Aoyagi H, DiCosmo F, Tanaka H (2003) Production of cell wall accumulative enzymes using immobilized protoplasts of *Catharanthus roseus* in agarose gel. Biotechnol Lett 25:1687–1693

Miettinen K, Dong L, Navrot N, Schneider T, Burlat V, Pollier J, Woittiez L, der Krol VS, Lugan R, Ilc T, Verpoorte V, Oksman-Caldentey KM, Martinoia E, Bouwmeester H, Goossens A, Memelink J, Reichhart WD (2014) The seco-iridoid pathway from *Catharanthus roseus*. Nat Commun 5:3606

Minocha R, Minocha SC, Komamine A, Shortle WC (1990) Role of polyamines in DNA synthesis of Cathranthus roseus cells Grown in suspension culture. In: Flores HE, Arteca RN, Shannon JC (eds) Polyamines and ethylene: biochemistry, physiology, and interactions. Am Soc Plant Physiol pp 339-342

Minocha R, Minocha SC, Komamine A, Shortle WC (1991) Regulation of DNA synthesis and cell division by polyamines in *Catharanthus roseus* suspension cultures. Plant Cell Rep 10:126–130

Misawa M (1977) Production of natural substances by plant cell cultures described in Japanese patents. In: Barz W, Reinhard E, Zenk MH (eds) Plant tissue culture and its bio-technological application. Proceedings of the First International Congress on Medicinal Plant Research. Springer, Berlin, pp 17–26

Misra N, Luthra R, Kumar S (1996) Enzymology of indole alkaloid biosynthesis in *Catharanthus roseus*. Ind J Biochem Biophys 33:261–273

Misra N, Misra R, Mariam A, Yusuf K, Yusuf L (2014) Salicylic acid alters antioxidant and phenolics metabolism in *Catharanthus roseus* grown under salinity stress. Afr J Trad Complement Altern Med 11:118–125

Miura Y, Hirata K, Kurano N, Miyamoto K, Uchida K (1988) Formation of vinblastine in multiple shoot culture of *Catharanthus roseus*. Planta Med 54:18

Mohamed EH, Soad SE, Kamelia AES, Amira MH (2007) Response of *Catharanthus roseus* shoots to salinity and drought in relation to vincristine alkaloid content. Asian J Plant Sci 6:1223–1228

Mohan SCT, Priyadarshini GS, Balamurugan V (2015) GC-MS analysis of phytochemicals and hypoglycemic effect of *Catharathus roseus* in alloxan induced diabetic rats. Int J Pharm Sci Res 25:123–128

Mollers C, Sarkar S (1989) Regeneration of healthy plants from *Catharanthus roseus* infected with mycoplasma-like organisms through callus culture. Plant Sci 60:83–89

Monograph (2016) Thin-baw-ma-hnyo. *Catharanthus roseus*. Medicinal plants of Myanmar. Ministry of Health, Department of Traditional Medicine, pp 35–36

Moreno PRH, Heijden VDR, Verpoorte R (1993) Effect of terpenoid precursor feeding and elicitation on formation of indole alkaloids in cell suspension cultures of *Catharanthus roseus*. Plant Cell Rep 12:702–705

Moreno PRH, Heijden VDR, Verpoorte R (1995) Cell and tissue cultures of *Catharanthus roseus* (L.) Don. A literature survey II. Updating from 1988–1993. Plant Cell Tiss Org Cult 42:1–25

Moreno VOA, Coello-Coello J, Loyola-Vargas VM, Vazquez-Flota F (1999) Nutrient consumption and alkaloid accumulation in a hairy root line of *Catharanthus roseus*. Biotechnol Lett 21:1017–1021

Morgan JBA, Shanks JBV (2000) Determination of metabolic rate-limitations by precursor feeding in *Catharanthus roseus* hairy root cultures. J Biotechnol 79:137–145

Morrison EYSA, West M (1982) A preliminary study of the effects of some West Indian medicinal plants on blood sugar levels in the dog. West Indian Med J 31:194–197

Morris P (1986a) Regulation of product synthesis in cell cultures of *Catharanthus roseus*. Effect of temperature. Plant Cell Rep 5:427–429

Morris P (1986b) Regulation of product synthesis in cell cultures of *Catharanthus roseus*. II. Comparison of production media. Planta Med 52:121–126

Morton JF (1976) Periwinkle. In: Morton JF (ed) Major medicinal plants: botany, culture and uses. Charles C Thomas, Springfield, IL, pp 237–124

Mostofa M, Choudhury ME, Hossain MA, Islam MZ, Islam MS, Sumon MH (2007) Antidiabetic effects of *Catharanthus roseus*, *Azadirachta indica, Allium sativum* and glimepride in experimentally diabetic induced rat. Bangladesh J Vet Med 5:99–102

Mujib A, Ilah A, Aslam J, Fatima S, Siddiqui ZH, Maqsood M (2012) *Catharanthus roseus* alkaloids: application of biotechnology for improving yield. Plant Growth Regul 68:111–127

Mujib A, Ali M, Isah T, Dipti (2014) Somatic embryo mediated mass production of *Catharanthus roseus* in culture vessel (bioreactor)—a comparative study. Saudi J Biol Sci 21:442–449

Mukhapadhyay S, Cordell GA (1981) *Catharanthus* alkaloid XXXVI isolation of vinca leukoblastine (VLB) and periformyline from *Catharanthus* trichophyllus and pericyclivine from *Catharanthus roseus*. J Nat Prod 44:335–339

Muralidharan L (2015) Hypoglycemic and biochemical remedies of *Cathanthus roseus* (Linn) on Alloxan – induced diabetic rat and its antioxidant status in rat lenses. Int J Med Res Pharm Sci 2:1–6

Murata J, De Luca V (2005) Localization of tabersonine 16-hydroxylase and 16-OH tabersonine-16-O-methyltransferase to leaf epidermal cells defines them as a major site of precursor biosynthesis in the vindoline pathway in *Catharanthus roseus*. Plant J 44:581–594

Murugavel T, Akbarsha MA (1991) Anti-spermatogenetic effect of *Vinca rosea* Linn. Ind J Exp Biol 29:810–812

Mustafa NR, Verpoorte R (2007) Phenolic compounds in *Catharanthus roseus*. Phytochem Rev 6:243–258

Muthulakshmi S, Pandiyarajan V (2013) Influence of IAA on the vincristine content of *Catharanthus roseus* (L). G. Don. Asian J Plant Sci Res 3:81–87

Namedo A, Patil S, Fulzele DP (2002) Influence of fungal elicitors on production of ajamlicine by cell cultures of *Catharanthus roseus*. Biotechnol Process 18:159–162

Nammi S, Boini MK, Lodagala SD, Babu R, Behara S (2003) The juice of fresh leaves of *Catharanthus roseus* Linn. reduces blood glucose in normal and alloxan diabetic rabbits. BMC Complement Altern Med 3:4

Natarajan A, Zameer K, Ahmed SZ, Sundaresan S, Sivaraj A, Devi K, Senthil Kumar B (2012) Effect of aqueous flower extract of *Catharanthus roseus* on alloxan induced diabetes in male albino rats. Int J Pharm Sci Drug Res 4:150–153

Navitha A, Sheeba AH, Ramesh C, Banu S (2012) Hypoglycemic and diabetic activity of ethanolic extract of *Catharanthus pusillus* (Murray) G. Don. J Pharm 2:17–21

Nayak BS, Pereria PL (2006) *Catharanthus roseus* flower extract has wound-healing activity in Sprague Dawley rats. BMC Complement Altern Med 6:41

Negi NC, Bhatia MC (1956) Biological investigation of *Vinca rosea*. Ind J Pharma 18:73–76
Nejat N, Valdiani A, Cahill D, Tan Y, Maziah M, Abiri R (2015) Ornamental exterior versus therapeutic interior of Madagascar periwinkle (*Catharanthus roseus*): the two faces of a versatile herb. Sci World J 2015:1–19
Nguywen VD (1977) List of simple drugs and medicinal plants of value in Vietnam. Proceeding Seminar of the use of Medicinal Plants in Healthcare. Tokyo, WHO Regional Officer Manila, vol 13. pp 65–83
Noble RL (1990) The discovery of the vinca alkaloids-chemotherapeutic agents against cancer. Int J Biochem Cell Biol 68:1344–1351
Nuutila AM (1994) Bioreactor studies on hairy root cultures of *Catharanthus roseus*. Composition of three bioreactor types. Biotechnol Tech 8:61–66
Ohadoma SC, Michael HU (2011) Effect of co-administration of methanol leaf extract of *Catharnthus roseus* on the hypoglycemic activity of metformin and glibenclamide in rats. Asian Pac J Trop Med 4:475–477
Ouwerkerk PB, Hallard D, Verpoorte R, Memelink J (1999) Identification of UV-B responseive in the promoter of the tryptophane decarboxylase gene from *Catharanthus roseus*. Plant Mol Biol 41:491–503
Pan Q, Mustafa NR, Tang K, Choi YH, Verpoorte R (2016) Monoterpenoid indole alkaloids biosynthesis and its regulation in *Catharanthus roseus:* a literature review from genes to metabolites. Phytochem Rev 15:221–250
Pandey S, Bahadur AN, Kanungo VK, Tiwari U (2014) *In vitro* propagation of a medicinal plant *Catharanthus roseus* L. (G.) Don. Ind J Life Sci 4:125–128
Pandiangan D, Tilaar W, Nainggolan N (2013) Morphological changes of cell in relation to increased catharanthine content of *Catharanthus roseus* cell aggregate culture after Tryptophan treatment. J Basic Appl Sci 13:45–51
Park KH, Saimoto H, Nakagawa S (1989) Occurrence of brassinolide and castasterone in crown gall cells of *Catharanthus roseus*. Agric Biol Chem 53:805–812
Patel PJ, Ghosh JS (2010) Antimicrobial activity of *Catharanthus roseus*—a detailed study. Br J Clin Pharmacol Toxicol 1:40–44
Patel Y, Vadgama V, Baxi S, Tripathi CB (2011) Evaluation of hypolipidemic activity of leaf juice of *Catharanthus roseus* (Linn). G. Donn. in Guinea pigs. Acta Pol Pharm Drug Res 68:927–935
Pauw B, Duijin VB, Kijne JW, Memelink J (2004) Activation of the oxidative burst by yeast elicitor in *Catharatnhus roseus* cells occurs independenly of the activation of genes involved in alkaloid biosynthesis. Plant Mol Biol 55(6):797–805
Paynee GF, Paynee NN, Shuler ML (1988) Bioreactor considerations for secondary metabolite production from plant cell tissue culture; indole alkaloid from *Catharanthus roseus*. Biotechnol Bioeng 31:905–912
Perry LM (1980) Medicinal Plants of East and South East Asia, Attributed properties and Uses. MIT, London, p 732
Pietrosiuk A, Furmanowa M, Lata B (2007) *Catharanthus roseus*: micropropagation and in vitro techniques. Phytochem Rev 6:459–473
Posthouwer C, Verheijden TMS, Van Andel TR (2016) A rapid sustainability assessment of wild plant extraction on the Dutch Caribbean Island of St. Eustatius. Econ Bot 70:320. doi:10.1007/s12231-016-9356-9
Prasad A, Mathur P, Shrivastava M, Kumar D, Sharma E (2014) Larvicidal efficacy of *Catharanthus roseus* linn leaves and flowers against the malaria vector anopheles stephensi liston insecta: diptera: culicidae. Int J Recent Sci Res 5:1620–1623
Rahman AU (1982) Some approaches to the study of indigenous medicinal plants. Bull Islamic Med 2:562–568
Rai MK, Upadhyay S (1988) Screening of medicinal plants of Chhindwara district against Trichophyton mentagrophytes: A casual organism of Tineapedis. Hindustan Antibiot Bull 30:33–36
Rajas MCN, Cuellar MCA (1981) Comparative and microbiological studies of the alkaloids of *Catharanthus roseus* and other related compounds. Rev Cuba Farm 15:131–138

Ramani S, Jayabaskaran C (2008) Enhanced catharanthine and vindoline production in suspension cultures of *Catharanthus roseus* by ultraviolet-B light. J Mol Signal 3:9

Ramavat KG, Bhansall RR, Arya HC (1978) Shoot formation in *C. roseus* (L.) G. Don. Callus cultures. Curr Sci 47(3):93–96

Ramirez VR, Mostacero LJ, Garcia AE (1988) Vegetales empleados en medicina tradicional Noreruana. Banco Agrario Del Peru and NACL, Univ Trujillo, Trujill, Peru, p 54

Ramya V, Govindaraji K, Kannan N, Jayakumararaj R (2008) *In vitro* evaluation of antibacterial activity using crude extract of *Catharanthus roseus* L. (G.) Don. Ethnobot Leaflets 12:1067–1072

Rao SR, Ravishankar GA (2002) Plant cell cultures: chemical factories of secondary metabolites. Biotechnol Adv 20:101–153

Rasineni K, Bellamkonda R, Singareddy SR, Desireddy S (2010) Antihyperglycemic activity of *Catharanthus roseus* leaf powder in streptozotocin-induced diabetic rats. Pharmacogn Rev 2:195–201

Rasool N, Rızwan K, Zubaır M, Naveed K, Ahmed V (2011) Antioxidant potential of different extracts and fractions of *Catharanthus roseus* shoots. Int J Phytomed 3:108–114

Rho D, Andre G (1991) Growth and stoichiometry of a *Catharanthus roseus* cell suspension culture grown under nitrogen-limiting conditions. Biotechnol Bioeng 38:579–587

Rijhwani SK, Shanks JV (1998) Effect of elicitor dosage and exposure time on biosynthesis of indole alkaloids by *Catharanthus roseus* hairy root cultures. Biotechnol Prog 14:144–449

Rischer H, Oresic M, Seppanen-Laakso T, Katajamaa M, Lammertyn F, Ardiles-Diaz W (2006) Gene-to-metabolite networks for terpenoid indole alkaloid biosynthesis in *Catharanthus roseus* cells. Proc Natl Acad Sci U S A 103:5614–5619

Rohtas A (2001) Micropropagation of Salvandora oleoides Decne (Jhal) through tissue culture. M.Sc. thesis, Department of Botany, CCS Haryana Agricultural University, Hisar, India

Ross IA (2003) Medicinal Plants of the World 1.Chemical Constituents, Traditional and Modern and Medicinal Uses. 2nd ed., Humana Press. Springer Science + Business Media, New York, 175-177

Ross SA, Megalla SE, Bishay DW (1980) Studies for determining antibiotic substances in some Egyption plants. Part 1. Screening for antimicrobial activity. Fitoterapia 51:303–308

Ruffer M, Kan-Fan C, Husson H, Stockigt J, Zenk MH (1979) 4, 21-Dehydrogeissoschizine, an intermediate in heteroyohimbine alkaloid biosynthesis. J Chem Soc 22:1016–1018

Ruskin SR, Aruna SR (2014) *In-vitro in-vivo* antitumor activity of *Catharanthus roseus*. Int Res J Pharm App Sci 4:1–4

Sadowska A, Obidowska G, Szacho-Guchowicz M (1989) Effect of NPK fertilization upon production of crude material and alkaloid content of *Catharanthus roseus* (L.) G. Don. Biuletyn Instytutu Hodowli i Aklimatyzacji Roslin 170:55–63

Saifullah KS (2011) Callus induction and cell suspension culture production of *Catharanthus roseus* for biotransformation studies of (–)-caryophyllene oxide. Pak J Bot 43:467–473

Sain M, Sharma V (2013) *Catharanthus roseus* (an anti-cancerous drug yielding plant)—a review of potential therapeutic properties. J Int J Pure Appl Biosci 1:139–142

Sakano K, Kiyota S, Yazaki Y (1997) Acidification and alkalinization of culture medium by *Catharanthus roseus* cells is anoxic production of lactate a cause of cytoplasmic acidification. Plant Cell Physiol 38:1053–1059

Samara D (2015) Nature's healer "In Sri Lanka." http://exploresrilanka.lk/2015/12/natures-healer/

Sayeed MA, Jasmin MH, Sanker TC, Rahmann MM, Alam MF (2014) Antitumor activity of leaf extract of *Catharanthus roseus (*L.) G. Don. Plant Environ Dev 3:24–30

Schlatmann JE, Ten Hoopen HJG, Heijnen JJ (1992) Optimization of the medium composition for alkaloids production by *Catharanthus roseus* using statistical experimental designs. Med Fac Landbouw Universiteit Gent 7:1567–1156

Schlatmann JE, Moreno PRH, Ten Hoopen HJG (1994) Effect of oxygen and nutrient limitation on ajmalicine production and related enzyme activities in high density cultures of *Catharanthus roseus*. Biotechnol Bioeng 44:461–468

Schlatmann JE, Koolhaas CMA, Vinke JL, Ten Hoopen HJG, Heijnen JJ (1995) The role of glucose in ajmalicine production by *Catharanthus roseus* cell cultures. Biotechnol Bioeng 47:525–534

Schmeller T Wink M (1998) Utilization of alkaloids in modern medicine. In: Roberts MF, Wink M (eds) Alkaloids. Biochemistry, ecology and medicinal applications. Plenum Press, New York, pp 435–459

Schmelzer GH, Fakim A (2008) Plant resources of tropical Africa 11(1). Medicinal plants 1. PROTA foundation/backhuys publisher/CTA wageningen, Netherlands, pp 152–159

Schmidt B, Kutney J, Mayer L (1998) Anhydrovinblastine for the treatment of cervical and lung cancer. PCT Int. Appl. http://www.google.com/patent/EPO969839A1?cl=en

Schroder G, Unterbusch E, Kaltenbach M, Schmidt J, Strack D, De Luca V (1999) Light-induced cytochrome P450-dependent enzyme in indole alkaloid biosynthesis: Tabersonine 16-hydroxylase. FEBS Lett 458:497–102

Scragg AH, Ashton S, York A, Stepan-Sarkissian G, Grey D (1990) Growth of *Catharanthus roseus* suspension for maximum biomass and alkaloid accumulation. Enzyme Microb Technol 12:292–298

Seaforth CE (2006) Traditional medicines for modern times antidiabetic plants. In: Soumyanath A (ed) Antidiabetic plants in the Caribberian. Taylor & Francis Group, New York, p 314

Seitz HU, Eilert U, De Luca V (1989) Elicitor mediated induction of phenylalanine ammonia lyase and tryptophan decarboxylase: accumulation of phenols and indole alkaloids in cell suspension culture of *Catharanthus roseus*. Plant Cell Tiss Org Cult 18:71–78

Sharma G, Kumar M, Sharma S (2011) Studies on antibacterial activity and biochemical/biophysical properties of phytocystatin purified from *Catharanthus roseus* (Madagascar periwinkle): an evergreen subshrub commonly found in district Bijnor (UP). Adv Biosci Biotechnol 2:391–396.

Shaqha A, Khan WM, Salam MN, Azzi A, Chaudhary AA (2015) Anti-diabetic potential of *Catharanthus roseus* Linn. and its effect on the glucose transport gene (GLUT-2 and GLUT-4) in streptozotocin induced diabetic wistar rats. BMC Complement Altern Med 15:379

Shorti DS, Kelkar M, Deshmukh VK (1963) Investigation of the hyperglycemic properties of *Vinca rosea*, *Cassia auriculata* and *Eugenia jambolana*. Ind J Med Res 51:464–467

Siddiqui MJ, Ismail Z, Aisha AFA, Majid AMS (2010) Cytotoxic activity of *Catharanthus roseus* (Apocynaceae) crude extracts and pure compounds against human colorectal carcinoma cell line. Int J Pharmacol 6:43–47

Siddiqui ZH, Mujib A (2016) Accumulation of vincristine in calcium chloride elicitated *Catharanthus roseus* cultures. J Nat Prod 2:307–315

Siegel RK (1976) Herbal intoxication. Pscyhoactive effects from herbal cigarettes, tea and capsules. J Am Med Assoc 236:473–476

Singh SM, Vats PS, Suri R, Shyam MML, Kumari AR, Sridharan KS (2001) Effect of an antidiabetic extract of *Catharanthus roseus* on enzymic activities in streptozotoxin induced diabetic rats. J Ethnopharmacol 76:269–277

Singh R, Kharb P, Rani K (2011) Rapid micropropagation and callus induction of *Catharanthus roseus in vitro*. World J Agric Res 7:699–704

Singh A, Singh PK, Singh RK (2014) Antidiabetic and wound healing activity of *Catharanthus roseus* L. in streptozotocin-induced diabetic mice. Am J Phytomed Clin Ther 2:686–692

Smith JI, Smart M, Misawa WG, Kuurz W, Tallevi SG, Dicosmo F (1987) Increased accumulation of indole alkaloids by some cell lines of *Catharanthus roseus* in response to addition of vanadyl of vanadyl sulphate. Plant Cell Rep 6:142–145

Sottomayor M, Lopez-Serrano M, Dicosmo F, Ros Barcelo A (1998) Purification and characterization of alpha-3′, 4′-anhydrovinblastine synthase (peroxidase-like) from *Catharanthus roseus* (L.) G. Don. FEBS Lett 428:299–303

Spencer CF, Koniuszy FR, Rogers EF (1949) Survey of plants for antimalarial activity. Lloydia 10:145–174

Srivastava NK, Srivastava AK (2007) Influence of gibberellic acid on CO_2 metabolism, growth and production of alkaloids in *Catharanthus roseus*. Photosynthetica 45:156–160

Stevens L, Blom T, Verpoorte R (1993) Sucellular localization of tryptophan decarboxylase, strictosidine synthase and strictosidine glucosidase in suspension cultured cells of *Catharanthus roseus* and *Tabernaemontana divaricata*. Plant Cell Rep 12:573–576

Sutrisna E (2015) *Catharanthus roseus* (Tapak Dara): "A controversial" medicinal plant in Indonesia. Int J Res Ayurveda Pharm 6:629–633

Svoboda GH (1966) The role of the alkaloids of *Catharanthus roseus* (L.) G.don (*Vinca rosea*) and their derivatives in cancer chemotherapy Z Plants: The Potentials for Extracting Protein, Medicines, and Other Useful Chemicals-Workshop Proceedings. Indianapolis, India, pp 154–169

Svoboda GH, Blake DA (1975) The phytochemistry and pharmacology of *Catharanthus roseus*. In: Taylor WI, Farnsworth NR (eds) The Catharanthus alkaloids, botany, chemistry, pharmacology and clinical use, pp 45–83

Swanberg A, Dai W (2008) Plant regeneration of periwinkle (*Catharanthus roseus*) via organogenesis. J Hort Sci 43:832–836

Swanston-Flatt SK, Day C, Flatt PR, Gould BJ, Bailey CJ (1989) Glycemic effects of traditional European plant treatments for diabetes. Studies in normal and streptozotocin diabetic mice. Diabetes Res 10:69–73

Taafari IJ, Fowler MK, Thaman RR (2006). Traditional medicine of the Marshell islands. The women, the plants, the treatments *Catharanthus roseus* (L.) G. Don Madagascar periwinkle. IPS Publications, University of the South Pacific Suva, Fiji, pp 211–212

Taha HS, El-Bahr MK, Seif-El-Nasr MM (2009) In vitro studies on Egyptian *Catharanthus roseus* (L.). Effect of biotic and abiotic stress on indole alkaloids production. J Appl Sci Res 5:1826–1831

Taha HS, Shams KA, Nazif NM, El-Nasr SMM (2014) In vitro studies on Egyptian *Catharanthus roseus* (L.) G. Don. Impact of stirred reactor physical factors on achievement of cells proliferation and vincristine andvinblastine accumulation. Res J Pharm Biol Chem Sci 5:330–340

Tallevi SG, Dicosmo F (1988) Stimulation of indole alkaloid content in vanadium-treated *Catharanthus roseus* suspension cultures. Centre for Plant Biotechnology, Department of Botany, University of Toronto, Ontario M5S 1A1, Canada. Planta Med 54:149–152

Ten Hoopen HJG, Gulik VWM, Schlatmann JE, Moreno PRH, Vinke JL, Heijnen JJ, Vepoorte R (1994) Ajmalicine production by cell cultures of *Catharanthus roseus*: from shake flaske to bioreactor. Plant Cell Tiss Org Cult 38:85–89

Thingujam D, Rekha K, Megala J, Usha B (2015) Antioxidant and anticancer properties of *Catharanthus pusillus*. Int J Adv Chem Sci Appl 3:48–51

Thompson W (1976) Herbs that heal. J Roy Coll Gen Pract 26:365–370

Tiong SH, Looi CY, Hazrina H, Aditya A, Javad PM, Wong WF, Cheah SC, Rais MM, Awang K (2013) Antidiabetic and antioxidant properties of alkaloids from *Catharanthus roseus* (L.) G. Don. Molecules 18:9770

Toivonen L, Ojala M, Kauppinen V (1989) Studies on the optimization of growth and indole alkaloids production by hairy roots cultures of *Catharanthus roseus*. Biotechnol Bioeng 37:673–680

Tzin V, Galli G (2010) New insights into the shikimate and aromatic amino acids biosynthesis pathways in plants. Mol Plant 3:956–972

Valluri JV (2009) Bioreactor production of secondary metabolites from cell cultures of periwinkle and sandalwood. Methods Mol Biol 547:325–335

Vazquez-Flota F, De Carolis E, Alarco AM, De Luca V (1997) Molecular cloning and characterization of desacetoxyvindoline-4-hydroxylase, a 2-oxoglutarate dependent-dioxygenase involved in the biosynthesis of vindoline in *Catharanthus roseus* (L.) G. Don. Plant Mol Biol 34:935–948

Vazquez-Flota FA, De Luca V (1998) Developmental and light regulation of desacetoxyvindoline 4-hydroxylase in *Catharanthus roseus* (L.) G. Don. Evidence of a multilevel regulatory mechanism. Plant Physiol 17:1351–1361

Vazquez-Flota F, Hernandez-Dominguez E, Lourdes D, Miranda-Ham M, Monforte-Gonzalez M (2009) A differential response to chemical elicitors in *Catharanthus roseus in vitro* cultures. Biotechnol Lett 31:591–595

Veau B, Courtois M, Oudin A, Chenieux JC, Rideau M, Clastre M (2000) Cloning and expression of cDNAs encoding two enzymes of the MEP pathway in *Catharanthus roseus*. Biochim Biophys Acta 1517:159–116

Vega-Avila E, Cano-velasco JL, Alarcon-Anguilar FJ, Fajardo Ortiz MC, Almanza-perez JC, Roman Ramos R (2012) Hypoglycemic activity of aqueous extracts from *Catharanthus roseus*. Evid Based Complement Altern Med 2012:934528

Verma P, Mathur AK (2011) Direct shoot bud organogenesis and plant regeneration from pre-plasmolysed leaf explants in *Catharanthus roseus*. Plant Cell Tiss Org Cult 106:401–408

Verma A, Laakso I, Seppanen-Laakso T, Huhtikangas A, Marja-Liisa RML (2007) A simplified procedure for indole alkaloid extraction from *Catharanthus roseus* combined with a semi-synthetic production process for vinblastine. Molecules 12:1307–1315

Verma P, Mathur AK, Singh A, Srivastava A, Masood N, Luqman Upadhyaya M, Mathur A (2012a) Co-culturing of *Catharanthus roseus*, Vinca major and *Rauwolfia serpentine* cell suspensions in shake flask and bioreactor: production of a novel alkaloid with antioxidant potential. J Med Plant Res 6:4978–4988

Verma A, Singh R, Singh S (2012b) Improved alkaloid content in callus cultures of *Catharanthus roseus*. Bot Serb 36(2):123–130

Verma P, Mathur AK, Masood N, Luqman S, Shanker K (2013) Tryptophan over-producing cell suspensions of *Catharanthus roseus* (L) G. Don and their up-scaling in stirred tank bioreactor: detection of a phenolic compound with antioxidant potential. Protoplasma 250:371–380

Verpoorte R, Moreno PRH, Heijden R (1993) Secondary metabolismin plants and plant cell cultures of some terpenoids indole alkaloid producing plants. Abstract. XV International Botanical Congress, Yokohama, Japan

Verpoorte R, Heijden R, Moreno PRH (1997) Biosynthesis of terpenoid indole alkaloids in *Catharanthus roseus* cells. In: Cordell GA (ed) The alkaloids, vol 49. Academic, San Diego, pp 221–229

Virmani OP, Srivastava GN, Singh P (1978) *Catharanthus roseus*—the tropical periwinkle. Ind Drugs 15:231–252

Wagay SA, Dwivedi SD, Sharma DM, Mushtaq Ahmad JT (2013) Antimicrobial activity of *Catharanthus roseus*. Chem Mater Res 3:61–64

Wagner F, Vogelmann H (1977) Cultivation of plant tissue cultures in bioreactors and formation of secondary products. In: Barz W, Reinhard E, Zeuk MH (eds) Plant tissue culture and its biotechnological application. Springer, Berlin, pp 245–252

Wang S, Zheng Z, Weng Y, Yu Y, Zhang D, Fan W, Dai R, Hu Z (2004) Angiogenesis and antiangiogenesis activity of Chinese medicinal herbal extracts. Life Sci 74:2467–2478

Wang JY, Liu ZP, Liu L, Liu C (2008) Effects of NaCl on the growth and alkaloid content of *Catharanthus roseus* seedlings. J Appl Ecol 19:2143–2148

Washington DC (1983) Anticancer drug from the Madagasar periwinkle. The Potentials for Extracting Protein, Medicines and Other Useful Chemicals-Workshop Proceedings. Office of Technology Assessment, US Congress, pp 23–34

Webb LJ (1984) Guide to medicinal and poisonous plants of Queensland. Council for Scientific and Industrial Research Bull, Melbourne, p 232

Whitmer S, Verpoorte R, Canel C (1998) Influence of auxins on alkaloid accumulation by a transgenic cell line of *Catharanthus roseus*. Plant Cell Tiss Org Cult 53:135–141

Whitmer S, Heijden R, Verpoorte R (2002) Effect of precursorfeeding on alkaloid accumulation by a tryptophan decarboxylaseover-expressing transgenic cell line T22 of *Catharanthus roseus*. J Biotechnol 96:193–203

Widowati W, Mozef T, Risdian C, Yellianty Y (2013) Anticancer and free radical scavencing potency of *Catharanthus roseus*, *Dendrophthoe petandra, Piper betle* and *Curcuma mangga* extracts in breast cancer cell lines. Oxid Antioxid Med Sci 2:137–142

Yahia A, Kevers C, Gaspar T, Chenieux JC, Creche JR (1998) Cytokinins and ethylene stimulate indole alkaloid accumulation in cell suspension cultures of *Catharanthus roseus* by two distinct mechanisms. Plant Sci 133:9–15

Yang LL, Yen KY, Kiso Y (1987) Antihepatotoxic actions of formosan plant drugs. J Ethnopharmacol 19:103–110

Yao XG, Chen F, Li P, Quan L, Chen J, Yu L, Ding H, Li C, Chen L, Gao Z, Wan P, Hu L, Jiang H, Shen X (2013) Natural product vindoline stimulates insulin secretion and efficiently ameliorates glucose homeostasis in diabetic murine models. J Ethnopharmacol 150:285–297

Yuag F, Wang Q, Wang G, Zao J, Tian Y, Tang K (2011) An efficient based somatic embryogenesis based plant regeneration from the hypocotyls of *Catharanthus roseus*. Afr J Biotechnol 10:14786–14795

Yuan YJ, Hu ZD (1994) Effect of residual medium on *Catharanthus roseus* callus and suspension cell culture. Plant Physiol Commun 29:185–187

Zaguirre JC (1944) Guide notes of bed-size preparations of most common local (Philippines) medicinal plants

Zenk MK, El-Shage E, Arens H, Stockigt J, Weiler EW, Deus B (1977) Formation of the indole alkaloids serpentine and ajmalicine in cell suspension cultures of *Catharathus roseus*. In: Barz W, Reinhard E, Zeuk MH (eds) In plant tissue culture and its biotechnol applications. Springer, Berlin, pp 27–43

Zhang X, Brotherton JE, Widholm JM (2015) Co-expression of the tobacco anthranilate synthase β subunit with its feedback-insensitive α subunit as a selectiable marker that also markedly increases the free tryptophane content. In Vitro Cell Dev Biol 51:564–570

Zhao J, Verpoorte R (2007) Manipulating indole alkaloids production by catharathus roseus cell cultures in bioreactor: from biochemical processing to metabolic engineering. Phytochem Rev 6:435–457

Zhao J, Zhu WH, Hu Q (2000a) Improved alkaloid production in *Catharanthus roseus* suspension cell cultures by various chemicals. Biotechnol Lett 22:509–514

Zhao J, Zhu WH, Hu Q (2000b) Enhanced ajmaline production in *Catharanthus roseus* cell cultures by combined elicitor treatment: from shake flask to 20L airlift bioreactor. Biotechnol Lett 22:509–514

Zhao J, Zhu WH, Hu Q (2001a) Effects of light and plant growth regulators on the biosynthesis of vindoline and other indole alkaloids in *Catharanthus roseus* callus cultures. Plant Growth Regul 33:43–49

Zhao J, Hu Q, Guo YQ, Zhu WH (2001b) Effects of stress factors, bioregulators, and synthetic precursors on indole alkaloid production in compact callus clusters cultures of *Catharanthus roseus*. Appl Microbiol Biotechnol 55:693–698

Zhao J, Zhu WH, Hu YQ (2001c) Enhanced catharanthine production in *Catharanthus roseus* cell cultures by combined elicitor treatment in shake flasks and bioreactors. Enzyme Microb Technol 28:673–681

Zhao J, Hu Q, Guo YQ, Zhu WH (2001d) Elicitor-induced indole alkaloid biosynthesis in *Catharanthus roseus* cell cultures is related to Ca^{2+} influx and the oxidative burst. Plant Sci 16:423–431

Zhao J, Zhu W, Hu Q, Guo Y (2001e) Compact Callus cluster suspension cultures of *Catharanthus roseus* with enhanced indole alkaloid biosynthesis. In Vitro Cell Dev Biol 37:68–72

Zhao J, Zhu WH, Hu Q, He XW (2001f) Enhanced indole alkaloid production in suspension compact callus clusters of *Catharanthus roseus*: impacts of plant growth regulators and sucrose. Plant Growth Regul 33:33–41

Zhao J, Davis LC, Verpoorte R (2005) Elicitor signal transduction leading to production of plant secondary metabolites. Biotechnol Adv 23:283–333

Zhu J, Wang M, Wen W, Yu R (2015) Biosynthesis and regulation of terpenoid indole alkaloids in *Catharanthus roseus*. Pharmacogn Rev 9:24–28

Therapeutic Efficacy of *Catharanthus roseus* in Type 1 and Type 2 Diabetes Mellitus in Wistar Rats

Karuna Rasineni, Sudhakara Gujjala, Saisree Sagree, Mallaiah Putakala, Sasi Bhusana Rao Bongu, Ramesh Bellamkonda, Sreenivasa Reddy Singareddy, and Saralakumari Desireddy

Abstract This study was undertaken to systematically investigate the beneficial effects of *Catharanthus roseus (C. roseus)* in insulin-deficient and insulin-resistant conditions. Type 1 diabetes is a metabolic disorder due to insulin deficiency whereas insulin resistance is the prominent feature of Type 2, both types of diabetes are characterized by hyperglycemia. Chronic hyperglycemia by itself and by its associated oxidative stress plays an important role in the initiation and progression of diabetic complications. Due to increasing obesity, altered dietary habits and sedentary life style both in western and developing countries, the prevalence of both types of diabetes are growing at an exponential rate.

In conventional therapy, Type 1 diabetes is treated with exogenous insulin and Type 2 with oral hypoglycemic agents (sulfonylureas, biguanides, etc.). In modern medicine, no satisfactory effective therapy is available to cure Diabetes Mellitus (DM). Though insulin therapy is used for management of DM, there are several drawbacks like insulin resistance, anorexia nervosa, brain atrophy, and fatty liver after chronic treatment. Unfortunately, apart from having a number of side effects, none of the oral hypoglycemic agents have been successful in maintaining euglycemia and controlling long-term microvascular and macrovascular complications.

K. Rasineni, PhD
Department of Internal Medicine, University of Nebraska Medical Center, Omaha, NE, USA

S. Gujjala, PhD • S. Sagree, M.Sc • M. Putakala, PhD • S.B.R. Bongu, M.Sc
S. Desireddy, PhD (✉)
Department of Biochemistry, Sri Krishnadevaraya University,
Anantapur, Andhra Pradesh, India
e-mail: skumari1@yahoo.co.in

R. Bellamkonda, PhD
Department of Food Technology, Vikramasimhapuri University,
Nellore, Andhra Pradesh, India

S.R. Singareddy, PhD
Biochemistry Division, National Institution of Nutrition, Hyderabad, Telangana, India

M. Naeem et al. (eds.), *Catharanthus roseus*,
DOI 10.1007/978-3-319-51620-2_9

Moreover, these therapies only partially compensate for metabolic derangements seen in diabetes and do not necessarily correct the fundamental biochemical lesions. The yawning gap for additional agents to combat hyperglycemia and its accompanying complications presents an opening to revisit traditional antidiabetic plants. Many traditional plant treatments for diabetes exist. However, few have received scientific or medical scrutiny, and the World Health Organization has recommended that traditional plant treatments for diabetes warrant further evaluation.

C. roseus, belongs to family Apocynaceae, is a subshrub also known as Madagascar periwinkle, *Vinca rosea*, or *Lanchnera rosea* worldwide. The plant *C. roseus* has gained acceptance from the pharmaceutical industries as it is widely used as an infusion in different parts of world to treat diabetes. Results from animal studies support the claim that this plant shows antihyperglycemic and hypolipidemic activity. Although earlier reports indicate blood glucose lowering activity in alcoholic extracts of leaves, no information is available on the biochemical basis related to antidiabetic property of *C. roseus*.

The present study revealed the biochemical basis for the antidiabetic activity of *C. roseus* in streptozotocin-induced Type 1 DM and high fructose diet-induced insulin-resistant Type 2 diabetic rat models. In this study, a total of 48 male Wistar rats, aged 6–7 weeks were used. One-third of the experimental animals were made diabetic by a single intraperitoneal injection of freshly prepared Streptozotocin (STZ) (55 mg/kg body weight, in ice-cold 0.1 M citrate buffer, pH 4.5, in a volume of 0.1 mL per rat). Insulin resistance was induced in another one-third of experimental animals by feeding fructose-enriched diet (containing 66% fructose, 18% protein, 8% fat, 4% cellulose, 3% mineral and 1% vitamin mix). The remaining one-third of animals served as control and maintained on standard pellet diet. The rats were divided into six groups—control group (C-group), control rats treated with *C. roseus* (C + CR), STZ diabetic group (D-group), diabetic animals treated with *C. roseus* (D + CR), fructose-fed rats (Group-F), fructose-fed rats co-administered with *C. roseus* (F + CR).

Treatment with *C. roseus* to both STZ-induced Type 1 and fructose diet-induced Type 2 rats for 60 days, rectified the alterations in the activities of key enzymes of glycolytic, gluconeogenesis and polyol pathway, and intestinal disaccharides. Further, accumulation of lipids in hepatic and cardiac tissues, and enhanced oxidative stress in liver, pancreas, kidney, and heart tissues were brought to near-normal values in both the groups after treatment with *C. roseus.* The multiple beneficiary properties like antidiabetic, hpolipidemic, insulin sensitizing activities of *C. roseus* along with antioxidant potential offers an exciting opportunity to develop a novel and safe therapeutic approach for both Type 1 and Type 2 diabetes.

Keywords Antihyperglycemic activity • Hypolipidemic activity • Type 1 and 2 diabetes • Lipid peroxidation • Fructose diet - STZ - Oxidative stress • Wistar rats

1 Introduction

Diabetes mellitus (DM) is a metabolic disorder, characterized by hyperglycemia due to disturbance in carbohydrate, fat, and protein metabolism which is associated with absolute or relative deficiency in insulin secretion and/or action. DM is one of the world's oldest known diseases. The number of diabetics among adults (aged 20–79 years) was 185 million in 2010 which may increase to 439 million in the year 2030 (Shaw et al. 2010). As the numbers of people with DM multiply worldwide, it has been taking an ever increasing proportion of national and international health care budget. Diabetes can be classified into two types. Type 1 diabetes is a metabolic disorder due to insulin deficiency whereas insulin resistance (IR i.e., hyperglycemia and hyperinsulinemia) is the prominent feature of Type 2; both types of diabetes are characterized by hyperglycemia. Chronic hyperglycemia by itself and by its associated oxidative stress plays a key role in the initiation and progression of diabetic complications. Despite the role of genetic, age, sex and environmental factors, it is generally accepted that westernized lifestyle (characterized by high calorie intake i.e., fructose and physical inactivity) contributing greatly to the prevalence of both types of diabetes at an exponential level both in western and developing countries.

In many developing countries, ayurvedic medicine, in particular, the herbal medicine is sometimes the only affordable source of health care (Hamdan and Afifi 2004). Even the WHO approves the use of plant drugs for different diseases including diabetes mellitus (WHO 2002). The multifactorial pathogenicity of diabetes demands the need of multimodal therapeutic approach. The multiple activities of plant-based herbal preparations meant for diabetes offer huge scope for combating the threat of the diabetic epidemic. Many traditional plant treatments are used for diabetes throughout the world. Few medicinal plant treatments for diabetes received scientific scrutiny for which WHO has also recommended attention (WHO 1980).

Catharanthus roseus (Apocynaceae), commonly known as Madagascar periwinkle, *Vinca rosea* or *Lanchnera rosea* worldwide, in Sanskrit, Nityakalyani and Sadapushpi, in Telugu Billaganneru, in Hindi Sadaphul. It is a native of Madagascar and abundantly naturalized in many regions, particularly in arid coastal locations, and grown commercially for its medicinal uses in India, Africa, Southern Europe and Australia and is cultivated as an ornamental plant almost all parts of the tropical and subtropical world. *C. roseus* is one of the medicinal plants which has found mention in the folk medicinal literature as early as 2nd B.C. Traditionally, *C. roseus* has been used in folk medicine to treat diabetes, high blood pressure, tuberculosis, etc. The fresh juice from the flowers of *C. roseus* made into a tea and was used by Ayurvedic physician in India for external use to treat skin problems, dermatitis, eczema, and acne (Nayak et al. 2006). Sulphates of *C. roseus* alkaloids, vincristine and vinblastine, were widely used as chemotherapeutic agents against leukemia and Hodgkin's disease throughout the world (Ozgen et al. 2003). The closely related semisynthetic derivatives of vincamine, a major alkaloid of *C. roseus* were widely used as a medicinal agent, known as ethyl-apovincaminate or vinpocetine has vasodilating, hypoglycemic, and memory enhancing actions (Chattopadhya et al. 1991). *C. roseus* is widely used as an infusion in different parts of the world to treat diabetes (Alxeandrova et al. 2000; Heijden et al. 2004). The hot water decoctions of either the leaves or the whole plant of *C. roseus* were used for treatment of diabetes in several countries (Don 1999). The juices of fresh leaves of *C. roseus* have been reported to reduce blood glucose levels in normal and alloxan diabetic rabbits (Nammi et al. 2003). The antihyperglycemic activity of alcoholic extract of leaves (Chattopadhyay et al. 1991) and dichloromethane–methanol extracts of leaves, twigs, and whole plant (Somananth et al. 2001; Jayanthi et al. 2010; Ibrahim et al. 2011) and aqueous extract of flower (Natarajan et al. 2012) have been reported in animal model. The alkaloids, namely, vindoline, vindolidine, vindolicine, and vindolinine isolated from this plant possessed the hypoglycemic activity (Svoboda 1969). The extracts of *C. roseus* have been reported to have peroxisome proliferator receptor activating activity in cultured human cells (Rau et al. 2006). The leaf juice of *C. roseus* significantly reduced the serum total cholesterol and triglyceride in rats was well documented (Antia and Okokon 2005). As an antidiabetic remedy, it was believed to promote insulin production or to increase the body's utilization of sugars from food. Earlier studies on antihyperglycemic and antihyperlipidemic activity of *C. roseus* are fragmentary, and no studies are available on the efficacy of *C. roseus* in preventing IR. However, very little information is available on antioxidant activity of *C. roseus*.

Therefore, this study was undertaken to explore possible beneficial effects of *C. roseus* leaf powder in insulin-deficient and insulin-resistant conditions. In this study, STZ-induced diabetic rats (i.p. injection at a dose of 55 mg/kg b.wt, in 0.1 mL of 0.05 M citrate buffer, pH 4.5) represent the insulin-deficient model and chronic feeding of Wistar rats with fructose-rich diet (66%) served as a model for insulin resistance. Further, the biochemical basis for its antidiabetic property and protection against IR were investigated.

In India, water decoction of leaves and flowers of *C. roseus* are traditionally used by diabetic patients. Hence, this study was made by using aqueous leaf powder suspension of the plant.

2 Materials and Methods

2.1 Plant Material

Fresh mature leaves of *C. roseus* (white variety) were collected during September, 2006, from the University campus and taxonomically authenticated by the Department of Botany, Sri Krishnadevaraya University, Anantapur, AP. Leaves were shade dried and then grinded into fine powder. The leaf powder was suspended in distilled water prior to use.

2.2 Chemicals

Streptozotocin was obtained from Sigma Chemical Co., St. Louis, MO, USA. All other chemicals were of analytical grade and procured from SISCO Research Laboratories (P) Ltd., Mumbai, India; Merck laboratories, India. The fructose diet (contained 66 % fructose, 18 % protein, 8 % fat, 4 % cellulose, 3 % mineral, and 1 % vitamin mix) and standard pellet diet were procured from National Centre for Animal Science, National Institution Nutrition, Hyderabad, India.

2.3 Experimental Design

Male Albino Wistar rats (140–160 g) used for this study were procured from Sri Venkateswara Enterprises (Bangalore, India). The experimental work was carried out after approval by the Institutional Animal Ethical Committee (Regd. no. 470/01/a/CPCSEA). They were housed two per cage in an air-conditioned room (22 ± 2°C) with 12 h light/dark cycle and had free access to standard pellet diet and water. In this study, a total of 48 male Wistar albino rats were acclimatized to our animal house before induction of Type 1 diabetes/IR. One-third of experimental animals were made diabetic by STZ injection and maintained on standard pellet diet and used for further studies. Insulin resistance was induced in another one-third of experimental animals by feeding fructose-enriched diet throughout the experimental period. The remaining one-third of animals served as controls and maintained on standard pellet diet. Each set of animals control (C), STZ diabetic (D), fructose diet fed (F) were further subdivided into two groups, thus comprising a total of six groups—control group (C-group), control rats treated with *C. roseus* (C + CR-group), STZ diabetic group (D-group), STZ diabetic rats treated with *C. roseus* (D + CR-group), fructose-fed rats (F-group), and fructose-fed rats treated with *C. roseus* (F + CR-group). Groups C + CR, D + CR, and F + CR rats were administered *C. roseus* leaf powder suspension (100 mg/kg body weight in ~2 mL of water/day) through gastric intubation for 60 days.

The dose of *C. roseus* leaf powder in the study was based on preliminary experiments on dose-dependent antihyperglycemic effect in STZ-induced diabetic rats. A dose less than 100 mg/kg body weight was not found to be effective in rats. Blood was collected in Eppendorf tubes containing EDTA (10 mg/mL) from 12 h fasted rats by means of capillary tube through retinorbital flexus. Plasma was separated by centrifugation. Body weight and plasma glucose, insulin and lipids were monitored at 15 day intervals till the end of study.

2.4 Animal's Sacrifice and Organ Collection

After 60 days of treatment, all the rats were sacrificed following 12 h of fasting by cervical dislocation and immediately liver, pancreas, kidneys, small intestine, adipose, heart, and thigh muscle were removed and washed thoroughly with ice-cold

0.9% NaCl (saline). Each organ of every animal was suspended in 0.15 M KCl in polypropylene containers, sealed with parafilm, labeled carefully, and frozen at −80 °C until assays were carried out.

2.5 Phytochemical Screening and In Vitro Antioxidant Potential of Aqueous Leaf Suspension of *C. roseus*

The *C. roseus* leaf powder suspension in water was qualitatively screened for the presence of various phytochemical constituents using standard procedures (Harbone 1973; Brain and Turner 1975; Sofowora 1982). Qualitative estimation of total polyphenols (Singleton et al. 1999), flavonoids (Arvouet-Grand et al. 1994), and saponins (Hiai et al. 1976) were carried out.

2.6 In Vitro Screening of Antioxidant Potential of *C. roseus*

In vitro antioxidant potential of *C.roseus* was assessed by screening inhibition of lipid peroxidation induced by Fe^{3+}/ADP/ascorbate system in rat liver homogenate (Sugioka et al. 1987), reducing ability (Oyaizu 1986) and its scavenging capacity of various free radicals like hydroxyl (Gutteridge et al. 1981), hydrogen peroxide (Ruch et al. 1989), superoxide (Oktay et al. 1989), nitric oxide (Marcocci et al. 1994), and DPPH (Okada and Okada 1998) were studied and compared with a standard natural antioxidant curcumin.

2.7 Biochemical Estimations and Assays

Plasma glucose was estimated by the glucose oxidase–peroxidase method by using the Span diagnostic kit (Span Diagnostics Ltd., Surat, India). Plasma insulin was estimated by using the Radioimmunoassay kit (RIA K-1) from Bhabha Atomic Research Centre (Mumbai, India) according to the method of Yalow and Berson (1960). Homeostasis model assessment (HOMA), used as an index to measure the degree of IR, was calculated by the following formula: insulin (μU/mL) × glucose (mmol/l)/22.5 (Matthews et al. 1985).

2.8 Intestinal Disaccharidases

Intestinal mucosa was collected by following the procedure outlined by Ravinder et al. (1989). The intestinal disaccharidases, i.e., maltose, sucrose, and lactose were assayed by following the method outlined by Dahlqvist (1968).

2.9 Carbohydrate Metabolic Studies

Glycogen content was determined by using anthrone reagent as adapted by Carrol et al. (1956). The frozen liver, kidney, and muscle tissues were slowly thawed at 4 °C. Muscle tissue was powdered using liquid nitrogen. Ten percent tissue homogenate prepared in ice-cold 0.1 M Tris–HCl buffer (pH 7.4) was centrifuged at 12,000 rpm for 45 min in Sigma laboratory centrifuge 3K18 model, rotor no. 12150. The clear supernatant was used for assay of key enzymes of carbohydrate metabolism, i.e., hexokinase (HK) (Brandstrup et al. 1957), phosphofructokinase (PFK) (Sadava et al. 1997), pyruvate kinase (PK) (Sadava et al. 1997) of glycolysis; glucose-6-phosphatase (G6Pase) (King 1965), fructose-1,6-bisphosphatase (F1, 6BPase) (Sadava et al. 1997) of gluconeogenesis; glycogen phosphorylase (Sutherland 1955) of glycogenolysis; glucose-6-phosphate dehydrogenase (G6PDH) (Beutler et al. 1955) of pentose phosphate pathway; fructokinase (Adelman et al. 1966) of fructose metabolism and aldose reductase (AR) (Hayman and Kinoshita 1965), sorbitol dehydrogenase (SDH) (Galambos et al. 1963) of polyol pathway. Protein content in the tissue homogenate and supernatant was measured by the method of Lowry et al. (1951).

2.10 Transaminases

The same cytosolic fractions of liver and kidney used for assay of carbohydrate metabolic enzyme were used for the assay of transaminases aspartate transaminase (ALT) and alanine transaminase (AST) (Reitman and Frankel 1990).

2.11 Lipid Metabolic Studies

The extraction of lipids from tissues was carried out according to the procedure of Folch et al. (1957). Plasma and tissue lipids, i.e., total cholesterol (TC), triglycerides (TG) (Liquid Gold Diagnostic kit), HDL-cholesterol (Assmann et al. 1983), VLDL-cholesterol, and LDL-cholesterol were calculated by using the (Friedewald et al. 1972) formula

$$\text{VLDL-C} = TG/5$$

$$LDL\text{-C} = TC\text{-}TG/5 - HDL\text{-C}$$

Free fatty acids (FFA) (Duncombe 1963), inorganic phosphate (Fiske and Subbarow 1925), and phospholipids (Connerty et al. 1961) were analyzed.

2.12 Lipid Metabolic Enzymes

The frozen tissues were slowly thawed at 4 °C. Ten percent tissue homogenate prepared in ice-cold 0.1 M potassium phosphate buffer, pH 8.0 was centrifuged at 12,000 rpm for 45 min. The clear supernatant was carefully collected and used for assay of fatty acid synthase (FAS) (Gibson and Hubbard 1960), malic enzyme (Storey and Bailey 1978a, b), and lipoprotein lipase (LPL) (Quinn et al. 1982).

2.13 Oxidative Stress Markers and Antioxidants

Immediately after separation of liver and pancreas, 10% tissue homogenates were prepared in 0.15 M KCl using Potter-Elvehjem homogenizer at 4 °C. The whole homogenate was used for estimation of glutathione (GSH) (Ellman's 1959), lipid peroxidation (LPO) (Utley et al. 1967), and protein oxidation (PO) (Levine et al. 1990).

Ten percent tissues homogenate in 0.15 M KCl was prepared by using Potter-Elvehjem homogenizer at 4 °C and centrifuged in cold (4 °C) at 12,000 rpm for 45 min. The clear supernatant was used for assay of antioxidant enzymes, viz., superoxide dismutase (SOD) (Hollenberg 2003), catalase (CAT) (Islam et al. 2009), glutathione peroxidase (GPx) (Johnson et al. 2007), glutathione-S-transferase (GST) (Habig et al. 1974), and glutathione reductase (GR) (Pinto and Bartley 1969).

2.14 Statistical Analysis

The results were expressed as means ± S.E.M. Data were analyzed for significant differences using Duncan's Multiple Range (DMR) test at $P < 0.05$.

3 Results and Discussion

3.1 Phytochemicals

C. roseus was investigated for its phytochemical components and therapeutic efficacies from ancient time. The plant has enormous phytoconstituents of medical importance. The preliminary qualitative phytochemical analysis of aqueous leaf suspension of *C. roseus* revealed the presence of alkaloids, flavonoids, tannins, phenols, and saponins indicating the presence of pharmacologically important phytochemicals (Table 1a). In contrast to our observation, Ibrahim et al. (2011) reported the absence of flavonoids, tannins, saponins, proteins, and amino acids in

Table 1a Qualitative phytochemical profiles of *C. roseus*

Phytochemicals	Presence/absence	Phytochemicals	Presence/absence
Alkaloids	+	Coumarins	–
Anthocyanins	–	Aminoacids	+
Anthocyanidins	–	Flavanoids	+
Glycosides	–	Tannins	+
Antraquinones	–	Phenols	+
Carbohydrates	+	Saponins	+
Carboxylic acids	–	Steroids	–
Catecholic compounds	+	Proteins	+

"+" denotes presence and "–"denotes absence

Table 1b Quantitative phytochemical profile of *C. roseus*

Total polyphenolic compounds (mg gallic acid equivalents/g extract)	Flavanoids (mg quercetin equivalents/g extract)	Saponins (mg diosgenin equivalents/g extract)
30.8	11.4	2.0

dichloromethane: methanol (1:1) extract of whole plant with the presence of alkaloids and carbohydrates, whereas Monika and Vandhana (2013) reported that carbohydrates, flavonoids, saponins, and alkaloids as potentially active phytochemicals of *C. roseus*. The quantitative analysis (Table 1b) revealed that 1 g of *C. roseus* contained 30.8 mg gallic acid equivalents of total polyphenols, 11.4 mg quercetin equivalents of flavonoids and 2.0 mg diosgenin equivalents of saponins which are found to be effective antioxidants, besides having antidiabetic, hypolipidemic, hepato, and cardioprotective activities.

Numerous epidemiological studies suggest that herbs/diets rich in phytochemicals and antioxidants execute a protective role in health and disease (Vinson et al. 2001). Flavonoids, steroids, alkaloids, and saponins are reported as bioactive antidiabetic (Kapoor 1990; Bone 1996) and anti-atherogenic principles (Omolekan and Olaiya 2013).

Polyphenolic compounds also possess a variety of other biological activities such as reduction of plasma lipids, which might be due to upregulation of LDL receptor expression (Kuhn et al. 2004) and inhibition of hepatic lipid synthesis (Thériault et al. 2006). Polyphenols are the major plant compounds with antioxidant activity. This activity is believed to be mainly due to redox properties (Wichi 1988), which play an important role in adsorbing and neutralizing free radicals, quenching singlet and triplet oxygen, or decomposing peroxides. In addition, the antioxidant properties of phenolic compounds and flavonoids are also attributed to their redox properties and the ability to chelate metals (Rice-Evans et al. 1996).

Flavonoids, one of the main groups of phenolic compounds, are conjugated to carbohydrate moiety and have a variety of biological activities, such as antioxidative (Moure et al. 2001), radical scavenging (Dugas et al. 2000), anti-inflammatory

(Crespo et al. 1999), and antidepressant (Butterweek et al. 2000). Furthermore, flavonoids are found to possess cancer preventive effects. Flavonoids are reported to regenerate damaged β-cells in the alloxan (Chakravarthy et al. 1980) and STZ (Coskun et al. 2005)-induced diabetic rats. Antiobesity potential of quercetin a well-known flavonoid is attributed by its inhibitory effect on lipogenesis at the level of gene expression to its correction of disturbed lipid metabolism. Kobori et al. (2011) observed that dietary intake of quercetin reduced body weight gain as well as visceral fat accumulation and improved systemic parameters related to metabolic syndrome (hyperglycemia, hyperinsulinemia, and dyslipidemia) by its antioxidant potential. They have shown a variety of activities such antitumor, cholesterol lowering, immune potentiating, anticancer, and antioxidant properties (Blumert and Liu 2003).

Saponins are known antinutritional factors and possess hypocholesterolemic effect due to reduced uptake of certain nutrient including glucose and lipid especially cholesterol at the gut through intra-lumena physicochemical interaction. Saponins, polyphenolic compounds, and Gallic tannins possess potent inhibition of intestinal glucose transport by inhibiting sodium glucose co-transporter-1 (S-GLUT-1) of intestines (Murakami et al. 1996; Kobayashi et al. 2000). Hence, the plant extract contains the phytoconstituents which are reported to have beneficial effects in modulating metabolic alterations observed under IR and insulin-deficient conditions.

3.2 *In Vitro Antioxidant Studies*

The state of increased free radical activity under insulin-deficient and insulin-resistant conditions is postulated to be the cause for establishment of oxidative stress, one of the mechanisms underlying diabetes and diabetic complications. On the basis of phytochemical profile which indicates the presence of phytoconstituents with antioxidant potential like polyphenols, alkaloids, tannins, etc. prompted to investigate the antioxidant potential of aqueous leaf suspension of *C. roseus* by conducting in vitro and in vivo experiments.

In vitro screening for antioxidant potential and radical scavenging activities of *C. roseus* revealed anti-lipidperoxidative effect and scavenging ability against hydrogen peroxide, superoxide, nitric oxide, hydroxyl, and DPPH radical with reducing ability. The extract inhibited lipid peroxides generated by the induction of Fe^{3+}/ADP/ascorbate in rat liver homogenate by 08.8–58.3% from 15 μg to 350 μg/mL concentration of *C. roseus* in a dose-dependent manner as presented in Fig. 1A. This effect of *C. roseus* was compared with well-known naturally isolated antioxidant curcumin from *Curcuma longa* (Zingiberaceae) as a standard. The IC_{50} value for inhibition of LPO is 240.12 μg/mL.

Superoxide radical generated in PMS-NADH systems by oxidation of NADH was assayed by the reduction of NBT. As furnished in the Fig. 1B, inhibition of O_2 formation was observed (2.32–90.2%) at 12.5–1000 μg/mL of *C. roseus*.

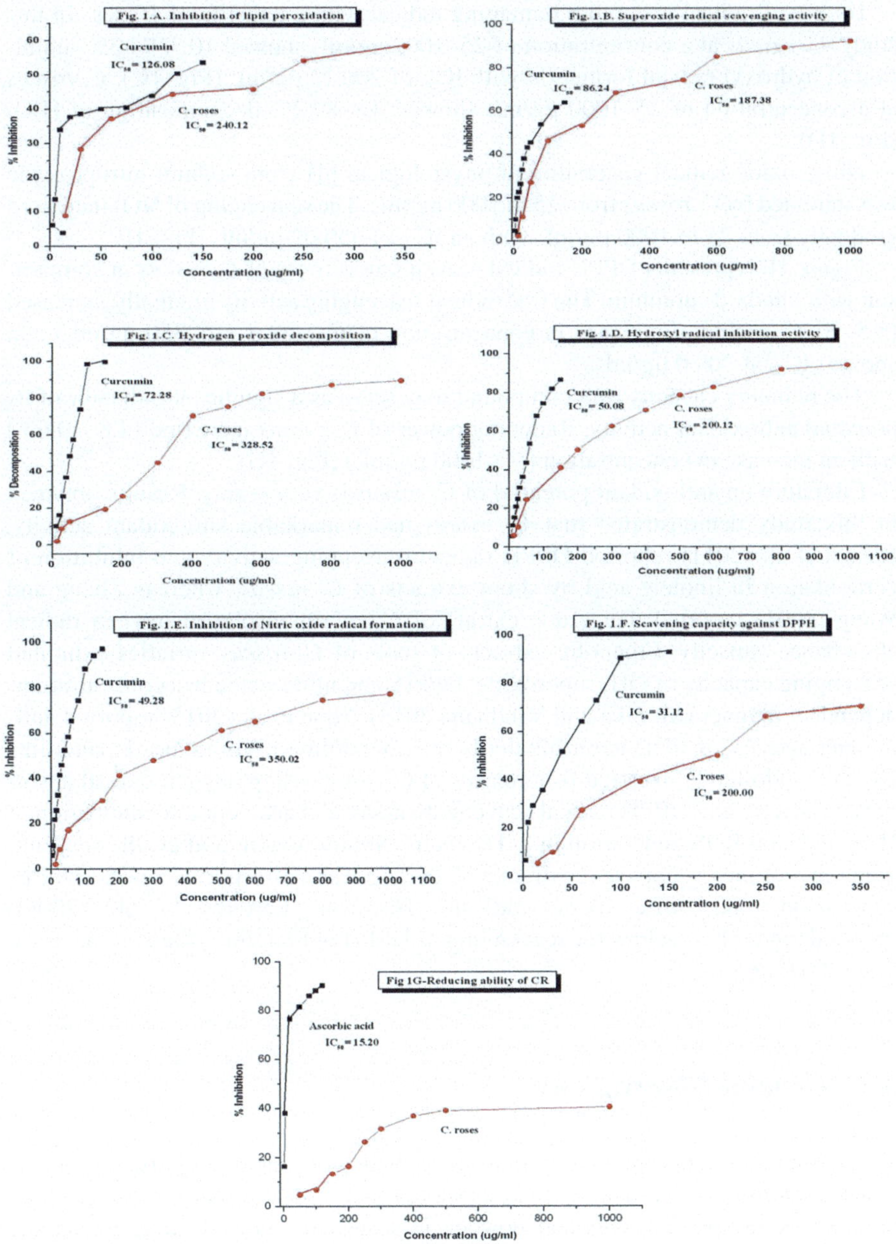

Fig. 1 Comparison of in vitro antioxidant potential of *C. roseus* with a standard natural antioxidant curcumin. (**A**) Inhibition of LPO, (**B**) superoxide radical scavenging activity, (**C**) hydroxyl radical inhibition activity, (**D**) hydrogen peroxide decomposition, (**E**) inhibition of nitric oxide radical formation, (**F**) scavenging capacity against DPPH and reducing the ability of *C. roseus* with a standard natural antioxidant ascorbic acid (**G**)

Hydroxyl radical is another damaging radical with a half-life of 10 sec. In this study, *C. roseus* at a concentration of 25–1000 μg/mL showed 10.3–94.3% inhibition of hydroxyl radical formation with IC_{50} of 200.12 μg/mL (Fig. 1C). *C. roseus* at a concentration of 25–1000 μg/mL showed 8.8–89.2% decomposition of H_2O_2 (Fig. 1D).

Nitric oxide radical generation at physiological pH from sodium nitroprusside was inhibited by *C. roseus* from 25 to 1000 μg/mL. The scavenging of $NO^{\cdot}$ increased gradually from 25 to 1000 μg/mL with an IC_{50} of 350.02 μg/mL (Fig. 1E).

Figure 1F represents DPPH radical scavenging activity of *C. roseus* in comparison with standard curcumin. The free radical scavenging activity gradually increased (5.5–70.4%) with an increase in concentration of *C. roseus* (15–350 μg/mL) and showed IC_{50} at 200.0 μg/mL.

The reducing capacity of a compound may serve as a significant indicator of its potential antioxidant activity. Reducing power of *C. roseus* increased (4.8–40.6%) with an increase in concentration (50–1000 μg/mL) (Fig. 1G)

Literature on antioxidant potential of *C. roseus* is very scanty. Results obtained in this study demonstrated that *C. roseus* had remarkable antioxidant activity. Rasool et al. (2011) reported DPPH radical scavenging activity and inhibition of peroxidation in linoleic acid by shoot extracts of *C. roseus,* whereas Zheng and Wang (2001) reported that crude extract of *C. roseus* displayed oxygen radical absorbance capacity. Ethanolic extracts of roots of *C. roseus* varieties exhibited scavenging capacity of OH, superoxide, DPPH, and nitric oxide in a concentration-dependent manner (Monika and Vandhana 2013). Soon et al. (2013) reported antioxidant activity of four major alkaloids, viz., vindoline I, vindoline II, vindoline III, and vindoline IV isolated from leaves of *C. roseus* using oxygen radical absorbance capacity and DPPH radical scavenging assay and also reported alleviation of H_2O_2-induced ROS generation in β-TC6 cells. Similar to our studies, the scavenging ability of the aqueous leaf extract of *C. roseus* was also reported against DPPH, superoxide, and nitric oxide radicals. However, Ferreres et al. (2008)) reported higher IC_{50} values for scavenging of DPPH (447 μg/mL) than our observation (200.0 μg/mL).

3.3 General Observations

No visible side effects and variation in animal behavior (respiratory distress, abnormal locomotion, and catalepsy) was observed in C + CR-group representing the nontoxic nature of *C. roseus* leaf powder. D-group rats showed the characteristic signs of diabetes such as polyuria, polydipsia, and polyphagia and failure to gain body weight. A significantly higher intake of food and water was observed in F-group from 10 days onwards of the experimental period compared to C-group. No significant variation was observed in the intake of food and water among the three groups: C, C + CR, and F + CR. Traditional medicines for the treatment of diabetes mellitus are probably based mainly on treatment of its obvious symptoms of pronounced thirst (polydipsia) and polyuria. Administration of *C. roseus* leaf powder suspension resulted in rectification of the signs of diabetes within 10 days of treatment in D + CR-group.

Our studies clearly demonstrated the antihyperglycemic activity of aqueous suspension of *C. roseus* in both STZ-induced diabetic and fructose diet-fed IR Wistar rats by rectifying the abnormalities of carbohydrate and lipid metabolisms under insulin-deficient and insulin-resistant conditions (Karuna and Saralakumari 2011; Karuna et al. 2010, 2013).

3.4 Carbohydrate Metabolic Studies

3.4.1 Body Weight, Plasma Glucose, Insulin, and HOMA

During the experimental period of 60 days, the mean body weight, plasma glucose, plasma insulin, and HOMA of the six experimental groups at 15-day interval are summarized in Fig. 2a–d respectively. STZ diabetic rats (D-group) showed a gradual increase in blood glucose level with gradual decrease in plasma insulin and body weights during experimental period. Fructose-fed rats (F-group) showed a gradual increase in plasma glucose, insulin, and body weight during experimental period. Thus, fructose-fed rats exhibited insulin resistance from its increased HOMA values during experimental period. There are reports showing an increase in energy intake, body weight, and adiposity with the consumption of high fructose diets both in humans and animals (Tordoff and Alleva 1990). Over weight and obese people have a much higher risk of developing Type 2 DM compared to those with healthy body weight. The results clearly indicate that the intensity of hyperglycemia is more prominent in STZ diabetic rats compared to fructose-fed rats due to decreased plasma insulin, whereas insulin resistance is more prominent in F-group compared to D-group.

Oral administration of *C. roseus* leaf powder suspension partially prevented the weight loss in STZ diabetic-treated rats (D + CR-group) and completely protected the fructose feed-induced weight gain in F + CR-group. The clinical symptoms of diabetes like polyphagia, polydipsia, and polyuria observed in the STZ diabetic rats were reversed within 15 days of *C. roseus* administration in D + CR-group. *C. roseus* administration resulted in gradual decrease in plasma glucose with a gradual increase in plasma insulin levels in D + CR-group. By the end of experimental period, plasma glucose levels of D + CR-group reached near-normal values but the observed increase in the plasma insulin level did not reach normal value as it was still significantly lesser than control rats. *C. roseus* treatment completely prevented the fructose-induced hyperglycemia and partially prevented hyperinsulinemia, resulting in a significantly decreased HOMA values in F + CR-group compared to F-group. Further C + CR rats showed normoglycemia with significantly lower plasma insulin levels than C-group indicating the beneficial effects of *C. roseus* in maintaining normoglycemia with lower insulin levels. Thus, C + CR rats are more insulin sensitive than C-group.

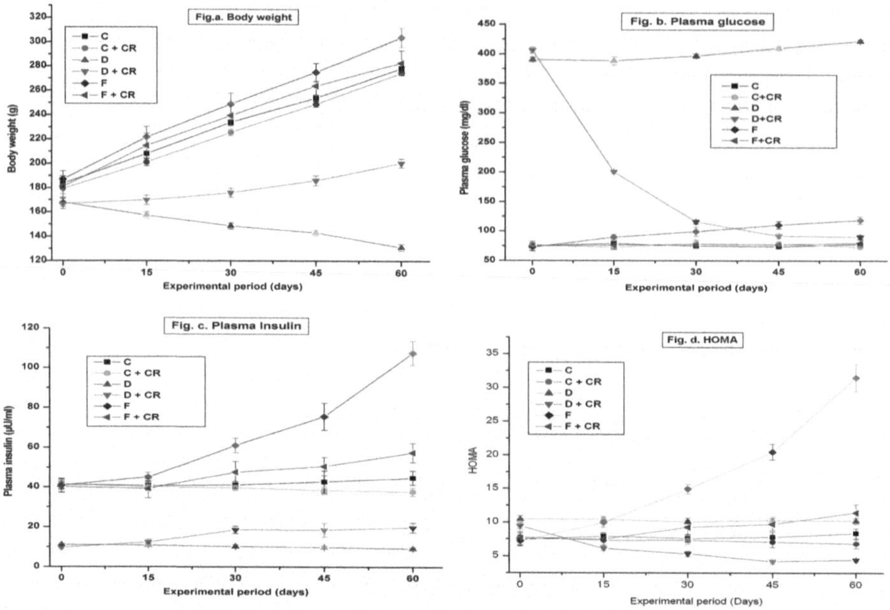

Fig. 2 Changes in body weight (**a**), fasting plasma glucose (**b**), fasting plasma insulin (**c**), and HOMA (**d**), of rats during the experimental period in group-C (control); group-C + CR (control treated with *C. roseus*); group-D (diabetic); group-D + CR (diabetic treated with *C. roseus*); group-F (fructose-fed rats); and group-F + CR (fructose fed treated with *C. roseus*). Values are mean ± SEM (*n*=8 animals in each group)

Similar to our studies, ethanolic extracts of leaves and flowers of *C. roseus* reported to show blood glucose lowering activity in hyperglycemic rats (Ghosh and Gupta 1982; Chattopadhya et al. 1991). Dichloromethane: methanol (1:1) extract of *C. roseus* leaves, twigs and whole plant showed antihyperglycemic activity along with prevention of body weight loss in alloxan (Jayanthi et al. 2010; Ibrahim et al. 2011) and STZ (Chattopadhyay et al. 1999; Singh et al. 2001)-induced diabetic rats. A complete protection against STZ challenge was reported by dichloromethane: methanol (1:1) extract prior treatment (500 mg/kg body weight for 30 days) (Chattopadhyay et al. 1999). However, the dose used in the above studies is five times higher than our study, i.e., 100 mg/kg body weight and it is organic solvent extract rather than aqueous leaf powder suspension. In contrast to our observations of normoglycemia of C + CR-group, the juice of fresh leaves of *C.roseus* showed blood glucose lowering activity both in alloxan diabetic and normal rabbits (Nammi et al. 2003). Soon et al. (2013) proposed the antidiabetic activity of alkaloid fractions isolated from leaves of *C. roseus* by promoting glucose uptake in pancreatic β-TC6 and myoblast C2C12 cells and also by inhibition of PTP-1B (a protein tyrosine phosphatase) a negative regulator of insulin signalling pathway.

3.4.2 Intestinal Disaccharidases

Diabetes mellitus either insulin deficient or insulin resistant is a state of nutrient starvation that frequently results in severe metabolic imbalances and pathological changes in many tissues including small intestine. Enhanced activities of intestinal disaccharidases, i.e., maltase, sucrase, and lactase were observed both in D-group (56.0, 15.2, and 26.1%) and F-group (12.6, 19.3, and 15.6%) rats compared to C-group, (Table 2) indicating the increased rate of digestion of disaccharides in these two metabolic conditions (insulin deficiency and insulin resistance). Therefore, enhanced activities of intestinal disaccharidases may play an important role in aggravating postprandial hyperglycemia. Diabetes stimulates the functional activity of the intestinal brush border membrane with enhancement of both hydrolytic enzyme activity (McAnuff-Harding et al. 2006) and membrane transport system (Hopfer 1975). Chronic diabetes enhances glucose transport by nonspecific increases in intestinal mass (Ferraris et al. 1993). *C. roseus* treatment significantly decreased the activities of intestinal disaccharides in D + CR and F + CR groups compared to D and F groups, respectively. This indicates the beneficial effects of *C. roseus* treatment in restoring the activities of intestinal disaccharidases to normal values. The α-glucosidase inhibitors are a new class of antihyperglycemic drugs that have a unique effect on the glycemic profile by delaying the absorption of disaccharides from intestine under hyperglycemic conditions. The α-glucosidase inhibitors not only delay carbohydrate absorption but they also alter the gastrointestinal hormonal axis which plays a role in the regulation of insulin secretion (Sunil and Sadekar 1999). Hence, acarbose, voglibose, and miglitol, the α-glucosidase inhibitors, are widely used either alone or in combination with insulin secretagogues in patient with Type 2 diabetes (Standl et al. 1999). Thus, antidiabetic property of *C. roseus* may also be due to its intestinal disaccharidase inhibitory activity. To our knowledge, this is the first report of *C. roseus* on intestinal disaccharidase inhibitory activity.

3.4.3 Key Enzymes of Carbohydrate Metabolic Pathways

The concept of over and underutilization of glucose by peripheral tissues under insulin-deficient and insulin-resistant conditions plays a pivotal role in establishment of hyperglycemia. Hence, in an attempt to gain an insight into the underlying biochemical mechanism of antihyperglycemic activity of *C. roseus* the carbohydrate metabolism was studied by assaying the activities of key enzymes of different pathways of metabolism, i.e., glycolysis, HMP shunt, gluconeogenesis, glycogenolysis, and polyol pathway in appropriate tissues of six experimental groups.

Table 2 Effect of *C. roseus* treatment on intestine disaccharidases

Disaccharidase (nmol of disaccharide hydrolyzed/min/mg protein)	C	C + CR	D	D + CR	F	F + CR
Maltase	3.96±0.04[a]	3.80±0.03[a]	6.18±0.09[b]	3.82±0.02[a]	4.46±0.08[c]	3.85±0.02[a]
Sucrase	3.73±0.02[a]	3.63±0.02[a]	4.20±0.04[b]	3.70±0.02[a]	4.45±0.14[b]	3.64±0.03[a]
Lactase	2.75±0.03[a]	2.65±0.03[a]	3.47±0.06[b]	2.84±0.05[a]	3.18±0.05[c]	2.67±0.03[a]

Values are mean ± S.E.M. (n=8 animals). Values with different superscripts within the row are significantly different at $P<0.05$ (Duncan's multiple range test)

Glycolysis

Glycolysis is the central pathway of glucose catabolism and metabolic energy. The data on activities of key glycolytic enzymes, i.e., HK, PFK, and PK of liver and muscle of six experimental groups are presented in Table 3. Regulation of flux through glycolysis is dependent on the tissue under consideration and nutritional and hormonal state of the tissue. D-group and F-group rats showed significantly decreased activities of HK and PFK the first two rate-limiting steps of glycolysis both in liver and muscle compared to C-group. The percentage decrease in HK and PFK in liver and muscle of D-group were 43.2 and 40.7%, 37.3 and 58.1% and in F-group are 27.7 and 35.3, 25.3, and 37.4% compared to control rats. But the decrease in these enzyme activities of F-group was significantly lower than D-group. Unlike HK and PFK, the activity of PK showed a different trend in D-group and F-group. It is significantly decreased in D-group both in liver (32.2%) and muscle (37.6%) compared to C-group and significantly enhanced both in the liver (22.5%) and muscle (7.3%) of F-group compared to C-group. In this study, the significantly decreased activities of key enzymes of glycolysis observed in STZ-induced diabetic rats indicated decreased operation of glycolysis under insulin-deficient condition both in the insulin-dependent tissue (skeletal muscle) and insulin-independent tissue (liver). Similar trend of decreased glycolytic enzyme activities was reported in various tissues of diabetic animals (Grover et al. 2002; Rathi et al. 2002; Ramesh et al. 2013). Decreased HK activity also indicates decreased availability of glucose-6-phosphate for HMP shunt operation under diabetic condition. Further, the decreased rate of glycolytic pathway in diabetic animals may be due to the competition for NAD between the sorbitol pathway and glycolysis (Williamson et al. 1993). Similar to our observation, decreased activities of HK and PFK with increased activity of PK in liver and muscle of fructose-fed rats were also reported (Rewana et al. 1993; Kannappa et al. 2006; Reddy et al. 2009) indicating decreased rates of glucose oxidation by glycolysis in these tissues. The prevailing hyperglycemia in fructose-fed conditions may lead to increased uptake of glucose into hepatocytes through insulin-independent glucose transporter GLUT-2. In addition, decreased operation of glycolysis is also responsible for building up of intracellular glucose concentration leading to glucotoxicity resulted in impaired insulin sensitivity in the liver. *C. roseus* treatment in D + CR-group prevented the diabetic-induced alterations in glycolytic enzyme activities. F + CR-group showed significantly enhanced muscle HK (64.7%) and PFK (55.9%) activities and significantly decreased hepatic and muscular PK activity (16.1 and 9.0%) compared to F-group. Studies of Weber and Convery (1966) indicated that insulin administration restored the decreased glycolytic enzyme activities of diabetic animals. Improvement in plasma insulin levels by *C. roseus* treatment in D + CR-group might be responsible for increased activity of glycolytic enzymes because of the transcriptional upregulation of glucokinase, PFK, and PK genes by insulin (Howard 1995). *C. roseus* administration to fructose-fed rats prevented the fructose feed-induced decrease in HK and PFK activity in muscle and the increase in PK activity in muscle and liver. Even though improvement was observed in the activities of hepatic glycolytic enzymes in F + CR-group

Table 3 Effect of *C. roseus* treatment on glycolytic enzymes of liver and muscle in STZ diabetic and fructose-fed IR rats

Parameter	Tissue	C	C + CR	D	D + CR	F	F + CR
Hexokinase (µmol of G6P formed/min/mg protein)	Liver	3.93±0.08[a]	4.052±0.03[a]	2.23±0.04[b]	3.80±0.12[a]	2.84±0.08[c]	3.00±0.08[c]
	Muscle	4.39±0.07[a]	4.52±0.08[a]	2.60±0.05[b]	4.53±0.07[a]	2.84±0.06[c]	4.68±0.27[a]
Phosphofructo kinase (µmol of F16Bisphosphate formed/min/mg protein)	Liver	3.35±0.05[a]	3.59±0.05[b]	2.10±0.05[c]	3.27±0.04[a]	2.50±0.04[d]	2.62±0.08[d]
	Muscle	4.17±0.07[a]	4.30±0.07[a]	1.75±0.02[b]	4.15±0.07[a]	2.61±0.07[c]	4.07±0.10[a]
Pyruvate kinase (µmol of NADH oxidized/min/mg protein)	Liver	2.57±0.04[a]	2.76±0.06[b]	1.74±0.02[c]	2.69±0.06[a]	3.15±0.05[d]	2.64±0.06[a]
	Muscle	2.87±0.03[a]	2.92±0.05[a]	1.79±0.04[b]	2.78±0.06[a]	3.08±0.04[c]	2.80±0.03[a]

Values are mean ± S.E.M. (n=8 animals). Values with different superscripts within the row are significantly different at $P<0.05$ (Duncan's multiple range test)

compared to F-group, the plant treatment does not restore the liver PK and PFK enzymes to normal level. The current results of glycolytic enzyme activities represent the beneficial effects of *C. roseus* treatment in rectifying the impairment in the insulin-mediated glucose transport observed in the muscle tissue of fructose-fed rats, whereas insulin resistance is partially rectified at hepatic level.

3.5 Fructose Metabolism

Fructose is rapidly phosphorylated by ATP in liver to form fructose-1-phosphate, catalyzed by the first enzyme of the fructose catabolic pathway i.e., fructokinase. This enzyme is virtually specific for fructose. Glucose is a general substrate for all body tissues while fructose has to be processed in the liver first and its intake represents a carbohydrate load targeted on the liver. The high capacity of this enzyme ensures the channeling of the majority of fructose into hepatic metabolism. Fructose is absorbed by intestinal epithelium and transported into hepatic portal vein. Then, all fructose absorbed flows through the liver initially. Fructokinase rapidly phosphorylates fructose to fructose-1-phosphate. FK activity in the liver of six experimental groups presented in Table 4 revealed no significant variation among C, C + CR, D, and D + CR groups as these groups are fed on normal chow diet. However, significantly increased hepatic fructokinase activity was observed both in F-group (33.1%) and F + CR-group (26.2%) compared to C-group. *C. roseus* treatment resulted in a slight but not significant decrease in hepatic fructokinase activity in F + CR-group compared to F-group. It was demonstrated that prolonged feeding is necessary for meaningful induction of enzyme (Fukuda et al. 1983). Mammalian liver is capable of adaptation to fluctuations in dietary fructose content by showing substantial inducibility of FK by fructose.

3.5.1 Gluconeogenesis

Glucose is produced through gluconeogenesis and glycogenolysis. Both phenomena are inhibited by insulin and enhanced by a deficiency of insulin action (Friedmann et al. 1967). Gluconeogenesis and glycolysis are reciprocally regulated so that one pathway is relatively inactive while the other is highly active.

Table 4 Effect of *C. roseus* treatment on hepatic fructokinase in STZ diabetic and fructose fed IR rats

Parameter	C	C + CR	D	D + CR	F	F + CR
Fructokinase (μmol NADH oxidized/min/mg protein)	1.09 ± 0.03[a]	1.04 ± 0.02[a]	0.99 ± 0.07[a]	1.17 ± 0.05[a]	1.45 ± 0.03[b]	1.38 ± 0.04[b]

Values are mean ± S.E.M. (*n*=8 animals). Values with different superscripts within the row are significantly different at $P<0.05$ (Duncan's multiple range test)

Gluconeogenesis mostly takes place in liver and to some extent in kidney. Increased hepatic glucose production is a major component of diabetes and insulin resistance-induced hyperglycemia. The data present in Table 5 revealed significantly increased activities of hepatic and renal F1,6BPase and G6Pase both in D and F groups compared to C-group. However, the enhancement in these enzyme activities is more prominent in D-group than F-group. The percent increase in hepatic and renal F1,6BPase and G6Pase activities of D-group are (76.9%, 63.6%) and (133.3%, 42.9%) and in F-group are (43.5%, 45.5%) and (59.5%, 23.2%), respectively, compared to C-group. Enhanced gluconeogenesis under diabetic condition may also be due to increased availability of gluconeogenic substrates by enhanced catabolism of proteins and lipids. *C. roseus* treatment for 60 days resulted in no significant alterations in the activities of hepatic and kidney F1,6BPase and G6Pase in C + CR-group compared to C-group. Thus, intensity of hyperglycemia in D and F groups are in correlation with the level of operation gluconeogenesis. *C. roseus*-treated diabetic (D + CR-group) and fructose-fed (F + CR-group) rats showed a significant decrease in these enzyme activities both in the liver and kidney when compared to D and F groups, respectively, and this decrease further resulted in bringing these enzyme activities to normal values. Thus, *C. roseus* treatment in D + CR and F + CR groups prevented the increased operation of gluconeogenesis seen under diabetic conditions.

3.5.2 Glycogen and Glycogen Phosphorylase

In animals, glycogen is the main storage form of carbohydrates present mainly in liver and muscle. The concentration of tissue glycogen depends upon the rate of glycogenesis and glycogenolysis. Glycogen stores of muscle and liver serve entirely different roles, i.e., in liver it serves as a glucose reserve for the maintenance of blood glucose homeostasis whereas in muscle it is a readily available source of glucose for its glycolysis. When compared to C-group, D and F group rats showed differential trend in regard to the changes in hepatic glycogen content and hepatic glycogen phosphorylase activity (Table 6). D-group showed significantly decreased glycogen content in liver (60.9%) and muscle (70.3%), with significantly enhanced hepatic glycogen phosphorylase activity (119.4%) whereas F-group showed significantly enhanced glycogen content in liver (24.7%) and muscle (18.6%) with significant decrease (29.7%) in the activity of glycogen phosphorylase when compared to C-group. It appears that fructose is a better substrate for glycogen synthesis than glucose. This can be explained by enhanced fructose metabolic production of fructose-1-phosphate by increased activity of FK. Dietary fructose is mainly metabolized in the liver. Fructose after converting to fructose-1 phosphate by FK, it is converted to dihydroxy acetone phosphate (DHAP) and glyceraldehyde- 3-phosphate. Increased concentration of these triose phosphates in the liver drives the glycogenesis towards glucose-6-phosphate, glucose-1-phosphate, and glycogen formation (Mayes 1993). Once liver glycogen is replenished, the intermediates of fructose metabolism are primarily directed towards TG synthesis (Parniak 1988). Our results

Table 5 Effect of *C. roseus* treatment on gluconeogenic enzymes in liver and kidney of STZ diabetic and fructose-fed IR rats

Parameter	Tissue	C	C + CR	D	D + CR	F	F + CR
Fructose-1, 6-bisphosphatase (nmol of F6P formed/min/mg protein)	Liver	0.39 ± 0.008^{a}	0.40 ± 0.001^{a}	0.69 ± 0.013^{b}	0.42 ± 0.010^{a}	0.56 ± 0.01^{c}	0.40 ± 0.007^{a}
	Kidney	0.11 ± 0.002^{a}	0.12 ± 0.005^{a}	0.18 ± 0.011^{b}	0.12 ± 0.003^{a}	0.16 ± 0.006^{c}	0.12 ± 0.005^{a}
Glucose-6-phosphatase (nmol of Pi formed/min/mg protein)	Liver	20.58 ± 0.53^{a}	21.37 ± 0.21^{a}	48.02 ± 2.50^{b}	22.00 ± 0.65^{a}	32.83 ± 1.35^{c}	21.51 ± 0.22^{a}
	Kidney	17.45 ± 0.27^{a}	17.32 ± 0.45^{a}	24.92 ± 0.82^{b}	18.25 ± 0.27^{a}	21.49 ± 0.37^{c}	18.06 ± 0.20^{a}

Values are mean ± S.E.M. (*n*=8 animals). Values with different superscripts within the row are significantly different at $P<0.05$ (Duncan's multiple range test)

Table 6 Effect of *C. roseus* treatment on glycogen and hepatic glycogen phosphorylase in STZ diabetic and fructose-fed IR rats

Parameter	Tissue	C	C + CR	D	D + CR	F	F + CR
Glycogen (mg glucose/g tissue)	Muscle	2.36 ± 0.08^{a}	2.48 ± 0.07^{a}	0.70 ± 0.02^{b}	1.43 ± 0.11^{c}	2.80 ± 0.13^{d}	2.51 ± 0.06^{e}
	Liver	29.24 ± 0.35^{a}	28.89 ± 10.70^{a}	11.42 ± 0.36^{b}	22.25 ± 0.59^{c}	36.53 ± 1.20^{d}	31.60 ± 0.79^{e}
Glycogen phosphorylase (μmol of Pi formed/min/mg protein)	Liver	0.131 ± 0.01^{a}	0.137 ± 0.01^{a}	0.288 ± 0.03^{b}	0.149 ± 0.01^{c}	0.092 ± 0.008^{d}	0.118 ± 0.007^{e}

Values are mean ± S.E.M. ($n=8$ animals). Values with different superscripts within the row are significantly different at $P<0.05$ (Duncan's multiple range test)

of enhanced glycogen content in fructose-fed rats are in accordance with earlier studies (Youn et al. 1987; Murakami et al. 1996; Reddy et al. 2009). However, a few contradictory observations of decrease hepatic glycogen content in fructose-fed rats also appeared in literature (Rajasekar and Anuradha 2007). In this study, the enhanced hepatic glycogen stores in fructose-fed rats can be explained by the observed decrease in the activity of glycogen phosphorylase. *C. roseus*-treated control rats (C + CR-group) showed no significant variation in the glycogen content of liver and muscle and hepatic glycogen phosphorylase when compared to C-group. Glycogen levels in various tissues especially in skeletal muscle is direct reflection of insulin activity as insulin promotes intracellular glycogen deposition by stimulating glycogen synthase and inhibiting glycogen phosphorylase (Ortmeyer et al. 1997). Two months of treatment with *C. roseus* partially prevented the depletion of glycogen in the liver and muscle with decreased hepatic glycogen phosphorylase of STZ diabetic rats. This may be due to increased circulatory insulin concentrations observed in the D + CR-group compared to D-group. Thus *C. roseus* partially corrected the altered glycogen metabolism observed in D-rats. Co-administration of *C. roseus* for 60 days partially prevented the fructose feeding-induced enhancement in the glycogen content of liver and muscle in F + CR-group

3.5.3 Polyol Pathway

Polyol pathway also known as sorbitol pathway basically involves the conversion of glucose to fructose via sorbitol. Aldose reductase, the first and rate-limiting step of polyol pathway, reduces glucose to sorbitol with the aid of co-factor NADPH and the second enzyme SDH with its co-factor NAD^+, converts sorbitol to fructose. Under hyperglycemic conditions, glucose disposal through polyol pathway tends to increase to 30% from normal 3%, and it plays a major role in the pathogenesis of diabetic complications as it is one of the major sources of oxidative stress. The product of this pathway, i.e., fructose and its metabolites, fructose-3-phosphate and 3-deoxyglucose, are more potent nonenzymatic glycation agents than glucose, the flux of glucose through the polyol pathway would increase advanced glycation end products (AGEs) formation.

The activities of AR and SDH in liver, pancreas, and heart tissues of six experimental groups are given in Table 7. In comparison with group-C, group-D showed a significant increase in the AR and SDH activities in liver (40.0%, 35.1%), pancreas (16.2%, 18.6%), and heart (22.9%, 12.4%) whereas F-group showed slight but not a significant increase in AR activity and a significant increase in SDH activities in liver (11.7%), pancreas (7.6%), and heart (8.2%). Thus, enhanced operation of polyol pathway is more prominent in D-group than F-group. *C.roseus* treatment resulted in the restoration of the enhanced activities of both AR and SDH to normal values in all tissues in D + CR-group. Except hepatic SDH, the activities of AR and SDH of F + CR-group are not significantly varied from C-group. Reports are available on polyol pathway inhibiting activity of many plant extracts and phytochemicals. Inhibition of aldose reductase has been shown to ameliorate vascular and other

Table 7 Effect of *C. roseus* treatment on aldose reductase and sorbitol dehydrogenase in STZ diabetic and fructose-fed IR rats

Parameter	Tissue	C	C + CR	D	D + CR	F	F + CR
Aldose reductase (μmol of NADPH oxidized/min/mg protein	Liver	0.87 ± 0.03^{a}	0.86 ± 0.04^{a}	1.22 ± 0.03^{b}	0.83 ± 0.04^{a}	0.94 ± 0.04^{a}	0.85 ± 0.02^{a}
	Pancreas	0.68 ± 0.02^{a}	0.67 ± 0.02^{a}	0.79 ± 0.02^{b}	0.69 ± 0.01^{a}	0.75 ± 0.04^{a}	0.69 ± 0.02^{a}
	Heart	0.48 ± 0.02^{a}	0.47 ± 0.01^{a}	0.59 ± 0.01^{b}	0.49 ± 0.02^{a}	0.55 ± 0.04^{a}	0.50 ± 0.02^{a}
Sorbitol dehydrogenase (μmol of NADH oxidized/min/mg protein)	Liver	4.93 ± 0.08^{a}	5.00 ± 0.03^{a}	6.66 ± 0.11^{b}	5.23 ± 0.10^{c}	5.51 ± 0.06^{d}	5.20 ± 0.08^{c}
	Pancreas	2.09 ± 0.03^{a}	2.06 ± 0.03^{a}	2.48 ± 0.06^{b}	2.11 ± 0.04^{a}	2.25 ± 0.03^{c}	2.07 ± 0.02^{a}
	Heart	2.17 ± 0.03^{a}	2.16 ± 0.03^{a}	2.44 ± 0.04^{b}	2.14 ± 0.03^{a}	2.35 ± 0.03^{c}	2.20 ± 0.01^{a}

Values are mean ± S.E.M. (n=8 animals). Values with different superscripts within the row are significantly different at $P<0.05$ (Duncan's multiple range test)

complications in diabetes (Greene et al. 1993; Williamson et al. 1993). Thus restoration of enhanced polyol pathway enzymes to normal levels by *C. roseus* supplementation in diabetic and fructose-fed animals (D + CR and F + CR) avoided the deleterious alterations due to enhanced operation of polyol pathway towards oxidative stress.

3.6 *Protein Metabolism*

3.6.1 Transaminases

To understand the beneficial effects of *C.rosues* treatment against alteration in protein catabolism under diabetic conditions, hepatic and renal transaminase (ALT and AST) activities are assessed in six experimental groups and represented in Table 8. The data revealed a significant increase in tissue transaminase activities both in D and F groups compared to C-group. An increase in transaminase activities was more pronounced in D-group than F-group. Tissue transaminase activities are increased in situations associated with enhanced amino acid catabolism and gluconeogenesis. The observed elevation of transaminase activities in liver and kidney of D-group and F-group rats is an indication of increased protein degradation and amino acid catabolism in these metabolic conditions, thus providing precursors for gluconeogenesis. In addition, enhanced nonenzymatic glycation of proteins under hyperglycemic conditions may decrease the half-life of proteins, thus contributing to the enhanced protein degradation (Vlassara and Palace 2002). More prominent enhancement in transaminase activities of diabetic rats compared to fructose-fed rats indicates more intensified protein degradation and amino acid catabolism under insulin-deficient condition than insulin-resistant condition. This is further supported by a decrease in body weight of D-group rats by decreased muscle mass. *C. roseus* administration for 60 days prevented the increased transaminase activities observed both in D + CR and F + CR groups. The restoration of transaminase activities of liver and kidney of D + CR and F + CR rats to their respective normal levels further strengthens the protective effect of *C. roseus* against diabetes-induced alterations in protein metabolism. In contrasts to our study, decreased activities of hepatic transaminases reported in diabetic rats are not ameliorated by *C. roseus* treatment (Singh et al. 2001).

3.7 *Lipid Metabolism*

3.7.1 Plasma Lipid Profile

The plasma lipid profiles of the six groups of animals at initial and final days of the experimental period are represented in Table 9. Both D and F groups showed dyslipidemia with respective significant increase in the plasma TC (13.2%, 62%), TG

Table 8 Effect of *C. roseus* treatment on tissue transaminases in STZ diabetic and fructose-fed IR rats

Parameter	Tissue	C	C + CR	D	D + CR	F	F + CR
ALT (µg of pyruvate formed/min/mg protein)	Liver	1.22±0.03[a]	1.20±0.01[a]	2.08±0.04[b]	1.28±0.03[a]	1.38±0.04[c]	1.26±0.02[a]
	Kidney	0.21±0.01[a]	0.22±0.01[a]	0.41±0.02[b]	0.26±0.02[c]	0.29±0.02[d]	0.22±0.01[a]
AST (µg of pyruvate formed/min/mg protein)	Liver	0.81±0.02[a]	0.82±0.02[a]	1.30±0.03[b]	0.86±0.03[a]	0.92±0.02[c]	0.78±0.01[a]
	Kidney	0.42±0.01[a]	0.38±0.03[a]	0.73±0.02[b]	0.44±0.01[a]	0.51±0.01[c]	0.44±0.01[a]

Values are mean ± S.E.M. (n=8 animals). Values with different superscripts within the row are significantly different at $P<0.05$ (Duncan's multiple range test)

Table 9 Effect of *C. roseus* treatment on plasma lipid profile in STZ diabetic and fructose-fed IR rats

Parameter	Days	C	C + CR	D	D + CR	F	F + CR
Total cholesterol (mg/dL)	Initial (0 day of experimentation)	65.26 ± 3.21^{a}	67.19 ± 2.67^{a}	72.67 ± 3.40^{b}	73.03 ± 2.86^{b}	68.60 ± 3.45^{a}	67.18 ± 4.67^{a}
	Final (60 days of experiment)	72.89 ± 3.20^{b}	62.21 ± 4.22^{a}	82.51 ± 5.86^{c}	75.60 ± 3.04^{a}	118.20 ± 6.19^{d}	76.27 ± 3.76^{b}
Triglycerides (mg/dL)	Initial	69.91 ± 3.31^{a}	75.01 ± 3.60^{a}	79.51 ± 2.68^{b}	76.89 ± 2.91^{b}	70.25 ± 4.81^{a}	73.88 ± 5.79^{a}
	Final	76.28 ± 4.68^{b}	68.78 ± 2.74^{a}	89.37 ± 5.60^{c}	77.46 ± 2.66^{b}	156.54 ± 8.80^{d}	80.48 ± 8.26^{b}
HDL-cholesterol (mg/dL)	Initial	33.71 ± 1.16^{a}	32.26 ± 0.78^{a}	30.41 ± 1.46^{b}	29.18 ± 1.92^{b}	33.35 ± 2.88^{a}	32.94 ± 1.60^{a}
	Final	29.43 ± 1.28^{b}	37.76 ± 0.89^{c}	26.10 ± 1.45^{d}	32.50 ± 0.77^{a}	25.91 ± 1.81^{d}	28.93 ± 1.55^{b}
LDL-cholesterol (mg/dL)	Initial	13.90 ± 0.66^{a}	14.99 ± 0.72^{a}	15.90 ± 0.53^{b}	15.21 ± 0.58^{b}	14.04 ± 0.96^{a}	14.77 ± 0.40^{a}
	Final	15.25 ± 0.52^{b}	13.75 ± 0.54^{a}	17.87 ± 1.12^{c}	15.49 ± 0.53^{b}	31.30 ± 1.71^{d}	16.09 ± 1.65^{b}
VLDL-cholesterol (mg/dL)	Initial	17.65 ± 3.73^{a}	19.93 ± 2.85^{a}	26.36 ± 2.97^{b}	28.83 ± 2.40^{b}	21.20 ± 3.10^{a}	22.47 ± 2.54^{a}
	Final	28.17 ± 2.93^{b}	10.76 ± 3.82^{c}	38.70 ± 4.27^{d}	27.60 ± 2.18^{b}	59.97 ± 4.36^{e}	31.05 ± 2.12^{b}
Atherogenic index	Initial	1.93 ± 0.20^{a}	2.05 ± 0.08^{a}	2.39 ± 0.22^{b}	2.56 ± 0.10^{b}	2.10 ± 0.13^{a}	2.13 ± 0.07^{a}
	Final	2.47 ± 0.17^{b}	1.64 ± 0.18^{c}	3.20 ± 0.21^{d}	2.32 ± 0.15^{b}	4.55 ± 0.20^{e}	2.63 ± 0.12^{b}

Values are mean ± S.E.M. (n=8 animals). Values with different superscripts within the row are significantly different from initial values of controls at $P<0.05$ (Duncan's multiple range test)

(17.2%, 105.0%), LDL-cholesterol (11.3%, 112.7%), and VLDL-cholesterol (17.2%, 104.5%) and a significant decrease in HDL-cholesterol (11.3%, 12.0%) compared to the C-group. The dyslipidemia observed in both D and F groups further resulted in a significant increase in the atherogenic index (27.9%, 80.0%). In contrast to hyperglycemia, the intensity of hyperlipidemia was more in insulin-resistant rats (F-group) than insulin-deficient rats (D-group) which may be due to the fact that fructose is more lipogenic than glucose. The abnormal high concentration of serum lipids in diabetes is mainly due to the increase in the mobilization of FFAs from the peripheral depots since insulin inhibits the hormone-sensitive lipase. High levels of plasma triacylglycerols is a well-established consequence of dietary fructose intake (Kelley et al. 2004). Enhanced hepatic lipogenesis, overproduction of VLDL, and impairment in their peripheral catabolism (Busserolles et al. 2002) are responsible for the observed dyslipidemia in fructose-fed rats.

C. roseus administration for 60 days improved the plasma lipid profile by lowering TC, TG, LDL-cholesterol, and VLDL-cholesterol and increasing HDL-cholesterol resulting in decreased atherogenic index in both D + CR and F + CR groups. Thus by the end of the experimental period, the lipid profile of D + CR and F + CR groups were not deviated from C-group indicating the protective effect of *C. roseus* against dyslipidemia. The beneficial anti-atherogenic effects of *C. roseus* are also evident from the healthy lipid profile of C + CR-group compared to C-group with 23.2% decrease in atherogenic index.

Natarajan et al. (2012) reported antihyperlipidemic activity of aqueous extract of flowers of *C. roseus* in alloxan-induced diabetic rats at a dose of 350 and 450 mg/kg body weight for 30 days with no significant effect at lower dose, i.e., 250 mg/kg body weight.

3.7.2 Tissue Lipids

The observed therapeutic effect of *C. roseus* treatment against the detected dyslipidemia with development of decreased insulin sensitivity in both diabetic models promoted to study the alterations in tissue lipids and lipid metabolic enzymes in these experimental groups. The data presented in Table 10 reveal the hepatic and heart tissue total cholesterol, triglycerides, phospholipids, and FFA of the six experimental groups. When compared to C-group, the percent increase in hepatic and cardiac TC, TG, phospholipids, and FFA are 12.8, 9.13, 5.7, and 15.2%, and 21.6, 17.9, 16.5, and 32.8%, respectively, in D-group and 21.6, 17.8, 51.7, and 32.2%, and 41.4, 21.6, 63.9, and 53.7%, respectively, in F-group. Thus, F-group showed higher lipid accumulation in both liver and heart tissues compared to STZ diabetic rats. It was well known that the process of lipid accumulation interferes with utilization of glucose through principles of Randle cycle and hepatic lipid accumulation seen both under insulin-deficient and insulin-resistant conditions may result in fatty liver. Over accumulation of TG in the liver and muscle may lead to overproduction of metabolites (ceramides and diacylglycerol) which induce a cascade of serine/tyrosine phosphorylation reactions that diminish the glucose transport activity and

Table 10 Effect of *C. roseus* treatment on tissue lipids in STZ diabetic and fructose-fed IR rats

Parameter	Tissue	C	C + CR	D	D + CR	F	F + CR
TC (mg/100 g tissue)	Liver	149.67±2.10[a]	137.41±3.01[b]	168.80±2.38[c]	149.16±2.32[a]	181.92±3.55[d]	152.37±2.18[a]
	Heart	62.53±1.95[a]	56.34±1.20[b]	70.17±1.35[c]	61.52±1.60[a]	88.41±3.73[d]	63.65±1.50[a]
TG (mg/100 g tissue)	Liver	156.61±2.28[a]	145.82±2.68[b]	170.97±2.22[c]	149.78±2.37[a]	184.48±4.55[d]	160.14±2.30[a]
	Heart	67.74±0.62[a]	62.85±0.90[b]	74.77±0.93[c]	68.52±1.24[a]	82.47±2.86[d]	69.82±1.10[a]
PLP (μg/100 g tissue)	Liver	887.6±13.06[a]	847.7±12.10[b]	938.5±17.08[c]	869.4±16.58[a]	1034.2±14.28[d]	892.4±9.40[a]
	Heart	447.5±5.14[a]	412.3±4.07[b]	634.4±5.00[c]	425.6±6.81[a]	733.1±6.10[d]	430.5±5.20[a]
FFA (mg/100 g tissue)	Liver	210.24±3.14[a]	191.49±4.16[b]	242.35±3.56[c]	205.66±3.66[a]	279.80±5.32[d]	212.38±3.89[a]
	Heart	122.50±2.69[a]	110.71±2.73[b]	159.81±2.41[c]	126.50±3.42[a]	188.44±2.23[d]	128.11±2.26[e]

Values are mean ± S.E.M. (n=8 animals). Values with different superscripts within the row are significantly different at $P<0.05$ (Duncan's multiple range test)

other events that desensitize insulin receptor signalling. A causal link of cardiac lipid accumulation and cardiac dysfunction has been proposed. The accumulation of neutral lipids in cardiac myocytes causes lipotoxic heart disease. Further, the lipotoxicity also plays a role in the development of contractile dysfunction. *C. roseus* administration for 60 days restored the tissue lipid fractions to normal values both in (D + CR) and (F + CR) groups. However, C + CR-group showed a slight but a significantly decreased level of tissue lipid fractions when compared to C-group.

In order to understand the contribution of lipogenic enzyme activities for the fatty liver in D and F groups and their alterations by *C. roseus* administration, the following hepatic lipogenic enzymes, i.e., FAS, malic enzyme, and G6PDH along with LPL of adipose tissue were assayed in six experimental groups.

3.7.3 Fatty Acid Synthase, G6PDH, Malic Enzyme, and Lipoprotein Lipase

Fatty acid synthase plays a central role in de novo lipogenesis in animals by catalyzing all the reactions in conversion of acetyl-CoA and malonyl-CoA to palmitate. Malic enzyme and G6PDH play a key role in generation of NADPH for lipogenesis. Lipoprotein lipase is an enzyme responsible for the hydrolysis of triacylglycerols from plasma lipoproteins, mainly chylomicrons and very low-density lipoproteins and its activity is influenced by nutritional and hormonal status and by environmental conditions. Adipose tissue LPL is the enzyme that initiates the entry of lipoprotein packaged fatty acids into adipose tissue for storage. The data on activity of hepatic G6PDH, malic enzyme, FAS, and lipoprotein lipase of visceral adipose tissue are furnished in Table 11. An opposite trend in the alteration of G6PDH malic enzyme and FAS activities were observed in STZ diabetic (D-group) and insulin-resistant rats (F-group) compared to C-group. In comparison with group-C, the activities of hepatic G6PDH, malic enzyme, and FAS are significantly lowered (54.3%, 18.3%, 21.2%) in group-D whereas significantly increased (32.5%, 37.4%, 25.3%) in group-F. The data clearly revealed the situation of a decreased lipogenesis under insulin-deficient condition and enhanced lipogenesis in IR condition. *C.roseus* treatment restored these enzyme activities to normal level in D + CR group. Co-administration of *C. roseus* along with fructose diet in F + CR group restored the hepatic FAS to normal level whereas malic enzymes was brought to near-normal value but significantly higher than group-C with no alteration in G6PDH from group-F. The abdominal adipose tissue was not visible in the D-group which is the reason for not assaying the LPL in D-group. A significantly decreased activity (11.7%) of LPL of adipose tissue was observed in F-group when compared to C-group. *C. roseus* administration for 60 days showed no significant variation in this enzyme activity in C + CR and D + CR groups. However, F + CR-group showed a significantly enhanced adipose LPL activity when compared to F and C groups. The observed very scanty adipose tissue in D-group indicates decreased de novo synthesis of lipids and increased lipolysis. However, an increase in the hepatic lipid levels in STZ diabetic rats may be due to increased uptake from the portal system as

Table 11 Effect of *C. roseus* treatment on lipid metabolic enzymes in STZ diabetic and fructose-fed IR rats

Parameter	Tissue	C	C + CR	D	D + CR	F	F + CR
Malic enzyme (μmol of NADP reduced/min/mg protein)	Liver	5.88 ± 0.17^{a}	6.13 ± 0.18^{a}	4.80 ± 0.06^{b}	5.68 ± 0.11^{a}	8.08 ± 0.13^{c}	6.92 ± 0.21^{d}
Fatty acid synthase (μmol of NADPH utilized/min/mg protein)	Liver	0.47 ± 0.02^{a}	0.45 ± 0.01^{a}	0.37 ± 0.02^{b}	0.42 ± 0.02^{a}	0.59 ± 0.02^{c}	0.50 ± 0.03^{a}
Lipoprotein lipase (μmol of PNP/min/mg protein)	Adipose	8.33 ± 0.13^{a}	8.25 ± 0.52^{a}	-	8.04 ± 0.19^{a}	7.35 ± 0.11^{b}	9.41 ± 0.16^{c}
Glucose-6-phosphate dehydrogenase (μmol of NADP reduced/min/mg protein)	Liver	2.43 ± 0.07^{a}	2.60 ± 0.04^{b}	1.11 ± 0.01^{c}	2.06 ± 0.05^{d}	3.22 ± 0.04^{e}	3.10 ± 0.06^{e}

Values are mean ± S.E.M. (*n*=8 animals). Values with different superscripts within the row are significantly different at $P<0.05$(Duncan's multiple range test)

shown earlier (Gupta et al. 1999) and not due to de novo synthesis. Thus, in this study the enhanced tissue lipids in fructose-fed rats can be explained by enhanced hepatic lipogenesis as reflected by the increased activities of FAS, malic enzyme, and G6PDH in the liver along with decreased activity of LPL of adipose tissue under fructose-fed condition .The therapeutic efficacy of *C. roseus* is evident from the restoration of lipid metabolic enzymes of D + CR-group to normal level. The protective role of *C. roseus* against fructose feed-induced lipid accumulation in tissues is also evident from preventing the enhanced lipogenesis observed under fructose-fed conditions by keeping the activity of key regulatory enzyme of lipolysis LPL of adipose and lipogenic enzymes of liver viz., FAS, malic enzyme, and G6PDH to the normal values. Thus, *C. roseus* treatment had given protection against diabetes-induced fatty liver and fructose feed-induced insulin resistance in target tissues by preventing lipid accumulation. However, these enzyme activities of C + CR-group did not deviate from C-group.

3.8 Oxidative Stress

The oxidative stress and resultant tissue damage are hallmark of chronic diseases and cell death, and diabetes is not an exception. Persistence hyperglycemia as observed in uncontrolled diabetes causes increased production of free radicals especially reactive oxygen species (ROS), for all tissues from glucose auto-oxidation and protein glycation (George and Mary 2004). In addition, hyperinsulinemia in insulin-resistant condition and enhanced FFA observed in diabetes and insulin-resistant conditions are also sources of free radicals which cause damage either directly affecting a specific molecule or indirectly by forming numerous toxic derivatives. Normalizing ROS generation not only reversed these changes, but also prevented the long-term complications of diabetes (Nishikawa et al. 2000). The observed efficient free radical scavenging capacity of *C. roseus* against different ROS and DPPH radicals from in vitro antioxidant screening experiments prompted to undertake in vivo studies. In order to understand the extent of oxidative stress and to assess the protective effect of *C. roseus* administration in insulin-deficient and insulin-resistant conditions, oxidative stress markers like lipid peroxidation, protein oxidation, and antioxidant status were assessed in tissues of six experimental groups.

3.8.1 Lipid Peroxidation, Protein Oxidation, and Reduced Glutathione

The extent of LPO and PO and the content of GSH in liver, pancreas, and heart of six experimental groups are summarized in Table 12. Both STZ diabetic (D-group) and insulin-resistant rats (F-group) showed significantly enhanced LPO in liver (47.3 and 22.0 %), pancreas (52.1 and 18.6%), and heart (32.1 and 24.9%) when compared to C-group. Thus under STZ-induced diabetic condition, the percent

Table 12 Effect of *C. roseus* treatment on tissue lipid peroxidation, protein oxidation, and reduced glutathione levels in six in STZ diabetic and fructose-fed IR rats

Parameter	Tissue	C	C + CR	D	D + CR	F	F + CR
Lipid peroxidation (nmol of MDA formed/ min/mg protein)	Liver	20.33±0.36^{a}	17.81±0.31^{b}	29.94±0.40^{c}	20.85±0.24^{a}	24.08±0.63^{d}	19.21±0.53^{a}
	Pancreas	7.41±0.13^{a}	6.366±0.22^{a}	11.27±0.37^{b}	8.43±0.18^{c}	8.79±0.22^{d}	7.67±0.37^{a}
	Heart	11.59±0.23^{a}	10.64±0.37^{b}	15.31±0.21^{c}	12.52±0.17^{a}	14.48±0.23^{d}	12.04±0.47^{a}
Protein oxidation (μmol of protein carbonyls/mg protein)	Liver	7.51±0.49^{a}	6.51±0.54^{b}	9.58±0.86^{c}	7.60±0.43^{a}	8.67±0.56^{d}	7.72±0.47^{a}
	Pancreas	3.11±0.05^{a}	3.02±0.10^{a}	4.04±0.11^{b}	3.30±0.52^{a}	3.71±0.06^{c}	3.23±0.06^{a}
	Heart	4.48±0.13^{a}	4.29±0.10 ^{a}	5.26±0.09^{b}	4.61±0.07^{a}	5.12±0.14^{c}	4.31±0.13^{a}
GSH (μg/mg protein)	Liver	5.55±0.20^{a}	6.33±0.20^{b}	3.92±0.32^{c}	5.11±0.16^{d}	4.78±0.25^{e}	5.58±0.18^{a}
	Pancreas	2.07±0.21^{a}	2.50±0.18^{b}	1.36±0.13^{c}	2.20±0.14^{a}	1.89±0.13^{a}	2.40±0.23^{d}
	Heart	3.72±0.31^{a}	4.16±0.034^{b}	2.12±0.30^{c}	3.62±0.31^{a}	3.22±0.22^{d}	3.59±0.15^{a}

Values are mean ± S.E.M. (n=8 animals). Values with different superscripts within the row are significantly different at $P<0.05$ (Duncan's multiple range test)

increase in LPO is more in pancreas than liver and heart whereas under insulin-resistant condition (F-group) percent increase in LPO is more in heart than liver and pancreas. The protein oxidation levels are also significantly increased in both D and F groups in liver (27.5 and 15.4%), pancreas (29.8%, 19.2%), and heart (10.1%, 7.1%) compared to C-group. GSH, the major portion of cellular nonprotein thiols, is an important antioxidant and considered as a major buffer in the cell. GSH functions directly in elimination of toxic peroxides and aldehydes and indirectly in maintaining Vit-C and Vit-E and SH-dependent enzymes in their reduced and functional state. Hence, the measurement of cellular GSH is considered as good assessment of antioxidant status. A significant decrease in the tissue GSH was observed both in D-group in liver (29.2%), pancreas (34.4%), and heart (42.9%) and F-group liver (13.7%) and heart (13.3%) compared to C-group. Unlike liver tissue, the pancreas of F-group showed no alterations in GSH content from C-group. The significantly decreased GSH content of D-group and F-group rats compared to control rats may be due to increased utilization of GSH or decreased synthesis because of enhanced oxidative stress. The data clearly indicates that the intensity of oxidative stress as reflected from increased LPO and PO with decreased GSH is comparatively more in insulin-deficient condition than insulin-resistant condition. Except pancreatic LPO and PO of D + CR-group *C. roseus* treatment for 60 days prevented the increased tissue LPO and PO and decreased tissue GSH content observed in STZ-induced diabetic and fructose-fed conditions by restoring to normal values in D + CR and F + CR groups. The significantly decreased pancreas LPO (25.2%) and increased hepatic GSH (30.2%) of D + CR-group compared to D-group did not reach the normal value. The beneficiary effect of *C. roseus* treatment is also reflected from tissue LPO and PO and GSH status of C + CR by maintaining better antioxidant status compared to C-group.

Control of diabetes-induced oxidative stress could theoretically be used to reduce the severity of diabetic complications. Clinical trials with conventional antioxidants in diabetic patients are limited. In addition to the many antioxidants examined, a number of commonly used drugs have shown promising antioxidant activity in addition to their primary pharmacological activity. These drugs include thiazolidinediones, metformin and HMG-CoA reductase inhibitors (statins), and inhibitors of the renin-angiotensin system. The observed significant elevation of GSH content of the tissues of D + CR and F + CR-rats compared D and F-rats indicate that *C. roseus* might have either increased the biosynthesis of GSH or lowered the utilization of GSH due to decreased oxidative stress or both. Generally, antioxidant treatment can exert beneficial effects in diabetes, with preservation of in vivo β-cell function. Antioxidant treatment suppresses apoptosis in β-cells without changing the rate of β-cell proliferation supporting the hypothesis that in chronic hyperglycemia, apoptosis induced by oxidative stress causes reduction in β-cell mass (Kaneto et al. 1999; Wiernsperger 2003). Particularly in D + CR-group, *C. roseus* protection against STZ diabetic-induced depletion of GSH is also evident from improved insulin secretion of these animals compared to D-group thus indicating the suppression of β-cell apoptosis. Maintenance of ample concentrations of antioxidants seems to be necessary for efficient insulin action. Sandrine et al. (2005) showed that insulin resistance induced by high fructose diet in rats was associated with oxidative stress. Oxidative stress could also participate in the progression of insulin resistance since the incubation of adipocytes in the presence of H_2O_2 decreases the sensitivity of cells to insulin and glucose transport (Hansen et al. 1999). Positive correlation was found between the malondialdehyde (MDA) level and index of insulin sensitivity (Kocic et al. 2007). Thus, prevention of GSH depletion seen in fructose-fed rats by *C. roseus* supplementation may also be responsible for the enhanced insulin sensitivity observed in F + CR-group. Besides hyperglycemia, hypertriglyceridemia, and hyperinsulinemia, the susceptibility of tissues to oxidative stress may depend on lipid over load and alterations in lipid composition. Enhanced lipid accumulation observed in the tissues of fructose-fed rats may also contribute to increased lipid peroxide levels found in these rats. *C. roseus* supplementation for 60 days to STZ diabetic and fructose-fed insulin-resistant rats alleviated the lipid accumulation in the skeletal muscle, heart, and liver tissues. This may depend upon its TG lowering and insulin sensitizing effects. Further, its antihyperglycemic effect could bring a favorable metabolic environment avoiding the pro-oxidant conditions with reduced oxidative stress in *C. roseus*-treated STZ and fructose-fed rats.

Besides transport and de novo synthesis, level of GSH is also regulated by GSH redox cycle composed of GPx, GR, and G6PDH. The reduction of GSSG to GSH is catalyzed by GR in the presence of NADPH which is generated from pentose phosphate pathway. Peroxides produced in a cell can be detoxified by the action of GPx and CAT. GPx catalyzes the reduction of peroxides with GSH to form GSSG and the reduction product of H_2O_2. This enzyme is specific for its hydrogen donor GSH and nonspecific for the hydroperoxides ranging from H_2O_2 to organic peroxides. GSH-independent antioxidant enzymes viz., SOD and CAT are widely distributed in all animal cells. Superoxide dismutase, a Cu/Zn containing enzyme, is a major enzyme defense for aerobic cells by providing the primary catalytic cellular defense that protects cells and tissues against potentially destructive reactions of $O_2^{\bullet}$ and their derivatives. CAT catalyzes the reduction of hydrogen peroxide and protects the tissues from highly reactive hydroxyl radicals. CAT decomposes H_2O_2 without generation of free radicals by minimizing one electron transfer.

The activities of glutathione-dependent (GR, GPx, and GST) and glutathione-independent (SOD and CAT) antioxidant enzymes of liver, pancreas, and heart of six experimental groups are summarized in Tables 13 and 14. Both GSH-dependent and GSH-independent antioxidant enzyme activities are highest in liver, the main organ involved in detoxification of process and least in pancreas indicating its susceptibility to oxidative stress. Studies of Grankvist et al. (1981) and Sigfrid et al. (2004) revealed more vulnerability of β cells to oxidative stress due to relatively low levels of GPx and other protective enzymes compared to other cells. Moreover, studies of Tiedge et al. (1997) and Sigfrid et al. (2004) also revealed incapability of pancreatic islets and β-cell lines to increased antioxidant enzyme expression in response to cellular stress.

Data presented in Table 13 indicates significantly decreased activities of GR, GPx, and GST in the tissues of both D and F groups compared to C-group. When compared to control rats, the percent decrease in the activity of GR in liver, pancreas,

Table 13 Effect of *C. roseus* treatment on glutathione-dependent enzyme activities in STZ diabetic and fructose-fed IR rats

Parameter	Tissue	C	C + CR	D	D + CR	F	F + CR
Glutathione reductase (μmol of NADPH oxidized/min/mg protein)	Liver	34.04±0.61^{a}	35.38±0.77^{a}	29.02±0.58^{c}	33.67±0.42^{a}	30.87±0.86^{b}	34.33±0.42^{a}
	Pancreas	22.19±0.68^{a}	23.10±0.51^{a}	18.71±0.36^{b}	21.79±0.50^{a}	20.36±0.52^{b}	23.06±0.33^{a}
	Heart	26.20±0.34^{a}	26.70±0.61^{a}	20.76±0.52^{b}	27.11±0.81^{a}	23.42±1.65^{c}	26.62±0.53^{a}
Glutathione peroxidase (μg of GSH consumed/min mg protein)	Liver	9.45±0.15^{a}	9.76±0.13^{a}	7.83±0.33^{b}	9.82±0.18^{a}	8.09±0.14^{b}	10.00±0.51^{a}
	Pancreas	4.75±0.24^{a}	4.95±0.22^{a}	3.81±0.07^{b}	4.6±0.06^{a}	4.06±0.10^{b}	5.08±0.18^{a}
	Heart	6.02±0.20^{a}	5.91±0.22^{a}	4.94±0.26^{b}	5.75±0.14^{a}	5.19±0.05^{c}	5.89±0.11^{a}
Glutathione-S-transferase (mmol of CDNB-GSH conjugate formed/mi/mg protein)	Liver	0.232±.001^{a}	0.220±.001^{a}	0.159±.001^{b}	0.226±.006^{a}	0.199±.005^{c}	0.228±.006^{a}
	Pancreas	0.037±.001^{a}	0.039±.001^{a}	0.026±.001^{b}	0.036±.0005^{a}	0.032±.001^{c}	0.036±.001^{a}
	Heart	0.046±.001^{a}	0.045±.001^{a}	0.030±.0007^{b}	0.044±.001^{a}	0.033±.001^{b}	0.045±.001^{a}

Values are mean ± S.E.M. (n=8 animals). Values with different superscripts within the row are significantly different at P<0.05 (Duncan's multiple range test)

and heart are 14.7, 15.6, and 20.7% in STZ diabetic rats and 9.3, 8.2, and 10.6% in fructose-fed rats, respectively. D-group rats showed a significant decrease in the activities of GPx and GST in liver (17.1 and 31.4%), pancreas (19.7 and 29.7%), and heart (17.9 and 34.7%) compared to C-group. Like STZ diabetic rats and fructose-fed rats also showed decreased activities of GR and GST in liver (14.3 and 14.2%), pancreas (14.5 and 13.5%), and heart (13.7 and 28.2%), respectively, compared to C-group. Decreased GR activity contributes to the decreased regeneration of GSH from GSSG. Decreased activities of GPx and GST in D and F groups rats are justified from the decreased GSH content in these tissues. Lowered activities of GSH-dependent antioxidant enzymes are in accordance with earlier studies in STZ diabetic rats (Anuradha and Selvam 1993; Venkateswaran and Pari 2003) and IR rats (Srividhya and Anuradha 2002; Busserolles et al. 2002). The depletion in the activities of these enzymes in both insulin-deficient and insulin-resistant conditions may result in deleterious oxidative stress due to accumulation of toxic products. Heart tissues of both STZ diabetic and fructose-fed rats showed higher percent decrease in GSH-dependent oxidative enzymes when compared to liver and pancreas. Vulnerability of cardiac tissues to oxidative stress is because these antioxidant enzymes play a key role in the cell protection against the deleterious effects of the ROS. *C. roseus* treatment in D + CR and F + CR groups resulted in retaining these enzyme activities with no deviation from C-group. This clearly reveals the protective effect of *C. roseus* treatment against oxidative damage by keeping GSH-dependent antioxidant enzymes at normal level in *C. roseus*-treated insulin-deficient and insulin-resistant animal models.

Significantly decreased activities of SOD and CAT in all tissues of D and F groups are evident from the data presented in Table 14. The percent decrease in the SOD activity in liver, pancreas, and heart tissues are 29.0, 46.3, and 28.2%, respectively, in D-group and 19.1, 31.3, and 26.0%, respectively, in F-group compared to C-group and decrease in the CAT activity of D-group and F-group in liver, pancreas, and heart are 24.12, 39.5, and 35.5%, and 19.1, 29.8, and 24.0% compared to C-group. The percent decrease in SOD and CAT activities are prominent in pancreatic tissue compared to liver and heart both in STZ diabetic and fructose-fed rats. The significant enhancement in the activities of SOD and CAT of F + CR-group compared to F-group resulted in restoration of these enzyme activities to normal values. Similarly, *C. roseus* treatment to D + CR group also normalized SOD activity in liver, heart, and pancreatic tissues and CAT activity in pancreas. However, CAT activities in hepatic and cardiac tissue of D + CR-group are still significantly lower than C-group. Restoration of SOD activity in D + CR and F + CR groups reveals an efficient defense against the first line of oxidative stress, i.e., $\mathbf{O_2^{\bullet-}}$ radicals which are known to inactivate CAT. Xu et al. (1999) demonstrated that over expression of CAT in mouse islet cells had given protection against H_2O_2-induced oxidative stress resulted in normal insulin secretion and reduced the diabetogenic effect of STZ in vivo.

The results from this study indicate that both insulin-deficient and insulin-resistant conditions have increased oxidative stress and a compromised antioxidant defense system in the liver, pancreas, and heart. This increase in oxidative

Table 14 Effect of *C. roseus* treatment on glutathione-independent enzyme activities in STZ diabetic and fructose-fed IR rats

Parameter	Tissue	C	C + CR	D	D + CR	F	F + CR
Superoxide dismutase (units/mg protein)	Liver	45.30±1.31[a]	50.85±1.69[b]	32.16±1.89[c]	44.13±1.33[a]	36.66±1.45[d]	46.72±2.18[a]
	Pancreas	26.67±1.18[a]	31.48±1.13[b]	14.31±1.14[c]	25.93±0.98[a]	18.32±1.71[d]	29.20±1.31[a]
	Heart	29.28±1.31[a]	31.08±1.46[a]	21.00±1.32[b]	27.52±1.30[a]	21.66±1.36[b]	27.67±1.32[a]
Catalase (μmol of H_2O_2 consumed/min/mg protein)	Liver	99.84±4.40[a]	111.55±4.22[b]	75.75±3.77[c]	91.69±3.20[d]	80.81±2.92[e]	100.95±3.77[a]
	Pancreas	14.56±0.53[a]	16.82±0.41[b]	8.84±0.30[c]	15.46±0.28[a]	10.22±0.44[d]	15.05±0.43[a]
	Heart	25.75±0.92[a]	26.10±0.71[a]	16.65±0.44[b]	22.77±0.88[c]	19.56±0.47[d]	24.45±0.77[a]

Values are mean ± S.E.M. (n=8 animals). Values with different superscripts within the row are significantly different at $P<0.05$ (Duncan's multiple range test)

stress is prevented by *C. roseus* administration in both Type 1 and Type 2 DM animal models.

Various natural products have long been used in traditional medical systems for treating diabetes contains a wide range of antioxidants with a potent scavenging activity for ROS. Similar to *C. roseus*, our earlier studies revealed antioxidant potential of other plant extracts, viz., *Phyllanthus amarus, Tinospora cordifolia, Commiphora mukul, Moringa oleifera*, and *Caralluma fimbriata* against oxidative stress in different tissues of STZ diabetic (Ramesh et al. 2011; Sudhakara et al. 2012; Divi et al. 2012; Vijaya Bharathi et al. 2014; Sasi Bhusana Rao et al. 2016), fructose diet (Reddy et al. 2009; Mallaiah et al. 2015; Ramesh and Saralakumari 2012; Ramesh et al. 2015), and high-fat diet (Sudhakara et al. 2014, 2016a, b)-fed insulin-resistant rats. Thus, *C. roseus* administration was found to be beneficial in insulin-deficient rats by increasing the plasma insulin levels, which may be due to regeneration of the damaged β-cells in islets of pancreas and insulin-resistant condition by increasing the insulin sensitivity as reflected by controlling the hyperinsulinemia observed in insulin-resistant condition.

4 Conclusion

The unidirectional therapeutic approach in the management of Type 1 and Type 2 DM does not appear to be the way to address the multifactorial pathogenicity of DM. The beneficial multiple pharmacological activities of *C. roseus* like antihyperglycemic activity by manipulating carbohydrate metabolism through various mechanisms, like hypolipidemic activity and restoration of enhanced intestinal disaccharidases of diabetic animals and antioxidant potential offer an exciting opportunity to develop this into a novel therapeutic approach for both types of diabetes.

Acknowledgements We thank Dr. Ch. Appa Rao and Dr. E. G. T. V. Kumar, Department of Biochemistry, Sri Venkateswara University, Tirupati, Andhra Pradesh (AP), India, for their help in Insulin assay, National Center for Laboratory Animal Sciences, National Institute of Nutrition, Hyderabad, Telangana, India, for High Fructose Diet supply and Dr. P. Mohammed Akhtar, Department of Statistics, Sri Krishnadevaraya University, Anantapuramu, (AP), India, for his help in Statistical analysis. We thank CPCSEA for providing registration to the University (470/01/a/CPCSEA) for carrying out animal experimentation.

References

Adelman RC, Spolter PD, Weinhouse S (1966) Dietary and hormonal regulation of enzymes of fructose metabolism in rat liver. J Biol Chem 241:5467–5472

Alxeandrova R, Alxeandrova I, Velcheva M, Varadino T (2000) Phytoproducts and cancer. Exp Pathol Parasitol 4:15–25

Antia BS, Okokon JE (2005) Effect of leaf juice of *Catharnathus roseus* Linn on cholesterol, triglyceride and lipoproteins levels in normal rats. BMC Complement Altern Med 37:401–403
Anuradha CV, Selvam R (1993) Effect of oral methionine on tissue lipid peroxidation and antioxidants in alloxan induced diabetic rats. J Nutr Biochem 4:212–217
Arvouet-Grand A, Vennat B, Pouratt A, Legret P (1994) Standardization dun extra it de propolis at identification des principaux constituents. J Pharm Belg 49:462–468
Assmann G, Schriewer H, Schmitz G, Hagele EO (1983) Quantification of high density lipoprotein cholesterol by precipitation with phophotungstic acid-$MgCl_2$. Clin Chem 29:2026–2030
Beutler E, Dern RJ, Alving AS (1955) The hemolytic effect of primaquine. VI. An in vitro test for sensitivity of erythrocytes to primaquine. J Lab Clin Med 45:40–50
Blumert M, Liu J (2003) Jiaogulan (*Gynostemma pentaphyllum*), China's immortality herb, 3rd edn. Torchlight, Badge, CA, p 80
Bone K (1996) Clinical applications of ayruvedic and Chinese herbs. Monographs for the western herbal practitioner. Phytotherapy, Warwick, QLD, pp 137–141
Brain KR, Turner TD (1975) The practical evaluation of phyto pharmaceuticals. Wright-Scientechnica, John Wright and Sons, Bristol, Dorchester, p 144
Brandstrup N, Kirk JE, Bruni C (1957) Determination of hexokinase in tissues. J Gerontol 12:166–171
Busserolles J, Roc E, Gueux E, Mazur A, Grolier P, Rayssiguier Y (2002) Short-term consumption of a high-sucrose diet has a pro-oxidanteffect in rats. Br J Nutr 87:337–342
Butterweek V, Jurgenliemk G, Nahrsted A, Winterhoff H (2000) Flavonoids from *Hypericum perforatum* show antidepressant activity in the forced swimming test. Planta Med 66:3–6
Carrol NV, Longley RW, Roe JH (1956) The determination of glycogen in liver and muscle by use of anthrone reagent. J Biol Chem 220:583–587
Chakravarthy BK, Gupta S, Gambir SS, Gode KD (1980) Pancreatic beta cell regeneration. A novel ntidiabetic mechanism of *Pterocarpus marsupium* Roxb. Ind J Pharmacol 12:123–127
Chattopadhyay RR (1999) A comparative evaluation of some blood glucose lowering agents of plant origin. J Ethanopharmacol 67:367–372
Chattopadhya RR, Sarkar SK, Ganguli S, Banerjee RN, Basu TK (1991) Hypoglycemic and antihyperglycemic effect of leaves of *Vinca rosea* Linn. Ind J Physiol Pharmacol 35:145–151
Connerty V, Briggs AR, Eaton EH (1961) Simplified determination the lipid components of blood serum. Clin Chem Acta 7:37–53
Coskun O, Kanter M, Korkmaz A, Oter S (2005) Quercetin, a flavonoi antioxidant, prevents and protects streptozotocin-induced oxidative stress and β-cell damage in rat pancreas. Pharmacol Res 51:117–123
Crespo ME, Galvez J, Cruz T, Ocete MA, Zarzuelo A (1999) Antiinflammatory activity of diosmin and hesperidin in rat colitis induced by TNBS. Planta Med 65:651–653
Dahlqvist A (1968) Assay of intestinal disaccharidases. Anal Biochem 22:99–107
Divi SM, Rramesh B, Saralakumari D (2012) Evaluation of antidiabetic and antihyperlipedimic potential of aqueous extract of *moringa oleifera* in fructose fed insulin resistant and STZ induced diabetic Wistar rats: a comparative study. Asian J Pharm Clin Res 5:67–72
Don G (1999) *Catharanthus roseus*. In: Ross IA (Ed) Medicinal plants of the world. Human press, Totowa, NJ, pp. 109–118
Dugas AJ Jr, Castaneda-Acosta J, Bonin GC, Price KL, Fischer NH, Winston GW (2000) Evaluation of the total peroxyl radical-scavenging capacity of flavonoids: structure activity relationships. J Nat Prod 63:327–331
Duncombe WG (1963) The colorimetric microdetermination of long-chain fatty acids. Biochem J 88:7–10
Ellmans GL (1959) Tissue sulphydryl groups. Arch Biochem Biophys 82:70–77
Ferraris RP, Casirola DM, Vinnakota RR (1993) Dietary carbohydrate enhances intestinal sugar transport in diabetic mice. Diabetes 42:1579–1587
Ferreres F, Pereira DM, Valentão P, Andrade PB, Seabra RM, Sottomayor M (2008) New phenolic compounds and antioxidant potential of *Catharanthus roseus*. J Agric Food Chem 56:9967–9974

Fiske CH, Subbarow Y (1925) The colorimetric determination of phosphorous. J Biol Chem 66:375–400

Folch J, Lees M, Sloane-Stanley GH (1957) A simple method for the isolation and purification of total lipides from animal tissues. J Biol Chem 226:497–509

Friedewald WT, Levy RI, Fredrickson DS (1972) Estimation of the concentration of low-density lipoprotein cholesterol in plasma without use of the preparative ultracentrifuge. Clin Chem 18:499–502

Friedmann B, Goodman E, Sidney Weinhouse H (1967) Effects of insulin and fatty acids on gluconeogenesis in the rat. J Biol Chem 242:3620–3627

Fukuda H, Iritani N, Tanaka T (1983) Effect of high fructose diet on lipogenic enzymes and their substrate and effector levels in diabetic rats. J Nutr Sci Vitaminol 29:691–699

Galambos JT, Asada M, Shanks JZ (1963) The effect of intravenous alcohol on serum enzymes in patients with normal or diseased liver. Gastroenterology 44:267

George LK, Mary RL (2004) Hyperglycemia-induced oxidative stress in diabetic complications. Histochem Cell Biol 122:333–338

Ghosh RK, Gupta I (1982) Effect of *Vinca rosea* and *Ficus racemososus* on hyperglycemia in rats. Ind J Anim Health 19:145–148

Gibson DM, Hubbard DD (1960) Incorporation of malonyl CoA into fatty acids by liver in starvation and alloxan-diabetes. Biochem Biophys Res Commun 3:531–535

Grankvist K, Marklund SL, Taljedal LB (1981) Cu Zn-superoxide dismutase, Mn-superoxide dismutase, catalase and glutathione peroxidase in pancreatic islet and other tissues in the mouse. Biochem J 199:393–398

Greene DA, Sima AA, Stevens MJ, Feldman EL, Killen PD, Henry DN, Thomas T, Dananberg J, Lattimer SA (1993) Aldose reductase inhibitors: an approach to the treatment of diabetic nerve damage. Diabetes Metab Rev 9:189–217

Grover JK, Yadav S, Vats V (2002) Medicinal plants of India with antidiabetic potential. J Ethnopharmacol 81:81–1000

Gupta D, Raju J, Prakash J, Baquer NZ (1999) Change in the lipidprofile, lipogenic and related enzymes in the livers of experimental diabetic rats: effect of insulin and vanadate. Diabetes Res Clin Pract 46:1–7

Gutteridge JM, Rowley DA, Halliwell B (1981) Superoxide-dependent formation of hydroxyl radicals in the presence of iron salts. Detection of 'free' iron in biological systems by using bleomycin-dependent degradation of DNA. Biochem J 199:263–265

Habig WH, Pabst MJ, Jakoby WB (1974) Glutathione-S-transferases. The first enzymatic step in mercapturic acid formation. J Biol Chem 249:7130–7139

Hamdan II, Afifi FU (2004) Studies on the in vitro and in vivo hypoglycemic activities of some medicinal plants used in treatment of diabetes in Jordanian traditional medicine. J Ethnopharmacol 93:117–121

Hansen LL, Ikeda Y, Olsen GS (1999) Insulin signaling is inhibited by micromolar concentrations of H_2O_2. Evidence for a role of H_2O_2 in tumor necrosis factor alpha-mediated insulin resistance. J Biol Chem 274:25078–25084

Harbone JB (1973) Phytochemical methods. A guide to modern technique of plant analysis. Chapman and Hall, London, pp 182–201

Hayman S, Kinoshita JH (1965) Isolation and properties of lens aldose reductas. J Biol Chem 240:877–882

Heijden RV, Jacob DI, Soneijer W, Hallard D, Verpoorte R (2004) The Catharanthous alkaloids: pharmacognosy and biotechnology. Curr Med Chem 11:607–628

Hiai S, Oura H, Nakajima T (1976) Color reaction of somesapogenins and saponins with vanillin and sulfuric acid. Planta Med 29:116–122

Hollenberg NK (2003) Considerations for management of fluid dynamic issues associated with thiazolidinediones. Am J Med 115:S111–S115

Hopfer U (1975) Diabetes mellitus: changes in the transport properties of isolated intestinal microvillous membranes. Proc Natl Acad Sci USA 72:2027–2031

Howard CT (1995) Metabolic regulation of gene transcription in mammals. J Biol Chem 270:23235–23238

Ibrahim M, SughraMehjabeen S, Narsu ML (2011) Pharmacological evaluation of *Catharanthus roseus*. Int J Pharmaceut Appl 2:165–173

Islam MM, Hoque MA, Okuma E, Banu MN, Shimoishi Y, Nakamura Y, Murata Y (2009) Exogenous proline and glycinebetaine increase antioxidant enzyme activities and confer tolerance to cadmium stress in cultured tobacco cells. J Plant Physiol 166:1587–1597

Jayanthi M, Sowbala N, Rajalakshmi G, Kanagavalli U, Sivakumar V (2010) Study of anti-hyperglycemic effect of *Catharanthus roseus* in alloxan induced diabetic rats. Int J Pharm Pharmaceut Sci 2:114–116

Johnson RJ, Segal MS, Sautin Y, Nakagawa T, Feig DI, Kang DH, Gersch MS, Benner S, Sánchez-Lozada LG (2007) Potential role of sugar (fructose) in the epidemic of hypertension, obesity and the metabolic syndrome, diabetes, kidney disease, and cardiovascular disease. Am J Clin Nutr 86:899–906

Kaneto H, Kajimoto Y, Fujitani Y, Matsuoka T, Sakamoto K, Matsuhisa M (1999) Ooxidative stress induces P21 expression in pancreatic islet cells: possible implication in beta cell dysfunction. Diabetologia 42:1093–1097

Kannappa S, Jayaraman P, Ramachandra MK, Anuradha CV (2006) Cinnamon bark extract improves glucose metabolism and lipid profile in the fructose fed rats. Singapore Med J 47:858–863

Kapoor LD (1990) CRC handbook of ayurvedic medicinal plants. CRC, Boca Raton, FL, pp 200–201

Karuna R, Saralakumari D (2011) Preventive effect of *Catharanthus roseus* (Linn.) against high-fructose diet-induced insulin resistance and oxidative stress in male Wistar rats. J Diabetes Mellitus 3:63–70

Karuna R, Ramesh B, Sreenivasa S, Saralakumari D (2010) Antihyperglycemic activity of *Catharanthus roseus* leaf powder in streptozotocin-induced diabetic rats. Pharmacog Res 2:195–201

Karuna R, Bellamkonda R, Singareddy SR, Desireddy S (2013) Abnormalities in carbohydrade and lipid metabolism in high-fructose diet fed insulin resistant rats: amelioration by *Ctharanthus roseus* treatments. J Physiol Biochem 69:459–466

Kelley GL, Allan G, Azhar S (2004) High dietary fructose induces a hepatic stress response resulting in cholesterol and lipid dysregulation. Endocrinology 145:548–555

King J (1965) Practical clinical enzymology. Van Nostrand, London, pp 85–234

Kobayashi K, Saito Y, Nakazawa I, Yoshizaki F (2000) Screening of crude drugs for influence on amylase activity and postprandial blood glucose in mouse plasma. Biol Pharm Bull 23:1250–1253

Kobori M, Masumoto S, Akimoto Y, Oike H (2011) Chronic dietary intake of quercetin alleviates hepatic fat accumulation associated with consumption of a Western-style diet in C57/BL6J mice. Mol Nutr Food Res 55:530–540

Kocic R, Pavlovic D, Kocic G, Pesic M (2007) Susceptibility to oxidative stress, insulin resistance, and insulin secretory response in the development of diabetes from obesity. Vojnosanit Pregl 64:391–397

Kuhn DJ, Burns AC, Kazi A, Dou QP (2004) Direct inhibition of the ubiquitin-proteasome pathway by ester bond-containing green tea polyphenols is associated with increased expression of sterol regulatory element-binding protein 2 and LDL receptor. Biochim Biophys Acta 1682:1–10

Levine RL, Garland D, Oliver CN, Amici A, Climent I, Lenz AG, Ahn BW, Shultiel S, Standman ER (1990) Determation of carbonyl content in oxidatively modified proteins. Methods Enzymol 186:464–478

Lowry OH, Rosebrough NJ, Farr AL, Randall RJ (1951) Protein measurement with Folin-Phenol reagent. J Biol Chem 193:265–268

Mallaiah P, Sudhakara G, Srinivasulu N, Sasi Bhusana Rao B, Saralakumari D (2015) Assessment of in vitro antioxidant and phytochemical screening of aqueous extract of *Phyllanthus amarus*. Int J Pharm Pharmaceut Sci 7:439–445

Marcocci L, Packer L, Droy-Lefaix MT, Sekaki A, Gardes-Albert M (1994) Antioxidant action of *Ginkgo biloba* extract EGb 761. Method Enzymol 234:462–475

Matthews D, Hosker JP, Rudenski AS, Naylor BA, Treacher DF, Turner RC (1985) Homeostasis model assessment: insulin resistance and beta-cell function from fasting plasma glucose and insulin concentrations in man. Diabetologia 28:412–419

Mayes PA (1993) Intermediary metabolism of fructose. Am J Clin Nutr 58:745–746

McAnuff-Harding MA, Omoruyi FO, Asemota HN (2006) Intestinal disaccharidases and some renal enzymes in streptozotocin-induced diabetic rats fed sapogenin extract from bitter yam (*Dioscorea polygonoides*). Life Sci 78:2595–2600

Monika S, Vandhana S (2013) *Catharanthus roseus* (an anticancerous drug yielding plant). A review of potential therapeutic properties. Int J Pure Appl Biosci 1:139–142

Moure A, Cruz JM, Franco D, Domoanguez JM, Sineiro J, DomõÂnguez H, NuÂnÄez M, ParajoÂ JC (2001) Natural antioxidants from residual sources. Food Chem 72:145–171

Murakami N, Murakami T, Kadoya M, Matsuda H, Yamahara J, Yoshikawa M (1996) New hypoglycemic constituents in "gymnemic acid" from *Gymnema sylvestre*. Chem Pharm Bull 44:469–471

Nammi S, Boini MK, Lodagala SD, Behara RB (2003) The fresh leaves of *Catharanthus roseus* Linn. Reduces blood glucose in normal and alloxan diabetic rabbits. BMC Complement Altern Med 3:4

Natarajan A, Ahmed KSZ, Sundaresan S, Sivaraj A, Devi K, Senthil Kumar B (2012) Effect of aqueous flower extract of *Catharanthus roseus* on alloxan induced diabetes in male albino rats. Int J Pharm Sci Drug Res 4:150–153

Nayak BS, Lexley M, Pinto P (2006) *Catharanthus roseus* flower extract has wound-healing activity in Sprague Dawley rats. BMC Complement Altern Med 6:1–6

Nishikawa T, Edelstein D, Du XL, Yamagishi S, Matsumura T, Kaneda Y, Yorek MA, Beebe D, Oates PJ, Hammes HP, Giardino M, Brownlee M (2000) Normalizing mitochondrial superoxide production blocke three pathways of hyperglycaemic damage. Nature 404:787–790

Okada Y, Okada M (1998) Scavenging effect of water soluble proteins in broad beans on free radicals and active oxygen species. J Agric Food Chem 46:401–406

Oktay M, Gülçin I, Küfrevioglu OI (1989) Determination of in vitro antioxidant activity of fennel (*Foeniculum vulgare*) seed extracts. Lebensmittel-Wissenchaft Technol 36:263–271

Omolekan TO, Olaiya COO (2013) Anti-atherogenic activity of ethanol extract of *Buccholzia coriacea* seeds on hypercholesterolemic rats. Int J Biochem Res Rev 3:190–199

Ortmeyer HK, Bodkin NL, Hansen BC (1997) Insulin regulates liver glycogen synthase and glycogen phosphorylase activity reciprocally in *Rhesus monkeys*. Am J Physiol Endocrinol Metab 272:133–138

Oyaizu M (1986) Studies on products of browning reactions: antioxidative activities of products of browning reaction prepared from glucosamine. Jpn J Nutr 44:307–315

Ozgen U, Turkoz Y, Stout M, Ozugurlu F, Pelik F, Bulut Y, Aslan M, Ravindranath Y, Savasan S (2003) Degradation of vincristine by myeloperoxidase and hypochlorous acid in children with acute lymphoblastic leukemia. Leuk Res 27:1109–1113

Parniak MA (1988) Enhancement of glycogen concentrations in primary cultures of rat hepatocytes exposed to glucose and fructose. Biochem J 251:795–802

Pinto RE, Bartley W (1969) The effect of age and sex on glutathione reductase and glutathione peroxidase activities and on aerobic glutathione oxidation in rat liver homogenates. Biochem J 112:109–115

Quinn D, Shirai K, Jackson RL (1982) Lipoprotein lipase: mechanism of action and role in lipoprotein metabolism. Prog Lipid Res 22:35–78

Rajasekar P, Anuradha CV (2007) Fructose-induced hepatic gluconeogenesis: effect of L-carnitine. Life Sci 80:1176–1183

Ramesh B, Saralakumari D (2012) Antihyperglycemic, hypolipidemic and antioxidant activities of ethanolic extract of comminiphora mukul gum resin in fructose- fed male wistar rats. J Physiol Biochem 68:573–582

Ramesh B, Rasineni K, Sreenivas Reddy S, Ramesh Babu K, Ramatholisamma P, Appa Rao CH (2011) Antihyperglycemic and antioxidant activities of alcoholic extract of *Commiphora mukul* gum resin in streptozotocin induced diabetic rats. Pathophysiology 18:255–261

Ramesh B, Karuna R, Sreenivasa Reddy S, Sudhakara G, Saralakumari D (2013) Ethanolic extract of *Commiphora mukul* gum resin attenuates streptozotocin-induced alterations in carbohydrate and lipid metabolism in rats. EXCLI J 12:556–568

Ramesh B, Sainath SB, Karuna R, Sreenivasa Reddy S, Manjunatha B, Sudhakara G, Sasi Bhusana Rao B, Saralakumari D (2015) Effect of *Commiphora mukul* gum resin on hepatic and renal marker enzymes, lipid peroxidation and antioxidants status in pancreas and heart in fructose fed insulin resistant rats. Beni-Suef Univ J Basic Appl Sci 4:269–278

Rasool N, Rızwan K, Zubaır M, Naveed KUR, Imran I, Ahmed VU (2011) Antioxidant potential of different extracts and fractions of *Catharanthus roseus* shoots. Int J Phytomed 3:108–114

Rathi SS, Grover JK, Vats V (2002) Anti-hyperglycemic effects of *Momordica charantia* and *Mucuna pruriens* in experimental diabetes and their effect on key metabolic enzymes involved in carbohydrate metabolism. Phytother Res 16:236–243

Rau O, Wurqlics M, Dingermann T, Abdel-Tawab M, Schubert-Zsilavecz M (2006) Screening of herbal extracts for activation of the human peroxisome proliferator-activated receptor. Pharmazie 61:952–956

Ravinder P, Srinivasan K, Radhakrishnamurty R (1989) Biochemical toxicity of hexachlorocyclohexane and its Y-isomer in albino rats. Ind J Exp Biol 27:248–251

Reddy SS, Ramatholisamma P, Ramesh B, Baskar R, Saralakumari D (2009) Beneficiary effect of *Tinospora cordifolia* against high-fructose diet induced abnormalities in carbohydrate and lipid metabolism in Wistar rats. Horm Metab Res 41:741–746

Reitman S, Frankel S (1990) Practical biochemistry in clinical medicine. Am J Clin Pathol (1957) 25:56. In: Nath RL (ed), 2nd edn. Academic, Calcutta, pp 155–161

Rewana S, Clark K, Zhong S, Buison A, Chackunkal S, Catherine Jen KL (1993) Low dose fructose ingestion during gestation affects carbohydrate metabolism in rat dams and their offspring. J Nutr 123:2158–2165

Rice-Evans CA, Miller JM, Paganga G (1996) Structure-antioxidant activity relationship of flavonoids and phenolic acids. Free Radic Biol Med 20:933–956

Ruch R, Cheng SJ, Klaunig JE (1989) Prevention of cytotoxicity and inhibition of intracellular communication by antioxidant catechins isolated from Chinese green tea. Carcinogenesis 10:1003–1008

Sadava D, Alonso D, Hong H, Pettit Barrett D (1997) Effect of methadone addiction on glucose metabolism in rats. Gen Pharmacol 28:27–29

Sandrine D, Eleni P, Richard M, Caroline A, Theophile D, Jean-Paul C, Gerard C, Jacqueline A (2005) Involvement of oxidative stress and NADPH oxidase activation in the development of cardiovascular complications in a model of insulin resistance, the fructose-fed rats. Atherosclerosis 179:43–49

Sasi Bhusana Rao B, Saisree S, Vijaya Bharathi G, Mallaiah P, Srinivasulu N, Sudhakara G, Ramesh B, Saralakumari D (2016) Protective role of aqueous extract of *Phyllanthus amarus* on oxidativ stress in pancreas of streptozotocin induced diabetic male Wistar rats. J Exp Appl Anim Sci 2:23–30

Shaw JE, Sicree RA, Zimmet PZ (2010) Global estimates of the prevalence of diabetes for 2010 and 2030. Diabetes Res Clin Pract 87:4–14

Sigfrid LA, Cunningham JM, Beeharry N, Håkan Borg LA, Rosales Hernandez AL, Carlsson C, Bone AJ, Green IC (2004) Antioxidant enzyme activity and mRNA expression in the islets of Langerhans from the BB/S rat model of type 1 diabetes and an insulin-producing cell line. J Mol Med 82:325–335

Singh SN, Vats P, Suri S, Shyam R, Kumria MM, Ranganathan S, Sridharan KJ (2001) Effect of an antidiabetic extract of *Catharanthus roseus* on enzymic activities in streptozotocin induced diabetic rats. Ethnopharmacology 76:269–277

Singleton VL, Orthofer R, Lamuela-raventos RM (1999) Analysis of total phenols and other oxidation substrates and antioxidants by means of Folin-Ciocalteu reagent. Method Enzymol 299:152–178

Sofowora A (1982) Medicinal plants and traditional medicines in Africa. Wiley, New York, pp 6–65

Somananth S, Praveen V, Shoba S, Radhey S, Kumari MML, Ranganathan S, Sridhran K (2001) Effect of an antidiabetic extract of a *Catharanthus roseus* on enzymatic activities in streptozotocin induced diabetic rats. J Ethanopharmacol 76:269–277

Soon HT, Chung YL, Hazrina H, Aditya A, Mohammadjavad P, Won FW, Shiau-Chuen C, Mohd Rais M, Khalijah A (2013) Antidiabetic and antioxidant properties of alkaloids from *Catharanthus roseus* (L.) G Don. Molecules 18:9770–9784

Srividhya S, Anuradha CV (2002) Metformin improves liver antioxidant potential in rats fed a high-fructose diet. Asia Pacific J Clin Nutr 11:319–322

Standl E, Baumgartl HJ, Fuchtenbusch M, Stemplinger J (1999) Effect of acarbose on additional insulin therapy in type 2 diabetic patients with late failure of sulphonylurea therapy. Diabetes Obes Metab 1:215–220

Storey JM, Bailey E (1978a) Effect of streptozotocin diabetes and insulin administration on some liver enzyme activities in the post-weaning rat. Enzyme 23:382–387

Storey KB, Bailey E (1978b) Intracellular distribution of enzymes associated with lipogenesis and gluconeogenesis in the fat body of the adult cockroach, Periplaneta. Insect Biochem 8:125–131

Sudhakara G, Ramesh B, Mallaiah P, Sreenivasulu N, Saralakumari D (2012) Protective effect of ethanolic extract of *Commiphora mukul* gum resin against oxidative stress in the brain of strecptozotocin induced diabetic wistar male rats. EXCLI J 11:576–592

Sudhakara G, Mallaiah P, Sreenivasulu N, Sasi Bhusana Rao B, Rajendran R, Saralakumari D (2014) Beneficial effects of hydro-alcoholic extract of Caralluma fimbriata against high fat diet induced insulin resistance and oxidative stress in Wistar male rats. J Physiol Biochem 70:311–320

Sudhakara G, Malliah P, Srinivasulu N, Manjunath B, Rajendran R, Saralakumari D (2016a) Renoprotective effect of Caralumma fimbriata against high-fat diet induced oxidative stress in Wistar rats. J Food Drug Anal 24:586–593

Sudhakara G, Mallaiah P, Gangarapu V, Nukala S, Bellamkonda R, Ramaswamy R, Desireddy S (2016b) Protective effect of Caralluma fimbriata against high-fat diet induced testicular oxidative stress in rats. Biomed Pharmacother 83:167–176

Sugioka K, Shimosegawa Y, Nakno M (1987) Estrogen as natural antioxidants of membrane phospholipid peroxidation. FEBS Lett 210:37–39

Sunil C, Sadekar SV (1999) Unique concept of alpha-glucosidase inhibition. J Gen Med 11:63–66

Sutherland EW (1955) In: Colowick SP, Kaplan NO (eds) Methods in enzymology, vol 2. Academic, New York, pp 215–218

Svoboda GA (1969) The alkaloids of *Catharanthus roseus* G. Don (*Vinca rosea* L.) in cancer chemotherapy. In: Gunkel JE (ed) Current topics in plant science. Academic, New York, pp 303–335

Thériault M, Caillet S, Kermasha S, Lacroix M (2006) Antioxidant, antiradical and antimutagenic activities of phenolic compounds present in maple products. Food Chem 98:490–501

Tiedge M, Lortz S, Drinkgern A, Lenzen S (1997) Relation between antioxidant enzyme gene expression and antioxidative defense status of insulin-producing cells. Diabetes 46:1733–1742

Tordoff MG, Alleva AM (1990) Effect of drinking soda sweetened with aspartame or high-fructose corn syrup on food intake and body weight. Am J Clin Nutr 51:963–969

Utley HG, Bernheim F, Hochstein P (1967) Effect of sulfhydryl reagents on peroxidation in microsomes. Arch Biochem Biophys 118:29–32

Venkateswaran S, Pari L (2003) Effect of *Coccinia indica* leaves on antioxidant status in streptozotocin-induced diabetic rats. J Ethnopharmacol 84:163–168

Vijaya Bharathi G, Sasi Bhusana Rao B, Mallaiah P, Sreenivasulu N, Sudhakara G, Ramesh B, Saralakumari D (2014) Hematological and hepatoprotective effects of aqueous extract of *Phyllanthus amarus* in streptozotocin induced male Wistar rats. J Exp Appl Anim Sci 1:186–198

Vinson J, Su X, Zubik L, Bose P (2001) Phenol antioxidation quantity and quality in foods: fruit. J Agric Food Chem 49:5315–5321

Vlassara H, Palace MR (2002) Diabetes and advanced glycation end products. J Int Med 251:87–101

Weber G, Convery HJ (1966) Insulin: inducer of glucose 6-phosphate dehydrogenase. Life Sci 5:1139–1146

WHO (1980) Expert Committee on Diabetes mellitus. Technical Report Series, World Health Organization, Geneva, p 646

WHO (2002) Traditional Medicine Strategy 2002–2005. WHO, pp 1–6

Wichi HP (1988) Enhanced tumor development by butylated hydroxyanisole (BHA) from the prospective of effect on forestomach and oesophageal squamous epithelium. Food Chem Toxicol 26:717–723

Wiernsperger NF (2003) Oxidative stress as a therapeutic target in diabetes: revisiting the controversy. Diabetes Metab 29:579–585

Williamson JR, Chang K, Frangos M, Hasan KS, Ido Y, Kawamura T, Nyengaard JR, Van den Enden M, Kilo C, Tilton RG (1993) Hyperglycemic pseudohypoxia and diabetic complications. Diabetes 42:801–813

Xu B, Moritz JT, Epstein PN (1999) Overexpression of catalase provides partial protection to transgenic mouse β-cells. Free Radic Biol Med 27:830–837

Yalow RS, Berson SA (1960) Immunoassay of endogenous plasma insulin in man. J Clin Invest 39(7):1157–1175

Youn JH, Kaslow HR, Bergman RN (1987) Fructose effect to suppress hepatic glycogen degradation. J Biol Chem 262:11470–11477

Zheng W, Wang SY (2001) Antioxidant activity and phenolic compounds in selected herbs. J Agric Food Chem 49:5165–5170

Biotechnological Interventions to Modulate Terpenoid Indole Alkaloid Pathway in *Catharanthus roseus* Using In Vitro Tools and Approaches

Priyanka Verma, Shamshad Ahmad Khan, Varsha Parasharami, and Ajay Kumar Mathur

Abstract *Catharanthus roseus* plant is valued for harboring more than 130 bioactive terpenoid indole alkaloids (TIAs) including the two of its leaf derived bisindole alkaloids—vinblastine and vincristine which are indispensible constituents of anti-neoplastic drugs used in metastatic malignancy associated with acute lymphoblastic leukaemia's and Hodgkin's/Non-Hodgkin's lymphomas. The extremely low *in planta* occurrence of TIAs in *C. roseus* plants resulting in high commercial demand and exorbitant price have brought this herb in focus of an intense scientific scrutiny in last 30 years. Research efforts have so far advanced in two major directions: towards understanding the enzymology and genetic regulation of the concerned metabolic pathway(s) leading to TIAs biosynthesis in plant and; secondly, exploring the possibility of developing cell/tissue culture based platforms for in vitro TIAs production to meet the industry's demand. Designing plants, free from such metabolic constraints, can be a possible approach to enhance the production of plant based medicines. This subject of plant metabolic engineering is gaining lot of attention these days. Pathway manipulation using the modern tools of genetic engineering to over-express a limiting enzyme or to suppress the expression of an enzyme using a shared substrate of a branched pathway are attractive options of metabolic engineering for diverting the metabolic flux towards the synthesis of a desired end product. Knowledge, thus gained, indicates that TIAs biogenetic route is characterized by extensive metabolic cross-talk and shuttling of at least 35 intermediates synthesized via 30 enzymatic reactions occurring in four different types of tissues (epidermis, internal phloem parenchyma, idioblasts and leticifers) and five different sub-cellular compartments (cytosol, vacuole, thyllakoid membrane, nucleus and

P. Verma (✉) • S.A. Khan • V. Parasharami
Division of Biochemical Sciences, CSIR-National Chemical Laboratory,
Dr. Homi Bhabha Road, Pashan, Pune, Maharashtra 411008, India
e-mail: priyankaurbest@gmail.com

A.K. Mathur
Plant Biotechnology Division, Central Institute of Medicinal and Aromatic Plants (CSIR-CIMAP), Kukrail Picnic Spot Road, P.O. CIMAP, Lucknow, Uttar Pradesh 226015, India

M. Naeem et al. (eds.), *Catharanthus roseus*,
DOI 10.1007/978-3-319-51620-2_10

endoplasmic reticulum). The complexity is further compounded by extremely high level of recalcitrancy of *C. roseus* plant for regeneration and *Agrobacterium*-mediated genetic transformation for pathway engineering. As a consequence, all genetic modulation efforts so far made in *C. roseus* are confined to cell suspension and transformed hairy root cultures that lack the required level of cyto- and tissue-differentiation essential for the expression of entire TIAs pathway genes and enzymes. A perusal of published work in *C. roseus* clearly suggests that inspite of several pathway manipulation/engineering attempts, the level of TIAs production in cell/tissue/hairy root cultures of this herb could never be enhanced to the level of expectations. The enzymatic, developmental and environmental rigidity/complexities associated with the biosynthetic pathway of these alkaloids have often been cited as possible reasons for these disappointing outcomes Therefore, three major areas of investigation are in focused attention of *Catharanthus* researchers' the world over are: (1) how to select or design the starting cells or tissue(s) to realize the full potential of applying metabolic engineering tools for up-regulating the TIAs pathway in them; (2) how to overcome the strong recalcitrancy of *Catharanthus* plant tissues for de novo organogenesis and in vitro plant regeneration for whole plant-level expression of a transgene coding either for a limiting pathway enzyme or a transcription factor that can control the global expression of several pathway genes and; (3) how to overcome the inability of non-differentiated cell cultures to execute those pathway steps that are expressed only in specialized tissues/cells of *C. roseus* plants. Various biotechnological approaches and generation of novel tissue types have been discussed in the present chapter for the modulation and increased TIAs flux in *C. roseus*.

Keywords Gene expression • In vitro culture • Terpenoid indole alkaloids • Transcription factors

1 Introduction

Plants have been serving as an extremely valuable source for treatments and therapies since human existence. Thousands of plant species are used as medicines and most of the world population uses them to cure the acute and chronic health problems. Many of the present day modern medicines in the western world have been developed on the basis of our ancient knowledge of traditional plant-based medicines. However, the receptors and mechanisms of their action in human body are being identified only recently. Plants produce a large number of low-molecular weight organic compounds that appear to have little function in plants' own growth and development. These compounds, designated as "secondary metabolites", however perform several diverse ecological functions in plant's defense responses against pathogens, herbivores UV radiations and other environmental extremities (Zhong 2001). Many of these plant secondary metabolites have also been found to possess important biological activities of pharmacological importance for humans. Therefore, intensive studies have been carried out during last six decades to characterize the nature of plant secondary metabolites from the viewpoints of their chemical structures, biosynthesis, bio-activities and clinical utilities.

Attempts to produce these plant based molecules by synthetic chemistry route have largely been found unfruitful and uneconomical due to their complex chirality. Hence, entire commercial demand for several important drug molecules like paclitaxel, vincristine, vinblastine, podophyllotoxin, camptothecin, ergot, digoxin, artemisinin, pilocarpine, morphine, reserpine, hypericine, theophylin etc. is still being met from their source plants (Verma et al. 2012a; Zhong 2001). As majority of the plant secondary metabolites are synthesized in very low amount in a season- and age-dependent manner their agri-based production is often less than the size of their clinical demands resulting in high market cost (Verma et al. 2012a, 2015a). The tools of modern day plant biotechnology are, therefore, finding increasing application to fill these gaps. These advancements are collectively studied under the umbrella of a new subject discipline called 'metabolic engineering" (Verpoorte and Alfermann 2000). Metabolic engineering is broadly referred to the directed improvement of cellular properties through the modification of a specific biochemical reaction(s) or the introduction of new enzymatic capability in a plant cell, with or without the use of recombinant DNA technology. There are three basic goals of metabolic engineering in plants: more production of a specific desired molecule, less production of a specific unwanted compound, and the production of a novel compound (i.e. a molecule that is produced in nature, but not usually in the target host plant). Strategies adopted for achieving these goals may include the engineering of single steps in a biosynthetic pathway to increase the metabolic flux towards a target compounds, blocking of competitive pathways at branch points of a shared substrate/intermediates, or to introduce short cuts that divert the metabolic flux towards a different route. In addition to these specifically directed gene insertion approaches, strategies like selection of plant genotype with increased density of metabolite producing or accumulating cell types, reduced catabolism of desired molecule, cells with increased availability of precursor molecules drawn from primary metabolite pools and chemical/biotic elicitation of regulatory elements (transcription factors) of a pathway etc. are also studied under the subject of metabolic engineering. The single gene manipulation has only limited value, because the effects of modulating single enzymatic steps are often absorbed by the system in an attempt to restore homeostasis. On the other hand, targeting the multiple steps in the same pathway could help in controlling metabolic flux in a more predictable manner but it is difficult to achieve (Verpoorte and Alfermann 2000). This would involve up-regulating several consecutive enzymes in a pathway or up-regulating enzymes in one pathway while suppressing those in another competing pathway, or using regulatory genes (transcription factors) to establish multipoint control over one or more genes of a given pathway (Verpoorte and Alfermann 2000).

These basic considerations of a plant metabolic engineering programme invariably require the ready availability of several metabolic tools, techniques and material to reach to a successful outcome (Verpoorte and Alfermann 2000). These may include: (1) identification and cloning of a target gene; (2) preparation of a suitable gene construct; (3) development of an efficient method of gene insertion into the host cell; (4) optimization of the regeneration protocol to recover plants from transformed tissue; (5) molecular and functional characterization of transgene to ensure its stable expression; (6) transfer of introduced gene to elite cultivars through conventional breeding, if required and; (7) evaluation of transgene expression for the envisaged task in the backdrop of bio-safety considerations.

Catharanthus roseus—The Madagascar periwinkle plant has a well-documented history of its use in the treatment of several physiological and metabolic disorders, particularly the various types of neoplasmas (Johnson et al. 1963). In traditional systems of medicines, the leaf, seed, flower and root of this plant are frequently used for treating diabetes, hypertension, menorrhagia and tumor growth. In modern systems of medicines the plant is commercially valued for the strong anti-neoplastic efficacy of two of its leaf-derived dimeric terpenoid indole alkaloids (TIAs) namely vincristine and vinblastine and two root-derived monomeric alkaloids—ajmalicine and serpentine that are in wide clinical usage to treat hypertension and other circulatory disorders (Verpoorte et al. 1997, 2002; Van der Heijden et al. 2004; Duarte et al. 2010). *Catharanthus* alkaloids are chemically classified as "terpenoid indole alkaloids (TIAs)" or "monoterpene indole alkaloids (MIAs)" because of the presence of an indole ring attached to a terpenoidal skeleton (El-Sayed and Verpoorte 2007). Among all the TIAs present in *C. roseus* plant, maximum clinical attention from a pharmacological point of view has been drawn by vincristine (VCR) and vinblastine (VLB) that occupy an indispensible place in most of the chemotherapy treatments against Hodgkin's lymphomas, childhood leukaemia, lymphosarcoma, neuroblastoma, and carcinoma of breast and lungs (Neuss and Neuss 1990; Arora et al. 2010). VLB and VCR, sold in the market by the name of velban and oncovin respectively, have so far defied the rules of synthetic chemistry for their chemical synthesis on a commercially and economically viable scale due to their multifaceted structural chirality (Hughes and Shanks 2002) and hence, *C. roseus* plants represent the sole bio-resource for their production; *albeit* in very low amount (Kumar et al. 2007). The *Catharanthus* alkaloids manifest their anti-cancerous activity by arresting tumor cell proliferation by binding to tubulin ("end-capping effect") and thereby disrupting the spindle assembly (Verma et al. 2012a). Both VCR and VLB are M-phase cell cycle-cycle mitotic inhibitors and are effective at sub-micromolar concentration (10nM–1μM). At higher concentration (>10 μM) they cause tubulin aggregation and results in the formation of tubulin para-crystals (Foye 1995). Due to the lack of cross confrontation with DNA-alkylating drugs, these alkaloids are preferentially included into the combination chemotherapy regimens (Van Der Heijden et al. 2004). Because of their strong cyto-toxic nature the *C. roseus* plant produces these drug molecules in extremely low amounts (<0.0002%) as a part of its auto-defense mechanism that results in high cost of extraction. Nearly 500–750 kg of dried leaves are required to yield just 1.0 g of VLB.

Vincristine, vinblastine and their derivatives form an essential ingredient of chemotherapy regimens for cancer treatment. They are either used singly or in combination with other medicines. All TIAs in clinical uses are administered intravenously and they are eventually metabolized in liver. Peripheral neuropathy, excessive hair loss, hyponatremia and constipation are some of the harmful side effects associated with these drug molecules. Some of the semi-synthetic derivatives of VLB and VCR such as vinorelbine and vinflunine have been developed to improve their therapeutic index with minimum side effects (Wilson et al. 1999). Both of these derivatives have been found useful in the treatment of non-small cell lung and metastatic breast and bladder cancers (Dipierre et al. 1991; Bennouna et al. 2006; Mano 2006). These alkaloids are frequently incorporated into combination chemotherapy because of their lack of cross resistance with DNA-alkylating drugs and their different mode of action (Van Der Heijden et al. 2004). Some of the clinical attributes of four of the most important TIAs found in *Catharanthus roseus* plants are described below:

- *VINBLASTINE* ($C_{46}H_{56}N_4O_9$): Vinblastine (VLB) is the generic name of the alkaloid formerly known as vincaleukoblastine. It is a colorless compound while its sulphate derivative is slightly yellow and hygroscopic. It is soluble in water and methanol. In plants VLB is formed as a chemical analogue of vincristine. It is an integral part of several chemo-therapeutic formulations including the most famous ABVD (Adriamycin, Bleomycin, VLB, Decarbazine) regimen which is highly effective in the treatment of metastatic testicular cancer, Kaposi's sarcoma and breast carcinoma. VLB can be safely stored at room temperature in an inert environment and is clinically administered in encapsulated form in multilamellar liposomes.
- *VINCRISTINE* ($C_{46}H_{56}N_4O_{10}$): Vincristine (VCR), also known as leurocristine, is sold under the brand name Oncovin after it was approved by FDA for clinical use in 1963 (Farnsworth 1985). In most commercial preparations VCR appears as a colorless fluid. VCR binds to tubulin dimer and disturbs the microtubule structure which in turn arrests mitosis at metaphase stage. It affects all rapidly dividing cell types including intestinal epithelium and bone marrow. VCR is an essential constituent of drug formulations like CHOP (Cyclophosphhamide, Hydroxydoxorubicin, VCR, Prednisone) and MOPP (Mechlo-rethamine, VCR, Procarbazine, Prednisone) that are strong anti-neoplastic regimens against non-Hodgkin's lymphoma, Hodgkin's lymphoma and lymphoblastic leukemia.
- *VINDESINE* ($C_{43}H_{55}N_5O_7$): Vindesine (VDS) is commercially available as white powder with the trade name Eldisine or Fildesin. It is particularly effective against melanoma, lung cancer, breast cancer and uterine cancers.
- *VINORELBINE* ($C_{46}H_{58}N_4O_9$): Vinorelbine (VNLB, VRL) was the first semi-synthetic derivative made with vindoline and catharanthine that reached the drug market in the name of Nevelbine). It is a colorless fluid with least side effects amongst all TIAs-based anti-cancer regimens. It is most effective against cell lung cancers.

The extremely low *in planta* occurrence of TIAs in *C. roseus* plants resulting in their short supply, high commercial demand and exorbitant price have brought them in focus of intense scientific scrutiny in last 30 years with sole intention of increasing their production. As a consequence, research in this area has proceeded in two major directions. Firstly, towards the understanding of their biosynthesis in plant, including the enzymology and genetic regulation of the concerned metabolic pathway(s) and secondly, towards exploring the possibility of developing cell/tissue culture based platforms for in vitro TIAs production. The wealth of information gathered so far has made TIAs biogenetic pathway as one of the most well dissected and understood metabolic routes at the level of enzymes and corresponding genes in plants.

2 Metabolic Engineering and Emergence of *C. roseus* as Model System for In Vitro Alkaloid Pathway Modulation

In a nut shell plant metabolic engineering takes into account the understanding of the architecture of a metabolic pathway to identify the major regulatory/limiting steps and then try to overcome these limitation by various biological engineering tools, including the omics approaches. Three conditions must be satisfied in a plant (cell) before a significant increment in the yield of a desired metabolite can be expected to occur. They are: (1) assured supply of starting precursor(s) and/or limiting intermediates, particularly those that lie at the interface of primary and secondary metabolisms; (2) optimal induction of genes/enzymes of an intended pathway at the right time and right place and; (3) availability of a suitable sink to store the synthesized product. Though initial two of these considerations are key to the success of any pathway modulation effort, better understanding of the dynamics of storage and catabolism of a targeted molecules within the plant/cell not only add towards its final productivity but also help in choosing the most suitable downstream process for its extraction and recovery.

Our understanding of metabolic pathway networks in plants, the enzymology and the genes involved at various steps of the metabolite flux within and between the pathways and, the temporal and spatial distribution of a given metabolite(s) within the plant body or its sub-cellular components has now reached to a level where engineering of a target pathway for obtaining higher yields of a given phytomolecule in homo- to hetero-logous expression systems is becoming a theoretical reality. Plant alkaloids (that have provided the maximum numbers of drug molecules to the pharma world), anthocyanins, terpenes and glycosides are occupying the centre stage of these ongoing metabolic engineering efforts. Plant cell and tissue culture techniques are integral components of these engineering enterprises in two major ways: (1) either the cultured tissues are being used as an alternate production platform as such or, (2) they are being used to provide requisite interface for genetic engineering for the hyper-expression of a limiting metabolic step or for the silencing of a branch point sub-way to block the diversion of a limiting pathway

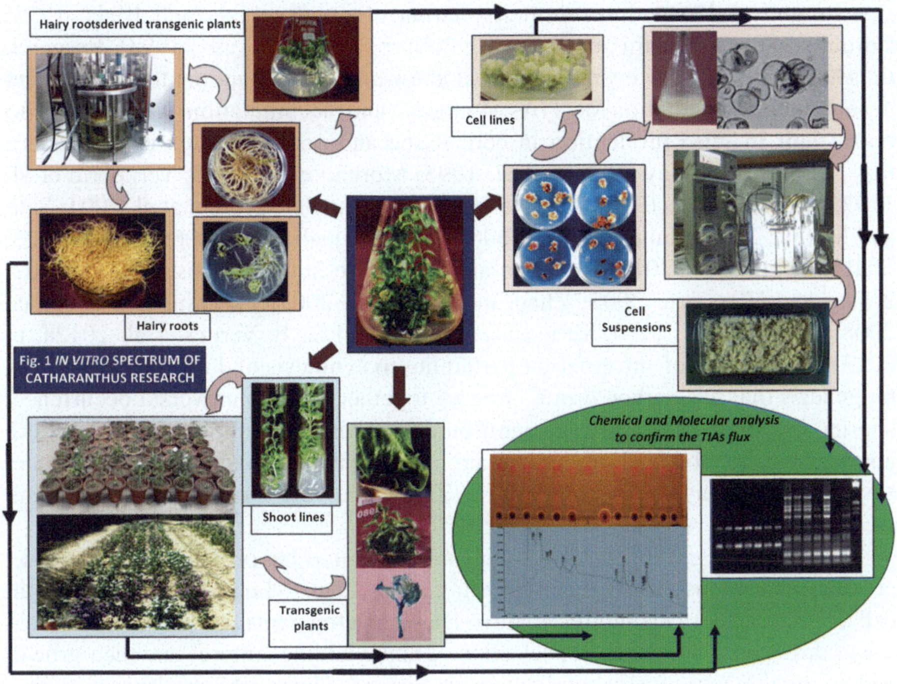

Fig. 1 In vitro spectrum of *Catharanthus* research

intermediate towards non-productive route. Cell or tissue cultures with a congenial physiological and biochemical backgrounds for the expression of a given pathway to produce a desired metabolite are also proving useful in avoiding other issues associated with their agri-based production such as: restricted geographical availability of source plant; seasonal fluctuations in yield; minimization of batch-to-batch variation in quality; avoidance of neighboring cell interference and complications related to long distant transport and segregation of metabolic pools; prolongation of pathway steps that occur for very short time in plants: easy up-scaling in bioreactor and; lesser complications in extractions and industrial down-stream processing. Figure 1 shows a spectrum of in vitro approaches that have potential application in enhancing the TIAs flux in *C. roseus*.

Agri-based production of *Catharanthus* alkaloids for pharma industry is highly expensive because of their low *in planta* occurrence leading to very high extraction cost. With a current market price vincristine and vinblastine represents some of the least abundant and most expensive plant natural products in use in the drug sector today. The great pharmaceutical value of *C. roseus* as a powerful anti-cancerous herb on one hand and high market price and clinical demand of its alkaloids on the other hand, has brought it under intensive scientific scrutiny since 1950s. The initial research efforts made in this plant species were more focused towards elucidation of the TIAs pathway, enzymology associated with their biogenesis and understanding of their pharmacology (Svoboda and Blake 1975; Cordell 1980; Ganapathi and Kargi 1990; Verpoorte et al. 1991, 1993). Research in last two decades has been more directed towards understanding the genetic and developmental regulation of TIAs synthesis and accumulation with an aim to boost their in vitro production in cell, tissue and organ cultures using modern tools of biotechnology (Meijer et al. 1993; Moreno et al. 1995; Verpoorte et al. 1997; St-Pierre et al. 1999; Di Flore et al. 2004; van der Heijden et al. 2004; Seth and Mathur 2005; Mahroug et al. 2006, 2007; Pasquali et al. 2006; Rischer 2006; Shukla et al. 2006, 2010; El-Sayed and Verpoorte 2007; Pietrosuik et al. 2007; Zarate and Verpoorte 2007; Zhao and Verpoorte 2007; Facchini and De Luca 2008; Shukla et al. 2010; Verma and Mathur 2011a, b; Verma et al. 2012a, b, 2013). The wealth of information pertaining to synthesis and regulation of TIAs biogenesis that has gathered in *C. roseus,* together with its universal occurrence, shorter breeding cycle, low chromosome number, amenability to cell culture approaches and, above all high commercial interest have made this plant a model system for pathway engineering efforts (Zarate and Verpoorte 2007; Facchini and De Luca 2008).

The insights regarding TIAs metabolism in *C. roseus* has prompted several workers to apply various metabolic engineering approaches to boost their production in cells, tissues and organ cultures in vitro. These developments are advancing in three major directions. These are: (1) selection of high yielding cultures and their growth and production optimizations following precursor feeding and biotic/abiotic elicitations; (2) Standardization of protocols for deliberate pathway gene insertion for hyper-expression of a limiting enzyme; (3) employment of transgenic hairy root cultures and their bioreactor up-scaling for metabolite production (Leckie et al. 1991; Fulzele and Heble 1994; Hirata et al. 1994; Hughes et al. 2004; ten Hoopen et al. 1994; Schlatmann et al. 1995a, b; Canel et al. 1998; van der Fits et al. 2001; Zhao and Zhu Wei-Hua 2001a, b; Whitmer et al. 2002b; Di Flore et al. 2004; Choi et al. 2004; Pasquali et al. 2006; Pietrosuik et al. 2007; Zarate and Verpoorte 2007; Zhao and Verpoorte 2007; Guirimand et al. 2009; Verma and Mathur 2011a, b; Verma et al. 2012a, b, 2013).

Following major gaps in *C. roseus* research were therefore identified that has been answered through the in vitro biotechnological interventions in recent pasts

- Majority of pathway modulation work in *C. roseus* has been carried out with wild type cells and tissue types. Very little attention was paid to generate mutant cell/tissue types with ideal physiological back-ground for pathway modulation, particularly with respect to adequate availability of TIAs precursors drawn from shikimate and terpenoid pathways of primary metabolic pools. Tryptophan availability in particular is crucial because of its least abundance in plant cells and higher demand for synthesis of auxins for growth-sustaining metabolic functions. Failure of efforts to get desired outcomes with exogenous feeding of these precursors in wild type cells might have been a consequence of their inability to accommodate these molecules in cells due to their strong feedback inhibitory actions on other metabolic pathways.

- Inability of in vitro grown heterotrophic cells/tissue cultures to express NMT enzyme involved in vindoline synthesis from tabersonine because of the absence of a functional chloroplast system under heterotrophic mode of in vitro growth. Efforts to select photo-autotrophic cells were not made.
- Absence of D4H and DAT transcripts in undifferentiated cell or hairy root cultures because of lack of special laticiferous and idioblast tissues.
- Poor sink and holding capacity of cells/tissues to store anti-mitotic bisindole alkaloids VLB and VCR.
- Lack of direct plant regeneration protocols to facilitate whole-plant expression of engineered pathway genes following genetic transformation.
- Limited know-how's on bioreactor designing and operation for *C. roseus* cells and hairy root tissues.
- Limited understanding of transcription factors and transporter proteins associated with TIAs biosynthesis.
- Lack of sufficient knowledge on catabolism of TIAs in cells

3 Terpenoid Indole Alkaloids Pathway: A Snap Shot

The biosynthesis of MIAs in *C. roseus* typically represents the elegant complexities of a plant secondary metabolic pathway. The entire multi-step pathway rigidly follows a complex and highly compartmentalized metabolic route which is strictly regulated by several developmental and environmental factors (Fig. 2). From a holistic perspective, the biosynthesis of TIAs in *C. roseus* proceeds via 30 co-ordinately regulated enzymatic steps involving at least 35 known intermediates (van der Heijden et al. 2004; Facchini and De Luca 2008). Consequently, 30 biogenetic and two regulatory genes have also been identified for their close association with this pathway. Out of a total of 42 cDNA clones identified in major TIAs producing plants, 25 belong to *C. roseus* (Guirimand et al. 2010). The entire pathway has been shown to precede with involvement of four discrete cell/tissue types (epidermis, internal phloem parenchyma, idioblasts) and participation of five intra-cellular compartments namely cytosol, vacuole, thylakoid membrane, nucleus, and endoplasmic reticulum (De Luca and Cutler 1987; Facchini 2001; El-Sayed and Verpoorte 2007; Mahroug et al. 2007; Facchini and De Luca 2008; Verma et al. 2012a). All TIAs are biosynthesised from a central precursor molecule- strictosidine, which is a condensation product of an indole ring donor tryptamine and a terpenoid moiety donor secologanin. The tryptamine and secologanin undergo a condensation reaction to form strictosidine in the cell vacuole. Strictosidine that heralds the first indication of a switch over of carbon flux from primary to secondary metabolism during TIAs synthesis. Strictosidine, the universal precursor of all subsequent MIAs is further de-glucosylated by the enzyme strictosidine β-glucoxidase to form an unstable aglycon that give rise to cathenamine which is a branch point intermediate that can be routed towards the synthesis of several different types of monomeric alkaloids like catharanthine, ajmalicine, and tabersonine. While

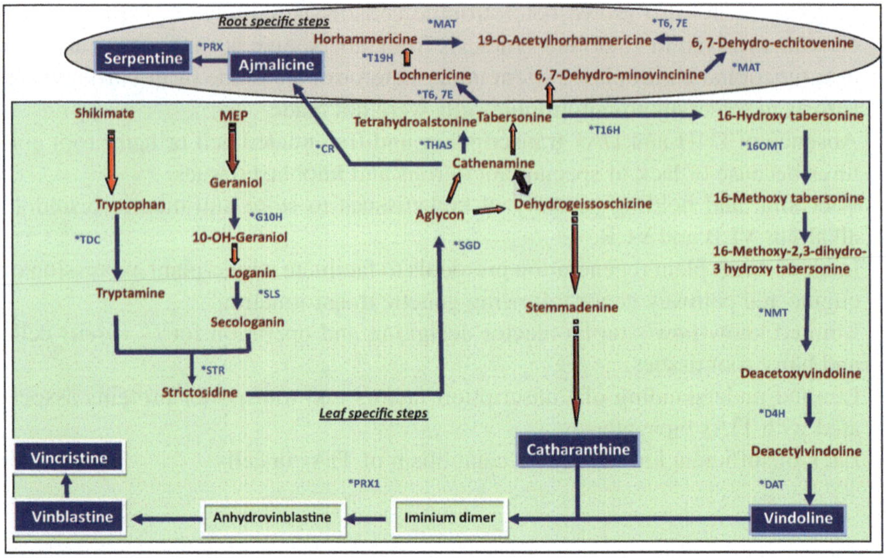

Fig. 2 TIAs pathway operating in *C. roseus*. *TDC* tryptophan decarboxylase, *G10H* geraniol-10-hydroxylase, *SLS* secologanin synthase, *STR* strictosidine synthase, *SGD* strictosidine β glucosidase, *T16H* tabersonine 16-hydroxylase, *16 OMT* 16-hydroxytabersonine-16-O-methyltransferase, *NMT* *N*-methyltransferase, *D4H* deacetoxyvindoline-4-hydroxylase, *DAT* deacetylvindoline 4-*O*-acetyltransferase, *T6,7E* tabersonine-6,7-epoxidase, *T19H* tabersonine 19 hydroxylase, *MAT* minovincinine-19-*O*-acetyltransferase, *CR* cathenamine reductase, *PRX1* peroxidase, *THAS* tetrahydroalstonine reductase

ajmalicine is further converted into serpentine in the vacuoles by the action of peroxidase, tabersonine is either converted into lochnericine and then to horhammericine in roots or is diverted towards catharanthine synthesis via dehydrogeissoschizine and stemmadenine as major intermediates.

While very little is known about the catharanthine sub-way (Loyola-Vargas et al. 2007), the vindoline route is fairly well dissected at the level of associated enzymes and genes (Schroder et al. 1999; St-Pierre and De Luca 1995; Vazquez-Flota et al. 1997; St-Pierre et al. 1999; Laflamme et al. 2001). Catharanthine has been shown to be exclusively excluded from epidermis and stored in surface wax layer of the leaves (Roepke et al. 2010) as a defense strategy against insect and microbial infestation. Tabersonine is also converted into vindoline in leaves through seven well-characterized enzymatic steps (Dethier and De Luca 1993; Vazquez-Flota et al. 1997; Luca and Laflamme 2001; Pasquali et al. 2006; Campous-Tamayoet al. 2008; Pan et al. 2015). The NMT is the only enzyme of this pathway whose activity is not affected by light, whereas D4H transcript abundance increased significantly upon eposing the plants to light (St. Pierre et al. 1998, Vasquez-Flota and De Luca 1998). Diversion of flux from cathenamine-derived intermediate tabersonine to vindoline is facilitated by seven enzymatic steps in the aerial tissues of *C. roseus* plants. The biochemistry of this seven-step route involves three hydroxylations and one each of *O*-methylation, *N*-methylation and *O*-acetylation reactions. The tabersonine to 16-hydroxy tabersonine conversion is catalyzed by a cytochrome P450-dependent tabersonine-16-hydroxylase (T16H) enzyme, followed by its methylation by 16-hydroxytabersonine-16-*O*-methyl transferase (16OMT) which also requires 5-adenosyl-L-methionine as a co-substrate (Levac et al. 2008). The conversion of 16-methoxy tabersonine to 16 methoxy-2, 3-dihydrotabersonine then occurs via an uncharacterised oxidation step. The subsequent step involves a thyllakoid associated N-methyltransferase (NMT) to obtain deacetoxyvindoline. The last two biogenetic reactions are catalyzed by a light-regulated deacetoxyvindoline-4-hydroxylase (D4H) and deacetylvindoline-4-*O*-acetyltransferase (DAT) enzymes that are expressed only in special idioblast/laticifer cells in leaves. (De Carolis et al. 1990; St-Pierre et al. 1999; Campous-Tamayo et al. 2008; Shukla et al. 2010; Guirimand et al. 2011). In contrast to aerial tissues, the roots of *C. roseus* plants operate another sub-way from tabersonine to lochnericine by the catalytic action of tabersonine-6, 7 epoxidase (T6,7E; Rodriguez et al. 2003). Lochnericine can then be converted into horhammericine by tabersonine 19-hydroxylase. Alternatively, tabersonine can also be routed towards horhammericine via 6,7 dehydrominovincine and both minovincinine and/or horhammericine can then be acetylated to yield 19-O-acetylhorhammericine by the enzyme minovincinine 19-hydroxy-*O*-acetyl transferase (MAT) which is localized in cortical cells of growing root tips. Finally, the coupling of monomeric TIA catharanthine and vindoline resulting in the formation of vinblastine and vincristine in leaves mark the termination of TIAs biogenetic pathway in *C. roseus*. Since vincristine and vinblastine are highly antimitotic molecules due to their inhibitory action on spindle assembly during cell division, this coupling reaction exclusively takes place in the cell vacuole as a measure of cellular containment. The dimerization reaction between catharanthine and vindoline is facilitated by class III basic peroxidase (Prx1; Sottomayor et al. 1998; Sottomayor and Ros Barcelo 2003; Kumar et al. 2007; Costa et al. 2008).

On the basis of cellular involvements during TIAs synthesis, the internal phloem parenchyma (IPAP) cells present in the periphery of stem pith are primary locations for expression of early pathway genes (Burlat et al. 2004) while the leaf epidermis harbors expression of SLS, TDC and STR gene products (St-Pierre et al. 1999) and idioblast and laticifer cells embedded in the palisade tissues of leaves. In root tissue TDC, STR and MAT transcripts are localized in protoderm and cortical cells around root apical meristem (Laflamme et al. 2001; Moreno-Valenzuela et al. 2003). T16H and SGD expressed in the epidermis, whereas ORCA3 and an AP2/ERF type of transcription factor were expressed in all four cell types (Murata and De Luca 2005). G10H expression is confined to vascular tissue (Burlat et al. 2004). TDC activity was maximum in the epidermal cells, while DAT activity was detected in the whole leaf. NMT activity was found in the whole leaf extract.

4 TIA Pathway Engineering: Current Status

(a) The Generation of Novel Tissue with Enhanced Precursor Availability

The strategies to engineer the TIAs pathway in *C. roseus* revolves around the generation of novel cell and tissue types characterized by sufficient precursor availability, over- expression of limiting enzymes, activation of regulatory transcription factors, silencing of competitive pathways to flux the intermediates in desired direction and optimized growth parameters (Canel et al. 1998; van der Fits and Memelink 2000; Hughes et al. 2004; van der Heijden et al. 2004; Zarate and Verpoorte 2007; Verma et al. 2012c). The materials being generated through these novel approaches now await their up scaling in bioreactors to assess their commercial utility (Zhao and Verpoorte 2007). In attempts to improve the production of the valuable alkaloids such as vincristine and vinblastine, several studies on *C. roseus* reported also the accumulation of phenolic compounds upon biotic and/or abiotic stress. The accumulation of phenolics may also affect other secondary metabolite pathways including the alkaloid pathways, as plant defence is a complex system. Elucidation of the pathways and understanding their regulation are important for metabolic engineering to improve the production of desired metabolites (Mustafa and Verpoorte 2007). The ability of plant cells to resist 5-MT induced toxicity has often been ascribed to their ability to over synthesize tryptophan to dilute the inhibitory effect of the analogue. This is generally achieved via *a* relaxed feedback inhibition of the enzyme anthranilate synthase involved in the biosynthesis of tryptophan from chorismate, the end product of shikimate pathway (Meijer et al. 1993; Scott et al. 1979; Radwanski and Last 1995; Seth and Mathur 2005). It is pertinent to recall that tryptophan is one of the least abundant amino acid present in plants but is the sole donor of indole ring for the synthesis of auxins, glucosinolates, phytoalexins and terpenoid indole alkaloids. Exogenous supplementation of tryptophan to enhance alkaloid synthesis in *C. roseus* has not been successful because it down regulates several other metabolic pathways via activation of another key regulatory enzyme chorismate mutase that push chorismate towards prephenate metabolism (Radwanski and Last 1995; Galili and Hofgen 2002; Whitmer et al. 2002b). Hughes et al. (2004) have shown that TIA accumulation in a transgenic hairy root line of *C. roseus* could be enhanced through larger tryptophan availability achieved via over-expression of an *Arabidopsis* feedback resistant anthranilate synthase gene in them.

To generate such novel tissue types with larger availability of tryptophan, the approach of selecting variants resistant to inhibitory stress of the tryptophan analogue namely 5-methyltryptophan was adopted in the studies conducted by Verma et al. (2012c, 2013) . The basis for adopting such an approach was its successful implementation in several earlier studies aimed to enhance the nutritional quality of many agricultural crops in terms of the improved content of essential dietary amino acids like tryptophan, arginine, lysine, phenylalanine, tyrosine, methionine, threonine etc. in them (Brotherton et al. 1996; Kisaka et al. 1996; Kim et al. 2004; Galili et al. 2005). Hyper-accumulation of targeted amino acid in cells resistant to corresponding analogue is a preferred strategy that a variant cell normally evolves to counter-balance the analogue inhibition by competing it out through a dilution effect during protein synthesis. Cultured plant cells selected for resistance to 5-methyltryptophan (5MT) have been shown to have elevated levels of free tryptophan. The key mechanism of 5MT resistance is reported to be the altered (relaxed/reduced) feedback inhibition of anthranilate synthase which is the key feedback control enzyme in the tryptophan biosynthesis from chorismate (Kang and Kameya, 1995; Kisaka et al. 1996). Anthranilate synthase (AS) has two subunits, ASα and ASβ, the former of which catalyzes the conversion of chorismate to anthranilate and is susceptible to feedback inhibition. 5MT resistant cultures have a mutated OASα1gene that code for ASα subunit OASα1D, in which aspartate-323 is replaced with asparagine making it thereby insensitive to feedback inhibition by tryptophan (Wakasa et al. 1999). Though not verified through enzymatic assays, the 5MT tolerant callus and multiple shoot lines selected by Verma et al. (2012c, 2013) have devised this strategy as was evident by hyper tryptophan accumulation in them with comparable growth in comparison to wild line cultures maintained on analogue-free medium, as proposed earlier by Seth and Mathur (2005) for 5MT tolerant callus mutants of *C. roseus*.

Verma et al. (2013) have selected five cell suspension lines of *C. roseus* resistant to 5-methyl tryptophan and characterized on the basis of growth, free tryptophan content and terpenoid indole alkaloid accumulation. Experimental parameters to scale up the most productive variant cell lines in 5–7 L air-lift stirred tank bioreactors were also standardized. Crude alkaloid extract of the cells grown in shake flask and this bioreactor batch also showed the formation of yellow-colored crystals which upon 1HNMR and ESI-MS analysis indicated a phenolic identity. This crude alkaloid extract of bioreactor-harvested cells containing this compound at 50 μg/mL concentration registered 65.21, 17.75, 97.0, 100 % more total antioxidant capacity, reducing power, total phenolic content, and ferric-reducing antioxidant power, respectively, when compared with that of extracts of cells grown in shake flask cultures. The latter, however, showed 57.47% better radical scavenging activity (DPPH) than the bioreactor-harvested cells. Another study of Verma et al. (2012c) showed in vitro selection of ten 5-methyltryprophan (5-MT)-resistant multiple shoot culture lines in three genotypes of *C. roseus*. The variant shoot lines displayed a differential threshold tolerance limit against the analogue stress, ranged from 20 to 70 mg/L 5-MT in the medium. The rooted shoots of 5-MT-tolerant lines were successfully acclimatized under glasshouse environment wherein they grew normally and set seeds. Flowering twigs or leaves excised from 1-year-old glasshouse grown plants of 5-MT variant lines upon postharvest in vivo elicitation with 30 mg/L 5-MT or 5.0 mg/L tryptophan registered an eight-to-tenfold increment in their vindoline content within 24–48 h.

The data obtained with respect to LD_{50} dose of 5MT stress for the wild line calli or multiple shoots also showed a tissue- and genotype-specific trend which was in agreement with reported doses specified for *C. roseus* (Seth and Mathur 2005), *Oryza sativa* (Kisaka et al. 1996; Kim et al. 2004) and Zeal Mays (Kang and Kameya 1995). Derivatization of an efficient selection scheme to isolate 5MT tolerant variant cultures was another highlight of the present study. The scheme has facilitated the recovery of altered phenotype in shortest possible time with efficient elimination of escapees, habituants and fall positives. Though the developed scheme has provided ample scope to isolate variants resistant to sub-lethal to supra-lethal stress of 5MT selection pressure, but it was found that variants selected for resistance to analogue stress around LD_{50} dose were more stable and showed better capacity to gradually build-up their tolerance level with each subsequent repeat selection or sieving cycle. The employment of callus or multiple shoots as starting material for screening 5MT tolerant variants in the present work was also found to be a better choice over the usually employed cell suspensions in earlier work on in vitro isolation of biochemical mutants. Problems associated with cell suspensions like cell aggregation, chromosomal instability and developmental asynchrony etc. were easily avoided. Though compact callus morphology generally hinders the screening of deep-seated variant cells amongst a larger population of non-variant cells in the wild population, but it could be effectively overcome by dot-plating technique during initial recurrent selection cycles under sustained analogue stress as was done in this study. Shoot cultures in particular also proved a useful starting tissue in case of *C. roseus* because of their advanced differentiation level (that has more conducive developmental/biochemical levels of specialization for monitoring the flux of the over-produced precursors through TIAs pathway) and the ease with which 5MT tolerant rooted plants could be obtained from them to observe their in vivo behaviour with respect to TIAs productivity and profiles at the whole plant level. A more recent study of Verma et al. (2015b) where 66 plants raised via direct shoot bud organogenesis from pre-plasmolysed leaf explants of *C. roseus* were assessed under in vivo conditions for their physio-morpho traits, tryptophan metabolism, genetic fidelity and alkaloid profile, showed a strong positive correlation between tryptophan content and 5-methyltryptophan tolerance.

(b) Transgenic Approaches to Enhance the TIAs Flux

(i) *Random T-DNA Insertion and Enhanced Alkaloid Profile*

Agrobacterium tumefaciens has a unique ability to transfer genes into plant genomes. This ability has been utilized for plant genetic engineering. For successful plant genetic transformations, information regarding interaction of bacterium with host plant proteins and plant genome, plant defense signaling and molecular mechanism of T-DNA transfer is already known. The *rol* of *onco* genes corresponding to T-DNA are known to alter morphology of the plant or host plant secondary metabolism. In *C. roseus*, some reports are there where simple t-DNA insertion leads to alter alkaloid content significantly. Increased level of serpentine and ajmalicine was reported in hairy roots (Parr et al. 1988; Batra et al. 2004; Verma et al. 2012b) as well as whole plant transgenics (Verma et al. 2015c). Bhadra et al. (1993) reported threefold increase in vindoline production in hairy roots while stable vindoline production was also observed in shooty teratomas (O'Keef et al. 1997) and whole plant transgenics (Verma et al. 2015c). Hong et al. (2005) observed increased horhammericine in hairy roots. Vincristine and vinblastine are reported to be rarely found in hairy roots, but it is found in shooty teratomas (Begum 2011, Begum et al. 2009) and hairy roots (Zargar et al. 2010). In due course of transformation events, it is reported that few T-DNA genes are differentially lost. Severe affects on morphology, growth, biosynthetic pathway gene expression and production of specific secondary metabolites have been observed on loss of some ORFs. The injury caused by pathogens enhances the production of defense compounds (secondary metabolites). The mode of action involves the activation of plant defense machinery via following steps: (a) Detection of signal by the pathogen; (b) activation of H^+-ATPase; (c) enhanced Ca^{2+} influx within the cells from the intercellular spaces; (d) activation of calcium dependent protein kinase (CDPK); (e) lastly the activation of NADPH oxidase. NADPH oxidase contributed in the commencement of MAP kinases that in turn produce active oxygen radicals that resulted into the increment of secondary metabolites biosynthesis via enhanced transcription of defense genes. Other important route to enhance secondary metabolite synthesis has been mediated by jasmonic acid and salicylic acid signaling pathway.

(ii) *Differentiation based Alkaloid Production*

The diversification of TIAs continuum in cultured tissues is linked with differentiation stage essential for the complete TIAs pathway genes and enzymes expression, most specifically those involved in the late steps of the vindoline biosynthetic pathway. The vindoline synthesis in *C. roseus* shoot cultures could be successfully linked with expression of key enzyme deacetylvindoline acetyl CoA acetyl transferase (DAT) that catalyzes the last step of vindoline biosynthesis. These studies have successfully confirmed the earlier contentions that light dependent vindoline synthesis requires the presence of a particular cellular organization in the form of idioblasts functional thyllakoid system (Murata and De Luca 2005). TIAs biosynthesis in *C. roseus* was strongly regulated by differentiation of specialized cells and tissues, which the undifferentiated cultures normally lack, hairy roots were considered to be a better candidate for in vitro production because of their higher level of cellular differentiation and improved genetic or biochemical stability in culture. Jung et al. (1995) demonstrated an inter-convertible method of hairy roots and its cell suspensions meant for TIAs production and found that cell suspension initiated from hairy root derived callus had 60% less catharanthine than in the transformed roots. This was restored to original level of 1.5 mg/L when roots were again produced from cell suspension. Moreno-Valenzuela et al. (1998) also observed reduction in TDC and STR activities by 5 and 30% respectively in cell suspension when TIAs production in hairy roots versus cell suspension was compared.

(iii) *Efficient Regeneration/Transformation Protocols*

Catharanthus roseus is generally considered as a genetically recalcitrant plant species from pathway engineering angle due to the absence of an efficient direct regeneration protocol for transgenic plant production (van der Fits et al. 2001; Di Flore et al. 2004). Most of the genetic modulation efforts so far made in *C. roseus* (Zarate and Verpoorte 2007) are confined to cell suspension and transformed hairy root cultures due to the fact that they don't have requisite level of tissue or cyto-differentiation which is necessary for the expression of complete TIAs pathway genes and enzymes (Di Flore et al. 2004). Moreover, the hairy roots and cultured cells of *C. roseus* are shown to be highly recalcitrant towards de novo regeneration into transgenic plants. Authors (Verma and Mathur 2011a, b) represented the first report on development of protocol for proficient shoot bud induction directly emerging from the leaves of *C. roseus*.

The most important aspect of a regeneration protocol utilized to genetically modify the species is shoot bud's de novo direct origin from the leaf tissue (Sharma et al. 2005). Since these de novo shoot buds originates from single cells, they are considered as better prospect for stable transformation leading to non-chimeric transgenic plants in comparison to those arising from pre defined germ lines (somatic embryos) or from a pre-formed axillary or apical meristem (Newell 2000). This work also highlights that a pre-plasmolytic treatment of *C. roseus* leaves in CPW: 13% mannitol solution. A 60 min pre-plasmolytic treatment was effective for release of the meristematic cells from the developmental block and allowed them to advance through a complete shoot bud regeneration cycle leading to plant formation. This dehydration treatment influenced the organogenesis either through the explants rapid uptake of plant growth promotors or by reducing the cyto-toxic TIAs concentration (through ex-osmosis) in the region of the dividing cells involved in the de novo organogenesis from the leaf explants. The regeneration protocol optimized in this study could also be successfully employed for *A. tumefaciens* – mediated transgenic plant production in *C. roseus. A. tumefaciens* (strain-LBA4404) harboring a binary vector pBI121 (with GUS and ntp II genes) with p35SGUS-INT (having GUS Introns) was used. Various parameters such as culture age of bacterial, inoculum density of bacterial suspension, infection method and co-cultivation condition for transformation were successfully optimized for this transgenic protocol development. The flooding approach became more effective if done via a SAAT treatment for 60 s. The Sonication treatments extended more than 60 s were lethal for explants. SAAT is a recently developed technique for efficient *A. tumefaciens*-based genetic transformation of crops that generally resist the usual manual pricking or vacuum infiltration approaches of infection like cotton, Papaya, Chenopodium, Vigna, Soybean etc. (Trick and Finer 1997). This method involves the short exposure of ultrasound waves to the target plant tissue immersed in an *Agrobacterium* suspension. The technique efficiently overcomes certain barriers such as the host specificity and the inability of bacterium to reach proper meristematic cells if they are buried deep inside the target tissues (Trick and Finer 1997). Besides, the wounded tissue could also produce more of phenolic substances like acetosyringone that are required for enhancing the binding accessibility of *Agrobacterium* to the cell surface. Also the extent of physical tissue injury caused during the making of thousands of nano-sized pores by the high energy of ultrasound waves is low in comparison to manual pricking by bacterial filled sterilized needle. This helped in minimizing the callusing response from the explants. Using hypocotyls as explants, Wang et al. (2010) developed genetic transformation method via *A. tumefaciens* strain EHA105. The construct used was pCAMBIA2301 having GUS reporter gene with a neomycin phosphotransferase II gene (NTPII) as selectable marker. The Best results with 11% transformation frequency were obtained when 10 min sonication was applied to hypocotyls with 80 W. After that explants were subjected to 30 min *A. tumefaciens* infection and 2 day co-cultivation on 1/2 MS medium having 100 μM acetosyringone. To asses the potential of the given protocol deacetylvindoline-4-*O*-acetyltransferase (DAT), was over-expressed that lead to nine separate transgenic plants. Vindoline was found to be enhanced in the DAT over-expressing transgenic plants. For optimum transformation frequency, various parameters such as density of *Agrobacterium* and acetosyringone concentration, duration of co-cultivation, sonication dose and duration and dose of selection pressure i.e. kanamycin were optimized. In a recent report of Weaver et al. (2014), The Fast Agro-mediated Seedling Transformation (FAST) method was developed that involves the co-cultivation and transient transformation of young seedlings. In this particular study, *Agrobacterium* strongly induced ZCT1 and ORCA3. There are just two reports on regeneration of transgenic plants from transformed hairy roots of *C. roseus* (Choi et al. 2004; Verma et al. 2012b). The regeneration response was found to be greatly influenced by genotype and varied with different root clones. The regenerated shoots showed prolific rooting with extensive lateral branching and shortened internodes in the transgenic plants. PCR and Southern blotting analysis confirmed the retention of Ri-TL-DNA in these regenerants. Verma et al. (2012b) also found excessive flowering in the regenerated plants.

(iv) *Elicitation/Precursor/Inhibition based Studies*

Alternation/supplementations in medium recipes of transformed tissue of *C. roseus* resulted into the altered alkaloid profile. While total alkaloid content was found to be enhanced by low medium nutrients (Toivonen et al. 1992) and penicillin supplementation (Sim et al. 1994) in hairy roots, acetyl salicylic acid increased total alkaloid content in tumor cell suspensions (Godoy-Hernandez and Loyola-Vargas 1997). MES-buffered medium had a negative effect on lochnericine accumulation but enhanced tabersonine synthesis (Morgan and Shanks 2000), whereas fructose supplementation enhanced the catharanthine level in hairy roots (Jung et al. 1992). Effect of oxygenase inhibitors like 1-aminobenzotriazole, clotrimazole (ABT) and 2,5-pyridinedicarboxylic acid was studied to decipher the pathway around tabersonine. The ABT inhibited horhammericine formation while 2,5-pyridinedicarboxylic acid specifically inhibited lochnericine accumulation (Morgan and Shanks 1999). Verapamil and $CdCl_2$, that block the Ca^{2+} flux across the plasma membrane enhanced the total alkaloid content by 25% in hairy roots and their discharge into the medium by ten times (Moreno-Valenzuela et al. 2003). The specific Ca^{2+} chelator, EGTA, stimulated 90% of the total alkaloid secretion. Recently, Thakore et al. (2013) reported 98% increment in ajmalicine on treatment with TritonX-100 (0.1% v/v) and n-hexadecane (2% v/v). Metabolic engineering of the biosynthetic pathway of these TIAs have indicated that extremely low yields of the pharmaceutically important alkaloids can be ascribed to the limitation in the availability of these two precursor molecules from the primary pool (Whitmer et al. 2002a). Tryptophan and terpenoid intermediates feeding to transgenic lines over expressing TDC/STR showed high alkaloid production (Whitmer et al. 2002a). It can be clearly interpreted that the terpenoid branch of pathway is limiting. Exogenous feeding of tryptophan and/or terpenoid intermediates like geraniol or loganin, etc in STR over-expressing cell lines has also resulted in higher TIAs production (Whitmer et al. 2003), indicating that it is not only the availability of tryptophan and secologanin, but their actual utilization which is more limiting in guiding the metabolic flux towards TIAs metabolism. Elicitors are known to induce synthesis of valuable secondary metabolites as defense against pathogen and the consequent biosynthetic pathway. Till date so many biotic or abiotic elicitors have been tested to enhance the discharge of secondary metabolites in *C. roseus* hairy roots. Methyl jasmonates (MeJA) is known to be involved in signal transduction, escorting the reaction of the plant to different environmental signals. One of such reaction is biosynthesis of proteins and secondary metabolites. For some recent years, MeJAs was frequently used as elicitors to induce biosynthesis of secondary metabolism in many plant species, including *C. roseus* (Loyola-Vargas et al. 2007). TIA genes demonstrate considerable variation in the degree and duration of induction by MeJA. Methyl jasmonate (MeJA) an organic compound used in plant defense caused significant enhancement in the transcript of TIAs pathway genes, particularly catharanthine accumulation while sodium nitroprusside alone or in combination with MeJA caused spectacular reduction in catharanthine accumulation. It is also known to enhance type-1 protein prenyltransferase transcripts in hairy roots of *C. roseus*. MeJA treated hairy roots induced the disturbance in the integrity of mitochondrial membrane and a reduction in ATP biosynthesis (Ruiz-May et al. 2009). In *C. roseus*, Octadecanoid-Responsive *Catharanthus* AP2-domain (ORCA1, ORCA2 and ORCA3) transcription factors play crucial role in up-regulating many TIAs genes in response to MeJA treatment. Goklany et al. (2009) carried out a dose specific study of the MeJA treated hairy roots. Low doses of MeJA were found to be favoring TIAs biosynthesis by keeping ZCT (transcriptional repressors) level low (two to seven-fold) and ORCA (29–40) level high resulting into 8 to 15-fold increment in genes involved in TIA biosynthesis. While at high doses of MeJA, TIAs biosynthesis was inhibited. This was attributed to the increased level of ZCT (40-fold) expression in comparison to ORCA (13 to 19-fold) resulting into and nominal induction of the TIA biosynthetic genes (0 to 6-fold). This study directly pointed out the crucial role of exogenous MeJA supplementation in *C. roseus* in up-regulating TIA biosynthetic pathway.

(v) *Pathway Gene Over-Expression*

Since biosynthesis of all TIAs initiates from a common precursor molecule-strictosidine that is formed by the coupling of tryptophan derived tryptamine and secologanin, limited availability of these precursor molecules from the primary shikimate and secoiridoid pools has often been documented as the most serious problem associated with low TIAs productivity in *C. roseus*. While over-expression of the gene tryptophan decarboxylase (TDC), which codes for the enzyme that converts tryptophan into tryptamine, has alone not found effective for a concomitant increase in TIAs accumulation, the overexpression of strictosidine synthase (STR), either alone or along with TDC, was found more conducive for enhanced TIAs synthesis (Canel et al. 1998). Tryptophan pool of the cell is maintained by feedback inhibition of the enzyme anthranilate synthase. To avoid a negative feedback regulatory step-the biosynthesis of indole precursor of TIAs i.e. tryptamine, attempts were made to over-express TIA s pathway gene coding enzymes. Tryptophan decarxylase (TDC) mediated catalysis of tryptophan yields tryptamine. The expression of a feedback resistant anthranilate synthase (AS) α-subunit from *Arabidopsis thaliana* concurrently with the AS β-subunit and TDC gene in the hairy roots of *C. roseus* improved the metabolic flux of the indole precursors tryptamine towards anti-hypertensive drug ajmalicine with a corresponding decrease in lochnericine, horhammericine and tabersonine (Hughes et al. 2002; Hong et al. 2005; Peebles et al. 2005). Recently *C. roseus* TDC/STR-TDC over-expression in other Apocynaceae members *Rawvolfia serpentina* (Mehrotra et al. 2013) and *Vinca minor* (Verma et al. 2015d) resulted into increased ajmalicine/serpentine and vincamine production respectively. In the mevalonate pathway, HMG-COA is converted to mevalonate by the action of enzyme 3-hydroxy-3-methylglutaryl-COA reductase (HMGR). Mevalonate an IPP precursor, is considered to be a crucial enzyme in the synthesis of cytoplasmic isoprenoids in plants such as fitosterols (campesterol, stigmasterol, and sitosterol) and prenyl chains for someproteins, sugars, and lipids. Ayora-Talavera et al. (2002) reported five to seven times more serpentine in hairy roots on the over-expression of HMGR gene. The coupling of isopentenyl diphosphate (IPP) and dimethylallyl diphosphate (DMAPP) catalysed by GPP synthase (GPPS) to yield Geranyl diphosphate (GPP), the initial point of formation of terpene moiety. The transient over-expression of *A. majus* AmGPPS.SSU (Geranyl diphosphate smaller sub unit) in *C. roseus* leaves resulted in the enhancement of vindoline content (Rai et al. 2013). In a study, conducted by Peebles et al. (2011), DXS gene alone as well as along with G10H were over-expressed in hairy roots of *C. roseus*. Noteworthy enhancement in ajmalicine, serpentine, and lochnericine by 67, 26 and 49%, respectively were observed with concurrent decrease in tabersonine (66%) and hörhammericine (54%). When DXS and G10H were co-over-expressed, ajmalicine, lochnericine and tabersonine were significantly enhanced by 16, 31 and 13%, respectively. Another study conducted by Chang et al. (2014) reported DXR and STR or MECS and STR over-expression in hairy roots. Again it favoured the co-over-expression rather than single gene expression. Higher ajmalicine was registered by co-over-expression of DXR-STR or MECS-STR in comparison to the single gene over-expression of DXR, MECS, and STR. Deacetylvindoline-4-*O*-acetyltransferase (DAT), the final enzyme in the vindoline biosynthesis that catalyses de-acetylvindoline into vindoline, was also over-expressed with CaMV35S promoter in hairy roots of *C. roseus* (Magnotta et al. 2007). These workers also found no increment in the vindoline level in the transformed roots but instead, observed fourfolds increase in horhammericine to control lines. Enhanced horhammericine accumulation was attributed to the inhibitory activity of minovincinine-19-O-acetyltransferase (MAT) by the DAT protein as revealed by the enzymatic analysis. Over-expression of DAT gene was found to enhance vindoline production in transgenic plants (Wang et al. 2010). The two peroxidases CrPrx and CrPrx1 are characterized from *C. roseus* and belong to class III peroxidases. CrPrx and CrPrx1 are known to be apoplastic and vacuoler in nature, respectively. For determining their respective function *in planta*, these were expressed in *Nicotiana tabacum* (Kumar et al. 2011). It was interpreted that salt and dehydration stress was preside over by the vacuolar peroxidase while apoplastic peroxidase provide cold stress tolerance. Over-expression of apoplastic peroxidase CrPrx in *C. roseus* hairy root lines resulted into enhanced ajmalicine and serpentine production. Using biolistic approach, Guirimand et al. (2009) has given a protocol of transient expression in *C. roseus* cells which was further utilized in subcellular characterization of HDS and G10H to plastids/stromules and to ER membrane. A very recent report showed the D4H like gene but with different transcriptional expression profile (Zhou et al. 2014). The wealth of information gathered so far has made TIAs biogenetic pathway as one of the best dissected and understood metabolic routes at the enzymatic level and corresponding genes level in plants. Over expression of TIAs pathway genes in *C. roseus* cell cultures has so far not be up to the execution in significantly augmenting sustainable production of the preferred alkaloids because of their lower level of cellular and tissue differentiation. Our understanding of terpenoid indole alkaloids has matured to a much better level of enzymatic and genetic regulations where a cell can now be engineered via incorporation of transgenes for homologous as well as heterologous expression of an indigenous or novel metabolic step(s).

(vi) *Transcription Factors and TIA Pathway Regulation*

A general understanding of the major routes of biosynthesis for TIAs pathway has been gathered, but the complete expression of pathways is necessary by determining the master regulators i.e. transcription factors that can manipulate whole biosynthetic steps involved in the pathway. ORCAs, ZCTs and WRKY transcription factors are the three major classes that came out in *C. roseus* pathway modulation. Interaction between the jasmonate- and elicitor-responsive element and two jasmonate-responsive transcription factors known as ORCA2 and ORCA3 takes place in *C. roseus*. The two ORCAs fit in the APETALA2/ethylene response-factor (AP2/ERF) family of transcription factors. The numerous genes govering primary and secondary metabolism, together with the TDC and STR genes, were shown to be regulated by ORCA3. MeJA induced all metabolite biosynthetic genes, primary and secondary, along with ORCA3 itself, that were tested (Van der Fits and Memelink 2000). Though, ORCA3 did not regulate all MeJA induced genes. For example, MeJA greatly induced expression of G10H, but ORCA3 overexpressing cell lines did not show G10H expression, suggesting that regulation of this gene involves additional jasmonate-responsive transcription factors. Although numerous genes of primary as well as secondary metabolism are regulated by ORCA3, it does control all genes in TIA metabolism. The involvement of additional transcription factors is possible in managing these genes. ORCA2, which is another jasmonate-responsive AP2/ERF-domain transcription factor from *C. roseus*, is a good applicant for this purpose. As indicated by preliminary results, ORCA2 and ORCA3 have interrelated, but different, assembly of target genes which suggest difference in functions of ORCA proteins, and the necessity of both for the whole range of jasmonate-stimulated metabolic modifications. An increment in AS, TDC, STR and D4H transcripts was generated by ORCA3 over-expression but CRMYC2 and

G10H transcription remained unaffected. G10H and ORCA3 co-over-expression induced a significant increase in G10H transcripts. Further, a substantial increase in the production of strictosidine, vindoline, catharanthine and ajmalicine took place due to ORCA3 and G10H over-expression but effects on anhydrovinblastine and vinblastine levels were limited (Pan et al. 2012). Another class of transcription factors i.e. zinc finger proteins (ZCT1, ZCT2, and ZCT3) are known to be transcriptional repressors in *C. roseus* and belong to the family Cys_2/His_2-type (transcription factor IIIA-type) ZCT family. They inhibit TDC and STR promoters in vitro by combining in a sequence specific manner (Pauw et al. 2004). Soybean transcription factor GmMYBZ2 inhibits the biosynthesis of catharanthine in hairy roots as stated by Zhou et al. (2011). Bax, a mammalian pro-apoptotic element of the Bcl-2 family when expressed in plants, stimulates oversensitive reactions. It generates transcriptional activation of two important genes TDC and STR in TIAs biosynthetic pathway of *C. roseus* cells. It induces the production of defense-related protein PR1 in the cells. It can be clearly interpreted that the mouse Bax activates the defense responses in *C. roseus* cells and stimutes the induction of TIA pathway (Jun and Fang 2007). It is now known that transcription factor CrMYC2 belonging to the family of helix-loop-helix (bHLH) plays crucial role in the activation of ORCA3 (a MeJA-responsive TIA up-regulator transcription factor) gene. The early response of Jasmonates is the activation of the CrMYC2 gene that binds to the sequence of the ORCA3 JRE in vitro. Transient assays showed that it further transactivates expression of reporter gene. Fall in the CrMYC2 expression by RNA interference resulted in a significant decrease in the production of ORCA3 mRNA. A root expressing CrWRKY1, *C. roseus* WRKY transcription factor, is induced by phytohormones namely jasmonate, ethylene and gibberellic acid. In hairy roots, CrWRKY1 up-regulates some of TIA pathway genes including TDC as well as transcriptional repressors ZCT1 ZCT2, and ZCT3 while down regulates the transcriptional activators ORCA2, ORCA3, and CrMYC2. But when the dominant-repressive form of CrWRKY1 was expressed it resulted into the suppression of TDC and ZCTs and the induction of ORCA3 and CrMYC2. The increased TDC activity, accumulation of tryptamine, resistance against 4-methyl tryptophan inhibition of CrWRKY1 expressing hairy roots were the outcome of up-regulation of TDC gene. The CrWRKY1 hairy roots were showing three times more serpentine as compared to control roots (Suttipanta et al. 2011).

5 Roadmap for Future Research

Catharanthus roseus today occupies the central position in ongoing metabolic engineering efforts in medicinal plants. The entire multi-step biogenetic pathway of its extremely priced anti-hypertensive and anti-cancerous alkaloids is fairly very well dissected at biochemical and gene levels. In order to increase the volumetric yield of these pharma molecules for drug industry, cell and tissue cultures of *C. roseus* are being increasingly tested to provide their alternate production platforms. However, a rigid developmental regulation and involvement of different cell, tissues and organelles in the synthesis of these alkaloids have restricted the utility of these cultures. Concerning its beneficial nature endophytes and allied members have considerable potential as bio-control agents and plant-growth promoters. Special efforts should therefore be made to define the molecular and biochemical bases of symbioses and their physiological effects on plant Endophytes have been extremely valuable in understanding the orchestration of root innate immunity. The root colonization by *endophytes* results in an increase in plant growth, early flowering, higher seed yield, alteration in the secondary metabolites, and adaptation to abiotic and biotic stresses. So one of the potential area revolve around determining the effect of the biotic and abiotic elicitors in relation with localization and expression of important MIA pathway genes required for the synthesis of the rate limiting products of MIA pathway like Vindoline, Catharanthine, Tabersonine, Vinblastine and Vincristine. Particular emphasis will be over to find out, whether the elicitation affects the cell metabolic machinery via altering the functional site of gene products at different locations in the cell compartments by mapping the location and transfer of gene products. Heterologous gene expression is another emerging trend in recent *C. roseus* biology. A recent study of Meittinen et al. (2014) reconstituted expression of the eight genes encoding seco-iridoid pathway, together with two genes boosting precursor formation and two downstream alkaloid biosynthesis genes, in *Nicotiana benthamiana*, allows the heterologous production of the complex MIA strictosidine. VIGS and RNAi-mediated down-regulation of sub-pathways operating at the branch point nodes in the upstream steps of the TIAs metabolism up to tabersonine synthesis will be the potential targets. Photo-autotrophic cell cultures of *C. roseus* with functional chloroplast system in the idioblast and laticifer networks are one of the future line to be targeted. Such tissue can be potential targets for hyper-expressing vindoline pathway limiting steps and their further condensation with catharanthine to produce antineoplatic alkaloids. Efforts must continue to identify transcriptional activators and transcriptional repressors of TIAs pathway genes to generate novel tissue types with a better understanding of regulatory mechanisms.

Acknowledgement The authors wish to express their sincere thanks to Director CSIR- NCL, for providing the facilities. Grateful appreciation also goes to the Council of Scientific & Industrial Research (CSIR), New Delhi, India, for the financial support in the form of CSIR-SRA (Pool Scientist) to the senior author and DBT-SERB for Fast Track Young Scientist Award to second author.

References

Arora R, Malhotra P, Mathur AK, Mathur A, Govil CM, Ahuja PS (2010) Anticancer alkaloids of *Catharanthus roseus*: transition from traditional to modern medicine. In: Arora R (ed) Herbal medicine: a cancer chemopreventive and therapeutic perspective. Jaypee Brothers Medical Publishers, New Delhi, pp 292–310

Ayora-Talavera T, Chappell J, Lozoya-Gloria E, Loyola-Vargas VM (2002) Overexpression of *Catharanthus roseus* hairy roots of a truncated hamster 3-hydroxy-3-methylglutaryl-CoA reductase gene. Appl Biochem Biotechnol 97:135–145

Batra J, Dutta A, Singh D, Kumar S, Sen J (2004) Growth and terpenoid indole alkaloid production in *Catharanthus roseus* hairy root clones in relation to left and right termini-linked Ri T-DNA gene integration. Plant Cell Rep 23:148–154

Begum F (2011) Augmented production of vincristine in induced tetraploids of *Agrobacterium* transformed shooty teratomas of *Catharanthus roseus*. Med Plants Int J Phytomed Relat Ind 3:59–64

Begum F, Rao SSN, Rao K, Devi YP, Giri A, Giri CC (2009) Increased vincristine production from *Agrobacterium tumefaciens* C_{58} induced shooty teratomas of *Catharanthus roseus* G. Don. Nat Prod Res 23:973–981

Bennouna J, Breton JL, Taurani JM, Ottensmeier C, O'brien M, Kosmidis P, Huat TE, Pinel MC, Colin C, Douillard JY (2006) Vinflunine—an active chemotherapy for treatment of advanced non-small cell lung cancer previously treated with a platinum- based regimen: results of a phase II study. Br J Cancer 94:1383–1388

Bhadra R, Vani S, Shanks J (1993) Production of indole alkaloids by selected hairy root lines of *Catharanthus roseus*. Biotechnol Bioeng 41:581–592

Brotherton JE, Schechter S, Ranch JP, Widholm JM (1996) Inheritance and stability of 5-methyltryptophan resistance in *Datura innoxia* selected in vitro. Plant Cell Physiol 37:389–394

Burlat V, Oudin A, Courtois M, Rideau M, St-Pierre B (2004) Co-expression of three MEP pathway genes and geraniol 10-hydroxylase in internal phloem parenchyma of *Catharanthus roseus* implicates multicellular translocation of intermediates during the biosynthesis of monoterpene indole alkaloids and isoprenoid-derived primary metabolites. Plant J 38:131–141

Campous-Tamayo F, Hernandez-Dominguez E, Vazquez-Flota FA (2008) Vindoline formation in shoot cultures of *Catharanthus roseus* is synchronously activated with morphogenesis through the last biosynthetic step. Ann Bot 102:409–415

Canel C, Lopes-Cardoso MI, Whitmer S, Van der Fits L, Pasquali G, Van der Heijden R, Hoge JHC, Verpoorte R (1998) Effects of over-expression of strictosidine synthase and tryptophan decarboxylase on alkaloid production by cell cultures of *Catharanthus roseus*. Planta 205:414–419

Chang K, Qiu F, Chen M, Zeng L, Liu L, Yang C, Lan X, Wang Q, Liao Z (2014) Engineering the MEP pathway enhanced ajmalicine biosynthesis. Int Union Biochem Mol Biol 61:249–255

Choi PS, Kim YD, Choi KM, Chung HJ, Choi DW Jr (2004) Plant regeneration from hairy-root cultures transformed by infection with *Agrobacterium rhizogenes* in *Catharanthus roseus*. Plant Cell Rep 22:828–831

Cordell GA (1980) The botanical, chemical, biosynthetic and pharmacological aspects of *Catharanthus roseus* (L.) G. Don. (Apocynaceae). In: Woo WS, Han BH (eds) Recent advances in natural product research. Seoul National University Press, Seoul, pp 65–72

Costa MMR, Hilliou F, Duarte P, Pereira LG, Almeida I, Leech M, Memelink J, Barcelo AR, Sottomayor M (2008) Molecular cloning and characterization of a vacuolar class III peroxidise involved in the metabolism of anticancer alkaloids in *Catharanthus roseus*. Plant Physiol 146:403–417

De Carolis E, Chan F, Balsevich J, De Luca V (1990) Isolation and characterization of a 2-oxoglutarate dependent dioxygenase involved in the second-to-last step in vindoline biosynthesis. Plant Physiol 94:1323–1329

De Luca V, Cutler AJ (1987) Subcellular localization of enzymes involved in indole alkaloid biosynthesis in *Catharanthus roseus*. Plant Physiol 85:1099–1102

Dethier M, De Luca V (1993) Partial purification of an *N*-methyltransferase involved in vindoline biosynthesis in *Catharanthus roseus*. Phytochemistry 32:673–678

Di Flore S, Hoppmann V, Fischer R, Schillberg S (2004) Transient gene expression of recombinant terpenoid indole alkaloid enzymes in *Catharanthus roseus* leaves. Plant Mol Biol Rep 22:15–22

Dipierre A, Lemarie E, Dabouis G, Garnier G, Jacoulet P, Dalphin JC (1991) A phase II study of Navelbine (vinorelbine) in the treatment of non-small cell lung cancer. Am J Clin Oncol 14:115–119

Duarte P, Memelink J, Sottomayor M (2010) Fusions with fluorescent proteins for subcellular localization of enzymes involved in plant alkaloid biosynthesis. In: Fett-Netto A (ed) Methods in molecular biology, metabolic engineering of plant secondary pathways. Humana Press, New York, pp 275–290

El-Sayed M, Verpoorte R (2007) *Catharanthus* terpenoid indole alkaloids: biosynthesis and regulation. Phytochem Rev 6:277–305

Facchini PJ (2001) Alkaloid biosynthesis in plants: biochemistry, cell biology, molecular regulation, and metabolic engineering applications. Annu Rev Plant Mol Biol 52:29–66

Facchini PJ, De Luca V (2008) Opium poppy and Madagascar periwinkle: model non-model systems to investigate alkaloid biosynthesis in plants. Plant J 54:763–784

Farnsworth NR (1985) Plants and modern medicine: Where science and folklore meet. Traditional medicine. World Health Forum 6:76–80

Foye WO (1995) Cancer chemotherapeutic agents. American Chemical Society, Washington, DC, pp 349–353

Fulzele DP, Heble MR (1994) Large-scale cultivation of *Catharanthus roseus* cells production of ajmalicine in a 20-l airlift bioreactor. J Biotechnol 35:1–7

Galili G, Hofgen R (2002) Metabolic engineering of amino acids and storage proteins in plants. Metab Eng 4:3–11

Galili G, Amir R, Hoefgen R, Hesse H (2005) Improving the levels of essential amino acids and sulfur metabolites in plants. Biol Chem 386:817–831

Ganapathi G, Kargi F (1990) Recent advances in indole alkaloid production by *Catharanthus roseus* (Periwinkle). J Exp Bot 41:259–267

Godoy-Hernandez G, Loyola-Vargas VM (1997) Effect of acetylsalicylic acid on secondary metabolism of *Catharanthus roseus* tumor suspension cultures. Plant Cell Rep 16:287–290

Goklany S, Loring RH, Glick J, Lee-Parsons CWT (2009) Assessing the limitations to terpenoid indole alkaloid biosynthesis in *Catharanthus roseus* hairy root cultures through gene expression profiling and precursor feeding. Biotechnol Prog 25:1289–1296

Guirimand G, Burlat V, Oudin A, Lanoue A, St-Pierre B, Courdavault V (2009) Optimization of the transient transformation of *Catharanthus roseus* cells by particle bombardment and its application to the subcellular localization of hydroxymethylbutenyl 4-diphosphate synthase and geraniol 10-hydroxylase. Plant Cell Rep 28:1215–1234

Guirimand G, Courdavault V, St-Pierre B, Burlat V (2010) Biosynthesis and regulation of alkaloids. In: Pua EC, Davey MR (eds) Plant developmental biology—biotechnological perspective, vol 2. pp 139–160

Guirimand G, Guihur A, Pierre P, Hericourt F, Mahroug S, St-Pierre B, Burlat V, Courdavault V (2011) Spatial organization of the vindoline biosynthetic pathway in *Catharanthus roseus*. J Plant Physiol 168:519–628

Hirata K,Miyamoto K, Miura Y (1994) *Catharanthus roseus* L (Periwinkle): production of vindoline and catharanthine in multiple shoot culture. In: Bajaj YPS (ed) Biotechnology in Agriculture and Forestry: Medicinal and Aromatic Plants I., vol 4. Springer, Berlin, pp 46–55

Hong SB, Peebles CAM, Shanks JV, San KY, Gibson SI (2005) Expression of the *Arabidopsis* feedback-insensitive anthranilate synthase holoenzyme and tryptophan decarboxylase genes in *Catharanthus roseus* hairy roots. J Biotechnol 122(1):28–38

Hughes RH, Shanks JV (2002) Metabolic engineering of plants for alkaloid production. Metab Eng 4:41–48

Hughes EH, Hong SB, Shanks JV, San KY, Gibson SI (2002) Characterization of an inducible promoter system in *Catharanthus roseus* hairy roots. Biotechnol Prog 18:1183–1186

Hughes EH, Hong SB, Gibson SI, Shanks JV, San KY (2004) Expression of a feedback-resistant anthranilate synthase in *Catharanthus roseus* hairy roots provides evidence for tight regulation of terpenoid indole alkaloid levels. Biotechnol Bioeng 86:718–727

Johnson S, Armstrong I, James G, Marvin G, Burnett JP (1963) The Vinca alkaloids: a new class of oncolytic agents. Lilly Laboratories for Research and Lilly Laboratories for Clinical Research, Indianapolis, IN, pp 1390–1427

Jun XM, Fang DJ (2007) Enhancing terpenoid indole alkaloid production by inducible expression of mammalian Bax in *Catharanthus roseus* cells. Sci China Ser C Life Sci 50:234–241

Jung KM, Kwak SS, Kim SW, Leo H, Choi CY, Liu JR (1992) Improvement of the catharanthine productivity in hairy root cultures of *Catharanthus roseus* by using monosaccharides as a carbon source. Biotechnol Lett 14:695–700

Jung KH, Kwak SS, Choi CY, Liu JR (1995) An interchangeable system of hairy root and cell suspension cultures of *Catharanthus* for indole alkaloid production. Plant Cel Rep 15:51–54

Kang KK, Kameya T (1995) Characterization of anthranilate synthetase and tryptophan synthase in a 5-methyl tryptophan resistant mutant (MR1) of *Zea mays* L. Breed Sci 45:321–325

Kim DS, Lee IS, Jang CS, Hyun DY, Seo YW, Lee YI (2004) Selection of 5-methyltryptophan resistant rice mutants from irradiated calli derived from embryos. Euphytica 135:9–19

Kisaka H, Kisaka M, Kameya T (1996) Characterization of interfamilier somatic hybrids between 5-methyltryptophan resistant (5-MT-resistant) Rice (*Oryza sativa* L.) and 5-MT-sensitive Carrot (*Daucas carota* L.); Expression of resistance to 5-MT by the somatic hybrids. Breed Sci 46:221–226

Kumar S, Dutta A, Sinha AK, Sen J (2007) Cloning, characterization and localization of a novel basic peroxidase gene from *Catharanthus roseus*. FEBS J 274:1290–1303

Kumar S, Jaggi M, Taneja J, Sinha AK (2011) Cloning and characterization of two new Class III peroxidase genes from *Catharanthus roseus*. Plant Physiol Biochem 49:404–412

Laflamme P, St Pierre B, De Luca V (2001) Molecular and biochemical analysis of a Madagascar periwinkle root-specific minovincinine-19-hydroxy-*O*-acetyltransferase. Plant Physiol 125:189–198

Leckie F, Scragg AH, Cliffe KC (1991) Effect of bioreactor design and agitator speed on the growth and alkaloid accumulation by cultures of *Catharanthus roseus*. Enzyme Microb Technol 13:296–305

Levac D, Murata J, Kim WS, De Luca V (2008) Application of carborundum abrasion for investigating leaf epidermis: molecular cloning of *Catharanthus roseus* 16 -hydroxy- tabersonine-16-*O*-methyltransferase. Plant J 53:225–223

Loyola-Vargas VM, Galaz-Avalos RM, Rodríguez-Ku JR (2007) *Catharanthus* biosynthetic enzymes: the road ahead. Phytochem Rev 6:307–339

Luca D, Laflamme P (2001) The expanding universe of alkaloid biosynthesis. Curr Opin Plant Biol 4:225–233

Magnotta M, Murata J, Chen J, De Luca V (2007) Expression of deacetylvindoline-4-O-acetyltransferase in *Catharanthus roseus* hairy roots. Phytochemistry 68:1922–1931

Mahroug S, Courdavault V, Thiersault M, St-Pierre B, Burlat V (2006) Epidermis is a pivotal site of at least four secondary metabolic pathways in *Catharanthus roseus* aerial organs. Planta 223:1191–1200

Mahroug S, Burlat V, St-Pierre B (2007) Cellular and sub-cellular organization of the monoterpenoid indole alkaloid pathway in *Catharanthus roseus*. Phytochem Rev 6:363–381

Mano M (2006) Vinorelbine in the management of breast cancer: new perspective, revived role in the era of targeted therapy. Cancer Treat Rev 32:106–118

Mehrotra S, Prakash O, Khan F, Kukreja AK (2013) Efficiency of neural network-based combinatorial model predicting optimal culture conditions for maximum biomass yields in hairy root cultures. Plant Cell Rep 32:309–317

Meijer JJ, Ten Hoopen HJG, Luyben KCAM (1993) Effects of hydrodynamic stress on cultured plant cells: a literature survey. Enzyme Microb Technol 15:234–238

Meittinen K et al (2014) The seco-iridoid pathway from *Catharanthus roseus*. Nat Commun 5:3606. doi:10.1038/ncomms4606

Moreno PRH, Van der Heijden R, Verpoorte R (1995) Cell and tissue cultures of *Catharanthus roseus*: a literature survey II. Updating from 1988 to 1993. Plant Cell Tiss Org Cult 42:1–25

Moreno-Valenzuela O, Galaz-Avalos RM, Minero-Garcoaa Y, Loyola-Vargas VM (1998) Effect of differentiation on regulation of indole alkaloid production in *Catharanthus roseus* hairy root. Plant Cell Rep 18:99–104

Moreno-Valenzuela OA, Minero-García Y, Chan W, Mayer-Geraldo E, Carbajal E, Loyola-Vargas VM (2003) Increase in the indole alkaloid production and its excretion into the culture medium by calcium antagonists in *Catharanthus roseus* hairy roots. Biotechnol Lett 25:1345–1349

Morgan JA, Shanks JV (1999) Inhibitor studies of tabersonine metabolism in *C. roseus* hairy roots. Phytochemistry 51:61–68

Morgan JA, Shanks JV (2000) Determination of metabolic rate-limitations by precursor feeding in *Catharanthus roseus* hairy root cultures. J Biotechnol 79:137–145

Murata J, De Luca V (2005) Localization of tabersonine 16-hydroxylase and 16-OH-tabersonine-16-*O*-methyltransferase to leaf epidermal cells defines them as a major site of precursor biosynthesis in the vindoline pathway in *Catharanthus roseus*. Plant J 44:581–594

Mustafa NR, Verpoorte R (2007) Phenolic compounds in *Catharanthus roseus*. Phytochem Rev 6:243–258

Neuss CL, Neuss MN (1990) Therapeutic use of bisindole alkaloids from *Catharanthus*. In: Brossi A, Suffness M (eds) The alkaloids, vol 37. Academic, San Diego, pp 229–240

Newell CA (2000) Plant transformation technology: developments and applications. Mol Biotechnol 16:53–65

O'Keef BR, Mahady GB, Gills JJ, Beecher CWW (1997) Stable vindoline production in transformed cell culture of *Catharanthus roseus*. J Nat Prod 60:261–264

Pan Q, Wang Q, Yuan F, Xing S, Zhao J, Choi YH, Verpoorte R, Tian Y, Wang G, Tang K (2012) Overexpression of ORCA3 and G10H in *Catharanthus roseus* plants regulated alkaloid biosynthesis and metabolism revealed by NMR-metabolomics. PLoS One 7, e43038

Pan Q, Rianika N, Mustafa Tang K, Choi YH, Verpoorte R (2015) Monoterpenoid indole alkaloids biosynthesis and its regulation in *Catharanthus roseus*: a literature review from genes to metabolites. Phytochem Rev 15:221. doi:10.1007/s111c1-015-9406-4

Parr AJ, Peerless ACJ, Hamill JD, Walton NJ, Robins RJ, Rhodes MJC (1988) Alkaloid production by transformed root cultures of *Catharanthus roseus*. Plant Cell Rep 7:309–312

Pasquali G, Porto DD, Fett-Neto AG (2006) Metabolic engineering of cell cultures versus whole plant complexity in production of bioactive monoterpene indole alkaloids: recent progress related to old dilemma. J Biosci Bioeng 101:287–296

Pauw B, Hilliou FAO, Martin VS, Chatel G, Wolf CJFD, Champion A, Pre M, Duijn BV, Kijne JW, Fits LVD, Memelink J (2004) Zinc finger proteins act as transcriptional repressors of alkaloid biosynthesis genes in *Catharanthus roseus*. J Biol Chem 279:52940–52948

Peebles CAM, Hong SB, Gibson SI, Shanks JV, San KY (2005) Transient effects of over-expressing anthranilate synthase a and b subunits in *Catharanthus roseus* hairy roots. Biotechnol Prog 21:1572–1576

Peebles CAM, Sander GW, Hughes EH, Peacock R, Shanks JV, San KY (2011) The expression of 1-deoxy-D-xylulose synthase and geraniol-10-hydroxylase or anthranilate synthase increases terpenoid indole alkaloid accumulation in *Catharanthus roseus* hairy roots. Metab Eng 13:234–240

Pietrosuik A, Furmanova M, Lata B (2007) *Catharanthus roseus*: micropropagation and *in vitro* techniques. Phytochem Rev 6:459–473

Radwanski ER, Last RL (1995) Tryptophan biosynthesis and metabolism: Biochemical and molecular genetics. Plant Cell 7:921–934

Rai A, Smita SS, Singh AK, Shanker K, Nagegowda DA (2013) Heteromeric and homomeric geranyl diphosphate synthases from *Catharanthus roseus* and their role in monoterpene indole alkaloid biosynthesis. Mol Plant 6:1531–1549

Rischer H (2006) Gene-to-metabolite networks for terpenoid indole alkaloid biosynthesis in *Catharanthus roseus* cells. Proc Natl Acad Sci USA 103:5614–5619

Rodriguez S, Compagnon V, Crouch NP, St-Pierre B, De Luca V (2003) Jasmonate- induced epoxidation of tabersonine by a Cytochrome P-450 in hairy root cultures of *Catharanthus roseus*. Phytochemistry 64:401–409

Roepke J, Salim V, Wu M, Thamm AMK, Murata J, Ploss K, Boland W, De Luca V (2010) Vinca drug components accumulate exclusively in leaf exudates of Madagascar periwinkle. Proc Natl Acad Sci U S A 107:15287–15292

Ruiz-May E, Galaz-Avalos RM, Loyola-Vargas VM (2009) Differential secretion and accumulation of terpene indole alkaloids in hairy roots of *Catharanthus roseus* treated with methyl jasmonate. Mol Biotechnol 41:278–285

Schlatmann JE, Koolhaas CMA, Vinke JL, Ten Hoopen HJG, Heijnen JJ (1995a) The role of glucose in ajmalicine production by *Catharanthus roseus* cell cultures. Biotechnol Bioeng 47:525–534

Schlatmann JE, Moreno PRH, Selles M, Vinke JL, Ten Hoopen HJG, Verpoorte R, Heijnen JJ (1995b) Two stage batch process for the production of ajmalicine by *Catharanthus roseus*: the link between growth and production stage. Biotechnol Bioeng 47:53–59

Schroder G, Unterbusch E, Kaltenbach M, Schmidt J, Strack D, De Luca V, Schroder J (1999) Light-induced cytochrome P450-dependent enzyme in indole alkaloid biosynthesis: Tabersonine 16-hydroxylase. FEBS Lett 458:97–102

Scott AI, Mizukami H, Lee S-L (1979) Characterization of a 5-methyltryptophan resistant strain of *Catharanthus roseus* cultured cells. Phytochemistry 18:795–798

Seth R, Mathur AK (2005) Selection of 5-methyltryptophan resistant callus lines with improved metabolic flux towards terpenoid indole alkaloid synthesis in *Catharanthus roseus*. Curr Sci 89:544–548

Sharma KK, Bhatnagar-Mathur PB, Thorpe TA (2005) Genetic transformation technology: status and problems. In Vitro Cell Dev Biol Plant 41:102–112

Shukla AK, Shasany AK, Gupta MM, Khanuja SPS (2006) Transcriptome analysis in *Catharanthus roseus* leaves and roots for comparative terpenoid indole alkaloid profiles. J Exp Bot 57:3921–3932

Shukla A, Shasany AK, Verma RK, Gupta MM, Mathur AK, Khanuja SPS (2010) Influence of cellular differentiation and elicitation on intermediate and late steps of terpenoid indole alkaloid biosynthesis in *Catharanthus roseus*. Protoplasma 242:35–47

Sim SJ, Chang HN, Liu JR, Jung KH (1994) Production and secretion of indole alkaloids in hairy root cultures of *Catharanthus roseus*: effects of *in situ* adsorption, fungal elicitation and permeabilization. J Ferment Bioeng 78:229–234

Sottomayor M, Ros Barcelo A (2003) Peroxidase from *Catharanthus roseus* (L.) G. Don and the biosynthesis of alpha-3′,4′-anhydrovinblastine: a specific role for a multifunctional enzyme. Protoplasma 222:97–10

Sottomayor M, Lopez-Serrano M, DiCosmo F, Ros-Barcelo A (1998) Purification and characterization of α-3′,4′-anhydrovinblastine synthase (peroxidase-like) from *Catharanthus roseus* (L.) G. Don. FEBS Lett 428:299–303

St-Pierre B, De Luca V (1995) A cytochrome P-450 monooxygenase catalyzes the first step in the conversion of tabersonine to vindoline in *Catharanthus roseus*. Plant Physiol 109:131–139

St-Pierre B, Laflamme P, Alarco AM, De Luca V (1998) The terminal O-acetyltransferase involved in vindoline biosynthesis defines a new class of proteins responsible for coenzyme A-dependent acyl transfer. Plant J 14:703–713

St-Pierre B, Vazquez-Flota FA, De Luca V (1999) Multicellular compartmentation of *Catharanthus roseus* alkaloid biosynthesis predicts intercellular translocation of a pathway intermediate. Plant Cell 11:887–900

Suttipanta N, Pattanaik S, Kulshrestha M, Patra B, Singh SK, Yuan L (2011) The transcription factor CrWRKY1 positively regulates the terpenoid indole alkaloid biosynthesis in *Catharanthus roseus*. Plant Physiol 157:2081–2093

Svoboda GH, Blake DA (1975) The phytochemistry and pharmacology of *Catharanthus roseus* (L.) G. Don. In: Taylor WI, Farnsworth NR (eds) The *Catharanthus* alkaloids. Marcel Dekker, New York, pp 45–83

Ten Hoopen HJG, Van Gulik WM, Schlatmann JE, Moreno PRH, Vinke JL, Heijnen JJ, Verpoorte R (1994) Ajmalicine production by cell cultures of *Catharanthus roseus*: from shake flask to bioreactor. Plant Cell Tiss Org Cult 38:85–91

Thakore D, Srivastava AK, Sinha AK (2013) Yield enhancement strategies for enhancement of indole alkaloids in hairy root cultures of *Catharanthus roseus*. Int J Chem Eng Appl 4:153–156

Toivonen L, Laakso S, Rosenqvist H (1992) The effect of temperature on hairy root cultures of *Catharanthus roseus*: growth, indole alkaloid accumulation and membrane lipid composition. Plant Cell Rep 11:395–399

Trick HN, Finer JJ (1997) SAAT: sonication assisted *Agrobacterium*-mediated transformation. Transgenic Res 6:329–336

Van der Fits L, Memelink J (2000) ORCA3, a jasmonate-responsive transcriptional regulator of plant primary and secondary metabolism. Science 289:295–297

Van der Fits L, Hilliou F, Memelink J (2001) T-DNA activation tagging as a tool to isolate regulators of a metabolic pathway from a genetically non-tractable plant species. Transgenic Res 10:513–521

Van der Heijden R, Jabos DJ, Snoeijer W, Hallard D, Verpoorte R (2004) The *Catharanthus* alkaloids: pharmacognosy and biotechnology. Curr Med Chem 11:1241–1253

Vasquez-Flota F, De Luca V (1998) Jasmonate modulates development and light regulated alkaloid biosynthesis in *Catharanthus roseus*. Phytochemistry 49:395–402

Vazquez-Flota F, De Carolis E, Alarco A-M, De Luca V (1997) Molecular cloning and characterization of desacetoxyvindoline 4-hydroxylase, a 2-oxoglutarate dependent dioxygenase involved in the biosynthesis of vindoline in *Catharanthus roseus* (L.) G. Don. Plant Mol Biol 34:935–948

Verma P, Mathur AK (2011a) Direct shoot bud organogenesis and plant regeneration from leaf explants in *Catharanthus roseus*. Plant Cell Tiss Org Cult 106:401–408

Verma P, Mathur AK (2011b) *Agrobacterium tumefaciens* mediated transgenic plant production via direct shoot bud organogenesis from from pre-plasmolyzed leaf explants of *Catharanthus roseus*. Biotechnol Lett 33:1053–1060

Verma P, Mathur AK, Shanker K (2012a) Growth, alkaloid production, rol genes integration, bioreactor up-scaling and plant regeneration studies in hairy root lines of *Catharanthus roseus*. Plant Biosyst 146:27–40

Verma P, Mathur AK, Shanker K (2012b) Increased availability of tryptophan in 5-methyltryptophan tolerant shoots of *Catharanthus roseus* and their post-harvest *in vivo* elicitation induces enhanced vindoline production. Appl Biochem Biotechnol 168:568–579

Verma P, Mathur AK, Srivastava A, Mathur A (2012c) Emerging trends in research on spatial and temporal organization of terpenoid indole alkaloids pathway in *Catharanthus roseus*: a literature up-date. Protoplasma 249:255–268

Verma P, Mathur AK, Masood N, Luqman S, Shankar K (2013) Tryptophan over producing cell suspensions of *Catharanthus roseus* (L) G. Don and their up-scaling in stirred tank bioreactor: detection of phenolic compound with antioxidant potential. Protoplasma 250:371–380

Verma P, Mathur AK, Khan SA, Verma N, Sharma A (2015a) Transgenic studies for modulating terpenoid indole alkaloids pathway in *Catharanthus roseus*: present status and future options. Phytochem Rev. doi:10.1007/s11101-015-9447-8

Verma P, Khan SA, Mathur AK, Srivastava A, Shanker K (2015b) Tryptophan metabolism and evaluation of morphological, biochemical and molecular variations in a field grown plant population derived via direct adventitious shoot bud regeneration from pre-plasmolysed leaves of *Catharanthus roseus*. Plant Cell Tiss Org Cult 123:357–375

Verma P, Sharma A, Khan SA, Mathur AK, Shanker K (2015c) Morpho-genetic and chemical stability of long term maintained Agrobacterium mediated transgenic *Catharanthus roseus* plants. Nat Prod Res 29:315–320

Verma P, Sharma A, Khan SA, Shanker K, Mathur AK (2015d) Over-expression of *Catharanthus roseus* tryptophan decarboxylase and strictosidine synthase in rol gene integrated transgenic cell suspensions of *Vinca minor*. Protoplasma 252:373–381

Verpoorte R, Alfermann AW (2000) Metabolic engineering of plant secondary metabolism. In: Verpoorte R, Alfermann AW (eds). Kluwer Academic, Dordrecht

Verpoorte R, Van der Heijden R, Van Gulik W M, Ten Hoopen H J G (1991) Plant biotechnology for the production of alkaloids: present status and prospects. In: Brossi A (ed) The alkaloids, vol 40. Academic, San Diego, pp 1–187

Verpoorte R, Van der Heijden R, Schripsema J (1993) Plant cell biotechnology for the production of alkaloids: present status and prospects. J Nat Prod 56:186–207

Verpoorte R, van der Heijden R, Moreno PRH (1997) Biosynthesis of terpenoid indole alkaloids in *Catharanthus roseus* cells. In: Cordell GA (ed) The alkaloids, vol 49. Academic, San Diego, p 221

Verpoorte R, Contin A, Memelink J (2002) Biotechnology for the production of plant secondary metabolites. Phytochem Rev 1:13–25

Wakasa K, Tozawa Y, Terakawa T, Hasegawa H (1999) Gene encoding a-subunit of rice anthranilate synthase and DNA relating thereto. World Intellectual Property Organization 99/11800

Wang CT, Liu H, Gao XS, Zhang HX (2010) Overexpression of *G10H* and *ORCA3* in the hairy roots of *Catharanthus roseus* improves catharanthine production. Plant Cell Rep 29:887–894

Weaver J, Goklany S, Rizvi N, Cram EJ, Lee-Parson CWT (2014) Optimizing the transient fast agro-mediated seedling transformation (FAST) method in *Catharanthus roseus* seedlings. Plant Cell Rep 33:89–97

Whitmer S, Van der Heijden R, Verpoorte R (2002a) Effect of precursor feeding on alkaloid accumulation by a strictosidine synthase over-expressing transgenic cell line S1 of *Catharanthus roseus*. Plant Cell Tiss Org Cult 69:85–93

Whitmer S, Van der Heijden R, Verpoorte R (2002b) Effect of precursor feeding on alkaloid accumulation by a tryptophan decarboxylase overexpressing transgenic cell line T22 of *Catharanthus roseus*. J Biotechnol 96:193–203

Whitmer S, Canel C, van der Heijden R, Verpoorte R (2003) Long-term instability of alkaloid production by stably transformed cell lines of *Catharanthus roseus*. Plant Cell Tiss Org Cult 74:73–80

Wilson L, Pand D, Jordan MA (1999) Modulation of microtubule dynamics by drugs: a paradigm for the actions of cellular regulators. Cell Struct Funct 24:329–335

Zarate R, Verpoorte R (2007) Strategies for the genetic modification of the medicinal plant *Catharanthus roseus* (L.) G. Don. Phytochem Rev 6:475–491

Zargar M, Farahani F, Nabavi T (2010) Hairy roots production of transgenic *Catharanthus roseus* L. plants with *Agrobacterium rhizogenes* under *in vitro* conditions. J Med Plant Res 4:2199–2203

Zhao J, Verpoorte R (2007) Manipulating indole alkaloid production by *Catharanthus roseus* cell cultures in bioreactors: from biochemical processing to metabolic engineering. Phytochem Rev 6:435–457

Zhao J, Zhu Wei-Hua HQ (2001a) Selection of fungal elicitors to increase indole alkaloid accumulation in *Catharanthus roseus* suspension culture. Enzyme Microb Technol 28:666–672

Zhao J, Zhu Wei-Hua HQ (2001b) Enhanced catharanthine production in *Catharanthus roseus* cell cultures by combined elicitor treatment in shake flasks and bioreactors. Enzyme Microb Technol 28:673–681

Zhong JJ (2001) Biochemical engineering of the production of plant-specific secondary metabolites by cell cultures. Adv Biochem Eng Biotechnol 72:1–26

Zhou ML, Hou HL, Zhu XM, Shao JR, Wu YM, Tang YX (2011) Soybean transcription factor GmMYBZ2 represses catharanthine biosynthesis in hairy roots of *Catharanthus roseus*. Appl Microbiol Biotechnol 4:1095–1105

Zhou C, Zhang J, Zhao SJ, Hu ZB (2014) An active *Catharanthus roseus* desacetoxyvindoline-4-hydroxylase-like gene and its transcriptional regulatory profile. Bot Stud 55:29

Vincristine and Vinblastine Anticancer *Catharanthus* Alkaloids: Pharmacological Applications and Strategies for Yield Improvement

M. Masidur Alam, M. Naeem, M. Masroor A. Khan, and Moin Uddin

Abstract *Catharanthus roseus* L. is a potent medicinal plant belonging to Apocynaceae family. In a number of countries, different parts of it are traditionally used in the treatment of various diseases, e.g. diabetes, menstrual irregulations, hypertension, cancer, etc. The high added-value of this plant is because of its enormous pharmaceutical features, which are because of its more than 130 terpenoid indole alkaloids (TIAs), some of which exhibiting imperative pharmacological activities. The most striking biological activity investigated is the antitumour effect of dimeric alkaloids such as that of anhydrovinblastine, vinblastine and vincristine, which are under study either in preclinical phase or are being used presently. The great pharmacological importance of these indole alkaloids contrasts with their small amounts in the plant, making their extraction a very expensive process. To overcome this problem, researches have looked for alternative sources and have been trying the strategies to produce them in higher amounts. Using biotechnological approaches, intensive research on the biosynthesis of TIAs and on the regulation of their biochemical pathways has been developed with the aim to increase the production of these high added-value compounds. This chapter is focused on the pharmaceutical application of the antitumour alkaloids and on the various strategies which improve the production of these alkaloids; it also analyses the beneficial effects that these compounds exert on human health.

M.M. Alam (✉)
Department of Botany, Malda College, Malda, West Begal, India

Plant Physiology Section, Department of Botany, Aligarh Muslim University, Aligarh, Uttar Pradesh 202002, India
e-mail: alam281@gmail.com

M. Naeem • M.M.A. Khan
Plant Physiology Section, Department of Botany, Aligarh Muslim University, Aligarh, Uttar Pradesh 202002, India

M. Uddin
Botany Section, Women's College, Aligarh Muslim University, Aligarh, Uttar Pradesh 202002, India

M. Naeem et al. (eds.), *Catharanthus roseus*,
DOI 10.1007/978-3-319-51620-2_11

Keywords *Catharanthus roseus* • Apocynaceae • Terpene indole alkaloids • Vinblastine • Vincristine

1 Introduction

Natural products, beneficial in human healthcare, were the basis of the first pharmaceutical practice. These plant products continue to play an important role in modern therapy too. The term 'natural products' is applied to materials derived from plants, microorganisms, invertebrates and vertebrates. They may serve as raw materials for chemical or biological modification to new products, which may lead to new therapeutic agents for new organic synthesis and fundamental metabolic studies. Approximately, 25–50% of the current pharmaceuticals are derived from plant products that show lesser side effects compared to the synthetic drugs (Upadhyay 2011). In fact, the synthetic drugs, after getting inside the body, generate many biochemical alterations and cause cross-reactivity inside the body fluids in addition to widely inhibiting bio-membrane functioning in human beings. Today, consumers prefer herbal medicines over the synthetic drugs as they believe that the former may be safer than the later. This has resulted in 'explosion' of researches in the field of identification, distribution and variations of plant species, the development of new and improved tests useful in the therapeutic evaluation of drugs, new procedures for the isolation, separation, identification and structural elucidation, and exploration of new types of organic synthesis with regard to plant products. The National Cancer Institute (NCI) of the US Public Health Service has recognized the value of plants as the sources of potential anticancer agents. In 1960, NCI initiated a systematic effort to collect and screen the plants for anticancer properties in collaboration with United States Department of Agriculture (USDA). Between 1960 and 1982, some 35,000 plants were collected by USDA from more than 60 countries; later, their pharmaceutical constituents were screened by NCI against a range of animal tumour systems.

Out of several anticancer plants, periwinkle (*Catharanthus roseus*) is considered as a unique plant in cancer therapy. It accumulates in its leaves the dimeric terpenoid indole alkaloids vincristine (VCR) and vinblastine (VBL), which may be regarded as the first natural drugs used in prevention and cure of cancer; these alkaloids are still included among the most valuable chemotherapeutic agents in the treatment of human neoplasma (Sottomayor and Ros Barceló 2005). Both of these anticancer alkaloids disrupt the spindle formation in mitotically dividing cells, thereby preventing the uncontrolled growth of malignant tumours (Fig. 1).

Catharanthus roseus L. is a perennial tropical medicinal plant belonging to the family Apocynaceae, which comprises eight species, of which seven (*Catharanthus coriaceus*, *C. lanceus*, *C. longifolius*, *C. ovalis*, *C. roseus*, *C. scitulus* and *C. trichophyllus*) are endemic to Madagascar and one (*C. pusillus*) is reported from India. Basically, *C. roseus* is an ornamental plant with beautiful pink flowers and nicely shaped leaves. However, it is also recognized as an important remedial plant with enormous pharmaceutical uses in human healthcare. In fact, it is nothing less than a biochemical factory, producing more than 130 different TIAs, some of which exhibit

Fig. 1 Figures of vincristine and vinblastine

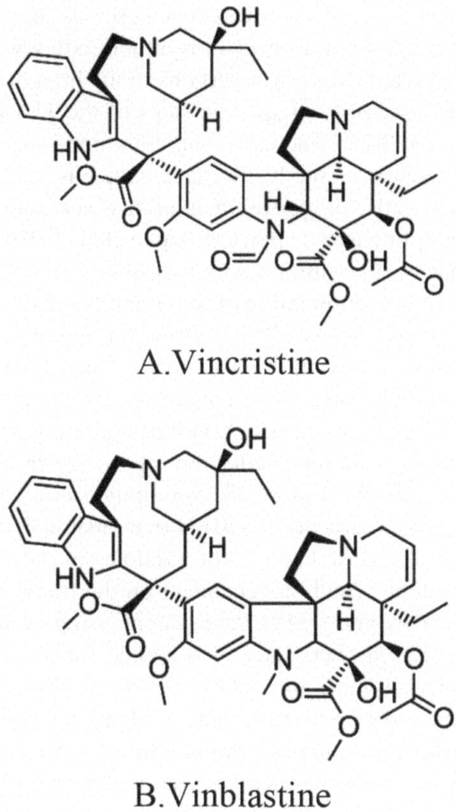

potent pharmacological activities in favour of human health (Van der Heijden et al. 2004). Vincristine and vinblastine alkaloids are commercial TIAs used in cancer chemotherapy, and together with many related semi-synthetic compounds are collectively named the *Vinca* alkaloids, as *Vinca rosea* is old name of *C. roseus*. Currently, vincristine, vinblastine, vinorelbine and vindesine have been used in clinical trials although only vincristine, vinblastine and vinorelbine have been approved for therapeutic exploitation in the United States (Rowinsky 2003). Fluorinated analogue of vinorelbine, called as vinflunine, has been approved for cancer chemotherapy in Europe (Bennouna et al. 2008; Schutz et al. 2011). Vincristine and vinblastine also exhibit strong antimicrobial activity (Grellier et al. 1999). Apart from anticancer alkaloids mentioned above, *C. roseus* also produces ajmalicine and serpentine alkaloids, which are monoterpenic indole compounds used as antihypertensive and anti-neuroinflammatory agents. Besides, yohimbine alkaloid of *C. roseus* is used in the treatments of erectile dysfunction, while vindolicine alkaloid is used for the development of antidiabetic drugs.

C. roseus anticancer compounds, vincristine and vinblastine, are derived from the coupling of catharanthine and vindoline alkaloids. Since there is very low recovery of these compounds from *C. roseus* plant with expensive extraction procedure, scientists have a large interest to increase their production through various biological techniques. Low levels of vincristine and vinblastine are mainly associated to the spatial separation of biosynthetic sites where these compounds are produced in the plant. Mainly, it pertains to the high degree of specialization in leaf cells where the assembly of specific steps of the TIA biosynthetic pathway occurs (Yu and de Luca 2013). Factually, catharanthine is accumulated almost exclusively in the wax exudates on the leaf surface, whereas vindoline is produced in specialized internal leaf cells, suggesting that an involvement of transport processes is needed for their coupling to take place (Roepke et al. 2010). Recently, an ABC transporter, CrTPT2, has been identified with its primary function in enhancing the transport; hence, its role in accumulation of catharanthine in the leaf epidermal surface has been expected (Yu and de Luca 2013). However, the physical separation of catharanthine and vindoline observed by Yu and De Luca (2013) is probably a limiting factor in very young leaves, where anhydro-vinblastine (AVBL) was actually shown to be absent (Naaranlahti et al. 1991), but definitely not in developed leaves, where the dimer AVBL was repeatedly reported to be in abundance (Balsevich and Bishop 1988; Goodbody et al. 1998; Sottomayor et al. 1998; Carqueijeiro et al. 2013). Further, Carqueijeiro et al. (2013) demonstrated that catharanthine, vindoline and AVBL were accumulated in the vacuoles of mesophyll cells by a specific proton antiport system, which is dependent on the transtonoplast pH gradient generated by V-H^+-ATPase and V-H^+-PPase system, using vacuoles isolated from leaves of adult plants.

Researchers have also looked for alternative sources and strategies to produce TIAs in high amounts in *C. roseus* plant. In fact, the low levels of the TIAs with anticancer activity found in plants have stimulated an intense interest in research efforts aiming to obtain in vitro *C. roseus* cultures with a higher production of these TIAs. Technologically, Zhao and Verpoorte (2007) showed that although *C. roseus* cells can be cultivated in bioreactors, the TIA biosynthesis is extremely low, which prevents their industrial production. To increase the anticancer alkaloids production, several approaches were tried (Zhao and Verpoorte 2007) using *C. roseus* cell cultures, using genetic modification or metabolic engineering, the most promising biotechnological tools for high production of these compounds (Van der Heijden et al. 2004) (Table 1).

The biosynthetic route for indole alkaloids has been studied by De-Luca and Cutler (1987). Because of the presence of cytotoxic *bis*-indole alkaloids of therapeutic importance, the production of vincristine, vinblastine and vindesine has become one of the main fields of interest in modern cell biotechnology (Zhao and Verpoorte 2007). The plant produces the active dimeric alkaloids in low concentrations (0.0005%), where nearly 500 kg of dry leaves of *C. roseus* are used to isolate 1 g of vinblastine (Van der Heijden et al. 2004) and 2 tons of macerated leaves produced 1 g of the alkaloid as the active principle, which is the amount required for the treatment of a child with leukaemia for 6 weeks (Karthikeyan et al. 2008). Because of the large number of alkaloids the *C. roseus* plant contains, the isolation of vincristine and vinblastine in the laboratory is very costly. Although all parts of the plant produce alkaloids (leaves, stems and roots) in different proportions (Soleimani et al. 2013), the maximum concentrations are found in the cortex of the roots, particularly when blooming (Jaleel et al. 2008). Wide arrays of different alkaloid subclasses have been identified, viz. vincosan, corynanthean, vallesiachotaman,

Table 1 Alkaloids isolated from different parts of *C. roseus* L. (Junaid et al. 2010)

Alkaloid	Extracted from (plant part)
β-Carboline	Leaf
Tryptamine,*N,N*-dimethyl	Cell suspension culture
Apparicine	Leaf, flower
Ammocalline	Plant extract, root
Anthirine	Plant extract, cell suspension culture
Akuammicine	Plant extract, leaf, root, callus culture, cell suspension culture, shoots
Iochrovicine	Leaf
Pericyclivine	Plant extract, leaf
Pleiocarpamine	Cell suspension culture
Cavincine	Plant extract, leaf, root, callus culture, hairy root
Iochnerine	Cell suspension culture
Tubotaiwine	Callus culture, cell suspension culture
Rosicine	Leaf
Catharanthine	Plant extract, leaf, flower, seedlings, callus culture, cell suspension culture, shoots
Tabersonine	Plant extract, leaf, seedlings, seed, callus culture, cell suspension culture
Venalstonine	Root
Akuammicine,12-Hydroxy	Cell suspension culture
Perivine	Plant extract, leaf, flower, root, callus culture, cell suspension culture
Vinervine	Cell suspension culture
Coronaridine	Flower
Vincadifformine	Cell suspension culture
Cyclolochnerine,21-Hydroxy	Callus culture, cell suspension culture, shoots, hairy root
Iochneridine	Leaf, callus culture, cell suspension culture, hairy root
Alstonine	Root, callus culture
Serpentine	Leaf, root, seedlings callus culture
Cathenamine	Plant extract
Vallesiachotamine	Callus culture, cell suspension culture
Isovallesiachotamine	Callus culture, cell suspension culture
Ajmalicine	Callus culture, cell suspension culture
Ajmalicine,19-*epi*,3-*iso*	Plant extract, callus culture, cell suspension culture
Ajmalicine, 3-*epi*	Plant extract, callus culture ,cell suspension culture
Akuammigine	Cell suspension culture
Akuammiline *O*-Deacetyl	Leaf, callus culture
Iochnericine	Plant extract, leaf, cell suspension culture
Minovincine	Plant extract
Preakuammicine	Seedlings

(continued)

Table 1 (continued)

Alkaloid	Extracted from (plant part)
Rosamine	Leaf
Tabersonine,19-Hydroxy	Cell suspension culture
Tetrahydroalstonine	Plant extract, flower, root, callus culture
Vindolinine, −Oxide	Plant extract, cell suspension culture
Vindolinine,19-*epi*,N-Oxide	Cell suspension culture
Fluorocarpamine, N-Oxide	Plant extract, leaf
Perividine	Plant extract
Isositsirikine, 19,20-Cis-16 (R)−	Plant extract, cell suspension culture
Isositsirikine, 19,20-Trans-16 (R)−	Plant extract, cell suspension culture
Isositsirikine, 19,20-Trans-16 (S)−	Plant extract, leaf, cell suspension culture
Minovincinine	Cell suspension culture
Sitsirikine	Plant extract, leaf, callus culture, cell suspension culture, shoots
Yohimbine	Plant extract, leaf, root, callus culture, cell suspension culture, hairy root
Sitsirikine,Dihydro-	Plant extract, leaf, root, callus culture, cell suspension culture
Perimivine	Plant extract, root
Tabersonine,11-Methoxy	Plant extract, flower
Almalicine, 7-Hydroxy-Indolenine	Callus culture
Ajmalicine *pseudo*-Indoxyl	Callus culture
Akuammiline,10-Hydroxy- Deacetyl	Callus culture
Epimisiline,19(S)	Hairy root
Horhammericine	Cell suspension culture, shoots
Mitraphylline	Flower, callus culture
Vincoline	Plant extract, leaf
Vindolinine	Plant extract, leaf, cell suspension
Vindolinine,19-*epi*	Plant extract, leaf, cell suspension culture
Vincolidine	Plant extract, leaf
Akuammine	Plant extract
Lochnerinine	Plant extract, leaf, cell suspension culture
Lochrovidine	Plant extract
Tabersonine,19-Hydroxy-11-Methoxy	Plant extract
Iochrovine	Plant extract
Vindolidine -Deacetyl-	Plant extract
Akuammiline	Plant extract, cell suspension culture
Horhammericine, 11-Methoxy	Cell suspension culture, shoots
Vincarodine	Plant extract, leaf
Vinosidine	Root
Vindoline, Deacetoxy-	Cell suspension culture, leaf, seedlings
Tabersonine,19-Acetoxy-11-Hydroxy-	Plant extract, leaf, cell suspension culture
Vindoline, Deacetyl-	Plant extract, leaf

(continued)

Table 1 (continued)

Alkaloid	Extracted from (plant part)
Iochnerinine	Leaf, root
Tabersonine, 19-Acetoxy-11- Methoxy	Cell suspension culture
Cathovaline	Leaf
Vindolidine	Plant extract, flower
Strictosidine Lactam	Cell suspension culture, shoots, hairy root
Vindoline	Plant extract, leaf, flower, seedlings, shoots
Akuammicine, Xylosyloxy-	Cell suspension culture
Strictosidine	Plant extract, leaf, root, seed, callus culture, cell suspension culture
Bannucine	Plant extract, leaf
Leurosivine	Leaf
Leurosine,17-Deacetoxy-	Plant extract
Vinblastine,4-Deacetoxy-	Plant extract, leaf
Vinblastine, Deacetyl-	Plant extract
Vinsedine	Seed
Leurosinine	Plant extract
Vinsedicine	Seed
Vinblastine,3,4-Anhydro-	Leaf, shoots
Vingramine	Seed
Vinblastine,4′-Deoxy-	Plant extract, leaf
Vinosidine	Plant extract
Vinblastine, N-Demethyl-	Plant extract
Vingrmine, Methyl-	Seed
Catharanthamine	Plant extract, leaf
Leurosine	Plant extract, leaf, shoots
Roseadine	Plant extract, leaf
Vincathicine	Plant extract, leaf
Roseamine	Plant extract
Vinblastine	Plant extract, leaf, flower, seedlings, cell suspension culture
Vinblastine,20′-*epi*-	Plant extract, leaf
Catharicine	Plant extract, leaf, flower
Catharine	Plant extract, leaf, shoot
Leurosine, 5′-Oxo-	Leaf
Carosine	Plant extract, leaf, flower
Leurosine,N B′-Oxide	Leaf
Vinamidine	Plant extract, leaf
Vincristine	Plant extract, leaf
Leurosidine, N B-Oxide	Plant extract
Vinblastine,14′-Hydroxy-	Plant extract
Vinblastine, 15’hydroxy-	Plant extract

(continued)

Table 1 (continued)

Alkaloid	Extracted from (plant part)
Neoleurocristine	Plant extract, leaf
Vindolidine	Plant extract
Leurosinone	Leaf
Neoleurosidine	Plant extract, leaf
Neoleurosidine, N B-Oxide	Plant extract, leaf
Vindolicine	Plant extract, leaf
Ammorosine	Root
Cathalanceine	Root
Cathindine	Leaf, root, cell suspension culture
Cavincidine	Plant extract, leaf, root, callus culture, cell suspension culture
Lochneririne	Leaf, root
Maandrosine	Plant extract, root
Perosine	Plant extract, leaf, root, callus culture
Rovindine	Plant extract, leaf
Vinaphamine	Plant extract, leaf
Vinaspine	Plant extract, leaf
Vincamicine	Plant extract, leaf

strychnan, aspidospermatan, plumeran, ibogan, eburnan and *bis*-indole alkaloids (Kisakurek and Hesse 1980). Up to 40 different *bis*-indole alkaloids have been found in *C. roseus*, many of which contain a moiety of plumerane (vindoline) and ibogaine (catharanthine). In relation to plant chemistry, *C. roseus* contains carbohydrates, flavonoids, saponins, phenol compounds, terpene indole alkaloids (Ataei-Azimi et al. 2008), anthocyanins, glucosides (Piovan and Filippini 2007), heart glycosides, steroids and mono-terpene glucosides (Van der Heijden et al. 2004). It has no tannins. Two flavonols have also been isolated and identified (Yadav et al. 2013) in addition to glycosidic flavonols that have been identified in seeds, stems, leaves and flowers of *C. roseus* (Ferreres et al. 2008). The extracts of the sprouts of *C. roseus* are used as a potential source of natural available antioxidants and with excellent pharmaceutical applications (Mallik et al. 2013).

The biological mechanism of action of the vincristine and vinblastine consists of the binding with the tubulin subunits of spindle apparatus during mitosis. These compounds inhibit the chromatin filaments drawn to their respective poles (Huxtable 1992), leading to the inhibition of cellular mitosis during the metaphase, and thereby starting the programmed cellular death or apoptosis (Leveque and Jehl 2007). Vincristine inhibits polymerization of the microtubules, producing an arrest in G2/M phase and inducing apoptosis (Casado et al. 2007). The semi-synthesis of vincristine and vinblastine, starting with the precursors and the organic synthesis (coupling of vindoline and catharanthine), is highly expensive and the production is poor in *C. roseus*. Therefore, alternative biotechnology strategies have been used to be able to increase the production of these secondary metabolites. They include the addition of biotic or abiotic inducers that stimulate the production of the metabolites in the biosynthesis pathway of the alkaloids.

2 Pharmacological Activities

The main secondary metabolites of *C. roseus* are terpene indole alkaloids (TIAs) with important applications in human medicine as mentioned above, which are presenting biological activities such as antitumour, anti-diabetes, anti-helminthic, anti-hypertensive, anti-diarrhoea, and antimicrobial actions and others. The Vinca alkaloids are generally known as compounds in the treatment of cancer (Moudi et al. 2013). These compounds repress cell growth because they alter the microtubular dynamics, which ultimately provokes apoptosis. Semi-synthetic compounds, similar to vincristine and vinblastine, have been developed to increase their therapeutic action (Nirmala et al. 2011). Vinblastine is used in particular for the treatment of Hodgkin's disease, besides lymphosarcoma, choriocarcinoma, neuroblastoma, carcinoma of breast and lung, and lymphocytic leukaemia (Junaid et al. 2010; Rai et al. 2014). Anhydrovinblastine, the direct precursor of vinblastine, also showed significant in vitro cytotoxic effect against human non-small cell lung cancer C4 besides that against human cervical carcinoma, human leukemic cells and A431 human carcinoma cells (Kutney et al. 2000). Vincristine is an oxidized form of vinblastine that arrests mitosis at metaphase and is very effective for treating acute lymphoblastic leukaemia in both children and adults. It is also used against Hodgkin's disease, Wilkins's tumour, neuroblastoma and reticulum cell sarcoma (Rowinsky 2003; Moudi et al. 2013). In addition, vincristine has also been used in the treatment of multiple non-malignant hematologic disorders like autoimmune and thrombotic thrombocytopenia, and hemolytic uremic syndrome sarcoma (Rowinsky 2003; Moudi et al. 2013). On the other hand, the cytotoxic effect of catharoseumine, which is a monoterpenic indole alkaloid isolated from the whole plant of *C. roseus*, was tested in different human tumour cell lines showing only a moderate cytotoxic effect against HL-60 cell line (Wang et al. 2012).

The antitumour alkaloids vincristine and vinblastine are used in malignant diseases; they are used in chemotherapy for leukaemia since they reduce the number of leukocytes in the blood (a high number of leukocytes indicate leukaemia) and in the treatment of Hodgkin's disease (Jaleel et al. 2008) characterized by being a monoclonal B cell neoplasia, and by the presence of abnormal cells called Reed–Sternberg cells (Jaffe et al. 2008). Vinblastine (vinblastine sulphate) is experimentally used for neoplasia treatment and for resistant pregnancy choriocarcinoma, a malignant neoplasia of the trophoblast, which is a highly aggressive and fetal lesion, since even when there is a timely diagnosis and it is appropriately treated with chemotherapy, it produces mortality in 10–15% of the cases (Cruz Ortíz et al. 2000). It is also effective in the treatment of advanced testicular tumours, breast cancer, Kaposi sarcoma and the Letterer–Siwe disease (Rocha and Leech 2002). Vincristine, formally known as leurocristine (vincristine sulphate), is used in leukaemia treatment in children. Vincristine is produced by the bonds of the terpene indole alkaloids: vindoline and catharanthine in the *C. roseus* plant (Evans et al. 2009). The use of the vinblastine and vincristine combined with chemotherapy has given 80% remission in Hodgkin's disease, 99% in acute lymphocytic leukaemia, 80% in Wilms' tumour in children,

70% in pregnancy corium cancer and the remission of 50% in Burkitt's lymphoma (Walts 2004; Amirjani, 2013). The indole alkaloid called amotin also has a strong anti-leukaemia activity (Taha et al. 2008).

At the neurologic level, the ajmalicine and serpentine are drugs used in treating depression and anxiety; these are also effective as anti-stress drugs (Taha et al. 2009; Hedhili et al. 2007). The supplements based on active ingredients of *C. roseus* such as vincamine are used for the prevention and treatment of cerebrovascular disorders and failures, vertigo, ischemic deficiencies and headaches, because they help oxygenate and increase brain glucose levels (Vas and Gulyas 2005); besides preventing abnormal clotting, they also increase the levels of serotonin, a brain neurotransmitter; the deficiency of serotonin produce schizophrenia, phobia, migraine and bulimia. On the other hand, vincamine is now known to increase the memory retention properties and it is effective in the treatment of vascular dementia (Gayatri and Chakravarthy 2013). Anhydrous vinblastine is used in the treatment of lung and cervix cancer (Kutney et al. 1998). The catharanthine isolated from *C. roseus* is cytotoxic in P-388 and KB human cancer cell lines (Wong et al. 2013). Furthermore, ajmalicine is used in the treatment of circulatory disorders and as an antihypertensive agent since it acts as an antagonist of the α1-adrenergic receptor, known as alpha blocker (Roquebert and Demichel 1984), with a preferential action on α2-adrenergic receptors (Chung et al. 2007).

2.1 How Do Alkaloids Execute Cancer Cell?

Drugs that interrupt mitotic progression, which are commonly referred to as 'antimitotics', are used extensively for the treatment of cancer. Currently, all such drugs that have been approved for clinical use target microtubules, with the taxanes and *Vinca* alkaloids showing much success against a number of cancers. Taxol (pacific yew tree), which is originally derived from the bark of *Taxus brevifolia*, is commonly used in the treatment of breast and ovarian cancers. *Vinca* alkaloids, such as vincristine, are often used in combination therapies to treat haematological malignancies (Jordan and Wilson 2004). Investigating the effects of these agents on microtubule dynamics has revealed much about their mechanism of action. The *Vinca* alkaloids interact with β-tubulin at a region adjacent to the GTP-binding site known as the *Vinca* domain (Rai and Wolff 1996). Within a concentration range that blocks proliferation, the *Vinca* alkaloids bind to tubulin at the plus-tip of microtubules. At the lower end of this range, this inhibits microtubule dynamics without altering polymer levels, whereas at higher concentrations, it induces microtubule depolymerization (Jordan et al. 1991). In both situations, mitotic spindle formation is disrupted, and cells therefore fail to complete a normal mitosis (Jordan et al. 1991). At very high concentrations (above 10 μM), vinca alkaloids can induce the aggregation of tubulin into paracrystals; however, this does not occur at clinically relevant concentrations (Jordan and Wilson 1999). Taxol steady the microtubules and dampens the dynamics of the polymer, thereby reducing depolymerization (Schiff et al. 1979). In mammalian cells, low concentrations of taxol stabilize microtubules, whereas higher concentrations increase polymerization (Jordan et al. 1993). Taxanes bind β-tubulin, but only when the monomer is incorporated into a microtubule. The binding site for taxol is on the inner face of the polymer, and the drug can bind the length of the polymer. Drug binding is thought to stabilize the structure of the polymer by inducing a conformational change, which enhances the affinity of the interaction between tubulin molecules (Nogales, 2000). Stabilization of microtubules by taxol binding prevents normal formation of mitotic spindles (Jordan et al. 1996). On entry into mitosis, chromosomes can attach taxol-stabilized microtubules; however, the lack of microtubule dynamics means that tension is not produced across sister chromatids (Kelling et al. 2003), and prevents correct chromosome bi-orientation. This leads to chronic activation of the spindle assembly checkpoint (SAC), which in turn leads to mitotic arrest (Musacchio and Salmon 2007). Although the mechanisms by which antimitotic drugs elicit a mitotic arrest are now well understood, relatively little is known about how cells respond to this prolonged cell-cycle delay. Recently, however, several studies have taken a new approach, using high content imaging and live-cell analysis to monitor the long-term behaviour of cells in response to antimitotic drugs. In this commentary, we focus on the recent studies and on how they have advanced our understanding of how cancer cells respond to antimitotic drugs, at least in cell culture. We also discuss the relevance of these new findings to the clinical use of both classical and novel antimitotic agents (Fig. 2).

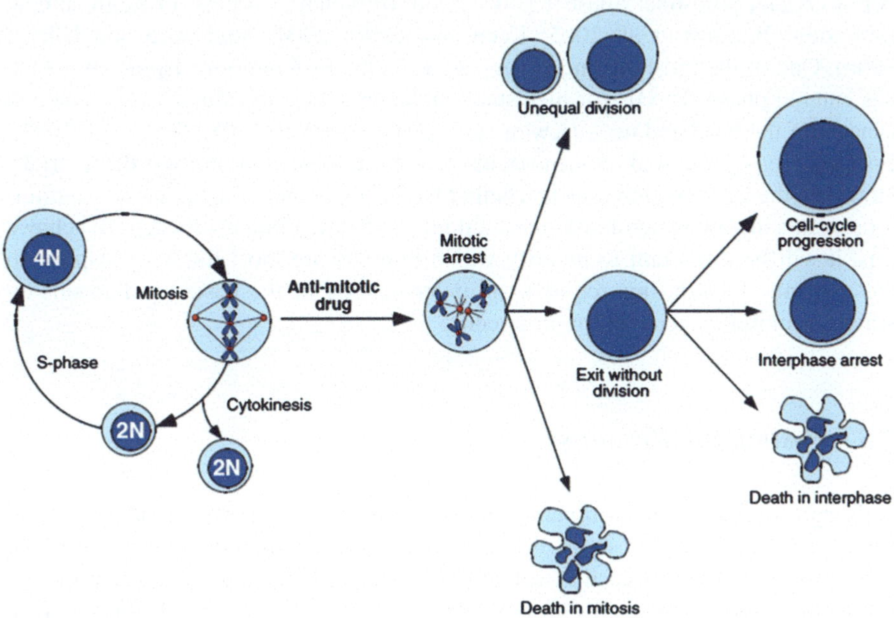

Fig. 2 Cell fate in response to antimitotic drug treatment. When cells are exposed to an antimitotic agent (*Vinca* alkaloids), they are arrested in mitosis due to chronic activation of the spindle assembly checkpoint. They then undergo one of several fates. Cells might die directly in mitosis, or divide unequally to produce aneuploid daughter cells. Alternatively, cells might exit mitosis without undergoing division. In this case, cells might then die in interphase, arrest in interphase indefinitely or enter additional cell cycles in the absence of division (from Gascoigne and Taylor 2009)

2.2 Clinical Use of Antimitotics

Although microtubule toxins have shown great success in the clinical therapy, two factors—namely resistance and toxicity—have limited their effectiveness. Patient resistance to classic antimitotic agents is commonly observed (McGrogan et al. 2008). Some patients respond well to treatment, but others rapidly acquire resistance and show little improvement. Toxicity is also a major limitation: in addition to killing tumour cells, antimitotic agents affect the division of normal cells, which therefore manifests as myelosuppression. Furthermore, because vincristine also disrupts microtubule dynamics in non-dividing cells such as peripheral neurons, neuropathies can also develop. Myelosuppression is reversible and therefore clinically manageable. By contrast, neurotoxicities are more problematic because they can often cause permanent damage (Rowinsky et al. 1993). To minimize neurological side effects, new agents are being developed that disrupt mitosis without interfering with microtubule dynamics in non-dividing cells. The rationale of this approach is that such drugs should prevent assembly of the mitotic spindles, and thereby retain antitumour activity, but they should not induce neuropathies. Frontrunners in this new class of therapeutics are inhibitors of the Eg5 kinesin, a motor protein that is required for the separation and movement of chromosomes to two opposite poles during mitosis. Agents are also being developed that inhibit mitotic kinases such as Aurora A and polo-like kinase 1 (Plk1), both of which also play a role in spindle formation (Bergnes et al. 2005; Keen and Taylor 2004; Strebhardt and Ullrich 2006). One of the most advanced new agents is the Eg5 inhibitor Ispinesib, which has entered phase II trials for metastatic melanoma, hepatocellular carcinoma and cancers of the head and neck (Kwon et al. 2008; Lee et al. 2008; Tang et al. 2008). Encouragingly, minimal neurotoxicity has been observed during these trials. However, the key question now is whether Ispinesib, or any of the other new antimitotic agents including taxol or *Vinca* alkaloids, will have clinical efficacy. To achieve this, it will be important to identify which tumours are most likely to respond to these agents. In turn, this requires an understanding of the basic mechanisms by which antimitotic agents kill tumour cells.

2.3 Postmitotic Response

Although the model of 'competing-networks' described above provides a useful framework by which to describe the decision of a cell to either die during mitosis or exit mitosis, it does not easily explain the variety of behaviours that are observed following mitotic exit—the postmitotic response. Several fates have been described for cells that exit mitosis in the presence of an antimitotic drug, including cell-cycle arrest, apoptosis and cell-cycle progression. The molecular factors that govern these fates are not well understood, but p53 protein appears to be involved in this regard. Substantial evidence supports the theory that p53 restrains cell-cycle progression following exit from a prolonged mitotic arrest (Lanni and Jacks 1998). It is unclear whether a p53-dependent response is induced during mitosis, or by a de novo signal that arises after mitotic exit. One possibility is that damage or stress that has accumulated during mitosis does not always trigger an apoptotic response during mitosis due to the inhibitory action of Cdk1 on caspase-9 (Allan and Clarke 2007). Following mitotic exit, and the loss of this Cdk1-mediated inhibition, the apoptotic threshold might fall to its interphase set-point. In turn, the pre-existing damage signal is then recognized, leading to execution of the apoptotic programme. Thus, it is likely that the fate of the cell in response to drug treatment is determined not only by events occurring during a mitotic arrest, but also by the consequences of these events after mitotic exit, as well as by additional signalling pathways that are active during interphase.

3 Approaches in the Enhancement of Pharmaceutically Active Compounds

Due to the pharmaceutical importance and the low content in the plant of vincristine and vinblastine, *Catharanthus roseus* turns into an important model system for biotechnological studies on plant secondary metabolism. Researchers are focusing their attention to enhance the alkaloids yield by various ways (chemically, enzymatically, synthetically or by cell culture method). The plant cell can be cultured at large scale (Verpoorte et al. 1991), but the yield of alkaloids production is too low and limits commercial applications. In recent times, however, two strategies have been commonly used for the enhancement of alkaloids.

(a) In vitro cultivation of shoot via organogenesis and somatic embryogenesis, callus or suspension by the optimization of media, phytohormones, temperature, pH, light, aeration, etc. In addition, high cell density culture, elicitor's treatment, mutagenesis, bioreactors and immobilization are also practised to improve alkaloids yield.
(b) Genetic engineering and overexpression of biosynthetic rate limiting enzymes in alkaloid biosynthesis pathways.

4 In Vitro Studies

In tissue culture, the response of culture has been influenced by a number of factors which in turn regulate alkaloids yield. The yield of alkaloids in suspension culture is directly influenced by the surrounding environmental conditions and genetic constitution of the concerned plant material. Over the years efforts have been made in numbers for optimization of culture media for better biomass and alkaloids production, some patents have also been filed (Van der Heijden et al. 1989; Ganapathi and

Kagri 1990; Moreno et al. 1995; Mujib et al. 2002). Carbon sources and inorganic compounds play a significant role in indole alkaloid production. It was earlier reported that nitrogen and phosphate both promoted growth but had an adverse effect on alkaloids yield (Knobloch and Berlin 1980; Van Gulik et al. 1993). The inhibitory effect of nitrogen on alkaloid production has not always been observed (Drapeau et al. 1987). The effect of nitrogen on alkaloids production is dependent on carbon availability to the cells which makes the carbon-to-nitrogen ratio (C/N ratio) an important factor to be taken into account. By the determination of the cellular C/N ratio, Rho and Andre (1991) identified three distinct growth phases: an active growth phase, an accumulation phase and a biomass decline phase (endogenous metabolism). They also noticed that phosphate (0.56 mM), nitrate (12.97 mM) and low concentration of ammonia were beneficial for maximum growth and increased alkaloids production. Similarly higher concentration of sucrose (carbon source) only enhanced the biomass; the optimized glucose (500 mM), ammonium and phosphate (0–12 mM) were previously used for higher alkaloids yield (Schlatmann et al. 1992). Medium composition and day's interval had direct effect on induction and accumulation of indole alkaloids (Junaid et al. 2009). A medium added with 6% sucrose is favourable for both biomass and alkaloids production in *Catharanthus* (Scragg et al. 1990). Liquid medium with 3–6% maltose was also found to be highly effective for production of somatic embryos (Junaid et al. 2006). It has been reported that agitated liquid media added with BAP (1.0 mg/L) was very productive for large-scale plant regeneration (Mujib et al. 1995). Alteration in macro- and micronutrient of MS medium (Murashige and Skoog 1962) has also been used to promote growth and subsequent alkaloid production (Smith et al. 1987a). Surface methodology (Tuominen et al. 1989) has been used for the rapid biomass growth and increase in ajmalicine production in hairy root cultures. Similar results in cell suspension culture have been noticed (Schlatmann et al. 1992). Hairy root culture is a unique system, often used for root-specific indole alkaloids production (Toivonen et al. 1992). Recently, Batra et al. (2004) have observed an increase in growth and in the yield of terpenoids indole alkaloids (ajmalicine and serpentine) when left and right termini-linked Ri T-DNA gene integration was made in hairy root cultures of *C. roseus*.

4.1 Phytohormones

The role of plant growth regulators in alkaloids production of *C. roseus* has been extensively studied, but the response varies with genetic makeup of the used explant, type and quantity of phytohormones (Ganapathi and Kagri 1990; Smith et al. 1987a). The cytokinin applied exogenously either alone or in combination with auxins to suspension cultured cells enhanced alkaloids accumulation in tumorous and non-tumorous cell lines (Kodja et al. 1989; Decendit et al. 1992). The enzyme peroxidase plays a significant role in alkaloids biosynthesis however, the addition of 2,4-D in the culture medium reduced the peroxidase activity (Liman et al. 1998).

Hirata et al. (1990) reported an increase in vindoline and catharanthine concentration by adding to the MS medium an amount of 0.1 mg/L of BAP and 0.1 mg/L of NAA. Exogenously supplied cytokinin increased ajmalicine and serpentine content in untransformed callus from cotyledons (Garnier et al. 1996). At the protein level, it was shown that endogenously produced cytokinin did not mimic the effect of exogenously applied cytokinin in *Catharanthus* (Carpin et al. 1997a), and they also noticed that the protein pattern of Ipt transgenic callus lines was insensitive to exogenously used cytokinin. A28 KD polypeptide and simultaneous ajmalicine accumulation was noted on omission of 2,4-D in medium and by the use of NaCl treatments. (Carpin et al. 1997b; Ouelhazi et al. 1993). In a separate study, Alam et al. (2012) worked out that the plants of two cultivars (Rosea and Alba) of *C. roseus* L. G. Don were sprayed with various PGRs viz. IAA, IBA, NAA, BAP, KIN, TDZ, GA_3, SA, HBR and TRIA, at the rate of 10^{-7} M at 60 days after planting (DAP). Application of HBR, KIN and GA_3 resulted in the ameliorative effects on total alkaloid content. Of the various PGRs, application of GA_3 increased vincristine content. It is reported that cultivar Rosea gave higher yield of foliage and roots and that of alkaloids compared to Alba (Idrees et al. 2010). Further, Alam (2013) studied the effect of GA_3 on the content of *C. roseus* alkaloids in addition to that on the yield of different plant parts like leaf, stem and root was taken into consideration separately. The content of total alkaloids (%) in all plant parts and the major alkaloids i.e. vincristine, vinblastine and vindoline were significantly increased by the influence of GA_3 treatments. The yield of total alkaloids in these plant parts was augmented accordingly owing to significant increase in total herbage yield of the plant. Further, he examined that EBL application positively influenced the content and yield of leaf and root alkaloid significantly. Further, Alam et al. (2016) find out the ameliorative effect of EBL on *C. roseus* leaf and root alkaloids content.

4.2 pH and Temperature of Culture Medium

In vitro biomass and alkaloid production are directly influenced by the pH values of the medium; pH values with a range of 5.5–6.5 did not have much effect on alkaloids yield. The value 5.5 was found as the optimum pH for serpentine production (Doller et al. 1976). It has been reported that alkaloids produced by suspension culture were stored in vacuole and simultaneously storage capacity changed in accordance with the changes of pH in the medium and vacuole (Neumann et al. 1983). Low and higher values of pH were used to release intracellular alkaloids into the culture medium (Asada and Shuler 1989). It is quite known that the optimized pH value (5.5–5.8) occasionally fluctuates during culture time and influences in vitro responses including alkaloid yield. For in vitro study temperature range from 20 to 30 °C has been considered the best for comparatively better biomass production and growth of cultures, but contradictory data have been reported about the alkaloids yield. Temperature in low range had inhibitory (Morris 1986), stimulatory (Courtosis and Guern 1980), or no effect (Scragg et al. 1988a, b) on alkaloid yield. In the tested cell lines under different temperature range (20, 25 and 30 °C), the highest serpentine production was recorded at 25 °C and no effect was recorded at temperature 17, 23 and 32 °C (Scragg et al. 1988a, b), while in hairy root culture low temperature enhanced the alkaloid yield (Toivonen et al. 1992).

4.3 Light and Aeration

Light is an important factor for both ex vitro and in vitro morphogenetic study. Its presence, absence, time and intensity directly influence the anabolic and catabolic processes, particularly with regard to secondary metabolism (Seibert and Kadkade 1980; Morris 1986). Most of the study of the effect of light was observed on serpentine and ajmalicine, where serpentine content was directly related to the intensity of light in *Catharanthus roseus* (Lounasmaa and Galambos 1989). Same was true for vindoline (De Luca et al. 1986); however, another alkaloid catharanthine was decreased in the absence of light. It has also been reported that light did not affect alkaloid yield but it affected the accumulation site (Drapeau et al. 1987). However, 15 h per day light exposure, instead of 24 h, improved the serpentine accumulation. On the contrary, dark-grown culture was much better in comparison to light grown regarding serpentine and ajmalicine content; in comparison to dark-grown culture, the alkaloid content decreased in light-grown culture from 79 to 14% regarding serpentine and from 78 to 18% regarding ajmalicine. Gradual transfer of dark-grown culture of *C. roseus* towards the light increased the serpentine content; however, continuous exposure of light decreased serpentine level (Scragg et al. 1988a, b). It has been optimized that 12 h is the best light period for better callus growth and alkaloid production (Hirata et al. 1990); however, dark period more than 12 h decreased the alkaloid contents. It was investigated that an increased chloroplast number and enhanced chlorophyll accumulation in response to light influenced the serpentine production (Loyola-Vargas et al. 1992). Besides, exposure of monochromic light such as blue (450 nm) or red (670 nm) did not affect growth and alkaloid accumulation; these variable wavelengths showed constant ajmalicine and serpentine synthesis which, however, decreased to some extent under white light (Hirata et al. 1990; Loyola-Vargas et al. 1992). Different types of gases, mainly CO_2 and ethylene, are usually evolved within the culture. In many cases, these gases reduce O_2 level in close vessels, inhibit plant culture growth as well as secondary metabolism. High dissolved oxygen and improved gaseous permeability at aerated condition stimulated secondary metabolism as observed by Schlatmann et al. (1994), as ajmalicine production was increased with high oxygen level. Improved oxidative metabolism at rich O_2 level is believed to be the reason for better product conversion. Aeration has been provided in culture to influence the alkaloids synthesis and to make it more efficient modern stirring devices have been employed along with traditional shake flask (Tom et al. 1991; Mohamed and Scragg 1990; Leckie et al. 1991; Lee and Shuler 1991). Different types of fermenters have also been used such as shikonin and ginseng; the two important secondary metabolites have been commercially produced by the use of fermenters. Several researchers (Paynee et al. 1988) have suggested the use of bioreactors in secondary metabolites production in plant cell culture of *C. roseus*. An impeller with a speed of 100 rpm was most appropriate for the accumulation of alkaloids; however, higher impeller speed increased the callus/suspension growth. Hoopen et al. (1994) studied the rate of ajmalicine production by using different vessels including shake flask and bioreactors. They found that biomass was not affected by different culture vessels; however, ajmalicine production was decreased with overfeeding of biomass in the shake flask and fermenter.

4.4 Elicitor's Effect

New groups of triggering factors, which are better known as elicitors, have been reported to stimulate the secondary metabolites (Eilert et al. 1986). The substance used as elicitors may be of biotic or abiotic in origin. Biotic elicitors include microbial filtrates (e.g. yeast, *Pythium* and other fungal filtrates), while abiotic elicitors comprise simple inorganic and organic molecules (e.g. vanadyl sulphate, oxalate, UV irradiation, etc.). It has been reported that addition of *Pythium aphanidermatum* filtrate -+*increased the accumulation of phenolic compounds instead of alkaloids production (Seitz et al. 1989). Effect of different concentrations of *Pythium vexans* extract was studied by Nef et al. (1991), who noticed that low elicitor concentration increased the serpentine production but no effect was observed on catharanthine yield. Addition of nicotinamide (8.2 mM) in *C. roseus* cell lines was used to enhance the anthocyanin accumulation (Berglund et al. 1993). The extract of *Pythium aphanidermatum* in hormone-free cell lines responded well and induced synthesis of enzymes (TDC and anthranilate synthase), which catalyse the biosynthesis of several intermediates and subsequent accumulation of tryptamine (Moreno et al. 1995). Several inorganic compounds (e.g. sodium chloride, potassium chloride and sorbitol) had also a positive effect on catharanthine accumulation (Smith et al. 1987b). Addition of vanadyl sulphate to cell suspension culture increased catharanthine, serpentine and tryptamine production but the event was concentration dependent (Tallevi and DiCosmo 1988); at 25 ppm, catharanthine and ajmalicine were primarily accumulated, and at 50–75 ppm, only tryptamine accumulation was noticed. Moreover, the effect of heavy metal was studied where addition of copper (200 μm) increased total indole alkaloid accumulation which was correlated with decreased tryptamine concentration. In addition, several stress factors (e.g. fungal elicitor, vanadyl sulphate and potassium chloride) were used and it was found that the alkaloids accumulation was concentration dependent (Kargi and Potts 1991). Adding the optimal concentration (29, 1.45 and 145 mg g^{-1} by dry weight) of fungal elicitor, vanadyl sulphate and potassium chloride into medium increased the alkaloids accumulation; however, higher concentration had toxic effects and resulted in the loss of cell viability. Twofold increase in alkaloids yield was noticed by adding tryptophan, fungal elicitor and vanadyl sulphate to the culture production medium

(Kargi and Ganapathi 1991). Exposure of 2,2-*azobis* dehydrochloride (AAPH, an oxidative stress agent) and UV-B irradiation to *C. roseus* culture increased the nicotinamide and trigolline contents (Berglund et al. 1996). Simultaneously, phenylalanine ammonia lysate (PAL) activity was also increased. However, the increase in PAL activity (caused by addition of 2 μm of AAPH) was prevented by 0.1 mm 3-amino benzomide, which is an inhibitor of poly-(ADP-ribose) polymerase. This suggests that nicotinamide and its metabolites function as signal transmitter in response to the oxidative stress, since poly-polymerase has defensive metabolic functions. In shoot culture of *C. roseus*, the level of vinblistine and leurosine were increased in response to irradiation with near ultraviolet light (370 nm) (Hirata et al. 1991; Hirata et al. 1992); however, catharanthine and vindoline content were decreased. Leaves were more sensitive to dimeric alkaloid accumulation in comparison to shoot; however, exposure of near ultraviolet irradiation to whole plant of *C. roseus* led to increased accumulation of dimeric alkaloids (Hirata et al. 1993). Yeast extract induced the transcription of the biosynthetic gene encoding strictosidine (STR) in the cultured *C. roseus* cells and alkalinization of the culture medium; the active principle from yeast extract was partially purified and found to be of proteinaceous in nature (Menke et al. 1999). Age of culture is very important factor for the elicitors to be effective (Ramos-Valdivia et al. 1997); addition of elicitors is preferred after a few days of inoculation of the culture when the cells are rapidly dividing.

4.5 Mutagenesis

Mutagenesis plays a potent role in the alteration of the genetic constitution, which leads to produce new varieties. *Penicillium* is the most classic example, with many other successful cases. Process of mutagenesis in diploid plants is very complex. Mutagenesis enhances alkaloids yield but the route of biosynthesis and the necessary regulation procedure are not elucidated yet clearly. Therefore, mutation at target site in duplicate genome is really difficult. In spite of several limitations in this process, scientists have used mutagens. Berlin (1982) noted accumulation of higher level of phenolics in some p-fluorophenylalanine resistant cell lines of *Nicotiana tabacum* and *N. glauca*. In case of *C. roseus,* he noticed that a tryptophan-analogue resistant mutant accumulated catharanthine in both growth and production medium. Similarly several research groups used X-rays to produce increased serpentine alkaloid. Beside these examples, some successful reports are available in other group of crops where mutagenesis improved metabolic accumulation. In order to increase secondary metabolites production, high cell density culture feeding has been attempted with or without much success. Ajmalicine production was very low when inoculum potential was increased to 2:8 from 1:9 mg/g. Moreover, low-density cultures increased alkaloids yields (Moreno et al. 1993). It has also been remarked that low oxygen level and inadequate nutrient uptake are among the possible causes for low metabolic accumulation during high cell density culture. Isolation and selection of superior lines from the heterogeneous cell populations help to improve the yield of alkaloids. These cells show genetic variability which was further diversified by the use of various mutagenic agents. Ajmalicine and serpentine level were increased in *C. roseus* by the selection of superior cell lines after mutagenesis (Zenk et al. 1977).

4.6 Bioreactor and Immobilization

In tissue culture, research for alkaloids production has been mainly focused on suspension culture, which requires a rotatory shaker. For large-scale production, however, large-sized culture vessel fermenter/bioreactor is most important. In both types of systems, a stirring device is provided for improved aeration (Drapeau et al. 1987; Kargi and Rosenberg 1987; Scragg et al. 1988a, b). In the device, there are several important vessels fitted with compressors, which provide filtered air. For plant culture growth and productivity, it is recommended that bioreactors with low shear-stress are much more suitable than those of high shear-stress. Bioreactors with improved mechanical designs are regularly introduced in bioreactors industry with innovated impeller, which helps to regulate shear agitation (Joilicoer et al. 1992). In *C. roseus,* immobilization of plants cells has been suggested for better accumulation of terpenoids (Hulst and Tramper 1989; Archambault et al. 1990). Immobilization not only maintains the cells viable for a longer period of time but also helps in extracellular alkaloids accumulation. Alginate-mediated immobilized cells enhanced the accumulation of tryptamide, ajmalicine and serpentine (Zenk et al. 1977; Majerus and Pareilleux 1986). The use of agar and agarose is found to be effective for long-term maintenance of cells. In the last few years, surface immobilization has been proposed using different types of matrices for large-scale production of alkaloids (Facchini et al. 1988; Facchini and DiCosmo 1990). In some other cases, negative influence of immobilization on cells was noticed (Archambault et al. 1990); gel or matrices, entrapment on polysaccharide sheet, is fairly successful in many plant systems and in *C. roseus*. Root of *C. roseus* contains a variety of secondary metabolites, which produce alkaloids. High rooting can be induced by genetic transformation using *Agrobacterium rhizogenes*. Induced roots grew with a faster rate in hormone-free medium with high accumulation of secondary metabolites in *C. roseus*. In transgenic *C. roseus* root, a significant increase in ajmalicine and catharanthine was noticed (Batra et al. 2004; Vazquez-Flota et al. 1997). Other groups used various types of bioreactors/fermenters to improve the growth of hairy roots, leading to better production of secondary metabolites (Davidou et al. 1989; Nuutila 1994). Although somatic embryogenesis (SE) has been reported in a wide variety of plant genera (Thorpe 1995; Mujib and Samaj 2006), it has been reported for the first time in *C. roseus* (Junaid et al. 2006). Earlier, a preliminarily study on plant regeneration from immature zygotic embryo was reported in *C. roseus* (Kim et al. 2004). The advantage of SE is that the initial cell populations can be used as a single cellular system and their genetic manipulation are easy and are similar to microorganisms.

5 Metabolic and Genetic Engineering in Alkaloids Biosynthesis

In alkaloids biosynthesis, the roles of several enzymes have been discussed in *C. roseus*: a few of them have been purified, identified, and characterized, and their encoding genes have also been cloned. The alkaloids biosynthesis is a very complex process that arises from the precursors tryptamine and secologanin. These two precursors are derived from two different pathways. Tryptamine is formed by the enzyme tryptophan decarboxylase (TDC), which has been reviewed earlier by various workers (Bentley 1990; Poulsen and Verpoorte 1992; Singh et al. 1991), while the strictosidine synthetase (SSS) helps in the coupling of tryptamine and secologanin to produce strictosidine (Madyastha and Coscia 1979; Inouye and Uesato 1986). The other enzymes such as geraniol 10-hydroxylase (G10H), NADPH-cytochrome P-450 reductase and anthranilate synthetase (AS) have the similar activities as TDC, which are involved in the biosynthesis of indole alkaloids (Poulsen et al. 1993). The TDC enzyme has been purified from cell suspension culture (Pennings et al. 1989) and ultimately its cDNA gene was established (Pasquali et al. 1992). The cytochrome P450 enzyme, geraniol-10-hydroxylase (G10H) and other enzymes have been studied extensively from intact plant of *C. roseus*. By HPLC study (Collu et al. 1999) and selection of a cell line with high G10H activity (Collu et al. 2001), the enzyme was purified to homogeneity (Collu et al. 1999). Based on the internal amino acid sequences obtained from the digested protein, gene was cloned and functionally expressed in yeast. The enzyme belongs to the CYP76B subfamily and is designated as CYP76B6. The activity of this enzyme was induced by treating the cells with the cytochrome P450 inducer Phenobarbital; it was decreased after treatment of the inhibitor ketoconazole (Contin et al. 1999). Besides, many other enzymes have been identified and characterized that metabolize strictosidine, which after undergoing several rearrangements produced cathenamine and ajmalicine (Hemscheidt and Zenk 1985; Stevens 1994). Another important enzyme is desacetoxyvindoline-4-hydroxylase (DAVH), active during vindoline biosynthesis; it was purified from intact plant of *C. roseus*. The native enzyme is a monomer and has a molecular weight of 45 KD with three isoforms (De Carolis and De Luca 1994). Recently, attention has been paid on the regulation of mevalonate biosynthesis that terminates with its end product strictosidine. Encoding genes and the enzymes of different steps of mevalonate pathway have been elucidated (Maldonado-Mendonza et al. 1992). After the formation of strictosidine, first step of alkaloid biosynthesis is the removal of sugar moiety from strictosidine to form an unstable aglycone. Two strictosidine β-glucosidases (SG) were partially purified and characterized from *C. roseus* cell cultures (Hemscheidt and Zenk 1980; Stevens 1994). Feeding of terpenoids precursors to *C. roseus* cell suspension cultures increased the alkaloids production (Naudascher et al. 1990; Facchini and DiCosmo 1991; Moreno et al. 1993). Addition of tryptophan (0.5 mM) to *C. roseus* cells resulted in high intracellular levels of tryptamine and an increase in STR activity but it did not influence ajmalicine accumulation much (Bongaerts 1998).

As in other feedback inhibitions, product accumulation depends upon the product degradation and this phenomenon has been reported in cell suspension culture of *C. roseus*. It is now known that the precursor for alkaloids (tryptophan to tryptamide) was located in the cytosol whereas the enzyme SSS was localized in the vacuole (Stevens et al. 1993).

5.1 Coupling Methods for Alkaloids Biosynthesis

The *bis*-indoles are derived from the coupling of vindoline and catharanthine. Catharanthine is thought to be derived from strictosidine via 4,21-dehydrogeissoschizine, stemmadenine and dehydrosecodine rout, while vindoline is derived from strictosidine via stemmadenine and tabersonine pathway. This pathway (transformation of tabersonine to vindoline) has got orderly six reactions (De Luca et al. 1986; Balsevich et al. 1986). The enzyme anhydrovinblastine synthase couples catharanthine and vindoline to yield AVBL, which was purified and characterized from *C. roseus* leaves. This heme protein has a molecular weight of 45 KD and shows the peroxidase activity. During this enzymatic coupling, both the monomers were incubated with cultured *C. roseus* cells at 30 °C at acidic pH (tris buffer 7.0). Only after 3 h the chemical reaction produced vinblistine and anhydrovinblastine as major products along with other dimeric alkaloids. Vindoline and catharanthine were also non-enzymatically coupled to the dihydropyridinium intermediate (DHPI) under near-UV light irradiation with a peak at 370 nm in the presence of flavin mononucleotide. Subsequently, DHPI can be reduced to anhydrovinblastin (AVBL) with an overall yield of 50%, based on initial amount of vindoline. Vinblastine content was further improved up to 50% by using various compounds as stimulants (Bede and DiCosmo 1992). Similarly, vincristine can be isolated from vinblastine by chemical conversion. Two routes are employed; first route is the isolation of *N*-deformyl-VCR, which was further converted into vincristine by formylation. The second method involves a formylation of the *C. roseus* extract in which conversion of *N*-deformyl–VLB to VCR takes place, after which the material is oxidized. In both cases, vincristine was purified by column chromatography and then sulphated. It was also reported that $MnCl_2$ and FMN/FAD stimulated coupling process. However, in the absence of *C. roseus* cell suspension enzymes, ferric acid stimulated coupling process. The production of vinblastine through enzymatic coupling pathway is thought to be highly efficient and is likely to be used commercially very soon. Vindoline and *bis*-indole alkaloids are accumulated only in green tissue and are not found in root and cell suspension cultures (Endo et al. 1987). The developmental regulation of TDS, SSC and the enzymes involved in late steps of vindoline biosynthesis has been studied extensively (De Luca and Cutler 1987; Fernandez et al. 1989). In seedlings of *C. roseus*, transcription of these enzymes was not under strong developmental control where enzymes activities were modulated by tissue specific or light dependent factors. The concentration of vindoline, catharanthine and 3’,4’-anhydrovinblastine (AVBL) are age-dependent

(Naaranlahti et al. 1991). Vinblastine was increased as seedlings matured, reaching a steady concentration when the plants became more than 3 months old. On an average, whole seedlings, young plants and mature plants contained 7, 11.5 and 12-μg/g dry weight VLB, respectively. After induction of shoot formation, the VLB contents increased rapidly to similar levels of in vitro seedlings (Datta and Srivastava 1997).

5.2 Subcellular Compartmentation

Subcellular compartmentation plays an important role in alkaloids metabolism. This process of metabolism involves the participation of plant cell to separate the enzyme from their substrates and end products. In this, alkaloids biosynthesis requires three cellular compartments, namely vacuole, cytosol and plastid (Meijer et al. 1993). The transformation of tryptophan into tryptamine takes place in cytosol (De Luca and Cutler 1987; Stevens et al. 1993) and that of SSS in vacuoles (Stevens et al. 1993; McKnight et al. 1991). SG was tightly bound to the tonoplast boundary (Stevens et al. 1993). Synthesis of strictosidine takes place inside the vacuole, which is later transported to the cytoplasm where its glucose moiety gets detached. Ajmalicine has the potentiality to move freely across the cell membrane and is accumulated into the vacuoles where it is converted into the serpentine using peroxidases (Blom et al. 1991); thus produced serpentine is stored in vacuole and cannot pass through the tonoplast. In cell suspension cultures, alkaloid accumulation seems to be restricted to certain cells (Asada and Shuler 1989). Permeability of cell plays a potent role to release plant products. There are several permeabilizing agents, like DMSO and Triton X-100, which are found to be very effective in *C. roseus* cell culture. Besides, for the release of secondary products, several other agents (e.g. chitosan, alginate beads, electroportion and ultra sonication) have been used with or without cell viability in other groups of plants. The cell membrane with active uptake mechanism has also been noticed in *C. roseus*. Most of the secondary products are generally accumulated intercellular; however, several compounds such as taxol and anthraquinones are identified in the media, which filtrate itself through membrane. For this extracellular product secretion, addition of resin XAD-7 enhanced the product adsorption in *Cinchona* (Ganapathi and Kagri 1990). The media provided with amberlite-type resin and XAD-7 resin adsorbed ajmalicine and catharanthine effectively in *C. roseus*.

6 Conclusion

C. roseus has been a research symbol because of the enormous number of phytochemical compounds, secondary metabolites and the therapeutic effects that they produce. The secondary metabolites of *C. roseus* are terpene indole alkaloids exhibited pharmacologic activity and with a number of applications in human medicine.

The plant has an extensive variety of properties: anticancer, anti-diabetes, anti-helminthic, antihypertensive, anti-diarrheic, antimicrobial, among others. The indole dimeric alkaloids, vinblastine and vincristine, have become important drugs in cancer chemotherapy due to their potent antitumour activity against several types of leukaemia and solid tumours. Remarkable example is vinblastine (a member of the iboga family of the indole alkaloids), which is produced by catharanthine, vindoline and catharanthine.

Because of the use of plants and in vitro cell cultures, the biosynthesis pathway has been determined, but not entirely clarified. Moreover, a considerable number of enzymes have been characterized and their particular cloned genes defined, with the production of the alkaloids being found as well regulated at the transcriptional level.

In order to increase the availability of alkaloids for therapeutic use, the production of biomass from in vitro cultures of calluses from leaves has taken place as biotechnological tool to augment the accumulation of alkaloids in *C. roseus*. The combination of various basal media as carbon sources, phytohormones and inducers of the biotic and abiotic type may provide positive ways for the rational technical development and the increase in production yields of several of these bioactive molecules in vitro. The in vitro cultures of calluses or cells in suspension could be used at a large industrial scale to obtain bioactive compounds that are of great significance in human health, and are envisaged as models to circumvent the limitations of other production systems.

References

Alam MM (2013) Influence of GA_3, Epibrassinolide and Kinetin on the performance of *Catharanthus roseus* L. with special reference to alkaloid production. PhD thesis, AMU, Aligarh, India

Alam MM, Naeem M, Idrees M, Khan MMA, Moinuddin (2012) Augmentation of photosynthesis, crop productivity, enzyme activities and alkaloids production in Sadabahar (*Catharanthus roseus* L.) through application of diverse plant growth regulators. J Crop Sci Biotechnol 15:117–129

Alam MM, Naeem M, Khan MMA (2016) Exploiting the Epibrassinolide as a plant growth promoter for augmenting the growth, physiological activities and alkaloids production in *Catharanthus roseus* L. J Med Plant Std 4:88–93

Allan LA, Clarke PR (2007) Phosphorylation of caspase-9 by CDK1/cyclin B1 protects mitotic cells against apoptosis. Mol Cell 26:301–310

Amirjani MR (2013) Effects of drought stress on the alkaloid contents and growth parameters of *Catharanthus roseus*. J Agric Biol Sci 8:745–750

Archambault J, Volesky B, Kurz WGM (1990) Production of indole alkaloids by surface immobilized *C. roseus* cells. Bioengineering 35:660–667

Asada M, Shuler ML (1989) Stimulation of ajmalicine production and extraction from *Catharanthus roseus*: effect of adsorption in situ, elicitors and alginate immobilization. Appl Microbiol Biotechnol 30:475–481

Ataei-Azimi A, Hashemloian BD, Ebrahimzade H, Majd A (2008) High in vitro production of anti-canceric indole alkaloid from periwinkle (*Catharanthus roseus*) tissue culture. Afr J Biotechnol 7:2834–2839

Balsevich J, Bishop G (1988) Distribution of catharanthine, vindoline and 3',4'-anhydrovinblastine in the aerial parts of some *Catharanthus roseus* plants and the significance thereof in relation to alkaloid production in cultured cells. In: Kurz WGW (ed) Primary and secondary metabolism of plant cell cultures II, 1st edn. Springer, Berlin, pp 149–153

Balsevich J, De-Luca V, Kurz WGM (1986) Altered alkaloid pattern in dark grown seedlings of *Catharanthus roseus*. The isolation and characterization of 4-desacetoxyvindoline: a novel indole alkaloid and proposed precursor of vindoline. Heterocycles 24:2415–2421

Batra J, Dutta A, Singh D, Kumar S, Sen J (2004) Growth and terpenoids indole alkaloid production in *Catharanthus roseus* hairy root clones in relation to left-and right- termini-linked Ri T-DNA gene integration. Plant Cell Rep 23:148–154

Bede JC, DiCosmo F (1992) Enzymatic synthesis of α-3', 4'-anhydrovinblastine: optimization and immobilization. Planta Med 58:576–579

Bennouna J, Delord JP, Campone M, Nguyen L (2008) Vinflunine: a new microtubule inhibitor agent. Clin Cancer Res 14:1625–1632

Bentley R (1990) The shikimate pathway—a metabolic tree with many branches. In: Fasman GD (ed) Critical reviews in biochemistry and molecular biology, vol 25. CRC, Boca Raton, FL, pp 307–384.

Berglund T, Ohlsson AB, Rydstrom J (1993) Nicotinamide increases glutathione and anthocyanins in tissue culture of *Catharanthus roseus*. J Plant Physiol 141:596–600

Berglund T, Kalbin G, Rydström J, Ohlsson AB (1996) UV-B and oxidative stress induced increase in nicotinamide and trigonalline and inhibition of defensive metabolism induction by poly (ADP-ribose) polymerase inhibitors in plant tissue. FEBS Lett 380:188–193

Bergnes G, Brejc K, Belmont L (2005) Mitotic kinesins: prospects for antimitotic drug discovery. Curr Top Med Chem 5:127–145

Berlin J (1982) Para-fluorophenylalanine resistant cell lines of tobacco. Z Pflanzenphysiol 97:309–311

Blom TJM, Sierra MI, Van Vliet TB, Franke-van Dijk ME, de Koning P, van Iren F, Verpoorte R, Libbenga KR (1991) Uptake and accumulation of Ajmalicine into isolated vacuoles of cultured cells of *Catharanthus roseus* (L). G. Don. and its conversion into serpentine. Planta 183:170–177

Bongaerts RJM (1998) The chorismate branching point in *Catharanthus roseus* : aspects of anthranilate synthase regulation in relation to indole alkaloide biosynthesis. Ph.D. Thesis, Leiden University

Carpin S, Garnier F, Andreu F, Chenieux JC, Rideau M, Hamdi S (1997a) Changes of both polypeptide pattern and sensitivity to cytokinin following transformation of periwinkle tissues with the isopentenyl transferase gene. Plant Physiol Biochem 35:603–609

Carpin S, Ouelhazi L, Filali M, Chenieux JC, Rideau M, Hamdi S (1997b) The relationship between the accumulation of a 28 KD polyleptide and that of indole alkaloids in *Catharanthus roseus* cell suspension cultures. J Plant Physiol 150:452–457

Carqueijeiro I, Noronha H, Duarte P, Geros H, Sottomayor M (2013) Vacuolar transport of the medicinal alkaloids from *Catharanthus roseus* is mediated by a proton-driven antiport. Plant Physiol 162:1486–1496

Casado P, Zuazua-Villar P, del Valle E, Martinez-Campa C, Lazo PS, Ramos S (2007) Vincristine regulates the phosphorylation of the antiapoptotic protein HSP27 in breast cancer cells. Cancer Lett 247:273–282

Chung IM, Ali M, Chun SC, Lee SJ, San KY, Christie AMP, Ahmad A (2007) New sesquiterpene glycosides from culture hairy roots of *Catharanthus roseus*. Chin J Chem 25:1695–1699

Collu G, Bink HHJ, Moreno PRH, Verpoorte R (1999) Determination of the activity of cytochrome p450 enzyme geraniol 10-hydroxylase in plants by high performance liquid chromatography. Phytochem Anal 10:314–318

Collu G, Unver N, Peltenburg-Looman AM, van der Heijden R, Verpoorte R, Memelink J (2001) Geraniol 10-hydroxylase, a cytochrome P450 enzyme involved in terpenoid indole alkaloid biosynthesis. FEBS Lett 508:215–220

Contin A, Collu G, Van Der Heijden R, Verpoorte R (1999) The effects of phenobarbital and ketoconazole on the alkaloid biosynthesis in *Catharanthus roseus* cell suspension cultures. Plant Physiol Biochem 37:139–144

Courtosis D, Guern J (1980) Temperature of *Catharanthus roseus* cells cultivated in liquid medium. Plant Sci Lett 17:473–482

Cruz Ortíz H, Rodríguez-Martínez HA, Román Bassaure E, Valle Gay A (2000) Coriocarcinoma gestacional con metástasis múltiples. Rev Fac Med UNAM 43:153–156

Datta A, Srivastava PS (1997) Variation in vinblastine production by *Catharanthus roseus* during in vivo and in vitro differentiation. Phytochemistry 46:135–137

Davidou E, Kan C, Quirion JC, Das BC, Husson HP (1989) Epiall-oyohimbine derivatives isolated from in-vitro hairy root cultures of Catharanthus trichophyllus. Phytochemistry 28:1383–1387

De Carolis E, De Luca V (1994) A novel 2-oxyglutarate dependent dioxygenase involved in vindoline biosynthesis; characterization, purification and kinetic properties. Plant Cell Tiss Org Cult 38:181–187

De Luca V, Cutler AJ (1987) Subcellular localization of enzymes involved in indole alkaloid biosynthesis in *Catharanthus roseus*. Plant Physiol 85:1099–1102

De Luca V, Balsevich J, Tyler RT, Eilert U, Panchuk BD, Kurz WGW (1986) Biosynthesis of indole alkaloids: developmental regulation of the biosynthetic pathway from tabersonine to vindoline in *Catharanthus roseus*. J Plant Physiol 125:147–156

Decendit A, Liu D, Ouelhazi L, Doireau P, Merillon JM, Rideau M (1992) Cytokinin enhanced accumulation of indole alkaloids in *Catharanthus roseus* cell culture: the factors affecting the cytokinin responses. Plant Cell Rep 11:400–403

Doller G, Alfermann AW, Reinhard E (1976) Production von indole alkaloiden in callus kulturen von *Catharanthus roseus*. Planta Med 30:14–20

Drapeau D, Blanch MW, Wike CR (1987) Ajmalicine, serpentine and Catharanthine accumulation in *Catharanthus roseus* bioreactor cultures. Planta Med 53:373–376

Eilert U, Constable F, Kurz WGW (1986) Elicitor stimulation of monoterpene indole alkaloids formation in suspension cultures of *C. roseus*. J Plant Physiol 126:11–22

Endo T, Goodbody A, Misawa M (1987) Alkaloid production in root and shoot cultures of *Catharanthus roseus*. Planta Med 53:479–482

Evans WC, Trease GE, Evans D (2009) Pharmacognosy, 16th edn. Elsevier, New York

Facchini PJ, DiCosmo F (1990) Immobilization of cultured *Catharanthus roseus* cells using fiberglass substratum. Appl Microbiol Biotechnol 33:36–42

Facchini PJ, DiCosmo F (1991) Secondary metabolites biosynthesis in cultured cells of *Catharanthus roseus* (L.) G. Don immobilized by adhesion of glass fibres. Appl Microbiol Biotechnol 35:382–392

Facchini PJ, Di Cosmo F, Radvanyi LG, Giguere Y (1988) Adhesion of *Catharanthus roseus* cells to surfaces: effect of substrate hydrophobocity. Biotechnol Bioeng 32:935–938

Fernandez JA, Oven TG, Kurz WGW, De Luca V (1989) Immunological detector and quantitation of tryptophan decarboxylase in developing *Catharanthus roseus* seedlings. Plant Physiol 91:79–84

Ferreres F, Pereira DM, Valentao P, Andrade PB, Seabra RM, Sottomayor M (2008) New phenolic compounds and antioxidant potential of *Catharanthus roseus*. J Agric Food Chem 56:9967–9974

Ganapathi G, Kagri F (1990) Recent advances in indole alkaloid production by *Catharanthus roseus* (Periwinkle). J Exp Bot 41:259–267

Garnier F, Carpin S, Label P, Creche J, Rideau M, Hamdi S (1996) Effect of cytokinin on alkaloids accumulation in periwinkle callus cultures transformed with a light inducible ipt gene. Plant Sci 120:47–55

Gascoigne KE, Taylor SS (2009) How do anti-mitotic drugs kill cancer cells? J Cell Sci 122:2579–2585

Gayatri CL, Chakravarthy R (2013) Micro propagation in *Catharanthus roseus*. Int J Innov Technol Explor Eng 2:194–196

Goodbody AE, Watson CD, Chapple CCS, Vukovic J, Misawa M (1998) Extraction of 3',4'-anhydrovinblastine from *Catharanthus roseus*. Phytochemistry 27:1713–1717

Grellier P, Sinou V, Garrea de Loubresse N, Bylen E, Boulard Y, Schrevel J (1999) Selective and reversible effects of vinca alkaloids on *Trypanosoma cruzi* epimastigote forms: Blockage of cytokinesis without inhibition of the organelle duplication. Cell Motil Cytoskeleton 42:36–47

Hedhili S, Courdavault V, Giglioli-Guivarch N, Gantet P (2007) Regulation of the terpene moiety biosynthesis of *Catharanthus roseus* terpene indole alkaloids. Phytochem Rev 6:341–351

Hemscheidt T, Zenk MH (1980) Glucosidases involved in indole alkaloid biosynthesis of *Catharanthus roseus* cell cultures. FEBS Lett 110:187–191

Hemscheidt T, Zenk MH (1985) Partial purification and characterization of a NADPH dependent tetrahydroalstonine synthase from *Catharanthus roseus* cell suspension cultures. Plant Cell Rep 4:216–219

Hirata K, Houriuchi M, Ando T, Miyamoto K, Miura Y (1990) Vindoline and catharanthine production in multiple shoot cultures of *Catharanthus roseus*. J Ferment Bioeng 70:193–195

Hirata K, Horiuchi M, Asada M, Miyamoto K, Miura Y (1991) Effect of near-ultraviolet light on alkaloid production in multiple shoot cultures of *Catharanthus roseus*. Planta Med 57:499–500

Hirata K, Horiuchn M, Asada M, Ando T, Miyamoto K, Miura Y (1992) Stimulation of dimeric alkaloid production by near-ultraviolet light in multiple shoot cultures of *Catharanthus roseus*. J Ferm Bioeng 74:222–225

Hirata K, Asada M, Yatani M, Miyamoto K, Miura Y (1993) Effects of near-ultraviolet light on alkaloid production in *Catharanthus roseus* plants. Planta Med 59:46–50

Hoopen HJG, Ten Gulik WMV, Schlatmann JE, Moreno PRH, Vinke JL, Verpoorte JJHR (1994) Ajmalicine production by cell cultures of *Catharanthus roseus:* from shake flask to bioreactor. Plant Cell Tiss Org Cult 38:85–91

Hulst AC, Tramper J (1989) Immobilized plant cells: a literature survey. Enzyme Microb Technol 11:546–558

Huxtable RJ (1992) The pharmacology of extinction. J Ethnopharmacol 37:1–11

Idrees M, Naeem M, Khan MMA (2010) The Superiority of cv 'rosea' over cv 'alba' of Periwinkle (*Catharanthus roseus* L.) in alkaloid production and other physiological attributes Turk. J Biol 34:81–88

Inouye H, Uesato S (1986) Biosynthesis of iridoids and secoiridoids. In: Herz W, Grisebach H, Kirby GW, Tamm CH (eds) Progress in the chemistry of natural products, vol 50. Springer, New York, pp 169–236.

Jaffe ES, Harris NL, Stein H, Isaacson PG (2008) Classification of lymphoid neoplasms: the microscope as a tool for disease discovery. Blood 112:4384–4399

Jaleel CA, Gopi R, Panneerselvam R (2008) Growth and photosynthetic pigments responses of two varieties of *Catharanthus roseus* to triadimefon treatment. C R Biol 331:272–277

Joilicoer M, Chavarie C, Carreau PJ, Archambault J (1992) Development of a helical ribbon impeller bioreactors for high density plant cell suspension cultures. Biotechnol Bioeng 39:511–521

Jordan MA, Wilson L (1999) The use and action of drugs in analyzing mitosis. Methods Cell Biol 61:267–295

Jordan MA, Wilson L (2004) Microtubules as a target for anticancer drugs. Nat Rev Cancer 4:253–265

Jordan MA, Thrower D, Wilson L (1991) Mechanism of inhibition of cell proliferation by Vinca alkaloids. Cancer Res 51:2212–2222

Jordan MA, Toso RJ, Thrower D, Wilson L (1993) Mechanism of mitotic block and inhibition of cell proliferation by taxol at low concentrations. Proc Natl Acad Sci U S A 90:9552–9556

Jordan MA, Wendell K, Gardiner S, Derry WB, Copp H, Wilson L (1996) Mitotic block induced in HeLa cells by low concentrations of paclitaxel (Taxol) results in abnormal mitotic exit and apoptotic cell death. Cancer Res 56:816–825

Junaid A, Bhatt MA, Mujib A, Sharma MP (2006) Somatic embryo proliferation maturation and germination in *Catharanthus roseus*. Plant Cell Tiss Org Cult 84:325–332

Junaid A, Mujib A, Nasim SA, Sharma MP (2009) Screening of vincristine in ex vitro and in vitro developed somatic embryo derived *Catharanthus roseus* (L) G. Don plantlets. Sci Hortic 119:325–329

Junaid A, Khan SH, Siddiqui ZH, Fatima Z, Maqsood M, Bhat MA, Nasim SA, Ilah A, Ahmad IZ, Khan SA, Mujib A, Sharma MP (2010) *Catharanthus roseus* (L.) G. Don. An important drug: it's applications and production. Pharmacie Globale: Int J Compr Pharm 1:1–16

Kargi F, Ganapathi B (1991) Effect of precursor stimulating agents on formation of indole alkaloids by *C. roseus* in a biofilm reactors. Enzyme Microb Technol 13:643–647

Kargi F, Potts P (1991) Effect of various stress factors on indole alkaloid formation by *C. roseus* (periwinkle) cells. Enzyme Microb Technol 13:760–763

Kargi F, Rosenberg MZ (1987) Plant cell bioreactors: present status and future trends. Biotechnol Prog 3:1–8

Karthikeyan B, Jaleel CA, Changxing Z, Joe MM, Srimannarayan J, Deiveekasundaram M (2008) The effect of AM fungi and phosphorous level on the biomass yield and ajmalicine production in *Catharanthus roseus*. Eurasia J BioSci 2:26–33

Keen N, Taylor S (2004) Aurora-kinase inhibitors as anticancer agents. Nat Rev Cancer 4:927–936

Kelling J, Sullivan K, Wilson L, Jordan MA (2003) Suppression of centromere dynamics by Taxol in living osteosarcoma cells. Cancer Res 63:2794–2801

Kim SW, In DS, Choi PS, Jang R (2004) Plant regeneration from immature zygotic embryo derived embryogenic calluses and cell suspension cultures of *Catharanthus roseus*. Plant Cell Tiss Org Cult 76:131–135

Kisakurek MV, Hesse M (1980) Chemotaxonomic studies of the Apocynaceae, Loganiaceae and Rubiaceae with reference to indole alkaloids. In: Phillipson JD, Zenk MH (eds) Indole and biogenetically related alkaloids. Academic, London, pp 11–26

Knobloch KH, Berlin J (1980) Influence of medium composition on the formation of secondary compounds in cell suspension cultures of *Catharanthus roseus* (L.) G. Don. Z Naturforsch 35:551–556

Kodja HP, Liu D, Merillon JM, Andreu F, Rideau M, Chenieux JC (1989) Stimulation par les cytokinines de l' accumulation d' alcaloides indoliques dans des suspension cellulaires de *Catharanthus roseus* G. Don. CR Acad Sci Paris Ser III 309:453–458

Kutney J, Mayer L, Schmidt B, inventors (1998) Anhydrovinblastine for the treatment of cervical and lung cancer. International Patent WO 199839004, 4 Mar 1998

Kutney J, Mayer L, Schmidt B (2000) Anhydrovinblastine for the treatment of cervical and lung cancer. EP0969839 A1, 2000

Kwon M, Godinho SA, Chandhok NS, Ganem NJ, Azioune A, Thery M, Pellman D (2008) Mechanisms to suppress multipolar divisions in cancer cells with extra centrosomes. Genes Dev 22:2189–2203

Lanni JS, Jacks T (1998) Characterization of the p53-dependent postmitotic checkpoint following spindle disruption. Mol Cell Biol 18:1055–1064

Leckie F, Scragg AH, Cliffe KC (1991) Effect of impeller design and speed on the large-scale cultivation of suspension cultures of *Catharanthus roseus*. Enzyme Microb Technol 13:801–810

Lee CWT, Shuler ML (1991) Different shake flask closure alter gas phase composition and ajmalicine production in *Catharanthus roseus* cell suspensions. Biotechnol Techniq 5:173–178

Lee CW, Belanger K, Rao SC, Petrella TM, Tozer RG, Wood L, Savage KJ, Eisenhauer EA, Synold TW, Wainman N. Seymour L (2008) A phase II study of ispinesib (SB-715992) in patients with metastatic or recurrent malignant melanoma: a National Cancer Institute of Canada Clinical Trials Group trial. Invest New Drugs 26:249–255

Leveque D, Jehl F (2007) Molecular pharmacokinetics of *Catharanthus* (vinca) alkaloids. J Clin Pharmacol 47:579–588

Liman F, Chahed K, Ouelhazi N, Ghrir R, Ouelhazi L (1998) Phytohormone regulation of isoperoxidase in *Catharanthus roseus* suspension cultures. Phytochemistry 49:1219–1225

Lounasmaa M, Galambos J (1989) Indole alkaloid production in *Catharanthus roseus* cell suspension cultures.In: Herz W, Crisebach H, Kirby GW, Tamm CH (eds) Progress in the chemistry of organic natural products, vol 55. Springer, New York, pp 89–115

Loyola-Vargas VM, Mende-Zeel M, Monforte-Gonzales M, Miranda-Ham ML (1992) Serpentine accumulation during greening in normal and tumor tissues of *Catharanthus roseus*. J Plant Physiol 140:213–217

Madyastha KM, Coscia CJ (1979) Enzymology of indole alkaloid biosynthesis. Rec Adv Phytochem 13:85–129

Majerus F, Pareilleux A (1986) Alkaloid accumulation in Ca-alginate entrapped cells of *Catharanthus roseus* using a limiting growth medium. Plant Cell Rep 5:302–305

Maldonado-Mendonza IE, Burnett RJ, Nessler CL (1992) Nucleotide sequence of a cDNA encoding 3 hydroxy-3-methylglutaryl coenzyme A reductase from *Catharanthus roseus*. Plant Physiol 100:1613–1614

Mallik J, Chowdhury HB, Al Farouq A, Das S (2013) Pharmacological profile of *Catharanthus roseus* (apocynaceae)—a detailed review. Asian J Pharm Res Dev 1:1–6

McGrogan BT, Gilmartin B, Carney DN, McCann A (2008) Taxanes, microtubules and chemoresistant breast cancer. Biochim Biophys Acta 1785:96–132

McKnight TD, Bergey DR, Burnett RJ, Nessler CL (1991) Expression of enzymatically active and correctly targeted strictosidine synthase in transgenic tobacco plants. Planta 185:148–152

Meijer AH, Verpoorte R, Hoge JHC (1993) Regulation of enzymes and genes involved in terpenoids indole alkaloid biosynthesis in *Catharanthus roseus*. J Plant Res 3:145–164

Menke FLH, Parchmann S, Mueller MJ, Kijne JW, Memelink J (1999) Involvement of the octadecanoid pathway and protein phosphorylation in fungal elicitor-induced expression of terpenoid indole alkaloid biosynthetic genes in *Catharanthus roseus*. Plant Physiol 4:1289–1286

Mohamed MTZ, Scragg AH (1990) A comparision of the growth and alkaloids production by suspension and immobilized *Catharanthus roseus* cells in 61 and 71 bioreactors. Immob Cell:335–342

Moreno PRH, Van der Heijden R, Verpoorte R (1993) Effect of terpenoids precursor feeding and elicitation on formation of indole alkaloids in cell suspension cultures of *Catharanthus roseus*. Plant Cell Rep 12:702–705

Moreno PRH, Van der Heijden R, Verpoorte R (1995) Cell and tissue cultures of *Catharanthus roseus;* a literature survey II. Updating from 1988–1993. Plant Cell Tiss Org Cult 42:1–25

Morris P (1986) Regulation of product synthesis in cell cultures of *Catharanthus roseus*. Effect of temperature. Plant Cell Rep 5:427–429

Moudi M, Go R, Yien CYS, Nazre M (2013) Vinca alkaloids. Int J Prev Med 4:1231–1235

Mujib A, Samaj J (2006) Somatic embryogenesis. Springer, Berlin

Mujib A, Das S, Dey S, Bhattacharya B (1995) Influence of agitation in in-vitro cultivation of *Catharanthus roseus* (L). G. Don multiple shoot. Phytomorphology 45:239–245

Mujib A, Ilah A, Gandotra N, Abdin MZ (2002) In vitro application to improve alkaloid yield in *Catharanthus roseus*. In: Govil JN, Kumar PA, Singh VK (eds) Biotechnology and genetic engineering of recent progress in medicinal plants, vol 4. Sci Tech, Houston, pp 415–440

Murashige T, Skoog F (1962) A revised medium for rapid growth and bioassays with tobacco tissue cultures. Physiol Plant 15:473–497

Musacchio A, Salmon ED (2007) The spindle-assembly checkpoint in space and time. Nat Rev Mol Cell Biol 8:379–393

Naaranlahti T, Auriola S, Lapinjoki S (1991) Growth related dimensation of vindoline and Catharanthine in *Catharanthus roseus* and effect of wounding on the process. Phytochemistry 36:1462–1463

Naudascher F, Doireau P, Thiersault M (1990) Influence de la disponibilite en precurseurs accumulation alcaloidique dans les cellules de *Catharanthus roseus* cultivees in vitro comparision entre suspensions en phase de crossence et suspensions en phase stationale. Les Colloq de FINRA 5:307–309

Nef C, Rio B, Chrestin H (1991) Induction of Catharanthine synthesis and stimulation of major indole alkaloids production *Catharanthus roseus* cells under non-growth altering treatment with Pythium vexons extracts. Plant Cell Rep 10:26–29

Neumann D, Kraus G, Heike M, Gröger D (1983) Indole alkaloid formation and storage in cell suspension cultures of *C. roseus*. Planta Med 48:20–23

Nirmala MJ, Samundeeswari A, Sankar PD (2011) Natural plant resources in anti-cancer therapy—a review. Res Plant Biol 1:1–14

Nogales E (2000) Structural insights into microtubule function. Annu Rev Biochem 69:277–302

Nuutila AM (1994) Bioreactor studies on hairy root cultures of *Catharanthus roseus*. Composition of three bioreactors types. Biotechnol Techniq 8:61–66

Ouelhazi L, Filali M, Decendit A, Chenieux JC, Rideau M (1993) Differential protein accumulation in zeatin-and 2,4-D-treated cells of *Catharanthus roseus* : correlation with indole alkaloid biosynthesis. Plant Physiol Biochem 31:421–431

Pasquali G, Goddijin OJ, De Wall A, Verpoorte R, Schilperoort RA, Hoge JH, Memelink J (1992) Coordinated regulation of two indole alkaloid biosynthetic genes from *Catharanthus roseus* by auxin and elicitors. Plant Mol Biol 18:1121–1131

Paynee GF, Paynee NN, Shuler ML (1988) Bioreactor considerations for secondary metabolite production from plant cell tissue culture; indole alkaloid from *Catharanthus roseus*. Biotechnol Bioeng 31:905–912

Pennings EJM, Van den Bosch RA, Van der Heijden R, Stevens LH, Duine JA, Verpoorte R (1989) Assay of strictosidine synthase from plant cells by high-performance liquid chromatography. Anal Biochem 176:412–415

Piovan A, Filippini R (2007) Anthocyanins in *Catharanthus roseus in vivo* and *in vitro*: a review. Phytochem Rev 6:235–242

Poulsen C, Verpoorte R (1992) Activities of chorismate utilizing enzymes and of enzymes involved in indole alkaloid biosynthesis in cell suspension cultures. Plant Physiol Biochem 30:105–113

Poulsen C, Bongaerts R, Verpoorte R (1993) Purification and characterization of anthranilate synthase from *Catharanthus roseus*. Eur J Biochem 212:431–440

Rai SS, Wolff J (1996) Localization of the vinblastine-binding site on beta-tubulin. J Biol Chem 271:14707–14711

Rai V, Tandon PK, Khatoon S (2014) Effect of chromium on antioxidant potential of *Catharanthus roseus* varieties and production of their anticancer alkaloids: vincristine and vinblastine. Biomed Res Int 2014:1–10

Ramos-Valdivia AC, van der Heijden R, Verpoorte R (1997) Elicitor-mediated induction of anthraquinone biosynthesis and regulation of isopentenyl diphosphate isomerase and farnesyl diphosphate synthase activities in cell suspension cultures of Cinchona robusta. Planta 203:155–161

Rho D, Andre G (1991) Growth and stoichiometry of a *Catharanthus roseus* cell suspension culture grown under nitrogen-limiting conditions. Biotechnol Bioeng 38:579–587

Rocha JP, Leech MJ (2002) Regulación transcripcional de genes involucrados en la producción de alcaloides indol-terpenoides en plántulas de *Catharanthus roseus*. Rev Colombiana Biotechnol 4:17–26

Roepke J, Salim V, Wu M, Thamm AM, Murata J, Ploss K, Boland W, de Luca V (2010) Vinca drug components accumulate exclusively in leaf exudates of Madagascar periwinkle. Proc Natl Acad Sci U S A 107:15287–15292

Roquebert J, Demichel P (1984) Inhibition of the alpha 1 and alpha 2-adrenoceptor-mediated pressor response in pithed rats by raubasine, tetrahydroalstonine and akuammigine. Eur J Pharmacol 106:203–205

Rowinsky E (2003) The Vinca alkaloids. In: Kufe DW, Pollock RE, Weichselbaum RR, Bast RC, Gansler TS, Holland JF, Frei E (eds) Cancer medicine, 6th edn, vol 2. BC Decker, Hamilton, ON

Rowinsky EK, Eisenhauer EA, Chaudhry V, Arbuck SG, Donehower RC (1993) Clinical toxicities encountered with paclitaxel (Taxol). Semin Oncol 20:1–15

Schiff PB, Fant J, Horwitz SB (1979) Promotion of microtubule assembly *in vitro* by taxol. Nature 277:665–667

Schlatmann JE, Ten Hoopen HJG, Heijnen JJ (1992) Optimization of the medium composition for alkaloids production by *Catharanthus roseus* using statistical experimental designs. Med Fac Landbouw Univ Gent 57:567–1569

Schlatmann JE, Moreno PRH, Vinke JL, Ten Hoopen HJ, Verpoorte R, Heijnen JJ (1994) Effect of oxygen and nutrient limitation on Ajmalicine production and related enzyme activities in high density cultures of *Catharanthus roseus*. Biotechnol Bioeng 44:461–468

Schutz FA, Bellmunt J, Rosenberg JE, Choueiri TK (2011) Vinflunine: Drug safety evaluation of this novel synthetic vinca alkaloid. Expert Opin Drug Saf 10:645–653

Scragg AH, Allan EJ, Leckie F (1988a) Effect of shear on the viability of plant cell suspensions. Enzyme Microb Technol 10:361–367

Scragg AH, Creswell R, Ashton S, York A, Bond P, Fowler MW (1988b) Growth and secondary product formation of a selected *Catharanthus roseus* cell line. Enzyme Microb Technol 10:532–536

Scragg AH, Ashton S, York A, Stepan-Sarkissian G, Grey D (1990) Growth of *Catharanthus roseus* suspension from maximum biomass and alkaloids accumulation. Enzyme Microb Technol 12:292–298

Seibert M, Kadkade PG (1980) Environmental factors. A. Light. In: Staba EJ (ed) Plant tissue culture as a source of biochemicals. CRC, Boca Raton, FL, pp 123–141

Seitz HU, Eilert U, De Luca V, Kurz WGW (1989) Elicitor mediated induction of phenylalanine ammonia lyase and tryptophan decarboxylase: accumulation of phenols and indole alkaloids in cell suspension culture of *Catharanthus roseus*. Plant Cell Tiss Org Cult 18:71–78

Singh BK, Siehl DL, Connelly JA (1991) Shikimate pathway: why does it mean so much for so many? In: Miflin BJ (ed) Oxford surveys of plant molecular biology, vol 7. pp 143–185.

Smith JI, Smart NJ, Kurz WGW, Misawa M (1987a) The use of organic and inorganic compounds to increase the accumulation of indole alkaloids in *Catharanthus roseus* (L). G. Don cell suspension cultures. J Exp Bot 38:1501–1506

Smith JL, Quesnel AA, Smart NJ, Misawa M (1987b) The development of single-stage growth and indole alkaloids production medium for *Catharanthus roseus* (L.) G. Don suspension cultures. Enzyme Microb Technol 9:466–469

Soleimani F, Zarghami R, Ebrahimzadeh M (2013) Effects of 2,4-D and kinetin concentrations on vinblastine and vincristine alkaloid contents in callus of periwinkle (*Catharanthus roseus*). Int J Agric Sci 3:759–765

Sottomayor M, Ros Barceló A (2005) The vinca alkaloids: from biosynthesis and accumulation in plant cells, to uptake, activity and metabolism in animal cells. In: Atta-ur Rahman (ed) Studies in natural products chemistry (bioactive natural products), vol 33. Elsevier Science, Amsterdam, pp 813–857

Sottomayor M, Lopéz-Serrano M, DiCosmo F, Ros Barceló A (1998) Purification and characterization of α-3',4'-anhydrovinblastine synthase (peroxidase-like) from *Catharanthus roseus* (L.) G. Don. FEBS Lett 428:299–303

Stevens LH (1994) Formation and conversion of strictosidine in the biosynthesis of monoterpenoid indole quinoline alkaloids. Ph.D. Thesis, Leiden University (121 p)

Stevens LH, Blom TJM, Verpoorte R (1993) Subcellular localization of tryptophan decarboxylase, strictosidine synthase and strictosidine glucosidase in suspension cultured cells of *Catharanthus roseus* and Tabernaemontana. Plant Cell Rep 12:573–576

Strebhardt K, Ullrich A (2006) Targeting polo-like kinase 1 for cancer therapy. Nat Rev Cancer 6:321–330

Taha HS, El Bahr MK, Seif-El-Nasr MM (2008) In vitro studies on egyptian *Catharanthus roseus* (L.) G. Don.: 1-Calli production, direct shootlets regeneration and alkaloids determination. J Appl Sci Res 4:1017–1022

Taha HS, El-Bahr MK, Seif-El-Nasr MM (2009) In vitro studies on Egyptain *Catharanthus roseus* L. G. Don. IV: manipulation of some amino acids as precursors for enhanced of indole alkaloids production in suspension cultures. Aust J Basic Appl Sci 3:3137–3144

Tallevi SG, DiCosmo F (1988) Stimulation of indole alkaloid content in vanadium- treated *Catharanthus roseus* suspension cultures. Planta Med 54(2):149–152

Tang PA, Siu LL, Chen EX, Hotte SJ, Chia S, Schwarz JK, Pond GR, Johnson C, Colevas AD, Synold TW, Vasist LS, Winquist E (2008) Phase II study of ispinesib in recurrent or metastatic squamous cell carcinoma of the head and neck. Invest New Drugs 26:257–264

Thorpe TA (1995) In vitro embryogenesis in plants. Kluwer, Dordrecht
Toivonen L, Laakso S, Rosenqvist H (1992) The effect of temperature on hairy root cultures of *Catharanthus roseus* growth. Indolealkaloid accumulation and membrane lipid composition of cell suspension cultures. Plant Cell Rep 11:395–399
Tom R, Jardin B, Chavarie C, Rho D, Archambault J (1991) Effect of culture process on alkaloid production by *Catharanthus roseus* cells. II. Immobilized cultures. J Biotechnol 21:21–42
Tuominen U, Toivonen L, Kauppinen V, Markkanen P, Björk L (1989) Studies on the growth and cardenolide production of *Digitalis lanata* tissue culture. Biotechnol Bioeng 33:558–562
Upadhyay RK (2011) Plant natural products: their pharmaceutical potential against disease and drug resistant microbial pathogens. J Pharm Res 4:1179–1185
Van der Heijden R, Verpoorte R, Ten Hoopen HJG (1989) Cell and tissue cultures of *Catharanthus roseus* (L.) G.Don: a literature survey. Plant Cell Tiss Org Cult 18:231–280
Van der Heijden R, Jacobs DI, Snoeijer W, Hallard D, Verpoorte R (2004) The *Catharanthus* alkaloids: pharmacognosy and biotechnology. Curr Med Chem 11:607–628
Van Gulik WM, Ten Hoopen HJG, Heijnen JJ (1993) structured model describing carbon and phosphate limited growth of *Catharanthus roseus* plant cell suspensions in batch and chemostat cultures. Biotechnol Bioeng 41:771–780
Vas A, Gulyas B (2005) Eburnamine derivatives and the brain. Med Res Rev 25:737–757
Vazquez-Flota F, DeCarolis E, Alarco AM, De Luca V (1997) Molecular cloning and characterization of desacetoxyvindoline-4-hydroxylase, a 2-oxoglutarate dependent-dioxygenase involved in the biosynthesis of vindoline in *Catharanthus roseus* (L.) G.Don. Plant Mol Biol 34:935–948
Verpoorte R, Van der Heijden R, Van Gulik WM, Ten Hoopen HJG (1991) Plant biotechnology for the production of alkaloids: present status and prospects. In: Brossi A (ed) The alkaloids, vol 40. Academic Press, San Diego, pp 1–187
Walts LR (2004) The herbal encyclopedia. A practical guide to the many uses of herbs. Universe, New York
Wang L, He HP, Di YT, Zhang Y, Hao XJ (2012) Catharoseumine, a new monoterpenoid indole alkaloid possessing a peroxy bridge from *Catharanthus roseus*. Tetrahedron Lett 53:1576–1578
Wong SK, Lim YY, Chan WCE (2013) Botany, uses, phytochemistry and pharmacology of selected apocynaceae species: a review. Pharmacogn Commun 3:2–11
Yadav PD, Bharadwaj NSP, Yedukondalu M, Methushala CH, Kumar AR (2013) Phytochemical evaluation of *Nyctanthes arbortristis*, *Nerium oleander* and *Catharanthus roseus*. Ind J Res Pharm Biotechnol 1:333–338
Yu F, de Luca V (2013) ATP-binding cassette transporter controls leaf surface secretion of anticancer drug components in *Catharanthus roseus*. Proc Natl Acad Sci U S A 110:15830–15835
Zenk MH, El-Shagi H, Arens H, Stockigt J, Weiler EW, Deus B (1977) Formation of the indole alkaloids serpentine and Ajmalicine in cell suspension culture of *Catharanthus roseus*. In: Barz W, Reinhard E, Zenk MH (eds) Plant tissue culture and its biotechnological application. Springer, Berlin, pp 27–44
Zhao J, Verpoorte R (2007) Manipulating indole alkaloid production by *Catharanthus roseus* cell cultures in bioreactors: from biochemical processing to metabolic engineering. Phytochem Rev 6:435–457

Role of PGRs in Anticancer Alkaloids (Vincristine and Vinblastine) Production

Jagjit Kaur, Apoorva Singh, Teena Pathak, and Kuldeep Kumar

Abstract The synthetic compounds used as medicines serve a number of side effects and therefore, plants are used as an alternative source these days. One of such plants is *Catharanthus roseus*. It produces nearly 130 alkaloids, out of which vincristine and vinblastine are two major anti-cancerous alkaloids. These are structurally similar but differ in their clinical action. In the natural habitat the amount of alkaloids produced is very low. So, the plant is grown in vitro on MS medium supplemented with different plant growth hormones such as auxins, cytokinins, methyl jasmonate (MeJA), ethephon, abscisic acid, salicylic acid (SA), chlormequat chloride (CCC), gibberellica acid (GA_3) and triazoles. Ethephon, SA, CCC, and triazoles positively affect the production of vincristine and vinblastine whereas abscisic acid and GA3 reduces their production. On the other hand, MeJA has no significant effect on their production. Different combinations of auxins and cytokinins have also been studied on the production of vincristine and vinblastine.

Keywords Vincristine • Vinblastine • *C. roseus* • Plant growth regulators

1 Introduction

Plants play an important role in human life by providing nutritional values and prevention of disorders and diseases by providing different value-added products to human healthcare system. Plants have always been a matter of interest to humans from prehistoric time but now the interest in plants has increased considerably at much a higher rate since three decades as their synthetic counterparts cause a lot of side effects (Dorais et al. 2001). The medicinal potential of the plants lies in the chemical substances (alkaloids, phenolic compounds, flavonoids, carbohydrates, tannins, and steroids) produced by them and their definite physiological action on

J. Kaur
Department of Biotechnology, Maharishi Markandeshwar University,
Mullana 133 207, Haryana, India

A. Singh • T. Pathak • K. Kumar (✉)
Department of Biotechnology, Multani Mal Modi College, Patiala 147 001, Punjab, India
e-mail: kuldeepbio@gmail.com

M. Naeem et al. (eds.), *Catharanthus roseus*,
DOI 10.1007/978-3-319-51620-2_12

the human body (Edeoga et al. 2005). Since ages, a large number of plants have been used as medicines and according to WHO (World Health Organization), nearly 80% of the population still relies on herbal medicines. This is because the herbal medicines offer certain advantages such as no side effects, easy availability and higher efficacy over the synthetic drugs. The fact that is needed to be mentioned is that 80% of the drugs used today find their origin from plant sources. The herbs used as plant based remedies serve different functions such as a model for the production of new synthetic compounds, the raw material base for semi-synthetic compounds and source of direct therapeutic agents (Dixon 2001).

Catharanthus roseus (L.) G. Don, also known as sadabhar or periwinkle, is an important medicinal plant known for its anti-cancerous, anti-hypersensitivity, anti-diabetic, anti-microbial and anti-oxidant properties. It is known to produce about 130 terpenoid indole alkaloids (TIAs) (van Der Heijden et al. 2004) and 25 out of these alkaloids are dimeric in nature (van Der Heijden et al. 2004). According to Trease and Evans (1989) secondary metabolites production occurs mainly as a result of defense mechanism during various stress conditions faced by the plants (Trease and Evans 1989). The secondary metabolites are used as fragrances, pesticides, pigments, food additives and drugs (Yadav and Yadav 2016).

The most commonly occurring dimeric alkaloids found in *C. roseus* are vincristine and vinblastine. Vincristine and vinblastine, produced in aerial parts of *C. roseus* (Aslam et al. 2010) are known to be anti-cancerous alkaloids and were used for the production of first natural anti-cancer drug (Costa et al. 2008). Vincristine helps in the treatment of reticulum cell sarcoma, acute leukemia, Wilkin's tumor, lymphoblastic leukemia, reticulum cell sarcoma, Hodgkin's disease and neuroblastoma. It is available under brand name Oncovin, the sulphate salt of vincristine (Aslam et al. 2010). On the other hand vinblastine is used for the treatment of choriocarcinoma, lymphosarcoma, Hodgkin's disease, neuroblastoma and chronic leukemia (Zhu et al. 2015). It is also available as its sulphate salt under brand name Velban. Although the exact anti-cancerous mechanism of both these alkaloids is unknown, it is said that they bind to tubulin and prevent the occurrence of metaphase in mitosis. As a result, the cancer cells do not divide due to the prevention of spindle formation in the cell (Creasey 1979). Both vincristine and vinblastine are similar in structure and mechanism of action but differ considerably in their clinical activity spectrum and toxicity (Owellen et al. 1977). The production of alkaloids can be increased using various elicitors, chemicals (Zhao et al. 2001) and growth regulators in plant tissue culture. Various plant growth regulators such as BA (6-Benzylaminopurine), NAA (1-Naphthalene Acetic Acid), kinetin, and 2,4-D (2,4-Dicholorophenoxyacetic acid) resulted into the induction of callus (Misawa 1994). For the increased production of vincristine in *C. roseus*, Kalidass and Mohan (2009), in their study, used MS (Murashige and Skoog) medium supplemented with kinetin and NAA and found that this combination induced callus development (Kalidass and Mohan 2009). Further, it was found that vincristine production increased when NAA-BA and 2,4-D-kinetin combinations were used (Kalidass et al. 2010).

2 Biosynthetic Pathways for Vincristine and Vinblastine

Vincristine and vinblastine are the two major bisindole alkaloids present in the plant constituting anticancerous properties and came out as the first natural drugs for cancer therapy (Costa et al. 2008). The production of vincristine and vinblastine occurs by the condensation of catharanthine and vindoline, originating from (+)-stemmadenine, a biosynthetic intermediate of terpenoid indole alkaloid (Shukla et al. 2006). The formation of vindoline (present in green parts of plant only) from (+)-stemmadenine via branch-point intermediate tabersonine, is catalyzed by the action of six enzymes (acetyl-CoA: 4-*O*-deacetylvindoline 4-O-acetyl-transferas [DAT], S-adenosyl-L-methionine: 16-methoxyvindoline 4-hydroxylase [NMT], hydroxylase, 16-hydroxy-tabersonine 16-*O*-methyltransferase [OMT], desacetoxyvindoline 4-hydroxylase [D4H], tabersonine 16-hydroxylase [T16H]) (de Luca et al. 1986). During the conditions of wounding, stress and senescence in *C. roseus*, ethylene is biosynthesized (via aminocyclopropane-1-carboxylate oxidase enzyme) which further catalysis the accumulation of vinblastine and vincristine production. Although the biosynthetic pathway for vincristine and vinblastine are not completely understood but the overall pathway can be established in three stages (Rischer et al. 2006; El-Sayed and Verpoorte 2007; Liu et al. 2007) (Fig. 1). First Stage is constituted by the biosynthetic formation of Tryptamine from the shikimate pathway and secologanin from the terpenoid pathway. This is followed by formation of monomeric alkaloids in the second stage. The trytamine and secologanin formed in the first stage combines and forms the strictosidine, which acts as the central precursor in the biosynthetic pathway of vinblastine and vincristine. Almost all terpenoid alkaloids are formed from strictosiddine which is a fusion product of derived tryptamine and secologanin under catalysis of strictosidine synthase (STR) (Treimer and Zenk 1979; Mizukami et al. 1979).

Strictosidine is further converted to monomeric alkaloids like vindoline and catharanthine. In the final stage Vindoline and Catharanthin monomeric alkaloids are coupled to form the bisindole alkaloids Vinblastine and Vincristine. The product resulting from the coupling is α-3′,4′-anhydrovinblastine, catalysed by the enzyme anhydrovinblastine synthase (AVLBS) (Sottomayor et al. 1998). α-3′,4′-anhydrovinblastine, is converted into vinblastine and then further converted into vincristine. However, the enzymes for catalysis of formation of vinblastine and for conversion of vinblastine to vincristine are still unknown and not yet isolated (Zhu et al. 2015). The conversion of vindoline to vinblastine and vincristine was confirmed by the artificial addition of vindoline (0.25 mM) to the production medium along with MeJA. It was found that cambial meristem cells (CMCs) contained complete set of enzymes that are responsible for the production of vinblastine and vincristine from vindoline (Zhang et al. 2015).

3 Effect of Different PGRs

The yields of secondary metabolites are less due to various extrinsic and intrinsic factors that affect the development, growth and secondary metabolite production. The common factors influencing the production of primary and secondary

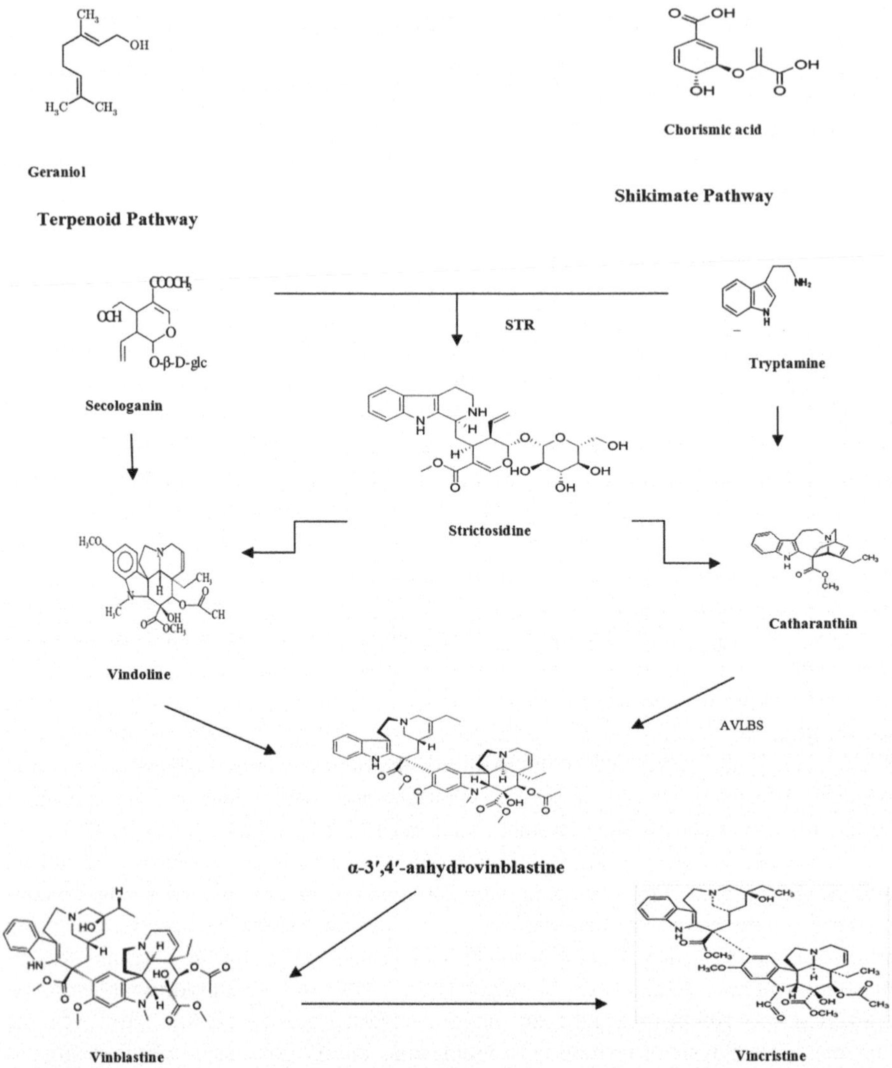

Fig. 1 Biosynthetic pathway for the production of vincristine and vinblastine

metabolites are phytohormones and PGRs (Muthulakshmi and Pandiyarajan 2013). For example, use of IAA (3-Indoleacetic acid) in medium for the growth of *Balanites aegyptiaca* increased the content of tannin, alkaloids and saponin (Mostafa and Abou Alhand 2011). Biotechnological approaches such as supplementing media with different PGRs can be used for the increased production of TIAs (Almagro et al. 2015). Different PGRs effect the production of alkaloids differently as shown in Table 1.

Table 1 Effect of PGRs on vincristine and vinblastine production

PGRs	Alkaloid production enhanced	Alkaloid production decreased
3-Indoleacetic acid (IAA)	Vincristine	
2,4-Dicholorophenoxyacetic acid (2,4-D)	Vincristine and vinblastine	
1-Naphthalene acetic acid (NAA)	Vincristine and vinblastine	
Methyl Jasmonate (MeJA)	No effect	No effect
Ethephon	Vinblastine	
Absciscic acid (ABA)		Vinblastine
Salicylic acid (SA)	Vinblastine	
Chlormequat chloride (CCC)	Vinblastine	
Giberllic acid (GA_3)		Vinblastine
Chromium (Cr)	Vincristine and vinblastine	
Triazole (Triadimefon)	Vincristine and vinblastine	

3.1 Auxins and Cytokinin

The callus growth, proliferation and alkaloid content is enhanced by auxins like 2,4-D. The combination of auxins and cytokinins is known to enhance callus growth and also increased alkaloid production (Verma et al. 2012). It has been reported that various combinations of cytokinins and auxins lead to the callus formation and proliferation (Kaya and Aki 2013). The seeds of *C. roseus* were treated with 150 and 200 ppm of IAA and it was observed that vincristine production was enhanced (Muthulakshmi and Pandiyarajan 2013). The low concentrations of 2,4-D enhanced the callus proliferation and alkaloid content in leaves whereas the callus proliferation was inhibited by the higher concentrations of 2,4-D however enhancement in alkaloid concentration was observed. During the growth phase alkaloid production is inhibited by 2,4-D (Arvy et al. 1994). It inhibits the expression of genes such as TDC gene, DXR (1-deoxy-D-xylulose-5-phosphate reductoisomerase) gene and DXS (1-deoxy-D-xylulose-5-phosphate synthase) gene (Pasquali et al. 1992; Hedhili et al. 2007). When the *C. roseus* was grown on MS medium supplemented with IAA (0.25–6.0 mg/L) callus growth was initiated followed by necrosis of callus. On the other hand BA and kinetin inhibited the growth of callus (Verma et al. 2012).

3.2 Methyl Jasmonate (MeJA)

With the increasing age of the seedling the methyl jasmonate (MeJA) decreased the biosynthesis of alkaloids. This is because MeJA is effective only during a particular stage of growth (seedling) after which it is no more effective (El-Sayed and

Verpoorte 2004). MeJA produces no significant response on the vinblastine production. Vinblastine is produced by condensation of vindoline and catharanthine. MeJA increased the contents of vindoline and catharanthine but did not regulate their condensation reaction. Therefore, there is no effect of MeJA on vinblastine content. The same was proved in another study conducted by Fernandez and his colleagues (Fernández-Pérez et al. 2013). Their study also showed that there was no effect of MeJA on the production of vincristine and vinblastine production (Fernández-Pérez et al. 2013). Addition of MeJA alongwith vindoline to the production media showed the increased conversion of vindoline to vinblastine and vincristine production (Zhang et al. 2015).

3.3 Ethephon

The introduction of ethephon in the medium during the growth of *C. roseus* increases the alkaloid biosynthesis. After treatment of medium with ethephon (0.1 mM) catharanthine and vindoline biosynthesis increased during the first 24–72 h. The application of ethephon had a lasting and rapid response on the monomeric alkaloids such as ajmalicine (Yahia et al. 1998), vindoline, tabersonine, serpentine and catharanthine (El-Sayed and Verpoorte 2004). Vinblastine biosynthesis is also increased after 72 h of application of ethephon followed by senescence of leaves of *C. roseus* (Naaranlahti et al. 1991). It was clear from the fact that anhydrovinblastine (direct precursor of vinblastine) was found in large amount in old leaves then young leaves (El-Sayed and Verpoorte 2005). Ethephon is converted to ethylene which results into the leaf senescence further triggering the synthesis of vinblastine (Wingler and Roitsch 2008).

3.4 Abscisic Acid (ABA)

Abscissic acid (ABA) reduces the accumulation of vinblastine, catharanthine and vindoline. When ABA is sprayed on the growing plants of *C. roseus* it resulted into 54% reduced accumulation of vinblastine at first 24 h but it came back to normal after 72 h (Pan et al. 2010).

3.5 Salicylic Acid (SA)

Salicylic acid (SA) influences the accumulation of alkaloids in *C. roseus* when the plants are treated with salicylic acid and in addition it serves an important component in the defense system of plants (Godoy-Hernandez and Loyola-Vargas 1997; El-Sayed and Verpoorte 2004; Dutta et al. 2007). In an experiment 0.1 mM SA was sprayed on the plantlets and significant increase in accumulation of catharanthine, vinblastine and vindoline was observed. The concentration of catharanthine was increased upto 42% after 24 h and the concentration of vinblastine increased upto 22% after 24 h. On the other hand vindoline concentration increased upto 22% after 24 h (Pan et al. 2010).

According to Idrees et al. (2011), foliar application of SA (10^{-5} M) reduced the damaging effect of salinity. Foliar spray of SA overcame the adverse effect of salinity by improving the content of vincristine (14.0%) and vinblastine (14.6%) in plants treated with 100 M NaCl.

3.6 Chlormequat Chloride (CCC)

Chlormequat chloride (CCC) showed a variable response on the alkaloid accumulation i.e. it enhanced the accumulation of vinblastine whereas slowly reduced the accumulation of vindoline and catharanthine. At lower concentrations of CCC (0.01 and 0.1 mM) the accumulation of vindoline and catharanthine was reduced after 48 h of incubation but had no effect on vinblastine accumulation. At higher concentration i.e. 1 mM there was no effect on the vindoline and catharanthine accumulation but the vinblastine accumulation was enhanced. It is surprising that CCC enhances the accumulation of vinblastine but reduces the accumulation of its precursors (vindoline and catharanthine). It is due to the fact that vindoline and catharanthine are synthesized in different parts of the plant and then transferred to vacuoles where they carry out the synthesis of vinblastine (Roytrakul and Verpoorte 2007). CCC does not significantly regulate the biosynthetic pathways involved in the biosynthesis of vindoline and catharanthine but regulates the pathway involved in biosynthesis of vinblastine from vindoline and catharanthine.

3.7 Gibberellic Acid (GA_3)

GA_3 has shown a negative impact on the accumulation of alkaloids in *C. roseus* plant. It was observed that when the plants were treated with 0.1 mM GA_3 the vinblastine concentration decreased to 15%, vindoline to 16% and catharanthine to 28% after 24 h (Pan et al. 2010).

Although, GA_3 treatment produced negative phenotypic response in total biomass production but positive response in content of total alkaloids in leaf, stem, and roots (Srivastava and Srivastava 2007). Misra et al. (2009) also reported that GA_3 has increased accumulation of total alkaloids in *C. roseus* plant. In another study, foliar application of GA_3 significantly increased the vincristine content (7.3%) in 'Rosea' cultivar of *C. roseus* (Alam et al. 2012).

3.8 *Chromium (Cr)*

Tissue and cell culturing has been used as a tool to enhance the production of alkaloids in plants using different growth regulators and nutrients. Yet very little work has been done to study the effect of heavy metals on alkaloid production. Chromium (Cr) is one of the commonly occurring heavy metals used in refractory steel production, leather industry and specialty chemical production. Due to excessive use of Cr in various industries its biomobility and bioavailability has increased greatly. Leather industry produces nearly 40% Cr of total industrial Cr (Barnhart 1997) and nearly 20,000–32,000 tons of Cr are released into the environment annually by tanning industries in India (Chandra et al. 1997). When the medium was supplemented with various concentrations of Cr it was found that the production of both vincristine and vinblastine was enhanced. It was observed that with 50 μM Cr nearly 69.42% increase in vincristine production was found, whereas vinblastine production was enhanced only by 2.29-fold (Rai et al. 2014).

3.9 *Triazoles*

Triazoles have the ability to alter the concentration of various PGRs such as cytokinins, ABA and gibberellic acid (Fletcher and Hofstra 1985). It also enhances root growth, inhibits shoot elongation and protects plants from environmental stresses (Davis et al. 1988; Fletcher and Hofstra 1988; Izumi et al. 1988). The protection from environmental stresses is due to the fact that triazoles increases the antioxidant potential and reduces the damage caused by free radicals (Fletcher and Hofstra 1988). Triazoles caused abiotic stress in the plants which leads to accumulation of alkaloids mainly in roots (Karadge and Gaikwad 2003). One example of triazole is triadimefon, it posses plant growth regulating properties alongwith fungicidal properties (Fletcher and Hofstra 1985, 1988). The application of triadimefon to the plants leads to the increased alkaloid concentration (Jaleel et al. 2006).

4 Conclusion

C. roseus is an important medicinal plant used for the treatment of various diseases. It posses different properties like anti-oxidant, anti-cancerous, anti-microbial and anti-hypersensitivity. PGRs effect differently on the production of vincristine and vinblastine i.e. some of the PGRs increases their production while others decrease it. Thus future holds a great deal in the field for the increased production of these alkaloids using different PGRs and also it can be elucidated if various metals have any regulatory role to play in increasing the production of these alkaloids.

References

Alam MM, Naeem M, Idrees M, Khan MMA, Moinuddin (2012) Augmentation of photosynthesis, crop productivity, enzyme activities and alkaloids production in *Catharanthus roseus* L. through application of diverse PGRs. J Crop Sci Biotechnol 15:53–57

Almagro L, Fernández-Pérez F, Pedreño MA (2015) Indole alkaloids from *Catharanthus roseus*: bioproduction and their effect on human health. Molecules 20:2973–3000

Arvy MP, Imbault N, Naudascher F, Thiersault M, Doireau P (1994) 2,4-D and alkaloid accumulation in periwinkle cell suspensions. Biochimie 76:410–416

Aslam J, Khan SH, Siddiqui ZH et al (2010) *Catharanthus roseus* (L.) G. Don. An important drug: it's applications and production. Pharm Glob 4(12):1–16

Barnhart J (1997) Occurrences, uses, and properties of chromium. Regul Toxicol Pharmacol 26(1):S3–S7

Chandra P, Sinha S, Rai UN (1997) Bioremediation of Cr from water and soil by vascular aquatic plants. In: Kruger EL, Anderson TA, Coats JR (eds) Phytoremediation of soil and water contaminants. ACS Symposium Series, vol 664. American Chemical Society, Washington, DC, pp 274–2782

Costa MM, Hilliou F, Duarte P, Pereira LG, Almeida I, Leech M, Memelink J, Barcelo AR, Sottomayor M (2008) Molecular cloning and characterization of a vacuolar class III peroxidase involved in the metabolism of anticancer alkaloids in *Catharanthus roseus*. Plant Physiol 146:403–417

Creasey WA (1979) The vinca alkaloids. In: Hahn FE (ed) Antibiotics, vol 5. Springer, New York, pp 414–438

Davis TD, Steffens GL, Sankhla N (1988) Triazole plant growth regulators. Hort Rev 10:63–105

de Luca D, Balsevich J, Tyler RT, Eilert U, Panchuk BD, Kurz WGW (1986) Biosynthesis of indole alkaloids: developmental regulation of the biosynthetic pathway from tabersonine to vindoline in *Catharanthus roseus*. J Plant Physiol 125:147–156

Dixon RA (2001) Natural products and plant disease resistance. Nature 411:843–847

Dorais M, Papadopoulos AP, Luo X, Leonhart S, Gosselin A, Pedneault K, Angers P, Gaudreau L (2001) Soilless greenhouse production of medicinal plants in north eastern Canada. Acta Hort 554:297–304

Dutta A, Singh D, Kumar S, Sen J (2007) Transcript profiling of terpenoid indole alkaloid pathway genes and regulators reveals strong expression of repressors in *Catharanthus roseus* cell cultures. Plant Cell Rep 26:907–915

Edeoga HO, Okwu DE, Mbabie BO (2005) Phytochemical constituents of some Nigerian medicinal plants. Afr J Biotechnol 4:685–688

El-Sayed M, Choi YH, Fréderichrich M, Roytrakul S, Verpoorte R (2004) Alkaloid accumulation in *Catharanthus roseus* cell suspension cultures fed with stemmadenine. Biotechnol Lett 26:793–798

El-Sayed M, Verpoorte R (2005) Methyljasmonate accelerates catabolism of monoterpenoid indole alkaloids in *Catharanthus roseus* during leaf processing Fitoterapia 76, 83–90

El-Sayed M, Verpoorte R (2007) Catharanthus terpenoid indole alkaloids: biosynthesis and regulation. Phytochem Rev 6:277–305

Fernández-Pérez F, Almagro L, Pedreño MA, Gómez Ros LV (2013) Synergistic and cytotoxic action of indole alkaloids produced from elicited cell cultures of *Catharanthus roseus*. Pharm Biol 51(3):304–310. doi:10.3109/13880209.2012.722646

Fletcher RA, Hofstra G (1985) Triadimefon, a plant multi-protectant. Plant Cell Physiol 26:775–780

Fletcher RA, Hofstra G (1988) Triazole as potential plant protectants. In: Berg D, Plempel M (eds) Sterol biosynthesis inhibitors. Ellis Horwood, Cambridge, England, pp 321–331

Godoy-Hernández GC, Loyola-Vargas VM (1997) Effect of acetylsalicylic acid on secondary metabolism of *Catharanthus roseus* tumor suspension cultures. Plant Cell Rep 16:287–290

Hedhili S, Courdavault V, Giglioli-Guivarc'h N, Gantet P (2007) Regulation of the terpene moiety biosynthesis of *Catharanthus roseus* terpene indole alkaloids. Phytochem Rev 6:341–351
Idrees M, Naeem M, Aftab T, Khan MMA, Moinuddin (2011) Salicylic acid mitigates stress by improving antioxidant defence system and enhances vincristine and vinblastine alkaloids production in periwinkle [*Catharanthus roseus* (L.) G. Don]. Acta Physiol Planta 33(3):987–999
Izumi K, Nakagawa S, Kobayashi M, Oshio H, Sakurai A, Takahashi N (1988) Levels of IAA, cytokinins, ABA and ethylene in rice plants as affected by a gibberellin biosynthesis inhibitor, uniconazole-P. Plant Cell Physiol 29:97–104
Jaleel CA, Gopi R, Lakshmanan GMA, Panneerselvam R (2006) Triadimefon induced changes in the antioxidant metabolism and ajmalicine production in *Catharanthus roseus* (L.) G. Don. Plant Sci 171(2):271–276
Kalidass C, Mohan VR (2009) *In vitro* rapid clonal propagation of *Phyllanthus urinaria* Linn. (Euphorbiaceae)—a medicinal plant. Researcher 1(4):56–61
Kalidass C, Ramasamy MV, Arjunan D (2010) Effect of auxin and cytokinin on vincristine production by callus cultures of *Catharanthus roseus* L. (Apocynaceae). Trop Subtrop Agroecosys 12:283–288
Karadge BA, Gaikwad PV (2003) Influence of NaCl salinity on growth and organic constituents of *Catharanthus roseus* G. Don. Ind J Plant Physiol 8:392–397
Kaya N, Aki C (2013) Effect of plant growth regulators on in vitro biomass changing in *Catharanthus roseus* (L) G Don. Ann Biol Res 4(4):164–168
Liu D, Jin H, Chen Y, Cui L, Ren W, Gong Y, Tang K (2007) Terpenoid indole alkaloids biosynthesis and metabolic engineering in *Catharanthus roseus*. J Integr Plant Biol 49:961–974
Misawa M (1994) Plant tissue culture: an alternative for production of useful metabolites (FAO Agricultural Services Bulletin). Bio International, Toronto, pp 18–19
Misra A, Srivastava NK, Srivastava AK, Khan A (2009) Influence of etherel and gibberellic acid on carbon metabolism, growth, and alkaloids accumulation in *Catharanthus roseus* L. Afr J Pharm Pharmacol 3:515–520
Mizukami H, Nordlöv H, Lee SL, Scott AI (1979) Purification and properties of strictosidine synthetase (an enzyme condensing tryptamine and secologanin) from *Catharanthus roseus* cultured cells. Biochemistry 18:3760–3763
Mostafa GG, Abou Alhand MF (2011) Effect of gibberellic acid and Indole 3-acetic acid on improving growth and accumulation of phytochemical composition in *Balanites aegyptiaca* plants. Am J Plant Physiol 6:36–43
Muthulakshmi S, Pandiyarajan V (2013) Influence of IAA on the vincristine content of *Catharanthus roseus* (L). G. Don. Asian J Plant Sci Res 3(4):81–87
Naaranlahti T, Auriola S, Lapinjoki SP (1991) Growth related dimerization of vindoline and catharanthine in *Catharanthus roseus* and effect of wounding on the process. Phytochemistry 30:1451–1453
Owellen RJ, Donigian DW, Hartke CA, Hains FO (1977) Correlation of biologic data with physicochemical properties among the vinca alkaloids and their congeners. Biochem Pharmacol 26:1213–1219
Pan Q, Chen Y, Wang Q, Yuan F, Xing S, Tian Y, Zhao J, Sun X, Tang K (2010) Effect of plant growth regulators on the biosynthesis of vinblastine, vindoline and catharanthine in *Catharanthus roseus*. Plant Growth Regul 60:133–141
Pasquali G, Goddijn OJM, de Waal A, Verpoorte R, Schilperoort RA, Hoge JH, Memelink J (1992) Coordinated regulation of two indole alkaloid biosynthetic genes from *Catharanthus roseus* by auxin and elicitors. Plant Mol Biol 18:1121–1131
Rai V, Tandon PK, Khatoon S (2014) Effect of chromium on antioxidant potential of *Catharanthus roseus* varieties and production of their anticancer alkaloids: vincristine and vinblastine. Biomed Res Int 2014: 934182. doi:10.1155/2014/934182
Rischer H, Oresic M, Seppänen-Laakso T, Katajamaa M, Lammertyn F, Ardiles-Diaz W et al (2006) Gene-to-metabolite networks for terpenoid indole alkaloid biosynthesis in *Catharanthus roseus* cells. Proc Natl Acad Sci U S A 103:5614–5619

Roytrakul S, Verpoorte R (2007) Role of vacuolar transporter proteins in plant secondary metabolism: *Catharanthus roseus* cell culture. Phytochem Rev 6:383–396

Shukla AK, Shasany AK, Gupta MM, Khanuja SPS (2006) Transcriptome analysis in *Catharanthus roseus* leaves and roots for comparative terpenoid indole alkaloid profiles. J Exp Bot 57(14):3921–3932

Sottomayor M, Lopez-Serrano M, DiCosmo F, Ros Barceló A (1998) Purification and characterization of alpha-3′,4′-anhydrovinblastine synthase (peroxidase-like) from *Catharanthus roseus* (L.) G. Don. FEBS Lett 428:299–303

Srivastava NK, Srivastava AK (2007) Influence of gibberellic acid on $^{14}CO_2$ metabolism, growth, and production of alkaloids in *Catharanthus roseus*. Photosynthetica 45:156–160

Trease GE, Evans GE (1989) Text book of pharmacognosy, 2nd edn. Bailliera Tindall, London

Treimer JF, Zenk MH (1979) Purification and properties of strictosidine synthase, the key enzyme in indole alkaloid formation. Eur J Biochem 101:225–233

van Der Heijden R, Jacobs DI, Snoeijer W, Hallard D, Verpoorte R (2004) The Catharanthus alkaloids: pharmacognosy and biotechnology. Curr Med Chem 11:607–628

Verma AK, Singh RR, Singh S (2012) Improved alkaloid content in callus culture of *Catharanthus roseus*. Bot Serbica 36(2):123–130

Wingler A, Roitsch T (2008) Metabolic regulation of leaf senescence: interactions of sugar signaling with biotic and abiotic stress responses. Plant Biol Suppl 1:50–62

Yadav R, Yadav N (2016) Secondary metabolites production in plant tissue culture: a review. Int Educ Sci Res J 2:63–64

Yahia A, Kevers C, Gaspar T, Chénieux JC, Rideau M, Crèche J (1998) Cytokinins and ethylene stimulate indole alkaloid accumulation in cell suspensions cultures of *Catharanthus roseus* by two distinct mechanisms. Plant Sci 133:9–15

Zhang W, Yang J, Zi J, Zhu J, Song L, Yu R (2015) Effects of adding vindoline and MeJA on production of vincristine and vinblastine, and transcription of their biosynthetic genes in the cultured CMCs of *Catharanthus roseus*. Nat Prod Commun 10:2095–2096

Zhao J, Zhu WH, Hu Q, He XW (2001) Enhanced indole alkaloid production in suspension compact callus clusters of *Catharanthus roseus:* impacts of plant growth regulators and sucrose. Plant Growth Regul 33(1):33–41

Zhu J, Wang M, Wen W, Yu R (2015) Biosynthesis and regulation of terpenoid indole alkaloids in *Catharanthus roseus*. Pharmacogn Rev 9:24–28

The Accumulation and Degradation of Alkaloids in *Catharanthus roseus* Supported by Various External Agents Under Different Environmental Conditions

Mohd. Idrees, Israr Ul Hassan, M. Naeem, Akbar Ali, Tariq Aftab, and M. Masroor A. Khan

Abstract *Catharanthus roseus* (L.) is a member of the family Apocynaceae, a rich source of indole alkaloids, used in several disorders like hypertension, diabetes, asthma, constipation, cancer, and menstrual problems. There are three common cultivars of *C. roseus*, namely, "Rosea," "Alba" and "Ocellata." four chief alkaloids in clinical use are vinblastine, vinorelbine, vincristine, and vindesine. Vincristine and vinblastine have widespread use in modern medicine as latent anticancer compounds. The physiologically important and anticancerous alkaloids, vinblastine and vincristine, are mostly present in leaves and antihypertensive alkaloids which originate in roots such as ajmalicine, serpentine, and reserpine. Accretion of alkaloids in *C. roseus* happens in response to various external and/or internal factors including elicitors or signal molecules. Secondary metabolites help plants in overcoming stress conditions by altering physiological processes. Environmental factors such as heat, moisture, light concentration, the source of water, mineral deposits, and CO_2 impact the growth of *C. roseus* and accumulation of metabolites (Marschner 1995). Dearth, high salinity, and very low temperature are environmental conditions that may have adverse effects on the growth and productivity of *C. roseus*. This chapter reviews the impact of various abiotic issues including dearth, salt, light, heavy metals, frost, heavy metels etc. on alkaloid accumulation in *C. roseus*. The main focus of the present review is on enhancement and degradation of alkaloids in *C. roseus* under different environmental conditions.

Keywords *C. roseus* • Alkaloids • Stress • Secondary metabolites

---

M. Idrees (✉) • I.U. Hassan
Department of Mathematics and Sciences, College of Arts and Applied Sciences, Dhofar University, PO Box 2509, Salalah 211, Sultanate of Oman
e-mail: midress@du.edu.om

M. Naeem • A. Ali • T. Aftab • M.M.A. Khan
Plant Physiology Section, Department of Botany, Aligarh Muslim University, Aligarh, Uttar Pradesh 202 002, India

M. Naeem et al. (eds.), *Catharanthus roseus*,
DOI 10.1007/978-3-319-51620-2_13

1 Introduction

Several internal elicitors and signal molecules caused by different types of stresses are responsible for gathering of secondary metabolites in plants. These metabolites have higher importance in medical and cosmetic industries (Chomel et al. 2016). The synthesis of these compounds is often low (less than 1% dry weight) and depends totally on the physiological and developmental stage of the plant (Rao and Ravishankar 2002). Plants accumulate several different types of natural products including drugs such as terpenes, alkaloids, flavonoids, flavones, ionones, steroids, phenolics etc. Various environmental factors including high and low temperature, drought, alkalinity, salinity, UV stress, and pathogen infection are potential threats to plant survival and development (Seigler 1998; Mohamed et al. 2007). Such stresses induced elicitors or signal molecules are responsible for the production and collection of secondary metabolites (Jaleel et al. 2007a, b). Stimulation has been broadly applied to proliferate the yield or to encourage de novo production of both primar and secondary metabolites in in vitro plant cell cultures (Dicosmo and Misawa 1985). Elicitors are used by various plant researcher to raise the production of both types of metabolites in different part of plants (Sudha and Ravishankar 2003; Karuppusamy 2009).

Environmental factors, such as attacks of various germs UV-irradiation, high light, acerbic, nutrient scarcities, malaise, and herbicide treatment increase the accretion of phenylpropanoids (Dixon and Paiva 1995). The nutrient content of soil has a remarkable influence on the accumulation of phenolic levels in plant tissues (Chalker-Scott and Fnchigami 1989). The accumulation of primary and secondary yields of plants is correlated to environmental and edaphic factors. The environmental conditions have impression on the metabolic pathways of natural products. *Catharanthus roseus* (Family: Apocynaceae) or previously known as *Vinca rosea*, Madagascar periwinkle is an herbaceous plant which probably originated from Madagascar. Periwinkle is a dicot, blossoming plant. It has a pentamerous flower. There are three cultivars of *C. roseus* L., which are different to each others on the basis of their flower colors (Idrees 2007), and two of them, namely, "rosea" (Pink) and "alba" (White), commonly flourish in northern part of India. *Catharanthus roseus* L. is a significant plant due to the presence of different types of indole alkaloids. The group of Catharanthus or Vinca alkaloids contains 130 terpenoid indole alkaloids. Alkaloids obtained from various parts of the plant are useful to treat various disorders like diabetes, malaise, malaria, throat infections, and chest complaints, menstrual cycles regulation, and as a euphoriant (van Der Heijden et al. 2004). There are several reports indicated that vincristine and vinblastine showed antineoplastic effect on different types of lymphoma and leukemia (Bates et al. 2013; Kang et al. 2007). These *Catharanthus* alkaloids are also applied to treat both malignant and nonmalignant cancers and in platelet and platelet-associated issues (Almagro et al. 2015). In the year 1985, scientists tested the extract of *C. roseus* against blood sugar level of rabbit and found that there was no effect on blood sugar but they observed a significant reduction in white blood cells. On the basis of this particular finding, several clinical trials were conducted to establish the fact that the extract of leaves extract of *C. roseus* proves useful in Hodgkin's disease (Panda 2005). Ajmalicine is a key monoterpene indole alkaloid found in this plant, the production of which has been used as a model system for improvement in the bio-process of several plant metabolites (Zhang et al. 2002). In this review article, we have reviewed literature regarding the influence of various environmental factors on plant secondary metabolites degradation as well as the production in *C. roseus*.

2 Effect of PGRs on *C. roseus*

Jaleel et al. (2008) explained effect of different plant growth regulators (paclobutrazol and gibberellic acid) and fungicide (*Pseudomonas fluorescens*) treatments on the growth characteristics of *C. roseus*. They observed that the total plant height increased with the age in the control, gibberellic acid (GA), and *P. fluorescens*-treated *Catharanthus roseus* plants, it reduced considerably under PBZ treatments. Further, Jaleel et al. (2011) reported that plant growth hormone like paclobutrazol and gibberellic acid affected various enzyme activities of *C. roseus* which are directly or indirectly related to alkaloid production. Another report published by Misra et al. (2014) suggested that salicylic acid induced growth, is directly linked with improvement of parameters like dry weight, moisture content, photosynthetic pigments, and soluble proteins. Salicylic acid has a promotive effect on phenylalanine ammonia-lyase (PAL) activity, which is ultimately culminated in increment of total soluble phenolics and lignin contents in both leaves and roots of *C. roseus*. Chen et al. (2013) reported that abscisic acid and nitric oxide enhanced accumulation of catharanthine in *C. roseus* treated with PB90, a protein promoter from phytophthora boehmeriae. A separate study, conducted by Xu et al. (2004) indicated that sodium nitroprusside at high concentration (10.0 and 20.0 mmol/L) enhanced catharanthine synthesis of *C. roseus* cells, but reduced cell growth. However, low doses of sodium nitoprusside (0.1 and 0.5 mmol/L) promoted the growth of *C. roseus* cells without any effect on accumulation of catharanthine. Foliar application of 300 μM melatonin to *C. roseus* leaves increased production of silver nanoparticles (AgNPs) (Sheshadri et al. 2015). They noticed that LC-MS/MS of leaf extract of *C. roseus* treated with melatonin indicated the presence of significant biomolecules such as 6-acetyl morphine (used as an analgesic), rauwolscine, and fisetin (anticancer compounds). Melatonin treated leaves improved antioxidant activity and total chlorophyll content. Thus, melatonin assists *C. roseus* to improve growth and therapeutic activities.

3 Effect of Salt Stress on *C. roseus*

Amirjani (2015) conducted an experiment to find out the effect of salt stress on seed germination of *C. roseus* and observed that increasing the NaCl concentrations of salt reduced germination percentage. Fresh and dry weights of treated plants showed a decrease with an increment of ascorbic acid content when treated by salt. Osman et al. (2007) investigated the response of *C. roseus* to salinity and drought for a period of 4 months. They recorded correlation between cellular proline level and capacity to survive to various environmental stresses. They also noticed that *C. roseus* accumulate more amino acid serine, methionine, and arginine in response to stressed condition. Jaleel et al. (2008a) conducted an experiment to study the effect of salinity on secondary metabolites accumulation in *C. roseus* and observed that amount of indole alkaloid contents was different with different level of soil salinity; higher amount of alkaloid contents was obtained from NaCl-treated plants than that of control. In another study, Jaleel et al. (2008b) observed that 80 mM NaCl reduced overall growth by decreasing the photosynthetic pigments protein, activities of antioxidant enzymes such as peroxidase (POX), superoxide dismutase (SOD), and polyphenol oxidase (PPO). The root alkaloid ajmalicine imroved by mild salt treatment. Further, they tested triadimefon as an amelioration agent to reduce the salt stress effect on the overall performance of *C. roseus*. They observed that 15 mg/L TDM reduced the inhibitory effects of NaCl stress by improving the growth of root, shoot, leaf area, chlorophyll, protein contents, and activities of antioxidant enzymes. The ajmalicine was increased in TDM-treated plants in comparison of both control and NaCl-treated plants. Results obtained by Jaleel et al. (2008b), indicated that the fungicide TDM have the capacity to reduce injurious free radicals, as a device of shielding plants against harmful oxidative stress from its surroundings and also improve the active components. Misra et al. (2014) in his study reported that salicylic acid showed growth stimulating property, which interrelated with the proliferation of dry weight, water content, chlorophyll and carotenoids contents, and soluble proteins. Salicylic acid has a preservative effect on the substantial rise in phenylalanine ammonia-lyase (PAL) activity, which is tracked by proliferation soluble phenolics and lignin contents in leaves and root of *C. roseus*. Salicylic acid (SA) also improves malondialdehyde content in leaves and root. The antioxidant enzymes such as catalase, glutathione reductase, glutathione-S-tranferase, superoxide dismutase, peroxidase as well as alkaloid yield amplified in all treatments. They concluded that application of SA can be used to treat salt stress, further, it enhanced the level of antioxidant and phenolics, and alkaloids in salt stress plant.

Idrees et al. (2011) used salicylic acid to ameliorate the adverse effects of salinity stress on periwinkle. They tested various doses of salicylic acid on this medicinal crop under different levels of salt stress. The particular dose of SA (10^{-5} M) reduced damage in growth caused by salinity. It not only enhanced the value of parameters related to growth but also reversed the effects of salinity. SA increased total alkaloid content under both normal and stress condition. Foliar spray of SA disabled the adverse effect of salinity and increased contents of vincristine (14.0%) and vinblastine (14.6%).

4 Effect of Drought Stress on *C. roseus*

Kim and van Iersel (2011) suggested that slowly developing drought stress increases photosynthetic acclimation of *C. roseus.* They postulated that plants act very differently to drought stress, depending on the rate at which it is imposed. Zhang et al. (2012) study the effect of water stress and nitrogen nutrition on alkaloids metabolism of *C. roseus*. They concluded that drought stress or supplementary nitrogen has influence on enzyme activities, decarboxylase and alkaloids production. The sufficient supply of nitrogen in low drought conditions favors the accumulation of important catharanthus alkaloids. Amirjani (2013) reported that various changes in physiological parameters like photosynthetic pigments, chlorophyll contents, rate of photosynthesis and transpiration, growth parameters, total alkaloid, vincristine and vinblastine in *C. roseus* grown under drought. He observed that drought has adverse effect on height, weight, and relative water content of *C. roseus*. A significant reduction in photosynthetic activity and transpiration rate was observed with increasing drought level. However, a significant increment was noted in data obtained for total alkaloid, vincristine and vinblastine content of *C. roseus* under drought.

5 Influence of Heavy Metal Stress on *C. roseus*

In a study, lanthanum nitrate was added to cell cultures of *Catharanthus roseus*, leading to a dose-dependent accumulation of alkaloids. Lanthanum nitrate promoted accumulation of alkaloids at low concentration, but it had an inhibitory effect at a higher concentration (Yuan and Hu 1993). Another study indicated that adding Ce_2O_3 and $CeCl_3$ to the medium of *C. roseus* enhanced the production of raubasine and adding Y_2O_3 and $NdCl_3$ increased the production of catharanthine (Asadian 2016). Ponarulselvam et al. (2012) tried to develop unique method for the green synthesis of silver nanoparticles by means of *C. roseus* leaves extracts. The leaves of *C. roseus* may be excellent tool for natural synthesis of silver nanoparticle which exhibits antiplasmodial activity against *P. falciparum*. The findings were helpful for the synthesis of biomedical and nanotechnology-based products (Ponarulselvam et al. 2012). Pandey et al. (2007) supplied concentrations of heavy metals like $CdCl_2$ and $PbCl_2$ to observe their bioaccumulation proficiency. Morphological changes like senescence of lower leaves and extensive chlorosis occurred after 4 days of transfer. In the case of $CdCl_2$ treatment, little effect was noticed on growth after 6 months, however, plants showed normal development and flowering in case of $PbCl_2$ treatment. Total alkaloids were found to be reduced in the roots of $CdCl_2$-treated plants. GA_3 to the $CdCl_2$-treated plants showed an increment in inter-nodal length and leaf area. AAS analyses of leaves of treated *C. roseus* showed 5–10% cadmium (Cd), but no traces of lead at all. Cd augmented the concentration of ajmalicine in culture medium by affecting tryptophan decarboxylase (TDC). Cd treatment enhanced

TDC transcript, the cellular tryptamine, and ajmalicine excretion (Zheng and Wu 2004). Feng et al. (2016) used perlite to study the effects of copper stress on the responses of *C. roseus* seedling growth, gene expression, and alkaloid production. The dry weights of root and leaf of *C. roseus* under exogenous Cu stress levels were increased under low concentration while decreased under high concentration.

6 Effect of Nutrient Stress on *C. roseus*

Nutrient deficiencies of nitrogen, phosphorus potassium, calcium, megnesium, silicon and boron may be cause of decrease in *Catharanthus* root's dry matter in deficient treatments. A decrease in soil potassium might be responsible for ajmalicine increment in roots of *C. roseus*. Nutrient deficiencies in N, P, Mg, and S reduced ajmalicine concentration in 55, 33, 22, and 26%, respectively, than that of complete treatment. Deficiencies in calcium and boron had no significant effect in ajmalicine concentration within the plant roots. Stafford and Fowler (1983) studied the effect of carbon and nitrogen on the growth and nutrient metabolism in cultures of *C. roseus*. They observed the influence of cultured cells under different nutritional level appeared in all aspects of growth; however few enzymes were unaffected. Cell viability continued at high level for many days after growth in both cultures. The possibility that protein degradation in nitrogen-limited batch cultures is under very stringent control is discussed. Meng-Yan et al. (2016) observed that three main alkaloids in plants decreased with reduced levels of nitrogen. The level of exogenous nitrate on vindoline and catharanthine in *C. roseus* was with more significant accumulation in the leaves at different leaf positions, but with little effect on the accumulation of vinblastine. In high nitrogen condition, plant synthesis and accumulation were at higher levels of free amino acids, which may provide the skeleton and promote the improvement of alkaloids in the partial synthesis. Gholamhosseinpour et al. (2011) significant variations among different levels of nitrogen in all measured parameters except the fresh and dry weights of shoot. Their results specified that the growth parameters (fresh and dry weights) of *C. roseus* improved with increasing the level of fertilization from 0 to 100 kg N ha^{-1} and then reduced at 150 kg N ha^{-1}. The concentration of vinblastine and vincristine alkaloids were measured by HPLC and TLC assays. The highest content and yield of vinblastine was detected in 150 kg N ha^{-1} while the lowest content and yield of vincristine was noticed in 150 kg N ha^{-1} using both approaches (HPLC and TLC) of analysis. The results showed that the content and yield of vinblastine and vincristine quantified with HPLC assay was higher than the samples determined by TLC method. The results delivered significant evidence of nitrogen effect on yield and alkaloid content that can be used to commercial production of periwinkle. Shimano and Ashihara (2006) observed that levels of adenine and guanine nucleotides, particularly ATP and GTP, were evidently low during Pi-starvation. There was an increment in the activity of RNase, DNase, 50- and 30-nucleotidases and acid phosphatase, which might be participated in the hydrolysis of nucleic acids and nucleotides. Accumulation of adenosine, adenine, guanosine, and guanine was observed during the long-term Pi-starvation. The activities of adenosine kinase, adenine phosphoribosyl transferase, and adenosine nucleosidase were maintained at a high level in long-term Pi-starved cells.

7 Conclusion

It is apparent that various external agents and environmental conditions affect growth and secondary metabolites production in *C. roseus*. Most importantly, wheather fluctuation, climate change, water availability and scarcity, salt and drought stress including several adverse soil conditions directly or indirectly influence alkaloids yield. However, *C. roseus* is a hardy nature plant that can tolerate unfavorable conditions for a short period. *Catharanthus roseus* exhibits morphological and molecular adaptations to different nutrient deficiency. The use of newly developed genetic tools with the structure and regulation of pathways for secondary metabolism serves as a basis for commercial production of alkaloid in *C. roseus*.

References

Almagro L, Fernández-Pérez F, Pedreño MA (2015) Indole alkaloids from *Catharanthus roseus*: bioproduction and their effect on human health. Molecules 20:2973–3000

Amirjani MR (2013) Effects of drought stress on the alkaloid contents and growth parameters of *Catharanthus roseus*. ARPN J Agric Biol Sci 8:10628–10632

Amirjani MR (2015) Effect of salinity stress on seed germination and antioxidative defense system of *Catharanthus roseus*. ARPN J Agric Biol Sci 10:163–171

Asadian AH (2016) A study on effect of earth elements on growth of medicinal plants. Int J Pharm Technol 8:10628–10632

Bates DJP, Danilov AV, Lowrey CH, Eastman A (2013) Vinblastine rapidly induces NOXA and acutely sensitizes primary chronic lymphocytic leukemia cells to ABT-737. Mol Cancer Ther 12:1504–1514

Chalker-Scott L, Fnchigami LH (1989) The role of phenolic compounds in plant stress responses; low temperature stress physiology in crops. In: Paul HL (ed). CRC, Boca Raton, FL

Chen Q, Chen Z, Lu L, Jin H, Sun L, Yu Q, Xu H, Yang F, Fu M, Li S, Wang H, Xu M (2013) Interaction between abscisic acid and nitric oxide in PB90-induced catharanthine biosynthesis of *Catharanthus roseus* cell suspension culture. Biotechnol Prog 29:994–1001

Chomel M, Guittonny-Larchevêque M, Fernandez C, Gallet C, DesRochers A, Paré D, Jackson BG, Baldy V (2016) Plant secondary metabolites: a key driver of litter decomposition and soil nutrient cyclin. J Ecol 104(6):1527–1541. doi:10.1111/1365-2745.12644

Dicosmo F, Misawa M (1985) Eliciting secondary metabolism in plant cell cultures. Trend Biotechnol 3:318–322

Dixon RA, Paiva NL (1995) Stress-induced phenylpropanoid metabolism. Plant Cell 7:1085–1097

Feng LIN, Mei-Jiao GAO, Bo-Fan YU, Hua-Feng CHEN (2016) Effects of copper stress on the alkaoids of *Catharanthus roseus*. Bull Bot Res 36:374–379

Freitas MSM, Gama MC, Monnerat PH, De Carvalho AJC, Lima TC, Vieira IJC (2016) Induced nutrient deficiencies in *Catharanthus roseus* impact ajmalicine bioproduction. J Plant Nutr 39:835–841

Gholamhosseinpour Z, Hemati K, Dorodian H, Bashiri-Sadr Z (2011) Effect of nitrogen fertilizer on yield and amount of alkaloids in periwinkle and determination of vinblastine and vincristine by HPLC and TLC. Plant Sci Res 3:4–9

Idrees M (2007) Performance of two cultivars of periwinkle 'Sadabahar' (*Catharanthus roseus* L.) with special reference of total alkaloid content. M.Phil. Dissertation, AMU, Aligarh

Idrees M, Naeem M, Aftab T, Khan MMA, Uddin M (2011) Salicylic acid mitigates salinity stress by improving antioxidant defence system and vincristine and vinblastine alkaloids production in *Catharanthus roseus* (L.) G. Don. Acta Physiol Planta 33:987–999

Jaleel CA, Gopi R, Manivannan P, Panneerselvam R (2007a) Antioxidative potentials as a protective mechanism in *Catharanthus roseus* (L.) G. Don. plants under salinity stress. Turk J Bot 31:245–251

Jaleel CA, Gopi R, Manivannan P, Panneerselvam R (2007b) Responses of antioxidant defense system of *Catharanthus roseus* (L.) G. Don. to paclobutrazol treatment under salinity. Acta Physiol Plant 29:205–209

Jaleel CA, Gopi R, Kishorekumar A, Manivannan P, Sankar B, Panneerselvam R (2008a) Interactive effects of triadimefon and salt stress on antioxidative status and ajmalicine accumulation in *Catharanthus roseus*. Acta Physiol Plant 30:287–292

Jaleel CA, Sankar B, Sridharan R, Panneerselvam R (2008b) Soil salinity alters growth, chlorophyll content, and secondary metabolite accumulation in *Catharanthus roseus*. Turk J Biol 32:79–83

Jaleel CA, Salem MA, Hasanuzzaman M, Nahar K (2011) Plant growth regulator interactions results enhancement of antioxidant enzymes in *Catharanthus roseus*. J Plant Interact 5:135–145

Kang MH, Kang YH, Szymanska B, Wilczynska-Kalak U, Sheard MA, Harned TM, Lock RB, Reynolds CP (2007) Activity of vincristine, L-ASP, and dexamethasone against acute lymphoblastic leukemia is enhanced by the BH3-mimetic ABT-737 in vitro and in vivo. Blood 110:2057–2066

Karuppusamy S (2009) A review on trends in production of secondary metabolites from higher plants by in vitro tissue, organ and cell cultures. J Med Plant Res 3:1222–1239

Kim J, van Iersel MW (2011) Slowly developing drought stress increases photosynthetic acclimation of *Catharanthus roseus*. Physiol Plant 143:166–177

Marschner H (1995) Mineral nutrition of higher plants. Academic, London

Meng-Yan ZHU, Bo-Fan YU, Hua-Feng CHEN (2016) Effects of nitrate nitrogen levels on physiological metabolism of *Catharanthus roseus*. Bull Bot Res 36:535–541

Misra N, Misra R, Mariam A, Yusuf K, Yusuf L (2014) Salicylic acid alters antioxidant and phenolics metabolism in *Catharanthus roseus* grown under salinity stress. Afr J Trad Complement Altern Med 11:118–125

Mohamed EH, Osman SS, El-Soud EKA, Hasan AM (2007) Response of *Catharanthus roseus* shoots to salinity and drought in relation to vincristine alkaloid content. Asian J Plant Sci 6:1223–1228

Panda H (2005) Periwinkle cultivation. In: Herbs cultivation & medicinal uses, 2nd edn. National Institute of Industrial Research, New Delhi.

Pandey S, Gupta K, Mukherjee AK (2007) Impact of cadmium and lead on *Catharanthus roseus*-a phytoremediation study. J Environ Biol 28:655–662

Ponarulselvam S, Panneerselvam C, Murugan K, Aarthi N, Kalimuthu K, Thangamani S (2012) Synthesis of silver nanoparticles using leaves of *Catharanthus roseus* Linn. G. Don and their antiplasmodial activities. Asian Pac J Trop Biomed 2:574–580

Rao SR, Ravishankar GA (2002) Plant cell cultures: chemical factories of secondary metabolites. Biotechnol Adv 20:101–153

Seigler DS (1998) Plant secondary metabolism. Chapman and Hall (Kluwer Academic), Boston, MA, p 711

Sheshadri SA, Sriram S, Balamurugan P, Anupriya R, Princy SA, Brindha P, Bindu S (2015) Melatonin improves bioreductant capacity and silver nanoparticles synthesis using *Catharanthus roseus* leaves. RSC Adv 5:47548–47554

Shimano F, Ashihara H (2006) Effect of long-term phosphate starvation on the levels and metabolism of purine nucleotides in suspension-cultured *Catharanthus roseus* cells. Phytochemistry 67:132–141

Stafford A, Fowler MW (1983) Effect of carbon and nitrogen growth limitation upon nutrient uptake and metabolism in batch cultures of *Catharanthus roseus* (L) G. Don. Plant Cell Tiss Org Cult 2:239–251

Sudha G, Ravishankar GA (2003) Elicitation of anthocyanin production in callus cultures of *Daucus carota* and involvement of calcium channel modulators. Curr Sci 84:775–779

van Der Heijden R, Jacobs DI, Snoeijer W, Hallard D, Verpoorte R (2004) The Catharanthus alkaloids: pharmacognosy and biotechnology. Curr Med Chem 11:607–628

Xu M, Dong J, Zhu M (2004) Effect of nitric oxide on catharanthine production and growth of *Catharanthus roseus* suspension cells. Biotechnol Bioeng 89:367–371

Yuan YJ, Hu ZD (1993) Effect of rare earth elements on plant cell culture of *Catharanthus roseus*. Acta Pharm Sinica B 3:20–24

Zhang W, Curtin C, Franco C (2002) Towards manipulation of postbiosynthetic events in secondary metabolism of plant cell cultures. Enzyme Microb Technol 30:688–696

Zhang N, Wen Q, Feng H, Cao R, Zhou X, Tagn J, Wu N (2012) Effects of water stress and nitrogen nutrition on regulation of *Catharanthus roseus* alkaloids metabolism. Zhongguo Zhong Yao Za Zhi 37:1346–1352

Zheng Z, Wu M (2004) Cadmium treatment enhances the production of alkaloid secondary metabolites in *Catharanthus roseus*. Plant Sci 166:507–514

Catharanthus roseus: Detoxification and Hepatic Protection of Aflatoxin B1

S. Patharajan, S. Bala Abirami, R. Elangomathavan, and S. Ramesh

Abstract Aflatoxins are a group of carcinogenic secondary metabolites produced by certain strains of *Aspergillus* species that are a serious food safety issue throughout the world. These fungi are common contaminants of groundnut, maize, rice, cottonseed, tree nuts, and other commodities. The major types of aflatoxins include B1, B2, G1, and G2, which have been detected as contaminants of crops in the field and during storage, transportation, and processing. Among the 20 different types of aflatoxins identified so far, aflatoxin B1 (AFB1) is considered the most toxic. The International Agency for Research on Cancer (IARC) classified aflatoxins as class I human carcinogens. Plant derivatives and secondary metabolites are being used for the treatment of liver cancer and hepatic protection. The World Health Organization estimates that approximately 80% of the world's inhabitants rely on traditional medicine for their primary health care. In this chapter, we discussed about the effectiveness of aqueous extracts of *Catharanthus roseus* medicinal plant in detoxification of AFB1. The results revealed that the aqueous extracts of *Catharanthus roseus* effectively reduced the growth of *Aspergillus flavus* and inhibited the maximum level of AFB1 production when compared to other medicinal plants. Degradation of AFB1 in the mixture with *C. roseus* plant extracts revealed a loss of over 90% of AFB1 suggesting that aqueous extract of *C. roseus* has the ability to degrade the AFB1 at maximum level. In addition to that, methanolic fraction of *C. roseus* leaf extract was evaluated against hepatocellular carcinoma induced by AFB1 in experimental mice. The results suggested that *C. roseus* could be able to protect the liver against the AFB1-induced oxidative damage in mice. This chapter covers the information regarding the risk of postharvest diseases and AFB1 contamination in agricultural products, detoxification, and hepatic protection of AFB1 by *C. roseus*.

Keywords Aflatoxin B1 • *C. roseus* • Hepatocellular carcinoma • Albino Wistar mice

---

S. Patharajan (✉) • S. Bala Abirami • R. Elangomathavan
Department of Biotechnology, PRIST University, Thanjavur 613403, Tamil Nadu, India
e-mail: spatharajan@yahoo.co.in

S. Ramesh
Department of Microbiology, PRIST University, Thanjavur 613403, Tamil Nadu, India

M. Naeem et al. (eds.), *Catharanthus roseus*,
DOI 10.1007/978-3-319-51620-2_14

1 Introduction

The understanding of the millennium development goal of reducing by half the number of people suffering from hunger by the year 2015 will require a significant increase in the amount of food grains produced in developing countries. However, food quality and safety issues resulting from aflatoxin B1 (AFB1) contamination present a serious obstacle to improving nutrition, enhancing agricultural production, and linking smallholder farmers to markets. The Food and Agriculture Organization (FAO) of the United Nations estimates that 25% of world food crops are affected while the Center for Disease Control (CDC) estimates that more than 4.5 billion people in the developing world are exposed to aflatoxins. About a quarter of the world population is at a high risk for toxin-related diseases nowadays. Aflatoxins are toxic metabolic substances produced by certain toxigenic strains of *Aspergillus* species growing in various feed and food commodities. They are the most potent hepatocarcinogens among all the known natural and synthetic compounds. Aflatoxins occurring in food commodities are secondary metabolites of filamentous fungi, which can contaminate many types of food crops throughout the food chain. Although hundreds of fungal toxins are known, a limited number of toxins are generally considered to play important roles in food safety (Shephard 2008; Reddy et al. 2010; Probst et al. 2014). AFB1 has been classified as a Class I human carcinogen, while Fuminosine B1 (FB1) and Ochratoxin A (OTA) have been classified as Class 2B (probable human) carcinogens by the International Agency for Research on Cancer (IARC 1993).

Mycotoxins are commonly produced by species of *Aspergillus, Penicillium,* and *Fusarium* (Chandra and Sarbhoy 1997; Masheshwar et al. 2009). Several strategies are used in controlling fungal growth and the mycotoxin biosynthesis in seeds, grains and foodstuff by chemical treatments, and food preservatives by physical and biological methods. These methods often require sophisticated equipment and expensive chemicals or reagents. Chemical control of fungi and mycotoxins also results in environmental pollution and health hazard, and affects the natural ecological balance (Yassin et al. 2011). Use of plant products and essential oils provides an opportunity to avoid synthetic chemical preservatives and fungicide risks (Mohammed et al. 2012). Medicinal plants have been contributed immensely to health care in Indian subcontinent. This is due in part to the recognition of the value of traditional medical systems, particularly in Asian origin, and the identification of medicinal plant from indigenous pharmacopoeias and traditional knowledge, which have significant healing power. Nature has been a source of medicinal agents for thousands of years and an impressive number of modern drugs have been isolated from natural sources; many of these isolations were based on the uses of agents in traditional medicine (Patharajan and Karthikeyan 2010). According to World Health Organization, medicinal plants would be the best source to obtain a variety of drugs. Therefore, such plants should be investigated for better understanding of their properties, safety, and efficacy (Nascimento et al. 2000).

The usage of plant derivatives for antifungal agents is common in developing countries before the advent of synthetic fungicides and due to relatively cost implication of imported fungicides (Galvano et al. 2001). In recent years, efforts have been devoted to the search for new antifungal materials from natural sources for food preservation (Juglal et al. 2002; Onyeagba et al. 2004; Boyraz and Ozcan 2005). Several edible botanicals like coriander (*Coriandrum sativum*), palak (*Spinacia oleracea*), Amaltas (*Cassia fistula*), and essential oils from clove, cardamom, almond, cinnamon, and *Eucalyptus* have been reported to have antifungal activity (Ferhout et al. 1999; Pradeep et al. 2003; Nayan and Shukla 2011). Afzal et al. (2010) reported that *A. sativum* has a wide antifungal spectrum, reached about 60–82% inhibition in the growth of seed-borne *Aspergillus* and *Penicillium* fungi. This was attributed to phytochemical properties of the garlic plant particularly allicin, which could decompose into several effective antimicrobial compounds such as diallyl sulfide, diallyl disulfide, diallyl trisulfide, allyl methyl trisulfide, dithiins, and ajoene (Salim 2011; Tagoe 2011). *C. roseus* is an important medicinal plant which contains more than 70 different types of alkaloids and chemotherapeutic agents that are effective in treating various types of cancers—breast cancer, lung cancer, uterine cancer, melanomas, Hodgkins and non-Hodgkins lymphoma. The anticancer drugs vincristine and vinblastine are synthesized from alkaloids of *C. roseus* (Balaabirami et al. 2016). This chapter reviewed the risk of postharvest diseases and AFB1 contamination in agricultural commodities, potential of botanicals with particular emphasis on *C. roseus* for the detoxification of AFB1 and hepatic protection.

2 Aflatoxin Contamination in Agricultural Products

India is an agricultural country with nearly three fourths of the people dependent on agriculture or rural economy. The great outstanding achievement of Indian agriculture since independence is the phenomenal growth of food grains output. Nearly 70% of the total production of food grains in India is retained at farm level where the unscientific and faulty storage conditions enhance the chances of fungal attack and thereby aflatoxin production. The decomposers of food grains, i.e., fungi and bacteria, are always present on food grains in dormant conditions (usually as spores) and grow under favorable climatic and other conditions. The fungal growth may cause a decrease in germ inability, discoloration of grain, heating and mustiness, loss in weight, biochemical changes, and production of toxins. All these changes may occur before the responsible fungi which could be detected on visual examination. The fungi produce a large number of aflatoxins in food grains and their products.

Aflatoxins are a group of highly toxic secondary metabolites of the fungi produced under certain favorable environmental conditions. Because of their potent toxic nature and fairly common occurrence under natural conditions, aflatoxins have attracted worldwide attention in the recent years. The problem of food contamination has been recognized throughout the recorded history and spoilage fungi are one of the major causes of food contamination (Ashiq 2015). Since historical times when primitive ancestors started out the cultivation and storage of crops, spoilage molds have shown their presence. Contamination of various commodities by fungi and the hazard of consuming the contaminated products have been noted worldwide (Pitt and Hocking 2009). Agricultural commodities with high mold content and high moisture content are susceptible to early spoilage and contamination. Once the agricultural product becomes infected in the field, fungi continue to propagate throughout the harvesting, storage, and processing stages as long as the environmental conditions are conducive to fungal growth. Every year massive amounts of food are wasted due to contamination by fungi that produce toxic metabolic products: the aflatoxins. To reduce the aflatoxin contamination of foods, several strategies have been investigated and that can be divided into natural, biological, chemical, and physical methods (Reddy et al. 2010). Over the years, efforts have been devoted to search for new antifungal materials from natural sources for food preservation (Boyraz and Ozcan 2005; Patharajan et al. 2011).

The inhibitory effects of plant extracts on aflatoxin synthesis have also been examined (Reddy et al. 2009). The oil of *Ocimum* species exhibited a broad range of activity against fungi, including human pathogens (Parida et al. 2014). Risk of aflatoxin contamination of food and feed increased due to environmental, agronomic, and socioeconomic factors. Temperature, food substrate, strain of the mold, and other environmental factors are some parameters that effect aflatoxin production. Growth of *Aspergillus* spp. and aflatoxin contaminations varied from location to location depending upon the agro-climatic and storage conditions. Preventing aflatoxin production at farm level is the best way to control aflatoxin contamination. Advances in molecular techniques and other decontamination methods could help to deal with these issues.

3 Detoxification of Aflatoxins with Plant Products

Postharvest losses are much more painful and costlier than preharvest losses both in terms of money. Postharvest losses involve parasitic diseases caused by fungi, bacteria and viruses, environmental factors, physiological and mechanical factors. Aflatoxin contamination of crops is a worldwide food safety concern and it occurs in maize, groundnut, wheat, sorghum, rice, etc. when the toxigenic fungi present under favorable conditions. An inhibitory effect of neem extracts on the biosynthesis of aflatoxins (groups B and G) in fungal mycelia was reported by Bhatnagar et al. (1990). More than 280 plant species have been investigated for their inhibitory effect on toxigenic Aspergilli and nearly 100 of these plants had some inhibitory effects on growth or toxin production by fungi (Montes-Belmont and Carvajal, 1998). Karapynar (1989) reported the inhibitory action of crude extracts from mint, sage, bay, anise, and ground red pepper on the growth of *A. parasiticus* and its aflatoxin production in vitro. Saxena and Mathela (1999) found antifungal activity of new compounds from *Nepeta leucophylla* and *N. clarkei* against *Aspergillus* sp. Mathela (1981) screened 12 terpenoids against the growth of *Aspergillus* species and found thymol and carvacrol to be more active than nystatin and talsutin. In another study, aflatoxin production by *A. parasiticus* was suppressed depending on the concentration of the plant aqueous extract added to the culture media at the time of spore inoculation.

Aflatoxin production in fungal mycelia grown for 96 h in culture media containing 50% *Azadirachta indica* leaf and seed extracts was inhibited by 90 and 65%, respectively (Razzaghi-Abyaneh et al. 2005). Mondall et al. (2009) studied the efficacy of different extracts of neem leaf on seed-borne fungi, *A. flavus*. In this study, the growth of the fungus was inhibited significantly and controlled with both alcoholic and water extracts. Efficacy of various concentrations of four plant extracts prepared from *Allium sativum*, *Azadirachta indica*, *Zingiber officinale*, and *Allium cepa* were studied on reduction of *A. flavus* on mustard. They found that garlic extract is most effective followed by neem (Latif et al. 2006). Srichana et al. (2009) studied the efficacy of the betel leaf extract on growth of *A. flavus* and it was found that the extract at 10,000 ppm completely inhibited the growth of this fungus. Hema et al. (2009) evaluated some of the South Indian spices and herbs against *A. flavus* and other fungi. They found that *Psidium guajava* is more effective on all tested fungi. In an another study by Satish et al. (2007), aqueous extracts of 52 plants from different families were tested for their antifungal potential against eight important species of *Aspergillus*. Among 52, 12 extracts have recorded significant antifungal activity against one or the other *Aspergillus* species tested. Similarly, Pundir and Jain (2010) studied the efficacy of 22 plant extracts against food-associated fungi and found that clove and ginger are more effective than other plant extracts. Awuah and Kpodo (1996) reported that the following plants *Ocimum gratissimum*, *Cymbopogon citratus*, *Xylopia aethiopica*, *Monodora myristica*, *Syzygium aromaticum*, *Cinnamomum verum*, and *Piper nigrum* are effective in inhibiting formation of non-sorbic acid, a precursor in the aflatoxin synthesis pathway. *Ocimum gratissimum* leaf powder has been successfully used in inhibiting mold development on stored soybean for 9 months (Awuah and Kpodo 1996). The powder extracts of *Cymbopogon citratus* inhibited the growth of fungi including toxigenic species such as *A. flavus* and *A. fumigatus* (Adegoke and Odesola 1996). Awuah and Ellis (2002) reported the effective use of leaf powders of *O. gratissimum* and cloves (*S. aromaticum*) combination with some packaging materials to protect groundnut kernels artificially inoculated with *A. parasiticus*. There have been a number of reports citing the inhibitory effects of onion extracts on *A. flavus* growth, and with an ether extract of onions, thio-propanol-S-oxide has been demonstrated to inhibit growth. Pepper extracts have been shown to reduce aflatoxin production in *A. parasiticus* and *A. flavus* (Ito et al. 1994). Large-scale application of different higher plant products like azadirachtin from *Azadirachta indica*, eugenol from *Syzygium aromaticum*, carvone from *Carum carvi*, and allyl isothiocyanate from mustard and horseradish oil have attracted the attention of microbiologists to other plant chemicals for use as antimicrobials (Singh et al. 2008). Such products from higher plants would most likely be biodegradable, renewable in nature, and perhaps safer to human health (Varma and Dubey 1999). Plant products, especially essential oils, are recognized as one of the most promising groups of natural compounds for the development of safer antifungal agents (Varma and Dubey 2001). Many reports are available for use of neem oil to control toxigenic fungi and their toxins. Plant essential oils from

Azadirachta indica and *Morinda lucida* were found to inhibit the growth of a toxigenic *A. flavus* and significantly reduced aflatoxin synthesis in inoculated maize grains (Bankole et al. 2006). Clove oil and its major component eugenol have been extensively used to control aflatoxigenic fungi and aflatoxins. Kumar et al. (2009) studied the efficacy of essential oil from *Mentha arvensis* L. to control storage molds of chickpea. The oil effectively reduced mycelial growth of *A. flavus*. During screening of essential oils for their antifungal activity against *A. flavus*, the essential oil of *Cymbopogon citratus* was found to exhibit fungal toxicity. Another extensive study, Tamil Selvi et al. (2003) demonstrated that *A. flavus* growth and AFB1 production were both inhibited by an essential oil containing mainly garcinol from the tropical shrub/tree *Garcinia indica* at 3000 ppm.

4 Importance and Medicinal Properties of *C. roseus*

C. roseus (*Vinca rosea*) is known as the common or Madagascar periwinkle. It is a perennial herb of the Apocynaceae family originally native to Madagascar. It measures about 2 ft height and has dark green glossy leaves and pale pink or white flowers. *C. roseus* is cultivated two common names on the basis of their flower colors. Pink: Rosea and White: Alba. *C. roseus* is an Indian originated herb which grows wild in the Indian subcontinent in south Asia. Traditionally, leaves of *C. roseus* are used as medicine for the treatment of following diseases such as menorrhagia, rheumatism, dyspepsia, indigestion, dysmenorrheal, diabetes, hypertension, cancer, menstrual disorders, skin disease, and bleeding diarrhea, and has sedative and antiviral properties. Generally it is known as *Vinca rosea*, *Ammocallis rosea,* and *Lochnera rosea*. Due to presence of various alkaloids in *C. roseus,* it has antihypertensive and antispasmodic properties. One of the important types of alkaloid is the vinblastine produced from *C. roseus* due to its antitumor activity and wide pharmaceutical use. *C. roseus* has been used for modern chemotherapeutic agent for their pain relieving properties.

Catharanthus is a genus of flowering plants in the dogbane family, Apocynaceae like genus vinca, they are commonly known as periwinkles and there are eight known species, and seven are endemic to Madagascar. Though one, *C. roseus* is widely naturalized around the world. The name Catharanthus comes from the Greek word, Pure flower.

4.1 Plant Usage

1. Used as medicine: In Ayurveda, the extracts of its roots and shoots are used against several disease including diabetes, malaria, etc.
2. Used as anticancer: The substances vinblastine and vincristine extracted from the plant are used in treatment of leukemia and Hodgkin's lymphoma.

Fig. 1 Growth habit

3. In plant pathology: *C. roseus* is used in plant pathology as an experimental host for phytoplasmas. This is because it is easy to infect with a large majority of phytoplasmas and also often has very distinctive symptoms such as phyllody and significant reduced leaf size.

Catharanthus roseus contains significant amounts of volatile and phenolic compounds, including caffeoylquinic acids and flavonol glycosides which are known for antioxidant activity. It has an important role in the body defense system that acting as antioxidants against reactive oxygen species (ROS), which are harmful by forming such products through normal cell aerobic respiration. Accumulation of free radicals can cause pathological conditions such as ischemia, asthma, arthritis, inflammation, neurodegeneration, Parkinson's, mongolism, aging process, and perhaps dementia.

The flower petals, seeds, and other parts of *C. roseus* (Fig. 1) exhibit antioxidant properties. Thus phenolic compounds have redox properties that act as reducing agents, hydrogen donors, singlet oxygen quenchers, or metal chelators. It has multiple applications in foods, cosmetics, and pharmaceutical industries. Besides antioxidant activity, these compounds exhibit antiallergic, anti-inflammatory, antimicrobial, antithrombotic, cardioprotective, and vasodilatory effects. This is influenced by number of geographical and environmental factors.

Natural antioxidants are the source of finding the potentially safe, cheap, and effective antioxidants. Phenolic acids, flavonoids, and alkaloids are the main active constituents in this plant. These active substances perform a number of protective functions in the human organism and are involved in important activities (Stanković et al. 2010, 2011; Quideau et al. 2011). Vinblastine and vincristine are the dimers, formed by the coupling of monoindole alkaloids such as catharanthine and vindoline (Koul et al. 2013) mainly present in aerial part of plants used to treat cancer. Vincristine and vinblastine require both aerial and root parts of a plant to be synthesized (De Luca and Laflamme 2001). *C. roseus* is a source of valuable alkaloids resembling those from *Rauwolfia* species. Roots of *Catharanthus* have more ajmalicine (vasodilating) and serpentine (hypertensive) than even *Rauwolfia serpentina.* They also possess reserpine (Mishra et al. 2001). The alkaloids possess hypotensive, sedative, and tranquilizing properties. The root bark contains the alkaloid alstonine which has been used traditionally for its calming effect and its ability to reduce blood pressure. Yohimbine (Procomil) is an alkaloid with stimulant and aphrodisiac effects found naturally in *Pausinystalia johimbe* (Millon et al. 2002). *C. roseus* also shows the presence of this compound along with another flavonoid hirsutidin (Piovan and Fillipini 2007).

Phytochemical screening of methanolic fraction of *C. roseus* leaf extract was performed and found the presence of alkaloid, carbohydrates, flavonoids, phenols, quinone, saponins, and tannins. The antioxidant activity of *C. roseus* was determined and the results indicated that the leaf extract possesses the highest antioxidant activity (Patharajan and Balaabirami 2014). Balaabirami et al. (2016) studied the efficacy of the various concentrations of aqueous extracts such as *C. roseus*, *Andrographis paniculata*, *Syzygium aromaticum* and *Ocimum gratissimum* for their potential in preventing AFB1 contamination. Among the plant extracts tested, *C. roseus* have the ability to inhibit the growth of aflatoxigenic fungi and degradation of AFB1. Patharajan et al. (2016) conducted the experiment on the evaluation of the methanolic fraction of *C. roseus* leaf extract against AFB1-induced hepatocellular carcinoma in albino mice. They observed the marked increase in lipid peroxidase level and concomitant decrease in enzymatic antioxidants levels in carcinoma-induced mice, whereas *C. roseus* treatment reversed the conditions to near normal level. Liver histopathology showed that *C. roseus* reduced the incidence of liver lesions, lymphocytic infiltrations, and hepatic necrosis induced by AFB1 in mice. These results suggested that *C. roseus* could protect the liver against the AFB1-induced oxidative damage in mice.

5 Biosynthesis of Terpenoid Indole Alkaloids (TIAs) in *C. roseus*

Antioxidants are radical scavengers which give protection to human body against free radicals by inhibiting the oxidizing chain reactions. When these substances are present at low concentration in body, they markedly delay or prevent the oxidation of an oxidizable substrate. These antioxidants always play important roles in delaying the development of chronic diseases such as cardiovascular diseases (CVD), cancer, atherosclerosis, inflammatory bowel syndrome, and Alzheimer's disease (Bozin et al. 2006). A variety of different alkaloids have present in *C. roseus* and more than 130 different compounds reported, including about 100 monoterpenoid indole alkaloids (Pereira et al. 2010). As an important medicinal plant, it has a good antioxidant potential throughout its parts under drought stress. Hence, if the compound has antioxidant potentials and phytochemical activity, it can be a good therapeutic agent for accelerating the wound healing process. The vinca alkaloids are also important for being cancer fighters. There are four major vinca alkaloids in clinical use: Vinblastine (VBL), vinorelbine (VRL), vincristine (VCR), and vindesine (VDS). VCR, VBL, and VRL have been approved for use in the United States. Vinflunine is also a new synthetic vinca alkaloid, which has been approved in

Europe for the treatment of second-line transitional cell carcinoma of the urothelium is being developed for other malignancies. Vinca alkaloids are the second-most-used class of cancer drugs and will stay among the original cancer therapies.

Many papers published in the biosynthetic pathway of terpenoid indole alkaloids (TIAs) in *C. roseus*, and the identification and characterization of the corresponding enzymes involved in the biosynthetic pathway (Zhu et al. 2014, 2015). Strictosidine is the central intermediate in the biosynthesis of different TIAs, which is formed by the condensation of secologanin and tryptamine. Secologanin is derived from terpenoid (isoprenoid) biosynthetic pathway, while tryptamine is derived from the indole biosynthetic pathway. Then various specific end products are produced by different routes during downstream process. Although many genes and corresponding enzymes have been characterized in the biosynthetic pathway. Our knowledge of the whole TIA biosynthetic pathway still remains largely unknown up to date. Full elucidation of TIA biosynthetic pathways is an important prerequisite to understand the regulation of the TIA biosynthesis in the medicinal plant and to produce valuable TIAs by synthetic biological technology.

The Madagascar periwinkle produces a large palette of Monoterpenoid Indole Alkaloids (MIAs), a class of complex alkaloids, including some of the most valuable plants natural products with precious therapeutical values. Evolutionary pressure on one of the hotspots of biodiversity has obviously turned this endemic Malagasy plant into an innovative alkaloid engine. *Catharanthus* is a unique taxon producing vinblastine and vincristine, heterodimeric MIAs with complex stereochemistry, and also manufactures more than 100 different MIAs, some shared with the Apocynaceae, Loganiaceae, and Rubiaceae members. For over 60 years, the quest for these powerful anticancer drugs has inspired biologists, chemists, and pharmacists to unravel the chemistry, biochemistry, therapeutic activity, cell and molecular biology of *C. roseus*. Recently, the "omics" technologies have fuelled rapid progress in deciphering the last secret of strictosidine biosynthesis, the central precursor opening biosynthetic routes to several thousand MIA compounds. *Catharanthus roseus* transcriptome, proteome, and metabolome databases, comprising organ-, tissue-, and cell-specific libraries, and other phytogenomic resources, were developed for instance by PhytoMetaSyn, Medicinal Plant Genomic Resources, and Smart Cell consortium. Tissue-specific library screening, orthology comparison in species with or without MIA-biochemical engines and clustering of gene expression profiles together with various functional validation strategies largely contributed to enrich the toolbox for plant synthetic biology and metabolic engineering of MIA biosynthesis (Dugé de Bernonville et al. 2014).

6 Aflatoxin and Hepatocellular Carcinoma

Cancer is one of the leading causes of adult deaths worldwide. In India, the International Agency for Research on Cancer estimated indirectly about 635,000 people died from cancer, representing about 8% of all estimated global cancer deaths and about 6% of all deaths in India (Karki et al. 2014). Hepatocellular carcinoma (HCC) is the most common liver cancer among men and women. HCC is the sixth most common cancer in the world and the third leading cause of cancer-related death, resulting in at least 500,000 deaths per year worldwide. HCC is a malignant tumor of liver parenchyma cells and the primary liver cancers are more prevalent among men than women. HCC is a major cause of morbidity and mortality and it is the seventh most common cancer worldwide and the third leading cause of cancer-related deaths complicating liver cirrhosis in most cases (Ferlay et al. 2010). Its incidence is increasing worldwide, ranging between 3 and 9% annually (Velazquez et al. 2003). Exposure to AFB1 is probably also an important contributor to the high incidence of liver cancer. Aflatoxins are a group of structurally related secondary fungal metabolites that are carcinogenic, hepatotoxic, teratogenic, and immunosuppressive. Reports are stating that even a single dose of aflatoxin is sufficient to induce liver cell tumors in rat (Angusbhakorn et al. 1990). For decades, it has been known that aflatoxin exposure causes liver cancer in humans and a variety of animal species. The International Agency for Research on Cancer has classified "naturally occurring mixes of aflatoxins" as a Group 1 human carcinogen. Concomitant exposure to aflatoxins and the hepatitis B virus (HBV) is common in developing countries and greatly increases HCC risk (Wu et al. 2013). Individuals with both exposures have a multiplicative greater risk of developing HCC than those exposed to aflatoxins or HBV alone (Groopman et al. 2008). A recent systematic review and meta-analysis determined that the risk of developing liver cancer was over six times higher in individuals with detectable aflatoxin biomarkers than in those without, over 11 times higher in individuals with chronic HBV infection than in those without, and over 73 times higher in individuals with both detectable aflatoxin biomarkers and HBV positivity compared with those with neither risk factor, a nearly perfectly multiplicative relationship (Liu et al. 2012). Two separate analyses have been conducted to estimate the global burden of liver cancer attributable to aflatoxins. Liu and Wu (2010) used a quantitative cancer risk assessment approach, using dose–response data for the relationship between aflatoxins and liver cancer risk in populations of HBV-negative and HBV-positive individuals and multiplying the corresponding cancer potency factors by aflatoxin exposure data for multiple nations worldwide (JECFA 1998; Henry et al. 1999). In their analysis that included about five billion individuals around the world (summing populations across nations for which aflatoxin data were available), they estimated that 25,200–155,000 liver cancer cases annually could be attributed to aflatoxin exposure.

In a follow-up study, Liu et al. (2012) used a different approach to estimate the global burden of cancer caused by aflatoxins: estimating population-attributable risk from a systematic review and meta-analysis of 17 epidemiological studies on aflatoxins, HBV, and liver cancer in Africa and Asia. It was estimated that about 23% (21–24%) of all HCC cases annually may be attributable to aflatoxins, for a total of up to 172,000 cases per year. Since liver cancer is the third leading cause of cancer deaths worldwide, and mortality rapidly follows diagnosis, the contribution of aflatoxins to this deadly cancer is significant.

Because aflatoxins are one of the most significant risk factors for liver cancer, one of the deadliest cancers worldwide, controlling its presence in the food supply is critical. It is possibly responsible for up to 172,000 liver cancer cases per year, most of which would result in mortality within 3 months of diagnosis. Possibly even more critical from a global public health standpoint is the link between aflatoxin exposure and childhood stunting, which can lead to a variety of adverse health conditions that last well beyond childhood. However, at the moment there is insufficient evidence for a quantitative risk assessment to evaluate exact daily doses of aflatoxins that lead to particular levels of risk or adverse health outcomes in children. Additionally, while aflatoxins may lead to immunomodulation, not enough information is currently known about how this leads to particular adverse health outcomes in humans. However, the human health evidence points to aflatoxins association with multiple adverse effects; hence, it is important to reduce human exposures to aflatoxins in the diet to the extent that feasible methods allow.

7 Plant Derivatives and Cancer Treatment

Natural products, especially plants, have been used for the treatment of various diseases for thousands of years. Evaluation of plants bearing efficiency in healing various diseases is growing in recent years. Innumerable biologically active compounds of plants are found to possess antibacterial properties. Practitioners of Ayurveda and Unani system of medicine regularly employ a large number of Indian medicinal plants as antibiotic agents and over the last 40 years, intensive efforts have been made to discover clinically used herbal-based antibacterial, antifungal, and anticancer drugs. The National Cancer Institute collected about 35,000 plant samples from 20 countries and has screened around 114,000 extracts for anticancer activity (Shoeb 2005).

Nature produces a variety of toxic compounds, which are often used as anticancer drugs. Up to now, there are at least 120 species of poisonous botanicals, animals, and minerals, of which more than half have been found to possess significant anticancer properties (Man et al. 2012). In spite of their clinical toxicity, they exhibit pharmacological effects and have been used as important traditional Chinese medicines for the different stages of cancer. Vincristine is a dimer-indo-alkaloid which is extracted from the leaves of *C. roseus*. It is effective to treat acute lymphocytic cell leukemia, odgkin disease, and non-Hodgkin disease clinically. The medicinal plants like *C. roseus*, *Podophyllum peltatum*, *Taxus brevifolia*, *Camptothecin acuminata*, *Cephalotaxus harringtonia*, *Viscum album*, *Ochrosia elliptica*, *Annona bullata*, *Asimina triloba*, and *Rhizoma zedoariae* are clinically proven for having anticancer activity. *C. roseus* is one of the best studied medicinal plants and recently, extract from periwinkle has been shown to be effective in the treatment of diabetes, high blood pressure, asthma, constipation, skin cancer, and Hodgkin's disease.

This book chapter will help to researchers and scholars to perform an in-depth study in this area as plant indicates the vast range of phytochemicals related to different season and agro-climatic zone.

8 Case Studies

8.1 Case Study I

Effect of methanolic extract of *C. roseus* on physiological enzymes in liver on control and experimental animals (Fig. 2)

The physiological marker enzymes like alanine transaminase (ALT) and alkaline phosphatase (ALP) were increasing in AFB1-induced group mice when compared with control group except aspartate transaminase (AST). After administration of *C. roseus* leaves extract to AFB1-induced group shows decreasing the enzyme level such as physiological enzyme and the values were near to control mice.

No significant changes were observed in *C. roseus* alone treated group of mice.

8.2 Case Study II

Effect of methanolic extract of *C. roseus* and AFB1 on liver lipid profile (Fig. 3)

The AFB1-induced group shows elevated level cholesterol, triglyceride, LDL, VLDL and decreased level of HDL compared with control group of mice. After treatment with *C. roseus* leaves extract significantly improved the altered level of cholesterol, triglyceride, LDL, VLDL and decreased level of HDL as compared with AFB1-induced group of mice. No detectable changes were observed in *C. roseus* treated control group of mice when compared to normal control group.

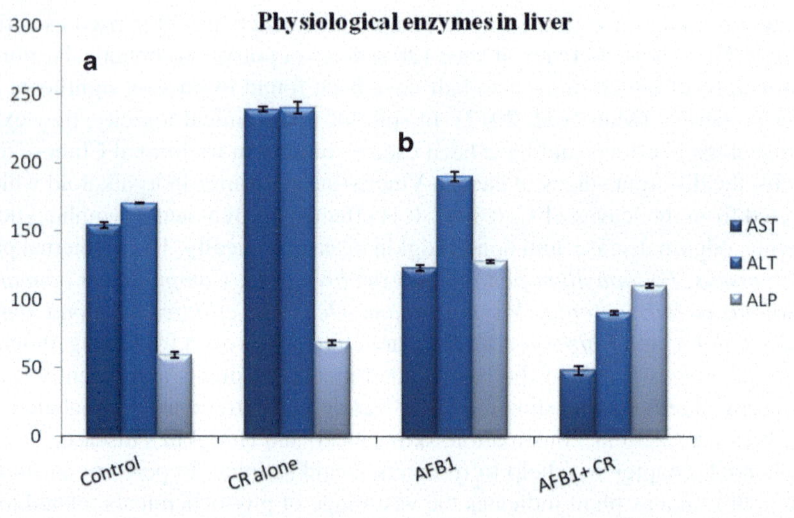

Fig. 2 Values of results are expressed as Mean ± SD for six mice. [a]$P < 0.05$ compared with control group of mice. [b]$P < 0.05$ compared with AFB1-induced group of mice, AST, ALT and ALP

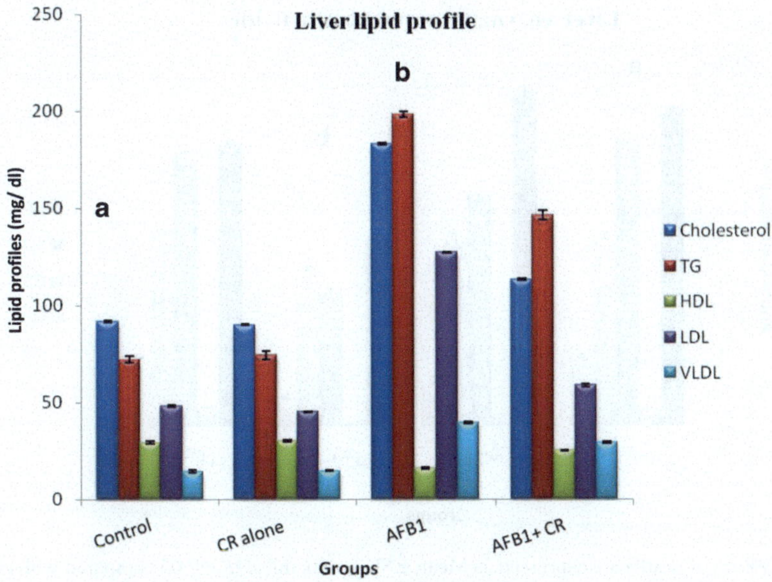

Fig. 3 Values of results are expressed as Mean ± SD for six mice. $^{a}P < 0.05$ compared with control group of mice. $^{b}P < 0.05$ compared with AFB1-induced group of mice, LDL = (Total cholesterol)—(HDL Cholesterol)—(Triglyceride/5), LDL cholesterol levels were expressed as mg/dl serum, VLDL = Triglycerides/5

8.3 *Case Study III*

Effect of methanolic leaf extract of *C. roseus* and AFB1 on liver enzymatic antioxidants (Fig. 4)

The levels of superoxide dismutase (SOD), catalase (CAT), glutathione peroxidase (GPx), glutathione-S-transferase (GST), and glutathione reductase (Gr) were significantly decreased ($P < 0.001$) in AFB1-induced group when compared to the control. Treatment with *C. roseus* extract recovered these decreased enzyme activities produced by AFB1 towards normalization compared with AFB1-induced group. No significant changes were observed in *C. roseus* alone treated group of mice.

9 Conclusion

Aflatoxins are very important because the contamination of aflatoxin poses serious problems in public health, agriculture, and economic aspects. Antifungal chemicals and pesticides have been used for preservation of stored grains. Because of health and economic considerations, natural plant products may be replaced by toxic chemicals and provide an alternative method to protect from AFB1 contaminations.

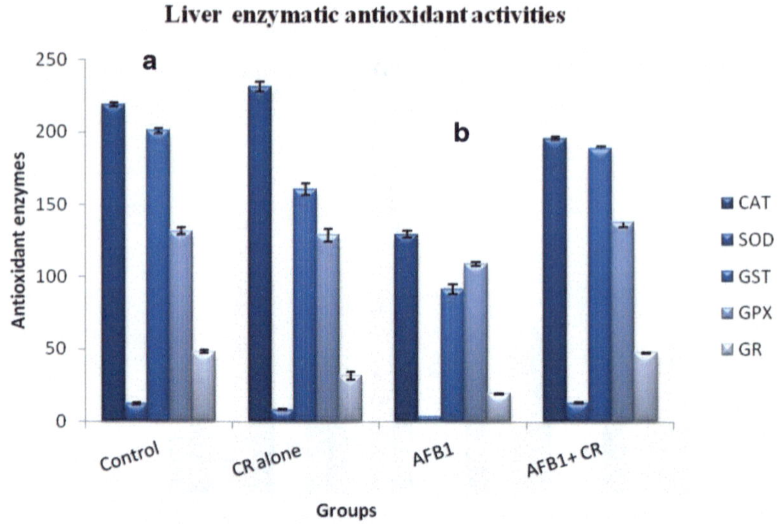

Fig. 4 Values of results are expressed as Mean ± SD for six mice. $^{a}P < 0.05$ compared with control group of mice. $^{b}P < 0.05$ compared with AFB1-induced group of mice, SOD, CAT, GPx, Gr, and GST

Many studies have confirmed the potentials of *C. roseus* for the degradation of AFB1 contamination and the protection of hepatocellular carcinoma. The methanolic fraction of *C. roseus* leaf extracts showed positive results for the protection of HCC induced by AFB1. Recently, more research has been focused on the role of alkaloids and flavonoids in cancer prevention because epidemiological investigations suggested that increased intake of fruits and vegetables are associated with the reduced risk of certain cancers.

Acknowledgements The authors are grateful to the authorities of PRIST University for the facilities and encouragements.

References

Adegoke GO, Odelusola BA (1996) Storage of maize and cowpea and inhibition of microbial agents of biodeterioration using the powder and essential oil of *Cymbopogon citratus*. Int Biodeter Biodegr 37:81–84

Afzal R, Mughal SM, Munir M, Kishwar S, Qureshi R, Arshad M, Laghari MK (2010) Mycoflora associated with seeds of different sunflower cultivars and its management. Pak J Bot 42:435–445

Angusbhakorn S, Get NP, Miyamoto M, Bharmarpravti N (1990) A single dose response effect of aflatoxin B1 on rapid liver cancer induction in two strains of rats. Int J Cancer 46:664–668

Ashiq S (2015) Natural occurrence of mycotoxins in food and feed: Pakistan perspective. Compr Rev Food Sci Food Saf 14:159–175

Awuah RT, Ellis WO (2002) Effects of some groundnut packaging methods and protection with *Ocimum* and *Syzygium* powders on kernel infection by fungi. Mycopathologia 154:29–36

Awuah RT, Kpodo KA (1996) High incidence of *Aspergillus flavus* and aflatoxins in stored groundnut in Ghana and the use of microbial assay to assess the inhibitory effects of plant extracts on aflatoxin synthesis. Mycopathologia 134:109–114

Balaabirami S, Elangomathavan R, Ramesh S, Patharajan S (2016) Inhibitory action on the growth of *Aspergillus flavus* and degradation of Aflatoxin B1 by medicinal herbs. Int J Sci Appl Res 3(3):35–43

Bankole SA, Ogunsanwo BM, Osho A, Adewuyi GO (2006) Fungal contamination and aflatoxin B1 of 'egusi' melon seeds in Nigeria. Food Control 17:814–818

Bhatnagar D, Zeringue HJ, Cormick SP (1990) Neem leaf extracts inhibit aflatoxin biosynthesis in *Aspergillus flavus* and *A. parasiticus*. In: Proceedings of the USDA neem workshop. US Department of Agriculture, Beltsville, pp 118–127

Boyraz N, Özcan M (2005) Antifungal effect of some spice hydrosols. Fitoterapia 76:661–665

Bozin B, Mimica-Dukic N, Simin N (2006) Characterization of the volatile composition of essential oils of some lamiaceae spices and the antimicrobial and antioxidant activities of the entire oils. J Agric Food Chem 54:1822–1828

Chandra R, Sarbhoy AK (1997) Production of Aflatoxins and Zearalenone by the toxigenic fungal isolates obtained from stored food grains of commercial crops. Indian Phytopathol 50:458–468

De Luca V, Laflamme P (2001) The expanding universe of alkaloid biosynthesis. Curr Opin Plant Biol 4:225–233

Dugé de Bernonville T, Clastre M, Besseau S, Oudin A, Burlat V, Glévarec G, Lanoue A, Papon N, Giglioli-Guivarc'h N, St-Pierre B, Courdavault V (2014) Phytochemical genomics of the Madagascar periwinkle: unravelling the last twists of the alkaloid engine. Phytochemistry 115:9–23

Ferhout H, Bohatier J, Guillot J, Chalchat JC (1999) Antifungal activity of selected essential oils, cinnamaldehyde and carvacrol against *Malassezia furfur* and *Candida albicans*. J Essential Oil Res 11:119–129

Ferlay J, Shin HR, Bray F, Forman D, Mathers C, Parkin DM (2010) Estimates of worldwide burden of cancer in 2008: Globocan 2008. Int J Cancer 127:2893–2917

Galvano F, Piva A, Ritieni A, Galvano G (2001) Dietary strategies to counteract the effects of mycotoxins. Rev J Food Protect 64:101–131

Groopman JD, Kensler TW, Wild CP (2008) Protective interventions to prevent aflatoxin-induced carcinogenesis in developing countries. Annu Rev Public Health 29:187–203

Hema R, Kumaravel S, Elanchezhiyan N (2009) Antimicrobial activity of some of the South-Indian spices and herbals against food pathogens. Global J Pharmacol 3:38–40

Henry SH, Bosch FX, Troxell TC, Bolger PM (1999) Public health: reducing liver cancer—global control of aflatoxin. Science 286(5449):2453–2454

International Agency for Research on Cancer (1993) Some naturally occurring substances: food items and constituents, heterocyclic aromatic amines and mycotoxins. In: IARC monographs on the evaluation of carcinogenic risks to humans, vol 56. IARC, Geneva, pp 489–521

Ito H, Chen H, Bunnak J (1994) Aflatoxin production by microorganisms of the *Aspergillus flavus* group in spices and the effect of irradiation. J Sci Food Agric 65:141–142

JECFA (1998) Safety evaluation of certain food additives and contaminants, WHO Food Additives Series 40, Aflatoxins. In: The forty-ninth meeting of the Joint FAO/WHO Expert Committee on Food Additives. World Health Organization, Geneva

Juglal S, Govinden R, Odhav B (2002) Spice oils for the control of co-occurring mycotoxin producing fungi. J Food Protect 65:683–687

Karki K, Singh KP, Kumar U, Mohan M (2014) Cancer scenario with future perspectives in Uttarakhand Region of India. Global J Res Anal 3(12):2277–8160

Karapynar M (1989) Inhibition effects of some spice agents on aflatoxigenic mould growths. In: Proceedings of the International Food Symposium, Apr 4–6, Bursa, Turkey, pp 129–137

Kumar A, Shukla R, Singh P, Singh AK, Dubey NK (2009) Use of essential oil from *Mentha arvensis* L. to control storage moulds and insects in stored chickpea. J Sci Food Agric 89:2643–2649

Latif MA, Kaoser A, Saleh M, Ashik M, Khan I, Rahman H, Hossain MA (2006) Efficacy of some plant extracts in controlling seed-borne fungal infections of mustard. Bangladesh J Microbiol 23:168–170

Liu Y, Wu F (2010) Global burden of aflatoxin-induced *Hepatocellular carcinoma*: a risk assessment. Environ Health Perspect 118:818–824

Liu Y, Chang CC, Marsh GM, Wu F (2012) Population attributable risk of aflatoxin-related liver cancer: systematic review and meta-analysis. Eur J Cancer 48:2125–2136

Man S, Gao W, Wei C, Liu C (2012) Anticancer drugs from traditional toxic Chinese medicines. Phytother Res 26(10):1449–1465

Masheshwar PK, Moharram SA, Janardhana GR (2009) Detection of fumonisin producing *Fusarium verticillioides* in paddy (*Oryza sativa*. L) using polymerase chain reaction (PCR). Braz J Microbiol 40:134–138

Mathela CS (1981) In vitro antifungal examination of some terpenoids. Proc Natl Acad Sci India 51:513–516

Koul M, Lakra NS, Chandra R, Chandra S (2013) *Catharanthus roseus* and prospects of its endophytes: a new avenue for production of bioactive metabolites. IJPSR 4(7):2705–2716

Millon MJ, Newman TA, Audinot V, Cussac D, Lejeune F, Nicolas JP, Cogé F, Galizzi JP, Boutin JA (2002) Agonist and antagonist actions of Yohimbine as compared to fluparoxan at alpha (2)—adrenergic receptors (AR). Synapse 35:79–95

Mishra P, Uniya GC, Sharma S (2001) Pattern of diversity for morphological and alkaloid yield related trades among the periwinkle *Catharathus roseus* accessions collected from in and around Indian Subcontinent. Genet Res Crop Evol 48:273–286

Mohamed AY, Mohamed AM, El-Rahim A, El-Samawaty MA (2012) Mycotoxins and non-fungicidal control of corn grain rotting fungi. J Plant Sci 7:96–104

Mondall NK, Mojumdar A, Chatterje SK, Banerjee A, Datta JK, Gupta S (2009) Antifungal activities and chemical characterization of neem leaf extracts on the growth of some selected fungal species *in vitro* culture medium. J Appl Sci Environ Manag 13:49–53

Montes-Belmont R, Carvajal M (1998) Control of *Aspergillus flavus* in maize with plant essential oils and their components. J Food Prot 61:616–619

Nascimento GGF, Lacatelli J, Freitas PC, Silva GL (2000) Antibacterial activity of plants extracts and phytochemical on antibiotic resistant bacteria. Braz J Microbiol 31:886–891

Nayan RB, Shukla VJ (2011) Antibacterial and antifungal activities from leaf extracts of *cassia fistula* L: an ethnomedicinal plant. J Adv Pharm Technol Res 2(2):104–109

Onyeagba RA, Ugbogu OC, Okeke CU, Iroakasi O (2004) Studies on the antimicrobial effects of garlic (*Allium sativum* Linn) ginger (*Zingiber officinale* Roscoe) and lime (*Citrus aurantifolia* Linn). Afr J Biotechnol 3:552–554

Parida R, Sandeep IS, Sethy BK, Sahoo S, Mohanty S (2014) Chemical composition, antioxidant and antimicrobial activities of essential oil from lime basil (*Ocimum americanum*): a potent source for natural antioxidant. Int J Pharm Pharm Sci 6(7):487–490

Patharajan S, Balaabirami S (2014) Antioxidant activity and phytochemical analysis of fractionated leaf extracts of *Catharanthus roseus*. Int J Pharm 1(2):138–143

Patharajan S, Karthikeyan V (2010) Integration of Indian traditional herbal medicines with modern technologies—an overview. Pharmacologyonline 1:964–978

Patharajan S, Balaabirami S, Elangomathavan R, Ramesh S (2016) Hepatoprotective effect of *Catharanthus roseus* extracts in experimental animals induced by Aflatoxin B1. (Submitted)

Patharajan S, Reddy KRN, Karthikeyan V, Lore A, Spadaro D, Gullino ML, Garibaldi A (2011) Potential of yeast antagonists on *invitro* biodegradation of ochratoxin A. Food Control 22:290–296

Pereira DM, Faria J, Gasper L, Ferreres F, Valentao P, Scotto mayor M (2010) Exploiting *Catharanthus roseus* roots: source of antioxidants. Food Chem 121:56–61

Piovan A, Fillipini R (2007) Anthocyanins in *Catharanthus roseus* in vivo and in vitro: a review. Phytochem Rev 6:235–242

Pitt JI, Hocking AD (2009) *Fungi and food spoilage*. Springer, New York, pp 1–2

Pradeep AG, Lokesh S, Rai VR (2003) Efficacy of some essential oils on seed mycoflora and seedling quality of some crop species saved by farmers. Adv Plant Sci 16:53–58

Probst C, Bandyopadhyay R, Cotty PJ (2014) Diversity of aflatoxin-producing fungi and their impact on food safety in sub-Saharan Africa. Int J Food Microbiol 174:113–122

Pundir RK, Jain P (2010) Antifungal activity of twenty two ethanolic plant extracts against food-associated fungi. J Pharm Res 3:506–510

Quideau S, Deffieux D, Douat-Casassus C, Pouysegu L (2011) Plant polyphenols: chemical properties, biological activities, and synthesis. Nat Prod 50:586–621

Razzaghi-Abyaneh M, Allameh A, Tiraihi T, Shams-Ghahfarokhi M, Ghorbanian M (2005) Morphological alterations in toxigenic *Aspergillus parasiticus* exposed to neem (*Azadirachta indica*) leaf and seed aqueous extracts. Mycopathologia 159:565–570

Reddy KRN, Reddy CS, Muralidharan K (2009) Potential of botanical and biocontrol agents on growth and aflatoxin production by *Aspergillus flavus* infecting rice grains. Food Control 20:173–178

Reddy KRN, Salleh B, Saad B, Abbas HK, Abel CA, Shier WT (2010) An overview of mycotoxin contamination in foods and its implications for human health. Toxin Rev 29(1):3–26

Salim AB (2011) Effect of some plant extracts on fungal and aflatoxin production. Int J Acad Res 3:116–120

Satish S, Mohana DC, Ranhavendra MP, Raveesha KA (2007) Antifungal activity of some plant extracts against important seed borne pathogens of *Aspergillus* sp. J Agricul Technol 3(1):109–119

Saxena J, Mathela CH (1999) Antifungal activity of new compounds from *Nepeta leucophylla* and *Nepeta clarkei*. Appl Environ Microbiol:702–704

Shephard GS (2008) Determination of mycotoxins in human foods. Chem Soc Rev 37:2468–2477

Shoeb M, Celik S, Jaspars M, Kumarasamy Y, MacManus S, Nahar L, Kong TLP, Sarker SD (2005) Isolation, structure elucidation and bioactivity of schischkiniin, a unique indole alkaloid from the seeds of *Centaurea schischkinii*. Tetrahedron 61:9001–9006

Singh P, Srivastava B, Ashok K, Dubey NK, Rajesh G (2008) Efficacy of essential oil of *Amomum subulatum* as a novel aflatoxin B_1 suppressor. J Herbs Spices Med Plants 14:208–218

Srichana D, Phumruang A, Chongkid B (2009) Inhibition effect of betel leaf extract on the growth of *Aspergillus flavus* and *Fusarium verticillioides*. Thammasat Int J Sci Tech 14:74–77

Stanković SM, Ćurčić GM, Žižić BJ, Topuzović DM, Solujić RS, Marković DS (2011) *Teucrium* plant species as natural sources of novel anticancer compounds: antiproliferative, proapoptotic and antioxidant properties. Int J Mol Sci 12:4190–4205

Stanković SM, Topuzović M, Solujić S, Mihailović V (2010) Antioxidant activity and concentration of phenols and flavonoids in the whole plant and plant parts of *Teucrium chamaedrys* L. var. *glanduliferum* Haussk. J Med Plant Res 4:2092–2098

Tagoe D, Baidoo S, Dadzie I, Kangah V, Nyarko H (2011) A comparison of the antimicrobial (Antifungal) properties of onion (*Allium cepa*), Ginger (*Zingiber officinal*) and garlic (*Allium sativum*) on *Aspergillus flavus*, *Aspergillus niger* and *Cladosporium herbarum* using organic and water base extraction methods. Res J Med Plant 5:281–287

Tamil Selvi A, Joseph GS, Jayaprakasha GK (2003) Inhibition of growth and aflatoxin production in *Aspergillus flavus* by *Garcinia indica* extract and its antioxidant activity. Food Microbiol 20:455–460

Varma J, Dubey NK (1999) Perspective of botanical and microbial products as pesticides of tomorrow. Curr Sci 76:172–179

Varma J, Dubey NK (2001) Efficacy of essential oils of *Caesulia axillaris* and *Mentha arvensis* against some storage pests causing biodeterioration of food commodities. Int J Food Microbiol 68:207–210

Velazquez E, Tournier HA, Mordujovich Bushiazzo P, Saavedra G, Schinella GR (2003) Antioxidant activity of Paraguayans plant extracts. Fitoterapia 74:91–97

Wu F, Stacy SL, Kensler TW (2013) Global risk assessment of aflatoxins in maize and peanuts: are regulatory standards adequately protective? Toxicol Sci 135(1):251–259

Yassin MA, El-Samawaty AMA, Moslem M, Bahkali A, Abd-Elsalam KA (2011) Fungal biota and occurrence of aflatoxigenic *Aspergillus* in postharvest corn grains. Fresen Environ Bull 20:903–909

Zhu J, Mingxuan W, Wei W, Rongmin Y (2015) Biosynthesis and regulation of terpenoid indole alkaloids in *Catharanthus roseus*. Pharmacogn Rev 9(17):24–28

Zhu X, Zeng X, Sun C, Chen S (2014) Biosynthetic pathway of terpenoid indole alkaloids in *Catharanthus roseus*. Front Med 8(3):285–293

Potential of *Catharanthus roseus* (L.) in Phytoremediation of Heavy Metals

V. Subhashini and A.V.V.S. Swamy

Abstract Phytoremediation is an environmental-friendly technology that exploits a plant's ability to remove contaminants for pollution prevention, control, and remediation from the environment. Plants are unique organisms equipped with remarkable metabolic and absorption capabilities as well as transport systems that can take up nutrients or contaminants selectively from the growth matrix, soil, or water. Plant species selection is a critical management decision for phytoremediation. Biosolutions are the best tools for all types of pollutions in future. Phytoremediation is one of the promising biosolutions for soil pollution. The earlier studies emphasize need for selecting more and more species for reclamation of soil quality through phytoremediation. The present study is an attempt to test the potential of the *Catharanthus roseus* species in the removal of heavy metals from the soil. The aim of the present study was to evaluate the metal accumulation capacity of the selected plant species. The heavy metals (lead, nickel, zinc, cadmium, and chromium) were used. Aqueous solutions of these metals added to the plant samples on alternate days for 60 days (2 months). After every 20 days, plant samples were collected from each pot, then dried in a hot air oven and powdered by a mortar and pestle. About 1 g of the powder from each part of the sample was taken for metal analysis by AAS. *Catharanthus roseus* was found to be a good accumulator of lead, nickel, zinc, cadmium and chromium. On consolidation of the results obtained the species can be recommended for the phytoextraction of lead, nickel, zinc, cadmium, and chromium contaminated soils.

Keywords Heavy metals • Phytoremediation • Bioconcentration factor • Translocation factor

V. Subhashini (✉) • A.V.V.S. Swamy (✉)
Department of Environmental Sciences, Acharya Nagarjuna University, Nagarjuna Nagar, Guntur, 522510 Andhra Pradesh, India
e-mail: subhashini2009@yahoo.com; arzavvsswamy1962@gmail.com

M. Naeem et al. (eds.), *Catharanthus roseus*,
DOI 10.1007/978-3-319-51620-2_15

1 Introduction

Heavy metal pollution in environment is mainly of anthropogenic origin and results from activities such as fossil fuels, vehicular emissions, industrial emissions, landfill leachates, fertilizers, sewage, and municipal wastes. As a consequence of the contamination from increasing mining and smelting activity, as well as excessive fertilizer application and wastewater irrigation in agriculture, emissions from the rapidly expanding industrial areas, mine tailings, disposal of high metal wastes, leaded gasoline and paints, land application of sewage sludge, pesticides, wastewater irrigation, coal combustion residues, spillage of petrochemicals and atmospheric deposition and heavy metal pollution of soils is becoming a more serious problem worldwide. (Nriagu 1979; Ensley 2000; McIntyre 2001; Bu Olayan and Thomas 2009; Kord et al. 2010; Zhang et al. 2010; Sekabira et al. 2011). All compartments of the biosphere are polluted by a variety of inorganic and organic pollutants as a result of anthropogenic activities and altered the normal biogeochemical cycling (Prasad 2003). Estimates suggested that plants can remove between 180 and 530 kg/ha of lead per year (Huang and Cunningham 1996; Blaylock et al. 1997), making remediation of sites contaminated up to 2500 mg/kg of lead possible in less than 10 years. Over recent decades, an annual worldwide release of heavy metals reached 22,000 t (metric ton) for cadmium, 939,000 t for copper, 783,000 t for lead, and 1,350,000 t for zinc (Chowdhury et al. 2015).

Phytoremediation technique is defined as the use of green plants to remove pollutants from the environment or to render them harmless (Raskin et al. 1994; Cunningham et al. 1995). Phytoremediation is a cost-effective plant-based approach of remediation that takes advantage of the ability of plants to concentrate elements and compounds from the environment and to metabolize various molecules in their tissues. Phytoremediation involves growing plants in a contaminated matrix, for a required growth period to remove contaminants from the matrix or facilitate immobilization or degradation (detoxification) of the pollutants. The plants can be subsequently harvested, processed, and disposed off in an environmentally sound manner. The mechanisms of phytoremediation involved in heavy metal remediation are limited to uptake, adsorption, transport and translocation, sequestration into vacuoles, hyperaccumulation and in some cases, volatilization (Salt et al. 1995; Meagher 2000).

A range of processes mediated by plants or algae are useful in treating environmental problems (Wang et al. 2006; Oh et al. 2013) are as follows:

- Phytosequestration—phytochemical complexation in the root zone, reduce the fraction of the contaminant that is bioavailable. Vacuolar storage in the root cells: contaminants can be sequestered into the vacuoles of root cells.
- Phytoextraction—uptake and concentration of substances from the environment into the plant biomass.
- Phytostabilization—reducing the mobility of substances in the environment, for example, by limiting the leaching of substances from the soil.

- Phytotransformation—chemical modification of environmental substances as a direct result of plant metabolism, often resulting in their inactivation, degradation (phytodegradation), or immobilization.
- Phytostimulation—enhancement of soil microbial activity for the degradation of contaminants, typically by organisms that associate with roots. This process is also known as rhizosphere degradation.
- Phytovolatilization—removal of substances from soil or water with release into the air, sometimes as a result of phytotransformation to more volatile and/or less polluting substances.
- Rhizofiltration—filtering water through a mass of roots to remove toxic substances or excess nutrients. The pollutants remain absorbed in or adsorbed to the roots.

Some important criteria involved in selection of plant species for phytoremediation are:

- The levels of tolerance with respect to metal known to exist at the site
- The level of adequacy of accumulation, translocation, and uptake potential of metals
- High growth rate and biomass yield
- Tolerance to water logging and extreme drought conditions
- Availability, habitat preference, e.g., terrestrial, aquatic, and semiaquatic, of the plants
- Tolerance to high pH and salinity

Natural sensitivity or tolerance of plants to accumulate metals is substantially affected by plant species and genotypes. In this context, plants can be divided into three groups.

1. Excluders: plants insensitive for uptake and accumulation of potentially toxic elements.
2. Indicators: majority of agricultural plants whose content of elements more or less linearly responds to increasing available content of trace elements in soil.
3. Accumulators: plants accumulating higher contents of elements in their tissues according to their increase in the soil. Hyper accumulators (extreme accumulators) plants can even prosper on contaminated soils and accumulate extremely high contents of trace elements in aboveground biomass (Baker 1987).

Heavy metals cannot be metabolized; only possible strategy to apply is their extraction from contaminated soil and transfer to the smaller volume of harvestable plants for their disposal (Padmavathiamma and Li 2007). Metals like Cadmium (Cd), Lead (Pb), Zinc (Zn), and Chromium (Cr) when present in high concentrations in soil exert potential toxic effects on overall growth and metabolism of plants and bioaccumulation of such toxic metals in the plant poses a risk to human and animal health. Heavy metal toxicity in plants is stunted growth, leaf chlorosis, and alteration in the activity of many key enzymes of various metabolic pathways. Heavy metals are toxic because they cause DNA damage and carcinogenic effects in animals and humans are probably caused by their mutagenic ability (Agarwal and

Sharma 2006; Arduini et al. 1996; Di Toppi and Gabrielli 1999; Knasmuller et al. 1998; Majer et al. 2002). Additionally, *Abutilon indicum, Catharanthus roseus,* and *Canna indica* species were tested for their ability of phytoextraction of Pb, Cd, and Cr from the contaminated soils. The experiments revealed that *Abutilon indicum* was a good accumulator of chromium, *Catharanthus roseus* was good accumulator of lead and chromium and *Canna indica* was good accumulator of chromium. The plant species was recommended for phytoextraction of lead and chromium contaminated soils (Subhashini and Swamy 2015).

2 Materials and Methods

The present study on "Potential of *Catharanthus roseus* in Phytoremediation of Heavy Metals" has been carried out during 2012–2014. *Catharanthus roseus* was selected for the present study to examine the potential to absorb the heavy metals from the soil and accumulate them in the above-ground and below-ground biomass. The plant species was affluent and native to the study area, i.e., Guntur District. The criteria followed for selection of species was their biomass, commonness, and tolerance to adverse climatic conditions (Fig. 1).

Systematic Position:	
Kingdom:	Plantae (Angiosperms, Eudicots, Asterids)
Order:	Gentianales
Family:	Apocynaceae
Genus:	*Catharanthus*
Species:	*roseus*
Binomial Name:	*Catharanthus roseus*

Catharanthus roseus (Periwinkle) is a species of *Catharanthus* native to Madagascar. Synonyms include *Vinca rosea* (the basionym), *Ammocallis rosea*, and *Lochnera rosea*; other English names occasionally used include Cape

Fig. 1 Plant and parts of *Catharanthus roseus* (L.) G. Don

Periwinkle, Rose Periwinkle, Rosy Periwinkle, and "Old-maid". It is also widely cultivated and is naturalized in subtropical and tropical areas of the world. It is an evergreen sub-shrub or herbaceous plant. The leaves are oval to oblong, 2.5–9 cm long and 1–3.5 cm broad, glossy green, hairless, with a pale midrib and a short petiole 1–1.8 cm long; they are arranged in opposite pairs. The flowers are white to dark pink with a darker red center. The fruit is a pair of follicles 2–4 cm long and 3 mm broad. The species has long been cultivated for herbal medicine and as an ornamental plant. In traditional Chinese medicine, extracts from it have been used to treat numerous diseases, including diabetes, malaria, and Hodgkin's disease. The substances vinblastine and vincristine extracted from the plant are used in the treatment of leukemia. Its alkaloids are hypotensive, sedative, and have tranquilizing properties and are anticancerous. It helps in relieving muscle pain, depression of central nervous system, and wasps stings. As an ornamental plant, it is appreciated for its hardiness in dry and nutritionally deficient conditions, popular in subtropical gardens. It is noted for its long flowering period, throughout the year in tropical conditions. Numerous cultivars have been selected, for variation in flower color (white, mauve, peach, scarlet, and reddish-orange), and also for tolerance of cooler growing conditions in temperate regions (Gamble 2008).

2.1 Selection of Plants for Experiment

The seedlings of the plants were selected from the vicinity of Acharya Nagarjuna University. Care was taken to select the seedlings of uniform size and devoid of any symptoms of disease. Ten seedlings were taken in each experimental pot. Ten such pots were taken in each experimental plant. Ten such pots were taken as one set. Out of the 12 such sets, two sets were considered as control (without addition of heavy metal solutions). Of the remaining 10 sets, two sets were used for each heavy metal. From these pots three harvests, i.e., on 20th day, 40th day, and 60th day, were taken for estimation of heavy metal accumulation in different parts of the plants such as leaves, stems and roots. The experimental pots were watered at regular intervals. For all control pots, only water without heavy metal was administered. Standard heavy metal solutions (1000 ppm) of Pb, Ni, Zn, Cd, and Cr were purchased from MERCK. These standard solutions were used as stock solution, from which 5 ppm diluted heavy metal working solutions were prepared by taking aliquots of stock solution. While administering heavy metal laden water, care was taken to prevent leaching from the pot. The alternate day water schedule was followed to prevent water logging condition in the pots. After every 20 days, plant samples were collected from each pot then dried in a hot air oven and powdered by a mortar and pestle.

2.2 Preparation of Samples for Wet Digestion

The powdered samples were subjected to acid digestion. Samples (1 g) of each part (leaves, stems, and roots) of the plant were weighed in digestion flasks and treated with 5 mL of concentrated HNO_3. One gram of the powdered plant material were weighed in separate digestion flasks and digested with HNO_3 and HCl in the ratio of 3:1. A blank sample was prepared applying 5 mL of HNO_3 into empty digestion flask. The digestion on hot plate at 110 °C for 3–4 h or continued till a clean solution was obtained. After cooling, the solution was filtered with Whatman NO. 42 filter paper. After filtering with Whatman No. 42 filter paper, the filtrate was analyzed for the metal contents in AAS.

2.3 Calculation of Bioconcentration Factor (BCF) and Translocation Factor (TF)

Heavy metals are currently of much environmental concern. They are harmful to humans, animals, and tend to bioaccumulate in the food chain. Phytoremediation is one of the promising methods for reclamation of soils contaminated with toxic metals by using hyper accumulator plants (Baker et al. 2000; Ghosh and Singh 2005; Lazaro et al. 2006). Under normal growing conditions, plants can potentially accumulate certain metal ions an order of magnitude greater than the surrounding medium (Kim et al. 2003).

2.4 Bioconcentration Factor (BCF)

Metal concentrations in plants vary with plant species. The concentration, transfer, and accumulation of metals from soil to roots and shoots was evaluated in terms of Biological Concentration Factor (BCF) and Translocation Factor (TF). Biological Concentration Factor (BCF) was calculated as metal concentration ratio of plant roots to soil (Yoon et al. 2006). The Bioconcentration Factor (BCF) of metals was used to determine the quantity of heavy metal absorbed by the plant from the soil. This is an index of the ability of the plant to accumulate a particular metal with respect to its concentration in the soil (Ghosh and Singh 2005).

2.5 Translocation Factor (TF)

Translocation Factor (TF) was described as ratio of heavy metals in plant shoot to that in plant root (Cui et al. 2007; Li et al. 2007). To evaluate the potential of the species for phytoextraction, the Translocation Factor (TF) was calculated. This ratio is an indication of the ability of the plant to translocate metals from the roots to the aerial parts of the plant. Metals that are accumulated by plants and largely stored in the roots of plants are indicated by TF values <1, with values greater indicating translocation to the aerial part of the plant (Yoon et al. 2006).

3 Results

3.1 Accumulation of Heavy Metals in Catharanthus roseus

3.1.1 Lead

Catharanthus roseus absorbed lead through root system in high quantities by 20th day itself, and the lead was translocated in a slow manner throughout the remaining period of experimentation. As a result by the 60th day, only 50% of the absorbed lead was translocated to stem and leaves. There was lowest accumulation in leaves (0.92 mg/kg) and highest in roots (67.33 mg/kg) with a total accumulation of lead (77.05 mg/kg) in the whole plant (Table 1). The results of lead accumulation in *Catharanthus roseus* during the experimental period revealed that the rate of translocation was meager (Fig. 2).

3.1.2 Nickel

Nickel accumulation in *Catharanthus roseus* was initially higher in stem followed by roots. The nickel accumulation was consistent in leaves from the beginning of the experiment. However, the 60th day the accumulation in roots increased by manifold and reached the highest followed by stem and leaves in that order (Table 2). Out of the 47.75 mg/kg of nickel accumulated in plants 20.63 mg/kg was retained in the roots, 16.63 mg/kg of nickel was accumulated in stem, and 10.49 mg/kg in leaves. Two thirds of the nickel remained in roots and stem. The results revealed that *Catharanthus roseus* is a good accumulator of nickel (Fig. 3).

Table 1 Accumulation of lead (mg/kg biomass) in different plant parts of *Catharanthus roseus* during the experimental period

Plant part	Control	20th day	40th day	60th day	Total accumulation
Leaf	24.03 ± 0.41	24.53 ± 0.15	24.5 ± 0.15	24.95 ± 0.08	0.92
Stem	60.69 ± 0.16	67.31 ± 0.18	68.09 ± 0.08	69.49 ± 0.17	8.79
Root	21.47 ± 0.17	84.32 ± 0.15	88.74 ± 0.07	88.81 ± 0.16	67.33
Total accumulation	106.19	176.16	181.33	183.25	77.05

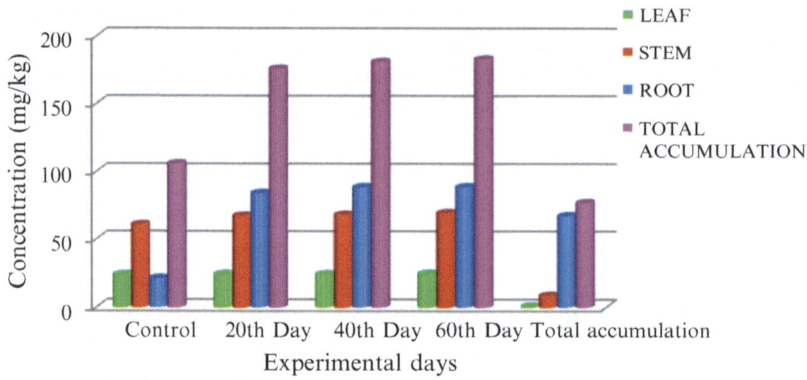

Fig. 2 Accumulation of lead in *Catharanthus roseus* during the experimental period

Table 2 Accumulation of nickel (mg/kg biomass) in different plant parts of *Catharanthus roseus* during the experimental period

Plant part	Control	20th day	40th day	60th day	Total accumulation
Leaf	2.09 ± 0.18	3.87 ± 0.15	6.18 ± 0.17	12.58 ± 0.08	10.49
Stem	5.39 ± 0.49	9 ± 0.17	9.21 ± 0.08	22.02 ± 0.17	16.63
Root	4.65 ± 0.16	5.94 ± 0.14	7.41 ± 0.07	25.28 ± 0.18	20.63
Total accumulation	12.12	18.82	22.8	59.87	47.75

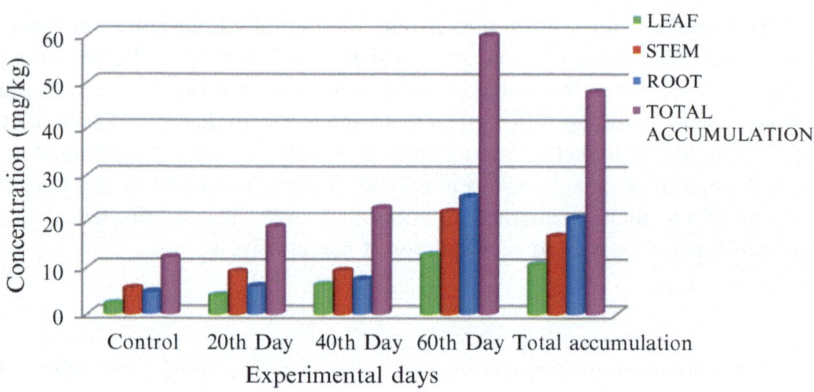

Fig. 3 Accumulation of nickel (mg/kg) in *Catharanthus roseus* during the experimental period

3.1.3 Zinc

The initial concentration of zinc was maximum in stem compared to roots and leaves. The increase of accumulation in roots and stem was consistent throughout the experimental period. However, from 40 to 60 day interval, the concentration of

Table 3 Accumulation of zinc (mg/kg biomass) in different plant parts of *Catharanthus roseus* during the experimental period

Plant Part	Control	20th day	40th day	60th day	Total accumulation
Leaf	40.62 ± 0.18	46.03 ± 0.15	46.99 ± 0.18	88.44 ± 0.09	47.82
Stem	84.83 ± 0.16	79.53 ± 0.17	81.66 ± 0.08	94.68 ± 0.17	9.85
Root	78.05 ± 0.15	87.79 ± 0.14	87.86 ± 0.07	87.98 ± 0.16	9.93
Total accumulation	203.5	213.35	216.51	271.1	67.60

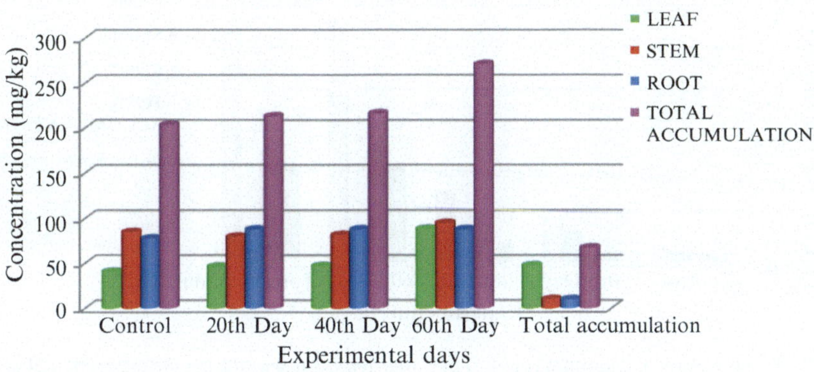

Fig. 4 Accumulation of zinc (mg/kg) in *Catharanthus roseus* during the experimental period

zinc increased to 88.44 mg/kg (from 46.99 mg/kg) in leaves. The accumulation in stem also increased from 81.66 to 94.68 mg/kg in stem. Overall, leaves recorded the highest accumulation of zinc followed by roots and stem (Table 3). The highest accumulation of zinc in leaves leaving low quantities of zinc in stem and roots reveal that maximum quantity of zinc is translocated up to leaves. Out of the total 67.60 mg/Kg of zinc absorbed, 9.93 mg/kg was translocated to root, 9.85 mg/kg to stem, and 47.82 mg/kg to leaves. The results revealed that *Catharanthus roseus* is a good accumulator of zinc (Fig. 4).

3.1.4 Cadmium

The concentrations of cadmium were very low in the initial stage. The accumulation of cadmium was the highest in the stem (13.17 mg/kg) followed by leaves (11.03 mg/kg) and the lowest in roots (3.25 mg/kg). The overall rate of accumulation in leaves, stem, and roots increased and consistently up to 40th day, and there was a sudden increase from 40 to 60th day. The results revealed that as the plant continued to grow the absorption of cadmium also increased but the cadmium absorbed by roots was completely translocated to stem and leaves. Out of the total accumulated

Table 4 Accumulation of cadmium (mg/kg) in different plant parts of *Catharanthus roseus* during the experimental period

Plant Part	Control	20th day	40th day	60th day	Total accumulation
Leaf	0.13 ± 0.18	0.97 ± 0.14	1.85 ± 0.16	11.16 ± 0.08	11.03
Stem	0.56 ± 0.16	0.79 ± 0.18	2.46 ± 0.08	13.73 ± 0.17	13.17
Root	0.84 ± 0.15	3.37 ± 0.19	3.81 ± 0.06	4.09 ± 0.18	3.25
Total accumulation	1.53	5.13	8.12	28.98	27.45

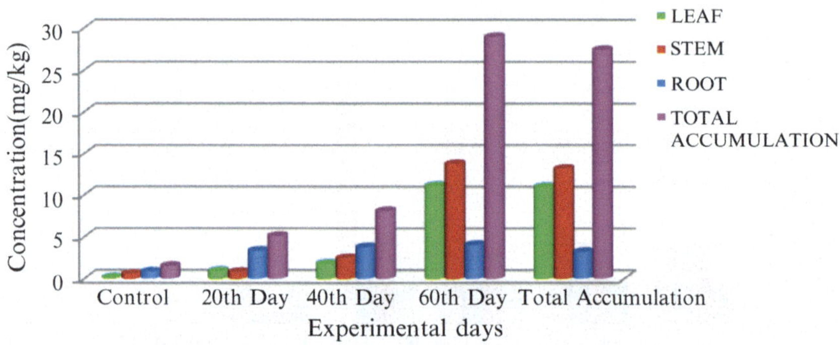

Fig. 5 Accumulation of cadmium (mg/kg) in *Catharanthus roseus* during the experimental period

27.45 mg/kg of cadmium roots retained only 3.25 mg/kg, and the remaining was translocated to stem (13.17 mg/kg) and leaves (11.03 mg/kg) (Table 4) (Fig. 5).

3.1.5 Chromium

Catharanthus roseus is a good accumulator of chromium. From the experiments conducted it is recorded that the roots have accumulated the highest chromium content (6.7–36.73 mg/kg), and then the absorption of chromium was consistent throughout. The accumulation in stem was very less compared to leaves and roots. Chromium accumulation in leaves was doubled from 20th to 40th day and threefold for 40th day to 60th day finally resulting in the highest accumulation of chromium in leaves followed by roots and stem, in that order (Table 5) (Fig. 6).

Catharanthus roseus showed a differential tendency of accumulation of different metals. The leaves have accumulated the lowest quantities of lead and nickel while zinc accumulated in the highest quantities (47.82 mg/kg) in leaves. The leaves showed a wide variation in affinity of accumulation, i.e., from 0.92 mg/kg of lead to 47.82 mg/kg of zinc. The total accumulation of metals showed a moderate range, the lowest being cadmium (27.44 mg/kg) followed by nickel (47.75 mg/kg), zinc (67.61 mg/kg), chromium (69.31 mg/kg), and lead (77.05 mg/kg) in the ascending order. Among the three plant parts, roots showed the highest accumulation

Table 5 Accumulation of chromium (mg/kg) in *Catharanthus roseus* during the experimental period

Plant part	Control	20th day	40th day	60th day	Total accumulation
Leaf	5.4 ± 0.18	6.52 ± 0.15	12.15 ± 0.16	37.13 ± 0.09	31.72
Stem	13.57 ± 0.16	19.49 ± 0.18	19.81 ± 0.08	20.27 ± 0.17	6.7
Root	6.07 ± 0.15	36.73 ± 0.14	36.57 ± 0.07	36.96 ± 0.16	30.89
Total accumulation	25.04	62.74	68.53	94.36	69.31

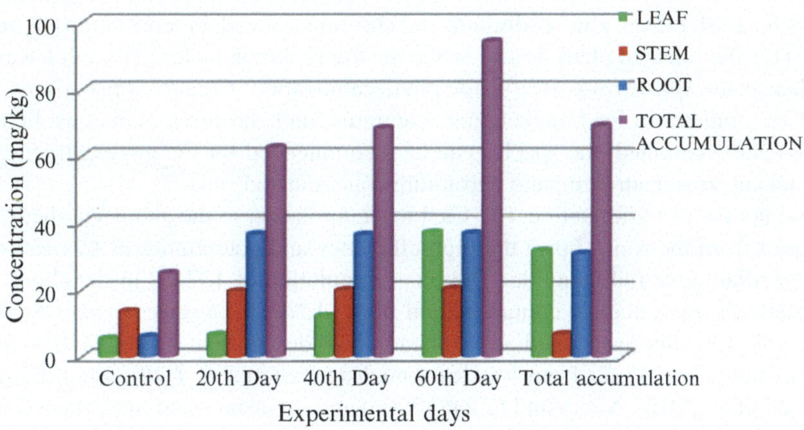

Fig. 6 Accumulation of chromium in *Catharanthus roseus* during the experimental period

of lead and nickel, while stem accumulated the highest quantities of cadmium, and no other metal accumulated in the highest quantity in stem compared with leaves and roots. Lead showed a very slow increase in leaves and stem from 20 to 40 and 40 to 60 while the roots showed a faster accumulation of lead during the first 20 days and then on the accumulation slowed down. Also it was evident most of the lead absorbed remained in the roots itself and the translocation was very poor. The strong root system and considerable biomass of the root system are favoring the accumulation of lead. *C. roseus* could well tolerate low amounts of chromium (and accumulate it to about 22% in leaves) and might, thus, prove useful in the reclamation and remediation of chromium contaminated soil and land (Ahmad and Mishra 2014) (Table 6).

Phytoextraction ensure elimination of the metal from the soil, as the plant absorbs the metals and store them in the roots, stem, and leaves. Phytostabilization ensures adsorption metals on roots, concentration or agglomeration or precipitation of metals in rhizosphere. Zinc and chromium was highly accumulated in leaves of *Catharanthus roseus*. Nickel and cadmium was highly accumulated in the stem of *Catharanthus roseus*. Lead was highly accumulated in roots of *Catharanthus roseus*. The TF value increases with increasing ability of the plant to translocate metals to stem and leaves. Thus, the plants showing high BCF and TF values (greater than one)

Table 6 Bioconcentration factor (BCF) and translocation factor (TF) of heavy metals in the experimental plant *Catharanthus roseus*

Metal	Bioconcentration factor	Translocation factor
Pb	8.13	0.15
Ni	5.66	1.31
Zn	5.17	5.8
Cd	74.99	7.43
Cr	5.1	1.24

are suitable for phytoextraction. While the plants showing TF value less than one can be used for phytostabilization. In the present study, the plants have shown varied BCF and TF values for each metal. *Catharanthus roseus* exhibited the highest BCF values for lead, nickel, zinc, cadmium, and chromium. Lead, bioconcentration factor (BCF) for the selected plant was >1 while the translocation factor (TF) value was <1 and hence the species was useful for phytostabilization. *Catharanthus roseus* was good accumulator of lead, nickel, zinc, cadmium, and chromium. On consolidation of the results obtained, the species can be recommended for the phytoextraction of lead, nickel, zinc, cadmium, and chromium contaminated soils.

The uptake of radionuclide-137 Cs, bioaccumulation in the shoot via the active transport from the root, shows the high efficiency and potentiality of *C. roseus* for the remediation of radionuclide. The bioaccumulation of 137 Cs in the shoot will remediate the radionuclide contamination from LLNW. *C. roseus* can also be made applicable for effective remediation of radionuclides present in the LLNW (phytoremediation of 137 Cs from low level nuclear waste using *Catharanthus roseus* (Fulekar et al. 2010). Misra and Gupta (2006) and Srivastava and Srivastava (2010) have evaluated the phytoremediation potential of *C. roseus* and study the effect of Cr stress on growth characteristics and alkaloid content of *C. roseus* and observed the production of over 100 alkaloids including vincristine, vinblastine, ajmaline, ajmalicine, and serpentine which are extremely important. Pandey et al. (2010) have studied the impact of cadmium and lead on *C. roseus.* Zheng and Wu (2004) have reported that cadmium treatment enhanced the production of alkaloid and secondary metabolites in *C. roseus.* It is known that under stress condition plants generally shift a major portion of their metabolic activities towards secondary metabolite synthesis, so an increase in alkaloid content was expected. In the present study, the total accumulation of metals showed a moderate range, the lowest being cadmium followed by Ni, Zi, Cr, and lead in that ascending order. Among the three plant parts root showed the highest accumulation of lead and nickel, while stem accumulated the highest quantities of Cd and no other metal accumulated in the highest quantity in stem compared with leaves and stem.

Huang and Cunningham (1996) and Blaylock et al. (1997) found that plants can remove between 180 and 530 kg/ha of Pb/year, making remediation of sites contaminated with up to 2500 mg/kg possible in fewer than 10 years. The highest BCF was recorded in soil polluted with 10 ppm Pb. This may be due to the fact that at moderately low concentration of lead in soil, plants tend to accumulate more metals than higher concentrations (Benzarti et al. 2008) while the lowest was recorded in soil polluted with 25 ppm. In the experimental plant *Catharanthus roseus, BCF values* of lead are >1 and TF values <1 are useful for phytostabilization, the species recommended for phytostabilization of lead contaminated soils *C. roseus* exhibited the highest BCF value and TF value for nickel. The plant species have good potential for phytoremediation of nickel contaminated soils. The chromium accumulation was the highest in leaves than stem and roots.

The ability of a plant to accumulate metals from contaminated soils was evaluated by the BCF, according to studies of Yadav et al. (2009). This study assumed that plants with BCF values >1 are accumulators, while plants with BCF values <1 are excluders (Baker 1981). Additionally, plants were classified as potential hyper accumulators if the BCF values were >10 (Ma et al. 2001). The present study has shown BCF values >1 for zinc, the species are potential species to be used for phytoextraction. The BCF and TF values for cadmium were >1 the study revealed that *C. roseus* was useful for phytoextraction of chromium. The function of zinc is to help a plant to produce chlorophyll. Leaves get discolored when the soil is deficient in zinc and plant growth is stunted (Kumar 1984). Zinc deficiency causes leaf discoloration called chlorosis tissue of the veins to turn yellow. Chlorosis by zinc deficiency usually affects the base of the leaf near the stem. Excess Zn can also give rise to manganese (Mn) and copper (Cu) deficiencies in plant shoots. Such deficiencies have been ascribed to a hindered transfer of these micronutrients from root to shoot. This hindrance is based on the fact that the Fe and Mn concentrations in plants grown in Zn-rich media are greater in the root than in the shoot (Ebbs and Kochian 1997). The BCF and TF values of *C. roseus* for chromium was >1, have good potential to be used as phytoextractor of chromium. The accumulation of zinc was the highest in the leaves of *C. roseus*. The accumulation in leaves was the highest followed by root and stem. In the experimental plant, nickel was highly accumulated in roots followed by stem and leaves.

Catharanthus roseus was good accumulator of lead, nickel, zinc, and chromium. The species can be recommended for the phytoextraction of lead, nickel, zinc, and chromium contaminated soils. Ni concentration is increasing in certain areas by human activities such as mining works, emission of smelters, burning of coal and oil, sewage, phosphate fertilizers, and pesticides (Gimeno Gatcia et al. 1996). Excess of Ni in soil causes various physiological alterations and diverse toxicity symptoms such as chlorosis and necrosis in different plant species, including rice (Zornoza et al. 1999; Rahman et al. 2005). In the present investigation, the Phytoremediation potential of *C. roseus* as evinced by its bioaccumulation coefficient (BAC) with respect to chromium, as well as the status and developmental activity profiles of enzymes POD and GST in leaves of plants grown in sludge amended soil were determined and evaluated. The ability of phytoremediation has commonly been characterized by a TF (Baker 1987; Kabata-Pendias and Pendias 1992; Yoon et al. 2006) which is defined as the ratio of the metal concentration in the shoots to that in the roots. Plants with TF values >1 are classified as high-efficiency plants for metal translocation from the roots to shoots (Ma et al. 2001).

Soil pH is a major factor influencing the availability of elements in the soil for plant uptake (Marschner 1995). Under acidic conditions, H^+ ions displace metal cations from the cation exchange complex (CEC) of soil components and cause metals to be released from sesquioxides and variable-charged clays to which they have been chemisorbed (McBride 1994). The retention of metals to soil organic matter is also weaker at low pH, resulting in more available metal in the soil solution for root absorption. Many metal cations are more soluble and available in the soil solution at low pH (below 5.5) including Cd, Cu, Hg, Ni, Pb, and Zn (McBride 1994; Blaylock and Huang 2000).

The accumulation of metal ions by root systems is a key function in terrestrial plants, which exhibit extensive ramifications through soil. Distribution of heavy metals in plant body depends upon availability and concentration of heavy metals as well as particular plant species and its population. For instance, roots usually show higher heavy metal concentration than shoots because they are the origin, which comes into contact with the toxic metals present in the soil (Breckle 1991).

4 Conclusions

Phytoremediation is defined as "the efficient use of plants to remove, detoxify or immobilize environmental contaminants in a growth matrix (soil, water, or sediments) through the natural biological, chemical, or physical activities and processes of the plants.". Plants are unique organisms equipped with remarkable metabolic and absorption capabilities, as well as transport systems that can take up nutrients or contaminants selectively from the growth matrix, soil, or water. The plants can be subsequently harvested, processed, and disposed. Many studies have reported the application of these technologies for the removal of metals such as lead, nickel, selenium, zinc, cadmium, chromium, arsenic, and manganese from the contaminated sites. The aim of the present study was to evaluate the metal accumulation capacity of the *C. roseus*. Aqueous solutions of the (lead, nickel, zinc, cadmium, and chromium) metals added to the plant samples, about 60 days (2 months). The species can be recommended for the phytoextraction of Pb, Ni, Zn, Cd, and Cr contaminated soils.

References

Agrawal V, Sharma K (2006) Phototoxic effects of Cu, Zn, Cd and Pb on in vitro regeneration and concomitant protein changes in *Holarrhena antidysentrica*. Biol Plant 50:307–310

Ahmad R, Mishra N (2014) Evaluation of Phytoremediation potential of *Catharanthus roseus* with respect to chromium contamination. Am J Plant Sci 5:2378–2388

Arduini I, Godbold DL, Onnis A (1996) Cadmium and copper uptake and distribution in Mediterranean tree seedlings. Physiol Plant 97:111–117

Baker AJM (1981) Accumulators and excluder-strategies in the response of plants to heavy metals. J Plant Nutr 3:643–654

Baker AJM (1987) Metal tolerance. New Phytol 106:93–111

Baker AJM, McGrath SP, Reeves RD, Smith JAC (2000) Metal hyperaccumulator plants: a review of the ecology and physiology of a biological resource for Phytoremediation of metal-polluted soils. In: Terry N, Banuelos GS (eds) Phytoremediation of contaminated soil and water Florida. Lewis Publishers, Boca Raton, pp 85–107

Benzarti S, Mohri S, Ono Y (2008) Plant response to heavy metal toxicity: comparative study between the Hyperaccumulator *Thlaspi caerulescens* (ecotype ganges) and nonaccumulator plants: lettuce, radish, and alfalfa. Environ Toxicol 23:607–616

Blaylock MJ, Salt DE, Dushenkov S, Zakharova O, Gussman C (1997) Enhanced accumulation of Pb in Indian mustard by soil-applied chelating agents. Environ Sci Technol 31:860–865

Blaylock MJ, Huang JW (2000). Phytoextraction of metals. In: I. Raskin and B.D. Ensley (eds) Phytoremediation of toxic metals using plants to clean up the environment. Wiley: New York. pp. 53–70

Breckle CW (1991) Growth under heavy metals. In: Waisel Y, Eshel A, Kafkafi U (eds) Plant roots: the hidden half. Marcel Dekker, New York, pp 351–373

Bu Olayan AH, Thomas BV (2009) Translocation and bioaccumulation of trace metals in desert plants of Kuwait Governorates. Res J Environ Sci 3(5):581–588

Chowdhury ASMHK, Das P, Sarkar I, Islam AL, Parvin F, Islam Z, Faris M, Cui S, Zhou Q, Chao L (2015) Potential hyper-accumulation of Pb, Zn, Cu and Cd in endurant plants distributed in an old smeltery, Northeast China. Environ Geol 51:1043–1048

Cui S, Zhou Q, Chao L (2007) Potential hyper-accumulation of Pb, Zn, Cu and Cd in endurant plants distributed in an old smeltery, northeast China. Environ Geol 51:1043–1048

Cunningham SD, Berti WR, Huang JW (1995) Phytoremediation of contaminated soils. Trends Biotechnol 13:393–397

Di Toppi LS, Gabbrielli R (1999) Response to cadmium in higher plants. Environ Exp Bot 41(2):105–130

Ebbs SD, Kochian LV (1997) Toxicity of zinc and copper to Brassica species: implications for phytoremediation. J Environ Qual 26:776–781

Ensley BD (2000) Phytoremediation for toxic metals—using plants to clean-up the environment. In: Raskin I, Ensley BD (eds) Rational for use of phytoremediation. Wiley, New York, pp 3–13

Fulekar MH, Singh A, Thorat V, Kaushik CP, Eapen S (2010) Phytoremediation of 137 Cs from low level nuclear waste using *Catharanthus roseus*. Indian J Pure Appl Phys 48:516–519

Gamble JS (2008) Flora of the Presidency of Madras. Bishen Singh Mahendra Pal Singh Publishers, Dehra Dun

Ghosh M, Singh SP (2005) A comparative study of cadmium phytoextraction by accumulator and weed species. Environ Pollut 133:365–371

Gimeno Gatcia E, Andreu V, Boluda R (1996) Heavy metals incidence in the application of inorganic fertilizers and pesticides to rice farming soils. Environ Pollut 92:19–25

Huang JW, Cunningham SD (1996) Lead phyto extraction: species variation in lead uptake and translocation. New Phytol 134:75–84

Kabata-Pendias A, Pendias H (1992) Trace Elements in Soils and Plants. CRC Press, Boca Raton, London

Kim B, Oh E, So J, Ahn Y, Koh S (2003) Plant terpene-induced expression of multiple aromatic ring hydroxylation oxygenase genes in *Rhodococcus* sp. strain T104. J Microbiol 41:349

Knasmuller S, Gottmann E, Steinkellner H, Fomin A, Pickl C, Paschke A, God R, Kundi M (1998) Detection of genotoxic effects of heavy metal contaminated soils with plant bioassays. Mutat Res 420:37–48

Kord B, Mataji A, Babaie S (2010) Pine (Pinus Elda rica Medw.) needles as indicator for heavy metals pollution. Int J Environ Sci Tech 7(1):79–84

Kumar N (1984) Effect of Zn X P fertilization on the growth of soyabean (*Glycin max* L.). M.Sc. thesis, C.S. Azad University of Agriculture and Technology, Kanpur, UP, India

Lazaro JD, Kidd PS, Martinez CM (2006) A phytogeochemical study of the Trasos Montes region (NE Portugal): possible species for plant-based soil remediation technologies. Sci Total Environ 354:265–277

Li MS, Luo YP, Su ZY (2007) Heavy metal concentrations in soils and plant accumulation in a restored manganese mineland in Guangxi, South China. Environ Pollut 147:168–175

Ma LQ, Komar KM, Tu C, Zhang W, Cai Y, Kennelley ED (2001) A fern that accumulates arsenic. Nature 409:579

Majer BJ, Tscherko D, Paschke A (2002) Effects of heavy metal contamination of soils on micronucleus induction in Tradescantia and on microbial enzyme activities: a comparative investigation. Mutat Res 515:111–124

Marschner H (1995) Mineral nutrition of higher plants, 2nd edn. Academic Press, London, p 889

McBride MB (1994) Environmental chemistry of soils. New York, Oxford University Press. p. 406

McIntyre T (2001) PhytoRem: A Global CD-ROM database of aquatic and terrestrial plants that sequester, accumulate, or hyperaccumulate heavy metals. Hull: Environment Canada

Meagher RB (2000) Phytoremediation of toxic elemental and organic pollutants. Curr Opin Plant Biol 3:153–162

Misra N, Gupta AK (2006) Effect of salinity and different nitrogen sources on the activity of antioxidant enzymes and indole alkaloid content in *Catharanthus roseus* seedlings. J Plant Physiol 163:11–18

Nriagu JO (1979) Production and uses of mercury. In: Nriagu JO (ed) The biogeochemistry of mercury in the environment. Elsevier/North Holl and Biomedical Press, Amsterdam

Oh K, Li T, Cheng HY, Xie Y, Yonemochi S (2013) Development of profitable phytoremediation of contaminated soils with biofuel crops. J Environ Protect 4:58–64

Padmavathiamma PK, Li LY (2007) Phytoremediation technology: hyperaccumulation metals in plants. Water Air Soil Pollut 184:105–126

Pandey S, Gupta K, Mukherjee AK (2010) Impact of cadmium and lead on *Catharanthus roseus -A* phytoremediation study. J Environ Biol 28:655–662

Prasad MNV (2003) Metal hyperaccumulators in plants- biodiversity prospecting for phytoremediation technology. Electron J Biotechnol 6:276–372

Rahman H, Sabreen S, Alam S, Kawai S (2005) Effects of nickel on growth and composition of metal micronutrients in barley plants grown in nutrient solution. J Plant Nutr 28:393–404

Raskin I, Nanda Kumar PBA, Dushenkov S, Salt DE (1994) Bioconcentration of heavy metals by plants. Curr Opin Biotechnol 5:285–290

Salt DE, Blaylock M, Kumar PBAN, Dushenkov V, Ensley BD, Chet L, Raskin L (1995) Phytoremediation: a novel strategy for the removal of toxic metals from the environment using plants. Biotechnology 13(2):468–474

Sekabira K, Oryem Origa H, Basamba TA, Mutumba G, Kakudidi E (2011) Application of algae in biomonitoring and phytoextraction of heavy metals contamination in urban stream water. Int J Environ Sci Tech 8:115–128

Srivastava NK, Srivastava AK (2010) Influence of some heavy metals on growth alkaloid content and composition in *Catharanthus roseus* L. Indian J Pharmaceut Sci 72:775–778

Subhashini V, Swamy AVVS (2015) Phytoremediation of lead, cadmium and chromium contaminated soils using selected weed plants. Acta Biologica Indica 4:205–212

Wang XJ, Chen L, Xia SQ, Zhao JF, Chovelon JM, Renault NJ (2006) Biosorption of Cu (II) and Pb (II) from aqueous solutions by dried activated sludge. Miner Eng 19:968–971

Yadav SK, Juwarkar AA, Kumar GP, Thawale PR, Singh SK, Chakrabarti T (2009) Bioaccumulation and Phytotranslocation of Arsenic, chromium and zinc by Jatropa curcas L, impact of dairy sludge and biofertilizer. Biores Technol 100(20):4616–4622

Yoon J, Cao X, Zhou Q, Ma LQ (2006) Accumulation of Pb, Cu, and Zn in native plants growing on a contaminated Florida site. Sci Total Environ 368:456–464

Zhang S, Chen M, Li T, Xu X, Deng L (2010) A newly found cadmium accumulator *Malva sinensis* Cavan. J Hazard Material 173:705–709

Zheng Z, Wu M (2004) Cadmium treatment enhances the production of alkaloid secondary metabolites of Catharanthus roseus. Plant Sci 166:507–514

Zornoza P, Robles S, Martin N (1999) Alleviation of nickel toxicity by ammonium supply to sunflower plants. Plant and Soil 208:221–226

Unraveling the Cumulative Effect of Soil-Applied Radiation-Processed Sodium Alginate and Polyacrylamide on Growth Attributes, Physiological Activities, and Alkaloids Production in Periwinkle [*Catharanthus roseus* (L.) G. Don]

Asfia Shabbir, Akbar Ali, Yawar Sadiq, Hassan Jaleel, Bilal Ahmad, M. Naeem, M. Masroor A. Khan, and Moin Uddin

Abstract Cancer has taken millions of lives in the passing decade alarming the scientists to urgently design the magic bullets for this jeopardy. *Catharanthus roseus* (L.) G. Don reportedly emerged as a remedy to this problem contributing, medicinal system, two of its paramount anticancer alkaloid constituents, vincristine and vinblastine. Recently, a new agro-technique has been evolved in which gamma-irradiated natural polysaccharides (e.g., alginate, carrageenan, and chitosan) are degraded into oligomers. When these oligomers are applied through foliar application to the plants, it proves as potent plant promoter. On the other hand, for quality soil structure, polyacrylamide (PAM) is used whose application to the soil contributes a great deal in terms of stability to soil and increased water holding capacity. A pot experiment was conducted to study the effect of soil-applied polyacrylamide (PAM) and irradiated sodium alginate (ISA) alone or in combination on *C. roseus*. The plants were harvested at 90 days after transplanting (DAP). Of the six treatments [(i) water sprayed control (Treatment 1), (ii) un-irradiated sodium alginate (UISA) (Treatment 2), (iii) irradiated sodium alginate (ISA) (Treatment 3), (iv) PAM (Treatment 4), (v) UISA + PAM (Treatment 5), and (vi) ISA + PAM (Treatment 6)] increased the values of most of the parameters significantly including fresh and dry weights of plants, contents of nitrogen, phosphorus and potassium, chlorophyll and carotenoids content, and activities of carbonic anhydrase and nitrate reductase in the leaves. The combined application of ISA + PAM also increased total content

A. Shabbir (✉) • A. Ali • Y. Sadiq • H. Jaleel • B. Ahmad • M. Naeem
M.M.A. Khan
Plant Physiology Section, Department of Botany, Aligarh Muslim University,
Aligarh, Uttar Pradesh 202002, India
e-mail: ashabbir164@gmail.com

M. Uddin
Botany Section, Women's College, Aligarh Muslim University,
Aligarh, Uttar Pradesh 202002, India

M. Naeem et al. (eds.), *Catharanthus roseus*,
DOI 10.1007/978-3-319-51620-2_16

and yield of leaf alkaloid by 14.1% and by 40.2% over the respective controls. Conclusively, the technique seems economically and ergonomically successful in increasing the expensive alkaloid production of *C. roseus*.

Keywords *Catharanthus roseus* • Alkaloids yield • Enzymes activity • Irradiated sodium alginate • Polyacrylamide

1 Introduction

The ongoing research, in the present times, is targeted at the exigency to fortify human health at all possible levels, therefore, keeping the diseases at the bay in the global context. Cancer, a deadly outbreak, sets the alarm ringing for researchers to take some imperative actions against this noxious menace. As per reports of Worldwide Cancer 2014, an estimated 14.1 million new cases of cancer occurred worldwide while 8.2 million people died from it during 2012 (Anonymous 2014). The American Cancer Society revealed deaths arising from cancer constitute 2–3% of the annual deaths worldwide. Plant kingdom exhibits endless biological and chemical diversity which eventually proves to be an extraordinary resource for the discovery of new antineoplastic drugs. Out of diverse plant species existing on earth, approximately 1000 species have anticancer potential (Bibi et al. 2012). Of the several medicinal plants, *Catharanthus roseus* (L.) G. Don commonly known as periwinkle, a member of Apocyanaceae family manifests the remedy of neoplasm (tumor or cancer) owing to its rich alkaloid content. In India, two varieties are commonly known, namely, "Rosea" and "Alba." The "Rosea" variety is reported to be superior over "Alba" in overall growth performance and alkaloids production (Idrees et al. 2010). The plant is a rich source of alkaloids which are distributed in all parts of the plant mainly in roots and leaves. The most important antineoplastic alkaloids, vinblastine and vincristine, are mainly present in the leaves, while the antihypertensive alkaloids such as ajmaline, serpentine, and reserpine are found in roots (Singh et al. 2000).

Recently, a novel agro-technique where gamma-rays irradiation is employed to degrade and lower down the molecular weight of some natural marine polysaccharides like alginates, chitosan, and carrageenan into small sized oligomers. These radiation-processed oligomers proved to be potent plant promoters when applied to plants through hydroponics or foliar sprays. Out of several natural polysaccharides, sodium alginate is a polysaccharide with its large quantity available in nature as structural part of brown algae *Sargassum* (Anthony et al. 2007). Irradiated sodium alginate (ISA) showed various biological effects on plants including enhanced seed germination, shoot elongation, and root growth (Yonemoto et al. 1993; Natsume et al. 1994; Hu et al. 2004; Idrees et al. 2012a; Ali et al. 2014). It also acts as the endogenous elicitor to promote the production of certain enzymes and plant growth (Akimoto et al. 1999; Khan et al. 2011; Sarfaraz et al. 2011; Naeem et al. 2011, 2015a; Aftab et al. 2013, 2016; Idrees et al. 2016). Besides, ISA has been reported to enhance the production of alkaloids and essential oil in opium poppy, lemongrass and periwinkle, respectively (Khan et al. 2011; Idrees et al. 2011, 2014; Naeem et al. 2015b).

Polyacrylamide (PAM), on the other hand, is a well-known soil conditioner found to improve the soil environment. It is grouped in a class of compounds formed by the polymerization of acrylamide (Barvenik 1994). The PAM application improves physical soil parameters such as hydraulic properties, infiltration rate, aeration, root penetration, and aggregate stability, thereby boosting up the plant establishment and growth rate (Flanagan et al. 2002; Hayat and Ali 2004; Sojka et al. 2007; Tümsava and Kara 2011; Keshavarz and Farahbakhsh 2012; Fazeli Rostampour et al. 2012; Gharache et al. 2013).

Considering the significance of *C. roseus* in the modern system of medicine, the importance that ISA holds as a plant growth promoter and the soil binding properties PAM unveil, an attempt was made to raise the cultivation of plants so as to amplify the production of its alkaloids on scientific lines. Therefore, a pot experiment was conducted to study the effect of ISA and PAM alone or in combination and to test whether any of them could accentuate the growth characteristics, physiological and biochemical parameters, and the content of alkaloids of periwinkle.

2 Materials and Methods

2.1 Chemicals

Sodium alginate was purchased from Sigma-Aldrich, USA. Polyacrylamide (PAM) was purchased from Wallace Laboratories, 365 Coral Circle, El Segundo, CA, USA.

2.2 Scanning Electron Microscopy (SEM) Analysis

The sodium alginate samples were examined for morphological structure using the Scanning Electron Microscope (Philips XL 30 ESEM, Jeol, Japan). The samples were coated with gold. Scanning electron microscopy and elemental analysis was performed for SEM analysis for un-irradiated as well as irradiated sodium alginate powder at the University Sophisticated Instrumentation Facility (USIF), Aligarh Muslim University, Aligarh, India. The SEM report clearly indicates that the gamma rays degraded the sodium alginate molecules by 1/12 to 1/10 times (approximately) (Fig. 1).

2.3 Plant Materials and Growth Conditions

Healthy seedlings of *C. roseus* cv. "Rosea" were obtained from Jawahar Horticulture Park, Aligarh, India. Uniform size of those seedlings were transplanted in earthen pots (25 cm diameter and 25 cm height) filled with 5 kg of homogenous mixture of

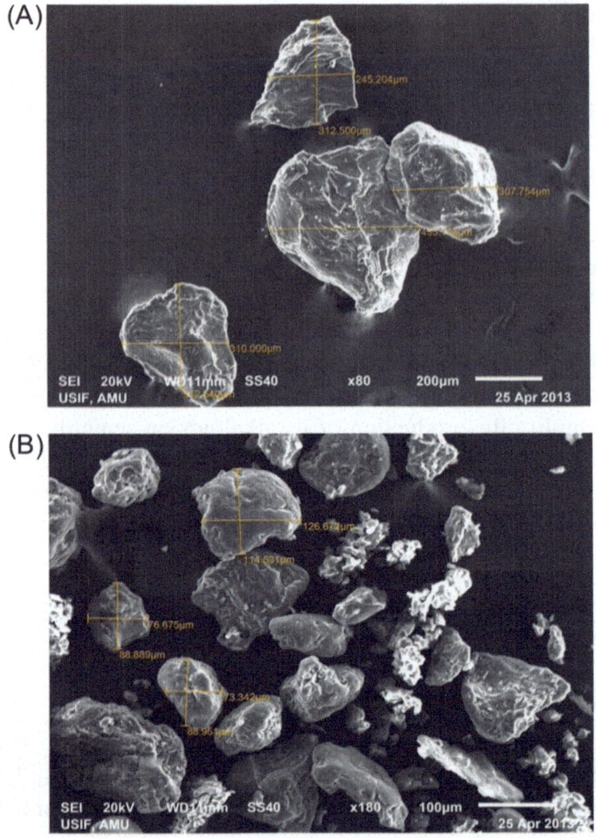

Fig. 1 (**A**) SEM images of un-irradiated sodium alginate. (**B**) SEM images of irradiated sodium alginate with gamma radiation

the field soil mixed with cow dung manure in the ratio of 4:1 and one seedling per pot was maintained. The soil samples were tested at the Government Soil Testing Laboratory, Quarsi Farm, Aligarh. Physicochemical characteristics of the soil were as follows: texture—sandy loam, pH (1:2)–6.6, E.C. (1:2)–0.48 m mhos cm^{-1}, available N, P, and K—98.5, 11.2, and 146.0 mg kg^{-1} soil, respectively.

2.4 Experimental Design, Growth, and Yield Analyses

A pot experiment was conducted according to simple randomized design in the natural conditions of Net-house at the Botany Department, A.M.U., Aligarh, India (27° 88′ N latitude, 78° 08′ E longitude, and 187.45 m altitude). Each treatment was replicated three times and treatments were applied through soil. Six treatments (water sprayed control (Treatment 1), un-irradiated sodium alginate (UISA) (Treatment 2), irradiated sodium alginate (ISA) (Treatment 3), PAM (Treatment 4), UISA + PAM (Treatment 5), and ISA + PAM (Treatment 6) were applied. UISA and

ISA were given at the rate of 10 mg kg^{-1} of soil and PAM was given at 40 mg kg^{-1} of soil. Standard agricultural practices were adopted during the whole period of crop growth and harvesting.

Growth and biochemical attributes of the crop were determined at 90 days after transplanting (DAP). Analysis of the crop was performed in terms of growth and other physiological attributes and total alkaloid content. At 90 DAP, plants from each treated pot were uprooted and washed carefully with tap water to remove all adhering foreign particles, then wiped off using blotting sheet and the fresh weights of plants were recorded thereafter. The plants were dried at 80 °C for 24 h using a hot air oven, and the dry weights of the plants were recorded subsequently. The above-ground plant height (shoot length) and root length were measured by meter scale. The total number of leaves per plant was counted in all the replicates of each treatment and the average leaf area was determined with the help of graph paper.

2.5 *Total Chlorophyll and Carotenoids Content*

The total chlorophyll content was estimated in fresh leaves collected from each pot by Mac Kinney (1941) method. Optical density (OD) was recorded at 645 and 663 nm wavelength for total chlorophyll content using a spectrophotometer (UV-1700 Shimadzu, Japan). Total carotenoids content was estimated by the method of Mac Lachlan and Zalik (1963). The OD was recorded at 480 and 510 nm wavelength. The contents of photosynthetic pigments were expressed as mg g^{-1} FW. The detailed procedure for the estimation of total chlorophyll content and total carotenoids content has previously been described by Idrees et al. (2012a) and Naeem et al. (2014).

2.6 *Activities of Nitrate Reductase (NR) and Carbonic Anhydrase (CA)*

Nitrate reductase (E.C. 1.7.1.1) activity was estimated by the intact tissue assay method developed by Jaworski (1971). The amount of nitrite formed was determined spectrophotometrically. The OD of the contents was recorded at 540 nm using the spectrophotometer. Activity of NR was expressed as nano moles of nitrite produced per gram fresh weight leaf tissue per hour nmol NO_2^{-1} g^{-1} FW h^{-1}.

The carbonic anhydrase (E.C. 4.2.1.1) activity in the fresh leaves was determined using the method described by Dwivedi and Randhawa (1974). The activity of CA was expressed as micromoles of CO_2 produced per kilogram of fresh leaf tissue per second (μmol CO_2 kg^{-1} FW s^{-1}). The detailed procedure for the estimation of activities of NR and CA can be followed from the previous works of Idrees et al. (2012a) and Naeem et al. (2014).

2.7 Estimation of N, P, and K Contents in Leaves

Leaf samples from each treatment were digested for the estimation of leaf-N, -P, and -K contents. The leaves were dried in a hot air oven at 100 °C for 24 h. The dried leaves were ground using a mortar and pestle and the leaf powder was sieved. The sieved leaf powder was used for the estimation of N, P, and K contents. The oven-dried leaf powder (100 mg) was carefully transferred into a digestion tube, to which 2 mL of AR (analytical reagent) grade concentrated sulfuric acid was added subsequently. The aliquot (peroxide-digested material), thus prepared, was used to estimate the percent N, P, and K contents in the leaves on dry weight basis.

2.7.1 Determination of N Content

Leaf-N content was estimated according to method of Lindner (1944) with slight modification by Novozamsky et al. (1983). The dried leaf-powder samples were digested with H_2SO_4 in the digestion tubes using temperature controlled Kjeldahl assembly. The OD (optical density) of the solution was recorded at 525 nm using the spectrophotometer.

2.7.2 Determination of P Content

The method of Fiske and Subbarow (1925), with slight modification by Rorison et al. (1993), was used to estimate the leaf-P content in the peroxide-digested material. The OD of the solution was recorded at 620 nm using the spectrophotometer.

2.7.3 Determination of K Content

Leaf-K content was determined according to Hald (1947), in the peroxide-digested material by a flame-photometer (Model, C150, AIMIL, India) with the help of emission spectra using specific filter.

The detailed procedure for the estimation of leaf-N, -P, and -K content can be followed from the previous work of Ali et al. (2014).

2.8 Total Alkaloids Content in Leaves

Total alkaloid content was estimated in leaves as described by Afaq and Tajuddin (1994). The method for the estimation of total alkaloids content has been described in detail by Naeem et al. (2015a). Total alkaloids content in leaves was calculated using the following formula:

$$\text{Total alkaloid content}(\%) = \frac{WA - \text{WE}}{WR} \times 100$$

where
WE = weight of empty porcelain dish (g)
WA = weight of porcelain dish after evaporation (g)
WR = weight of the dried powder (g)

2.9 Statistical Analysis

The data were analyzed statistically with SPSS-17 statistical software (SPSS Inc., Chicago, IL, USA). Mean values were statistically compared by Duncan's multiple range test (DMRT) at $P < 0.05\%$ level.

3 Result

The enhancement in the values for some of the parameters studied due to soil-applied ISA and PAM might be considered a valuable observation. Treatment 6 (ISA + PAM) gave the maximum value for most of the parameters studied. This was followed by the Treatment 5 (UISA + PAM). However, the control registered minimum value for most of the studied attributes.

3.1 Growth Attributes

The application of ISA had pronounced effect on various growth parameters. Thus, ISA in combination with PAM significantly affected the studied parameters. Application of ISA together with PAM gave maximum value for shoot fresh weight at 90 DAP. ISA + PAM recorded 21.7% higher value for shoot fresh weight per plant over the control (Table 1). A progressive increase in shoot dry weight was noted by the application of ISA combined with PAM. ISA + PAM proved optimum and enhanced shoot dry weight by 12.3% over the control. However, the effect of ISA alone was significantly similar in increasing shoot dry weight per plant (Table 1). Treatment 6 (ISA + PAM) also proved the best for fresh and dry weights of root exhibiting enhancement of 23.2 and 16.7%, respectively, over the control. The effect of ISA combined with PAM was found significant on the shoot length per plant. Treatment 6 (ISA + PAM) enhanced shoot length per plant by 25.5% compared to the control at 90 DAP (Table 2). Root length per plant was found significantly affected by combined application of ISA and PAM. The maximum root length was

Table 1 Effect of various doses of ISA (10 mg kg^{-1} soil) and PAM (10 mg kg^{-1} soil) alone or in combination on growth attributes of *Catharanthus roseus* L. G. Don. recorded at 90 DAP. Values represent the mean of three replicates ± standard error

Treatments/parameters	Treatment 1	Treatment 2	Treatment 3	Treatment 4	Treatment 5	Treatment 6
Shoot fresh weight (g)	38.87 ± 0.52^{e}	39.31 ± 0.58^{e}	42.56 ± 0.44^{d}	44.35 ± 0.32^{c}	44.47 ± 0.67^{b}	47.31 ± 0.29^{a}
Shoot dry weight (g)	8.48 ± 0.17^{d}	8.66 ± 0.31^{d}	9.27 ± 0.34^{c}	9.82 ± 0.19^{b}	10.00 ± 0.10^{b}	10.45 ± 0.15^{a}
Root fresh weight (g)	3.97 ± 0.02^{c}	3.99 ± 0.05^{c}	4.24 ± 0.11^{b}	4.09 ± 0.55^{c}	4.10 ± 0.46^{c}	4.46 ± 0.20^{a}
Root dry weight (g)	0.78 ± 0.09^{c}	0.80 ± 0.05^{c}	0.82 ± 0.11^{bc}	0.85 ± 0.06^{b}	0.87 ± 0.12^{b}	0.91 ± 0.19^{a}

Values of means within a row followed by the same letter are not significantly different according to Duncan's multiple range test ($P \leq 0.05$)

Table 2 Effect of ISA (10 mg kg^{-1} soil) and PAM (10 mg kg^{-1} soil) alone or in combination on growth attributes of *Catharanthus roseus* L. G. Don. recorded at 90 DAP. Values represent the mean of three replicates ± standard error

Treatments/parameters	Treatment 1	Treatment 2	Treatment 3	Treatment 4	Treatment 5	Treatment 6
Shoot length (cm)	27.03 ± 0.15^{d}	27.17 ± 0.12^{d}	32.47 ± 0.26^{bc}	31.83 ± 0.25^{c}	32.00 ± 0.26^{b}	33.93 ± 0.42^{a}
Root length (cm)	18.00 ± 0.25^{d}	18.20 ± 0.29^{d}	19.54 ± 0.12^{c}	21.29 ± 0.13^{b}	21.35 ± 0.12^{b}	21.96 ± 0.17^{a}
Number of leaves per plant	30.33 ± 2.08^{d}	31.33 ± 2.03^{d}	32.33 ± 0.33^{c}	34.67 ± 0.33^{b}	34.98 ± 0.58^{b}	39.00 ± 2.03^{a}
Average leaf area (cm^2)	6.67 ± 0.04^{d}	6.70 ± 0.13^{d}	7.27 ± 0.34^{c}	7.48 ± 0.13^{b}	7.64 ± 0.17^{b}	8.09 ± 0.30^{a}

Values of means within a row followed by the same letter are not significantly different according to Duncan's multiple range test ($P \leq 0.05$)

recorded in ISA + PAM treatment, which was 22.0% higher as compared with control (Table 2). The effect of ISA + PAM was also found significant on number of leaves and leaf area per plant. Treatment 6 (ISA + PAM) enhanced number of leaves and leaf area per plant by 28.6 and 21.3%, respectively, as compared with that of the control (Table 2).

3.2 *Physiological and Biochemical Parameters*

ISA in combination with PAM-treated plants had a noteworthy effect on physiological and biochemical parameters. The photosynthetic parameters, i.e., total chlorophyll and total carotenoids content registered an increase of 28.7 and 21.9%, respectively, as compared to the control. The enzymatic activities, i.e., nitrate reductase (NR) activity and carbonic anhydrase (CA) activity increased by 20.3 and 20.1% as compared to the control (Fig. 2). Leaf analysis of *C. roseus* revealed that application of ISA and PAM enhanced contents of leaf-N, -P, and -K (Fig. 3). Again Treatment 6 proved best and increased the leaf-N content by 15.9% over the

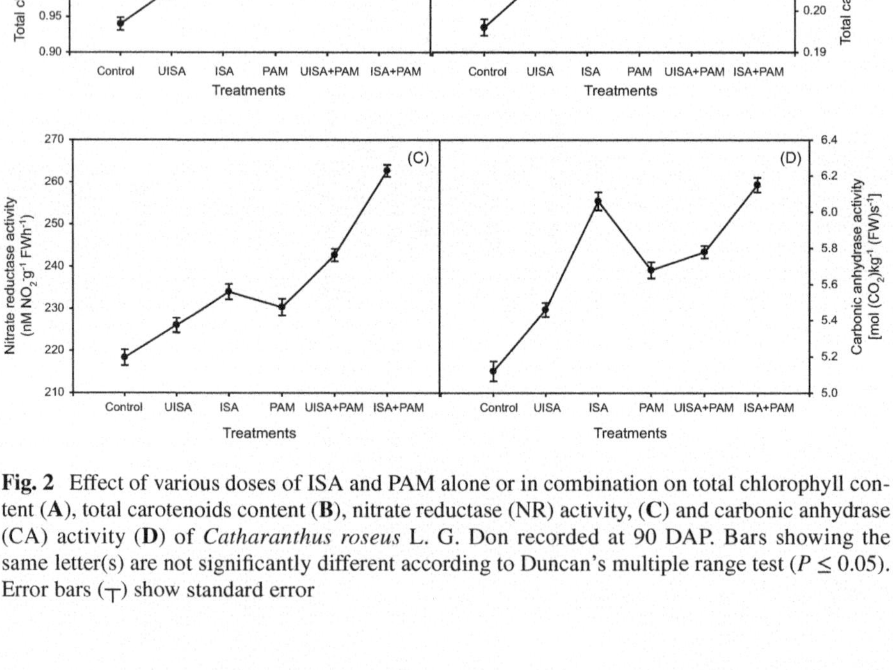

Fig. 2 Effect of various doses of ISA and PAM alone or in combination on total chlorophyll content (**A**), total carotenoids content (**B**), nitrate reductase (NR) activity, (**C**) and carbonic anhydrase (CA) activity (**D**) of *Catharanthus roseus* L. G. Don recorded at 90 DAP. Bars showing the same letter(s) are not significantly different according to Duncan's multiple range test ($P \leq 0.05$). Error bars (┬) show standard error

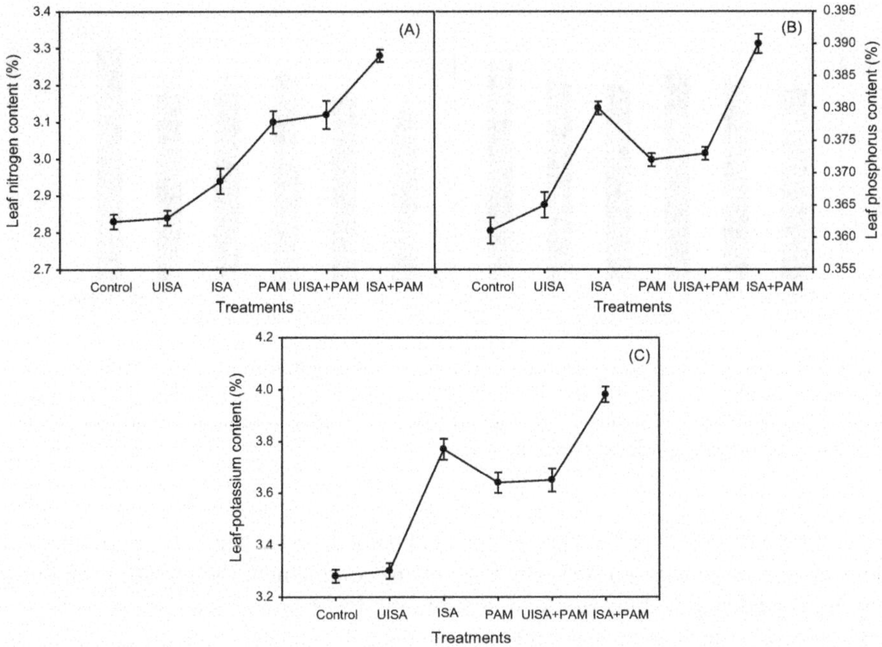

Fig. 3 Effect of various doses of ISA and PAM alone or in combination on leaf-nitrogen content (**A**), leaf phosphorus content (**B**), and leaf potassium activity (**C**) of *Catharanthus roseus* L. G. Don recorded at 90 DAP. Bars showing the same letter(s) are not significantly different according to Duncan's multiple range test ($P \leq 0.05$). Error bars (┬) show standard error

control (Fig. 3). The ISA + PAM also increased the leaf-P content by 8.0%. As compared to the control, leaf-K content, recorded with combination of ISA + PAM, was 21.3% higher (Fig. 3).

3.3 *Yield Attributes*

The leaf-alkaloid content and yield attributes were profoundly affected by the combined treatment (ISA + PAM). Treatment 6 ruled out all other treatments registering an increase of 14.1% in the total alkaloid content while 40.2% in total alkaloid yield as compared to the water sprayed control plants (Fig. 4).

4 Discussion

Growth attributes were found to be significantly affected by the application of ISA alone or in combination with PAM. Among various treatments applied, Treatment 6 (ISA + PAM) proved the best for all the growth parameters. The fact that

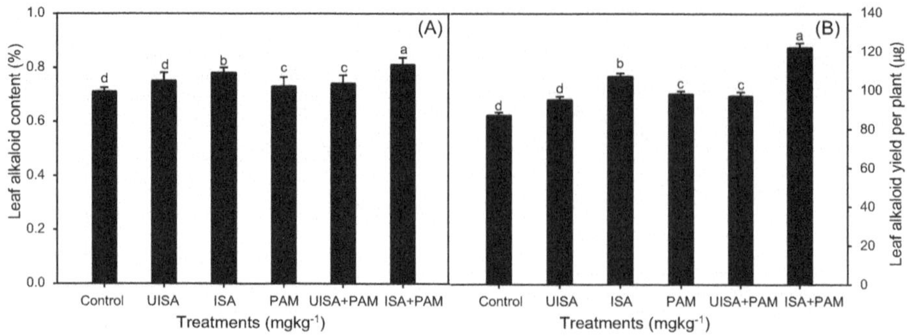

Fig. 4 Effect of various doses of ISA and PAM alone or in combination on leaf-alkaloid content (**A**) and leaf alkaloid yield per plant (**B**) of *Catharanthus roseus* L. G. Don recorded at 90 DAP. Bars showing the same letter(s) are not significantly different according to Duncan's multiple range test ($P \leq 0.05$). Error bars (┬) show standard error

combination dose proved the best is logical and could be related to the pertinent soil environment provided by PAM which seems to complement the growth promoting action of irradiated sodium alginate. Seemingly, PAM provides an impeccable soil environment by increasing water retention in the soil and improving the physical soil parameters such as hydraulic properties, infiltration rate, aeration, and root penetration thereby influencing the growth, physiology, and yield of the plant (Martin 1953; Azzam 1985; Wallace and Wallace 1986a, b; Danneels et al. 1994; Flanagan et al. 2002; El-Rehim 2006; Taban and Movahedi Naeini 2006; Sojka et al. 2007; Abedi-Koupai et al. 2008). Concurrently, the growth promoting activity of ISA shows cumulating effects with PAM and tends to increase the growth attributes maximally compared to other treatments. Although abundant reports are available regarding the enhanced growth and physiological activity of different plants through foliar application of degraded oligomers of sodium alginate, however, scientific literature on soil application of ISA is meagre. Takada-Oikawa et al. (2000) has reported that soil application of polysaccharides have stimulating effect on plant's growth in case of *Chrysanthemum corondrium*. Algam et al. (2010) also reported that chitosan solution applied through soil drenching proves ameliorative for tomato plant and increased all the growth parameters significantly.

The physiological and biochemical parameters were remarkably enhanced by Treatment 6 (ISA + PAM). The depiction of enhanced results is illustrated with Figs. 2 and 3. This can be synchronized with the previous records of El-Sayed et al. (1991) who reported photosynthetic pigments increased by hydrogel polymer. Since photosynthesis depends heavily on nutrient availability (El-Rehim 2006), PAM-mediated increased water retention and hence, increased nutrient availability might be attributed to the accentuated photosynthesis. Concurrently, ISA plays a critical role in enhancing photosynthetic activity which is evident from the previous works of Mollah et al. (2009), Khan et al. (2011), Sarfaraz et al. (2011), Aftab et al. (2013), and Idrees et al. (2012b).

Leaf analysis of *C. roseus* revealed the enhanced contents of N, P, and K in leaf due to ISA and PAM application (Fig. 3). Nitrogen being a constituent of chlorophyll molecule, application of nitrogen markedly promotes the synthesis of chlorophyll and other active photosynthetic pigments (Taiz et al. 2014). The positive effect of ISA + PAM on chlorophyll content might perhaps be attributed to the ISA + PAM-mediated increase in leaf-N content in this study. PAM and ISA individually have been reported to facilitate the efficient absorption and subsequent utilization of mineral nutrients in plants (El-Rehim 2006; Sarfaraz et al. 2011; Aftab et al. 2011). However, the combined effect of the applied treatments is worked out for the first time and it has encouraging results. Similar enhancement by ISA in leaf-N, -P, and -K contents has earlier been reported in various crops (Khan et al. 2011; Sarfaraz et al. 2011; Hashmi et al. 2012). PAM, on the other hand, is reported to have analogous results for leaf-N, -P, and -K as presented by deVarennes et al. (1999) in case of ryegrass (*Lolium perenne*) and Liu et al. (2013) in *Pinus pinaster*.

Carbonic anhydrase finds an active role in photosynthesis, which is evident by its presence in all photosynthesizing tissues. It catalyzes the reversible hydration of CO_2 to carbonic acid, thereby increasing the availability of CO_2 to RuBisCO in photosynthesis (Badger and Price 1994). Further, a probable reason for the enhancement of CA activity could be the ISA-mediated de novo synthesis of CA, which might involve transcription/translation of the genes associated as has been reported for other degraded natural polysaccharides (Knowles and Ries 1981). This can be corroborated to the findings of Tomoda et al. (1994), Natsume et al. (1994), Kume et al. (2002), and Luan et al. (2003), who reported that foliar spray of γ-irradiated sodium alginate enhanced the enzyme activities in different plants. Luan et al. (2003) also suggested a key role of γ-irradiated sodium alginate in enhancing the biological activity of the plants. The assumption that nitrate reductase is responsible for assimilation of nitrate and hence protein synthesis denotes that the increase in NR activity indicated by the application of γ-irradiated sodium alginate and PAM might have exerted a pivotal role in enhancing photosynthetic rate (Fig. 2c). Another report from Khan and Srivastava (1998) revealed about ISA-mediated increase in the membrane permeability to facilitate the absorption and utilization of mineral nutrients in addition to the improved transport of assimilates. Ultimately, these results can be considered as a justification to the increase in plant fresh and dry matter contents as a result of soil-applied γ-irradiated sodium alginate and PAM via improvement in the biochemical and physiological parameters studied in the treated plants.

The synergy of ISA and PAM continues to show up its escalatory upshot in terms of yield of the plant. El-Rehim (2006) underlined the dependence of all plantations on water retention in the soil so as to prevent the leaching of essential nutrients. He therefore reported that addition of organic material like PAM and sodium alginate increase water retention and hence, nutrient availability. The published reports also testify same characteristics of PAM to be beneficial for growth and yield of plants (Sojka et al. 2007; Keshavarz and Farahbakhsh 2012; Gharache et al. 2013). The possible explanation to this synergistic effect of ISA and PAM might perhaps be the apparent part played by PAM in assuring maximum nutrient availability as well as by ISA which is perhaps responsible for the elevated leaf-nitrogen content that might have promoted amino acid synthesis leading to the improved alkaloid content in the leaves. Khan et al. (2011) heralded an increase in the alkaloids production in case of opium poppy which goes in synchrony with the above result. The finding can also be corroborated with the recent reports of Hashmi et al. (2012) in fennel, Aftab et al. (2014) in artemisia, Ali et al. (2014) in eucalyptus, Naeem et al. (2014) in mint, and Idrees et al. (2014) in lemongrass.

5 Conclusion

Summing up the work in a nutshell, the results derived from the present study in an essay to accentuate the growth, physiology, and alkaloid production of *C. roseus* L. G. Don. were encouraging with combined application of ISA + PAM ruling out all other treatments. The application of this combined treatment will be highly cost-effective and economical for raising the production of such valuable alkaloids. However, a comprehensive study is still required for the mechanistic mode of action of alginate-derived oligomers and PAM on plants.

Acknowledgements The Bhabha Atomic Research Centre (BARC), Mumbai, Maharashtra, India, is highly acknowledged for its help in providing the γ-irradiated sodium alginate material for this investigation.

References

Abedi-Koupai J, Sohrab F, Swarbrick G (2008) Evaluation of hydrogels application on soil water retention characteristics. J Plant Nutr 31:317–331

Afaq SH, Tajuddin, Siddiqui MMH (1994) Standardization of herbal drugs. Publication Division, AMU, Aligarh

Aftab T, Khan MMA, Idrees M, Naeem M, Moinuddin, Hashmi N, Varshney L (2011) Enhancing the growth, photosynthetic capacity and artemisinin content in *Artemisia annua* L. by irradiated sodium alginate. Radiat Phys Chem 80:833–836

Aftab T, Naeem M, Idrees M, Khan MMA, Moinuddin, Varshney L (2013) Cumulative role of irradiated sodium alginate and nitrogen fertilizer on growth, biochemical processes and artemisinin production in *Artemisia annua*. Ind Crop Prod 50:874–881

Aftab T, Khan MMA, Naeem M, Idrees M, Siddiqui TO, Moinuddin, Varshney L (2014) Effect of irradiated sodium alginate and phosphorus on biomass and artemisinin production in *Artemisia annua*. Carbohydr Polym 110:396–404

Aftab T, Naeem M, Idrees M, Khan MMA, Uddin M, Varshney L (2016) Simultaneous use of irradiated sodium alginate and nitrogen and phosphorus fertilizers enhance growth, biomass and artemisinin biosynthesis in *Artemisia annua* L. J Appl Res Med Aromat Plants 3:186–194

Akimoto C, Aoyagi H, Tanaka H (1999) Endogenous elicitor-like effect of alginate on physiological activities of plant cells. Appl Microbiol Biotechnol 52:429–436

Algam S, Xie G, Li B, Yu S, Su T, Larsen J (2010) Effects of *Paenibacillus* strains and chitosan on plant growth promotion and control of *Ralstonia* wilt in tomato. J Plant Pathol 92:593–600

Ali A, Khan MMA, Moinuddin, Naeem M, Idrees M, Hashmi N, Dar TA, Varshney L (2014) Radiolytically depolymerised sodium alginate improves physiological activities, yield attributes and composition of essential oil of *Eucalyptus citriodora* Hook. Carbohydr Polym 112:134–144

Anonymous (2014) WHO Cancer Fact Sheet No. 297. http://www.who.int/mediacentre/factsheets/fs297/en/

Anthony J, Gabapathy A, Coothan KV, Streenivasan PP, Arjuna R, Palaninathan V (2007) Beneficial effects of sulphated polysaccharides from *Saragassum wightii* against mitochondrial alterations induced by cyclosporine A in rat kidney. Mol Nutr Food Res 51:1413–1422

Azzam RAI (1985) Tailoring polymeric gels for soil reclamation and hydroponics. Commun Soil Sci Plant Anal 16(10):1123–1138

Badger MR, Price GD (1994) The role of carbonic anhydrase in photosynthesis. Annu Rev Plant Physiol Plant Mol Biol 45:369–392

Barvenik FW (1994) Polyacrylamide characteristics related to soil application. Soil Sci 58:235–243

Bibi Y, Nisa S, Zia M, Waheed A, Ahmed S, Chaudhary MF (2012) In vitro cytotoxic activity of *Aesculus indica* against breast adeno carcinoma cell line (MCF-7) and phytochemical analysis. Pak J Pharm Sci 25:183–187

Danneels P, Beeckman T, Cotthem WV (1994) Effects of polyacrylamide and a fertilizer on germination and seedling growth of *Ammophila arenaria* (L.) link I at 5% soil moisture content. Biol Jb Dodonaea 61:84–98

deVarennes A, Torres MO, Conceição E, Vasconcelos E (1999) Effect of polyacrylate polymers with different counter ions on the growth and mineral composition of perennial ryegrass. J Plant Nutr 22:33–43

Dwivedi RS, Randhawa NS (1974) Evaluation of rapid test for the hidden hunger of zinc in plants. Plant Soil 40:445–451

El-Rehim HAA (2006) Characterization and possible agricultural application of polyacrylamide/sodium alginate crosslinked hydrogels prepared by ionizing radiation. J Appl Polym Sci 101:3572–3580

El-Sayed H, Kirkwood RC, Graham NB (1991) The effects of a hydrogel polymer on the growth of certain horticultural crops under saline conditions. J Exp Bot 42:891–899

Fazeli Rostampour M, Yarnia M, Rahimzadeh Khoee F (2012) Effect of polymer and irrigation regimes on dry matter yield and several physiological traits of forage sorghum. Afr J Biotechnol 11(48):10834–10840

Fiske CH, Subbarow Y (1925) The colorimetric determination of phosphorus. J Biol Chem 66:375–400

Flanagan DC, Chaudhari K, Norton LD (2002) Polyacrylamide soil amendment effects on runoff and sediment yield on steep slopes: part II. Natural rainfall conditions. Trans ASAE 45(5): 1339–1351

Gharache A, Kalidari AS, Rahimi MM, Khalatbari M (2013) Evaluation of the effect of different ranges super absorbent on quality and physiological characteristics of canola (*Brassica napus* L.) cvs Zarfam under water deficit stress. Tech J Eng Appl Sci 3:165–169

Hald PM (1947) The flame photometer for the measurement of sodium and potassium in biological materials. J Biol Chem 163:499–510

Hashmi N, Khan MMA, Moinuddin, Idrees M, Khan ZH, Ali A, Varshney L (2012) Depolymerized carrageenan ameliorates growth, physiological attributes, essential oil yield and active constituents of *Foeniculum vulgare* Mill. Carbohydr Polym 90:407–412

Hayat R, Ali S (2004) Water absorption by synthetic polymer (aquasorb) and its effect on soil properties and tomato yield. Int J Agric Biol 6:998–1002

Hu X, Jiang X, Hwang H, Liu S, Guan H (2004) Promotive effects of alginate derived oligosaccharide on maize seed germination. J Appl Phycol 16:73–76

Idrees M, Naeem M, Khan MMA (2010) The superiority of cv. 'rosea' over cv. 'alba' of periwinkle (*Catharanthus roseus* L.) in alkaloid production and other physiological attributes. Turk J Biol 32:81–88

Idrees M, Naeem M, Alam M, Aftab T, Hashmi N, Khan MMA, Moinuddin, Varshney L (2011) Utilizing the γ-irradiated sodium alginate as a plant growth promoter for enhancing the growth, physiological activities, and alkaloids production in *Catharanthus roseus* L. Agric Sci China 10(8):1213–1221

Idrees M, Naeem M, Aftab T, Hashmi N, Khan MMA, Moinuddin, Varshney L (2012a) Promotive effect of irradiated sodium alginate on seed germination characteristics of fennel (*Foeniculum vulgare* Mill.). J Stress Physiol Biochem 8(1):108–113

Idrees M, Moinuddin, Nasir S, Naeem M, Aftab T, Khan MMA, Varshney L (2012b) Gamma irradiated sodium alginate induced modulation of phosphoenolpyruvate carboxylase and production of essential oil and citral content of lemongrass. Ind Crop Prod 40:62–68

Idrees M, Dar TA, Naeem M, Aftab T, Khan MMA, Ali A, Moinuddin, Varshney L (2014) Effects of gamma-irradiated sodium alginate on lemongrass: field trials monitoring production of essential oil. Ind Crop Prod 63:269–275

Idrees M, Ul Hassan I, Ahmad NG, Naeem M, Tariq A, Khan MMA, Varshney L (2016) Functional activities and essential oil production in coriander plants supported with application of irradiated sodium alginate. Int J Appl Environ Sci 11:467–474

Jaworski EJ (1971) Nitrate reductase assay in intact plant tissues. Biochem Biophys Res Commun 43:1247–1279

Keshavarz L, Farahbakhsh H (2012) Effect of superabsorbent on physio-morphological traits and forage yield of millet (*Pennisetum americanum* L.) under different irrigation treatments. Int J Plant Anim Environ Sci 3(2):149–156

Khan MG, Srivastava HS (1998) Changes in growth and nitrogen assimilation in maize plants induced by NaCl and growth regulators. Biol Plant 41:93–99

Khan ZH, Khan MMA, Aftab T, Idrees M, Naeem M (2011) Influence of alginate oligosaccharides on the growth, yield and alkaloids production of opium poppy (*Papaver somniferum* L.). Front Agric China 5:122–127

Knowles NR, Ries SK (1981) Rapid growth and apparent total nitrogen increases in rice and corn plants following applications of triacontanol. Plant Physiol 68:1279–1284

Kume T, Nagasawa N, Yoshii F (2002) Utilization of carbohydrates by radiation processing. Radiat Phys Chem 63:625–627

Lindner RC (1944) Rapid analytical methods for some of the more common inorganic constituents of the plant tissues. Plant Physiol 19:76–89

Liu F, Ma F, Xing S, Du Z, Ma B, Jing D (2013) Effects of super-absorbent polymer on dry matter accumulation and nutrient uptake of *Pinus pinaster* container seedlings. J For Res 18: 220–227

Luan LQ, Hien NQ, Nagasawa N, Kume T, Yoshii F, Nakanishi TM (2003) Biological effect of radiation-degraded alginate on flower plants in tissue culture. Biotechnol Appl Biochem 38:283–288

Mac Kinney G (1941) Absorption of light by chlorophyll solutions. J Biol Chem 140:315–322

Mac Lachlan S, Zalik S (1963) Plastid structure, chlorophyll concentration and free amino acid composition of chlorophyll mutant of barley. Can J Bot 41:1053–1062

Martin WP (1953) Status report on soil conditioning chemicals. Soil Sci Soc Am J 17:1–9

Mollah MZI, Khan MA, Khan RA (2009) Effect of gamma irradiated sodium alginate on red amaranth (*Amaranthus cruentus* L.) as growth promoter. Radiat Phys Chem 78:61–64

Naeem M, Idrees M, Aftab T, Khan MMA, Moinuddin, Varshney L (2011) Irradiated sodium alginate improves plant growth, physiological activities and active constituents in *Mentha arvensis* L. J Appl Pharm Sci 02:28–35

Naeem M, Idrees M, Aftab T, Alam MM, Khan MMA, Uddin M, Varshney L (2014) Employing depolymerised sodium alginate, triacontanol and 28-homobrassinolide in enhancing physiological activities, production of essential oil and active components in *Mentha arvensis* L. Ind Crop Prod 55:272–279

Naeem M, Aftab T, Ansari AA, Idrees M, Ali A, Khan MMA, Uddin M, Varshney L (2015a) Radiolytically degraded sodium alginate enhances plant growth, physiological activities and alkaloids production in *Catharanthus roseus* L. J Radiat Res Appl Sci 8:606–616

Naeem M, Idrees M, Aftab T, Alam MM, Khan MMA, Uddin M, Varshney L (2015b) Radiation processed carrageenan improves plant growth, physiological activities and alkaloids production in *Catharanthus roseus L.* Adv Bot 2015:1–11. Article ID 150474

Natsume M, Kamao Y, Hirayan M, Adachi J (1994) Isolation and characterization of alginate derived oligosaccharides with root growth promoting activities. Carbohydr Res 258:187–197

Novozamsky I, Houba VJG, van Eck R, van Vark W (1983) A novel digestion technique for multielement plant analysis. Commun Soil Sci Plant Anal 14:239–248

Rorison IH, Spencer RE, Gupta PL (1993) Chemical analysis. In: Hendry GAE, Grime JP (eds) Methods in comparative plant ecology. Chapman and Hall, New York, pp 156–161

Sarfaraz A, Naeem M, Nasir S, Idrees M, Aftab T, Hashmi N, Khan MMA, Moinuddin, Varshney L (2011) An evaluation of the effects of irradiated sodium alginate on the growth, physiological activities and essential oil production of fennel (*Foeniculum vulgare* Mill.). J Med Plant Res 5:15–21

Singh DV, Maithy A, Verma RK, Gupta MM, Kumar S (2000) Simultaneous determination of *Catharanthus* alkaloids using reserved phase high performance liquid chromatography. J Liq Chromatogr Relat Technol 23:601–607

Sojka RE, Bjorneberg DL, Entry JA, Lentz RD, Orts W (2007) Polyacrylamide in agriculture and environmental land management. Adv Agron 92:75–162

Taban M, Movahedi Naeini SAR (2006) Effect of aqua sorbent and organic compost amendments on soil water retention and evaporation with different evaporation potential and soil texture. Commun Soil Sci Plant Anal 37(13–14):2031–2055

Taiz L, Zeiger E, Moller IM, Murphy A (2014) Plant physiology and development, 6th edn. Sinauer Associates, Inc., Sunderland

Takada-Oikawa N, Katoh N, Oshida T, Kawanabe S, Kaise T (2000) Studies on materials containing polysaccharides as soil amendments 1: effects on animal waste water purification in a potting experiment. J Mater Cycles Waste Manag 2:138–142

Tomoda T, Umemura K, Adachi T (1994) Promotion of barley root elongation under hypoxic conditions by alginate lyase lysate. Biosci Biotechnol Biochem 58:202–203

Tümsava Z, Kara A (2011) The effect of polyacrylamide (PAM) applications on infiltration, runoff and soil losses under simulated rainfall conditions. Afr J Biotechnol 10:2894–2903

Wallace A, Wallace GA (1986a) Effects of soil conditioners on emergence and growth of tomato, cotton and lettuce seedlings. Soil Sci 141(5):313–316

Wallace A, Wallace GA (1986b) Effects of very low rates of synthetic soil conditioners on soils. Soil Sci 141(5):324–327

Yonemoto Y, Tanaka H, Yamashita T, Kitabatake N, Ishida Y, Kimura A, Murata K (1993) Promotion of germination and shoot elongation of some plants by alginate oilgomers prepared with bacterial alginate lyase. J Ferment Bioeng 75:68–70

Catharanthus roseus: Cultivation Under Stress Conditions

Sanjeev Pandey

Abstract The Madagascar periwinkle or *Catharanthus roseus* L. G. Don of the family Apocyanaceae is a well-known aromatic plant that is cultivated in the gardens as ornamentals is now being considered for its wide-scale cultivation to accrue more than 125 types of alkaloids that it produces. All parts of this plant are medicinally important; the roots accumulate ajmalicine and serpentine, whereas the stem and the leaves are the source of vincristine and vinblastine. The latter two are indispensable in cancer chemotherapy. *Catharanthus roseus* is an extremely hardy plant and can grow at places where any other plant will normally succumb. It does not require too much of water and nutrition and in fact under stress condition the overall production of the alkaloids is increased. Most suitable temperature for its cultivation in open field is 20–30 °C. Deep sandy loam to loamy soil is preferable where water logging should be avoided. A precipitation of 100 cm per annum with little or no irrigation is most suitable for its growth. The plant is propagated either by their seeds or shoot cuttings. The method of sowing with a row-to-row gap of 45 cm and plant-to-plant distance of 30 cm should be maintained by thinning time to time. To avoid fungal attack, foliar sprays of carbendazime bavistin and benomyl is recommended. The plant is ready to harvest after 1 year of its growth. Either the whole plant could be uprooted and then separated and dried for processing or the bottom part could be left for ratooning.

Catharanthus could be a plant of choice under drought and salinity stress. The plant not only can withstand extreme drought and high salt concentration but the alkaloid biosynthesis also increases under such conditions.

Application of growth regulators like jasmonate and salicylic acid have positive effect on alkaloid biosynthesis. These growth substances also ameliorate the toxicity generated due to other stress factors like the salt stress or the heavy metal stress, etc. However, a minimum concentration of certain heavy metals, viz., cadmium, lead, nickel, etc. is also known to enhance alkaloid production.

S. Pandey (✉)
Department of Botany, Banwarilal Bhalotia College, Asansol 713303, WB, India
e-mail: spbot.bbc.asn@gmail.com

M. Naeem et al. (eds.), *Catharanthus roseus*,
DOI 10.1007/978-3-319-51620-2_17

Since the plant yield through field cultivation is not always satisfactory, in vitro cell and tissue culture practices have also been employed to have a more controlled growth and yield of the crop. In this regard, the callus and cell suspension cultures are more fashionable. Basic MS medium with various formulations have been used to get the maximum yield. Application of phytohormones in suitable combinations and implementation of osmotic stress as well as UV-B exposure have been possible to introduce in order to get the maximum yield in tissue culture as well.

Keywords Periwinkle • Indole alkaloids • Stress • Cultivation • Tissue culture • Yield

1 Introduction

Madagascar periwinkle or *Catharanthus roseus* L. G. Don of the family Apocyanaceae is a native of Central Madagascar and from there it has spread to the other parts of the world including India. The plant is widely cultivated in gardens and fields as ornamental plant and also for deriving life-saving drugs that it produces. Periwinkle is one of the few medicinal plants that has found mention in the folk medicinal books as early as 2nd BC. In India, it has been used to treat wasp stings with the juice of the leaves. In Hawaii, they prescribed an extract of the boiled plant to arrest bleeding. In Central America and parts of South America, people made a gargle to get relief from sore throats, chest ailments, and laryngitis. In Cuba, Puerto Rico, Jamaica, and other islands an extract of the flower was commonly administered as eyewash for the eyes of the infants, and in Africa leaves were used for menorrhagia and rheumatism (Dobelis 1997; Walts 2004).

The medicinal properties of *Catharanthus* are so diverse that very few plants have generated as much interest among the researcher as this plant has. The interest began in 1950s when researcher learnt about the "periwinkle tea" that was consumed in Jamaica as an antidiabetic medicine. The plant confers this property by increasing glucose utilization (Singh et al. 2001). Later it was discovered that the plant produces more than 125 monoterpenoid indole alkaloids in different organs of it (Groger 1985). The roots accumulate ajmalicine and serpentine that are used to control high blood pressure and other types of cardiovascular diseases. The leaves and stem are the source of dimeric alkaloids, vinblastine (VLB) and vincristine (VCR), that are indispensable parts of most anticancer chemotherapies (Jordon et al. 1991). There are more than 70 other alkaloids that have been isolated and characterized from the plant in addition to the above ones (Dobelis 1997; Walts 2004).

Catharanthus is normally a hardy plant that prefers to grow well at temperature 20–30 °C in soil with low water holding capacity and low fertility. This indicates the drought-resistant nature of the plant which makes it suitable for cultivation in dried arid regions of the world. Naturally, therefore, it is found mostly in tropical and subtropical regions of the world. High adaptability and the ability to grow on marginal lands makes this plant highly preferable to the farmers. Failure of the crop is minimum. Presence of alkaloids all over the plant body confers immunity to cattle-grazing and crop loss due to pilferage. The USA is the largest user of the raw material most of which it imports from Malagasy, India, and Mozambique. Hungary is also been one of the major consumers of the leaves. West Germany, Italy, the Netherlands, and the UK are interested in the roots (Aslam et al. 2010). In India, it is being grown in Tamilnadu, Karnataka, Andhra Pradesh, Telangana, Madhya Pradesh, Chhattisgarh, Gujarat, and Assam in an area of about more than 3000 ha (Farooqi, 2010).

Considering its growing economic value and ease of cultivation principally in degraded lands under stress condition, it is necessary to review the process of cultivation that could be adopted. It is also very important in terms of land utilization as the plant has prospects of cultivating in wastelands also.

1.1 Cultivation of Catharanthus

Catharanthus is mainly grown in the field for the most part of the world; however, in vitro production of *Catharanthus* biomass has also gained popularity in recent years. In this section, the field cultivation of the plant has been discussed first followed by the cultivation in culture -

1. ***Soil***—As mentioned earlier, *Catharanthus* is a hardy plant and grows well on a wide variety of soils except those which are alkaline and water logged. Deep sandy loam to loam soils of moderate fertility are preferred for large-scale cultivation of this plant. However, sandy soil with raw liquid cow-dung provides an aerated substratum to the crop, and it is preferred for its commercial cultivation (Karnick 1977). In South India, the plant is widely grown on red laterite soil and occasionally on black cotton soils.
2. ***Ecology***—The ecology of the plant says that it is naturally drought resistant and often grows in sandy locations near the sea coasts and river banks. It is salt tolerant and mostly found near sea level but occasionally it may grow up to an elevation of 1500 m. It can withstand drought but not severe heat. A rainfall of about 100 cm is most suitable for its growth. As far as irrigation and nutrition is concerned, it requires little irrigation or fertilizer. Too excessive watering may cause yellowing of leaves.
3. ***Propagation and Planting***—*Catharanthus roseus* is a perennial plant usually propagated by the seeds. However, vegetative propagation by cuttings is also very common. Seeds may remain dormant for several weeks after maturity and should be subjected to sowing after 2 months of their storage in refrigerated environment. The seeds remain viable for 3–5 years if stored properly. The seeds are sown in nurseries or sometimes direct sowing is also done. For direct sowing seeds are mixed with about 10 times their volume of sand and sown in monsoon in rows of 45 cm apart. When the plants are sufficiently grown, thinning is done to keep a

final distance between two plants to be 30 cm. Nursery sowing is said to be more economical than direct sowing. Seeds are sown in nurseries in February to March and when the seedlings are about 6–7 cm tall, they are transplanted to fields with plant-to-plant and row-to-row spacing with 30 cm and 45 cm, respectively. *Catharanthus roseus* can also be propagated vegetatively by greenwood or semi-ripe cuttings rooted in a closed container with bottom heat. The rooting can be induced by applying rooting powder in 4–5 weeks of their planting.

4. ***Management***—Though the plant does not require any special supply of fertilizers, a mixture of nitrogen, phosphorus, and potassium may be useful to increase yield. FYM (Farm yard manure) is also sometimes applied. Weeding must be done periodically. When the plants become too tall, pruning is necessary. To encourage branching, apical dominance has to be removed by decapitating the tip leaving merely 7–8 cm stem part above the ground level only. The side branches grow up always in opposite manner. If this pruning is not done, branching will start at a height of 20–30 cm that too in one sided way leading to imbalance in the weight of the plant and decumbent habit.
5. ***Diseases and Pests***—Several pathogens of fungal, viral, and mycoplasmal origin have been reported to attack *Catharanthus*. Among the fungal diseases twig blight/top-rot or dieback disease caused by *Phytophthora nicotinae* (Gill et al. 1977), *Pythium debaryanum*, *P. butleri*, *P. aphanidermatum*, and/or *Colletotrichum dermatum*; leaf spot due to *Alternaria tenussima*, *A. alternate*, *Rhizoctonia solani*, *Ophiolobus catharanthicola*, *Haplosporella marathwadensis*, *Gloemerilla cingulata*, and/or *Myrothecium rosedum* (Goyal and Pathak 1982); foot-rot and wilt caused by *Sclerotium rolfsii* and/or *Fusarium solani*. For controlling these diseases, foliar spray of the fungicides Captafol Foltaf 80 WP, Carbendazim Bavistin 50 WP, and Benomyl 50 WP are recommended (Kalra et al. 1991). A mycoplasmal disease causing chlorotic wilting and vein clearing together with several anatomical deviations is reported. These diseases are in turn transmitted by an insect *Onconetopia nigrans*. Again, leaf mosaic disease due to viruses also occurs due to which leaves develop irregular yellow patches, malformation, necrosis, and wilting. Other viral diseases are known to cause extreme shortening of the nodes and internodes with reduced fruit and seed formation. The mycoplasmal disease can be controlled by regular spray of oxytetracyclin, tetracycline, or nicotine sulphate. The viral disease can be overcome to some extent by GA_3 treatment that causes internode elongation and leaf widening.
6. ***Harvesting***—*Catharanthus roseus* is harvested after 12 months of sowing. Moreover, at flowering the alkaloid concentration is highest. Sometimes, the whole plants are uprooted first and then the roots are separated from stems and leaves. If only the leaves are to be harvested, the plants are left in the field for a ratoon crop. Ratooning is a process in which bottom stubble part is left with the roots which sprouts next season again and can give several cycles of crop if the process is repeated. The stems, leaves, and roots are separated, dried at low temperatures before being packaged for shipment.
7. ***Yield***—The yield of dried roots, stems, and leaves per hectare of irrigated land for the crop is 1.5, 1, and 3 tonnes, respectively.

1.2 Cytogenetic and Reproductive Variations of the Plant

Catharanthus roseus is diploid plant species with a karyotype comprising 16 chromosomes. These eight bivalents can be visualized at meiosis. Floral morphology of *Catharanthus* is conducive for both self-pollination and insect-mediated cross-pollination. The natural population of *C. roseus* possesses genetic variability to considerable extent that has been exploited to develop horticultural and drug-yielding varieties. In *Catharanthus*, 27 germplasm (indigenous and exotic) lines have been collected from a wide range of geographical areas and studied for 28 morpho-agronomic and chemical characters. Among these five promising lines of high root and foliage yield were selected for field trials. For comparative evaluation of economic characters and the correlation of this aspect indicate that an ideal plant type should have bushy growth producing higher biomass and having a much-branched root system.

Variations resulting from cross-breeding, induced mutagenesis, and polyploidy generation have been variously employed in this plant. Experimental hybrids have been obtained between *C. roseus* and *C. trichophyllous* (Sevestre-Rigouzzo et al. 1993). Significant heterosis has been observed (Levy et al. 1983) in terms of leaf and root yields in crosses involving three pure lines. However, no heterosis could be found at the level of ajmalicine production. On the other hand, an increased morphological difference as well as an increase in the alkaloid content was found in the leaves and roots of three induced mutants of *C. roseus* (Kulkarni et al. 1999).

Tetraploids and triploids have also been developed and characterized for this plant (Schnell 1941; Eigste and Tenney 1943). These autotetraploids were more tolerant to diseases like dieback, collar, and root-rot disease than their diploid counterpart. Certain strains of autotetraploids were observed to possess broader leaves and lower length/width ratio, reduced pollen fertility, larger pollens, shorter follicles with heavier seeds, lower leaf dry matter but higher alkaloid content. The problem with these higher ploidy strains are their high level of sterility and low seed viability (Janaki-Ammal and Bezbarauah 1963). However, selection of seeds for cultivation should be made considering specific objective of such cultivation and procurement must be made from government-authorized suppliers.

1.3 Factors Affecting Growth and Yield Under Field Conditions

1. ***Temperature***—Studies on the germination behavior of different cultivars of *Catharanthus roseus* have revealed that the optimum temperature for germination ranges from 20–27 °C (Mastalerz 1976; Styer and Laffe 1990; Choudhury and Gupta 1995). The lowest and highest limit of temperature beyond which germination ceases completely is 15 °C and 40 °C, respectively. In seeds of "Down Carpet" and "Little Bright Eye" varieties of *C. roseus*, a suppressed germination

percentage has been recorded at 15 °C day and 20 °C night temperatures. Optimum temperature of these two varieties is also ~25 °C (Blazich et al. 1995). Seeds of "*alba*" variety also show an ambient germination temperature of ~25 °C (Choudhury and Gupta 1995). In green house conditions, *Vinca* requires warm conditions with a temperature of about 22 °C at night. Under outdoor conditions, it is being agreed that a soil temperature of no less than 22 °C is to be maintained; however, it prefers a day air temperature of 26 °C and can grow well at a temperature as high as 35 °C. Temperature below 15 °C ceases flowering and at about 12 °C several problems like leaf rolling, chlorosis, leaf drop, and sometimes a purpling and bronzing effects on leaf could be observed.

2. ***Soil pH***—*Catharanthus* or *Vinca* shows a micronutrient deficiency when grown in soilless mixture with a pH above 6.3. A pH level of 5.6 is suggested for its better nutrient acquisition. The soil pH usually rises for 4–6 weeks after transplantation. So a continuous monitoring is necessary, and to keep it low acid forming fertilizers such as iron-sulphate may be applied time to time. Another problem of pH higher than 6.0 is *Thielaviopsis* infection that causes black-root rot disease.
3. ***Light***—Germination percentage of *Vinca* seeds are higher in darkness than in light (Ball 1991). In other words, *Vinca* is scotoblastic in nature. According to Mastalerz (1976) and Ball (1991), seeds should be covered by the sowing medium so that darkness is maintained while germination. After the seedlings have come out 1200–1500 fc light is essential. At maturity, *Catharanthus* grows naturally well in warm sunny days with a light requirement of above 2000 fc.

In a study with UV-B exposure under dark incubation, an overall increase in the expression of alkaloid biosynthetic genes was recorded through proteomics. However, these two factors have negative effects on chlorophyll biosynthesis and photosynthetic gene expression (Zhu et al. 2015).

4. ***Drought and Salinity***—*Catharanthus* grows well under drier environment. Excess watering in fact causes yellowing of leaves. According to Frischknecht et al. (1987), a mild water stress (Ψ of leaf = -1.3 MPa) though has no significant effect on the alkaloid content of *Catharanthus*, but an occasional life-saving irrigation gave more alkaloid. Singh et al. (2001) have recorded a higher value of total leaf and root alkaloids at low level of irrigation. There is in fact inverse relation between fresh and dry weight of leaves and alkaloid production with increased level of drought in the soil (Talha et al. 1975; Amirjani 2013).

Cheruth et al. (2008) studied the water use efficiency of the two varieties of *Catharanthus* under drought conditions. Two varieties, rosea and alba of *C. roseus* (L.) G. Don., were screened under two water regimes for their water use efficiencies. Drought stress was imposed at 60% field capacity from 30 to 70 days after sowing with 100% field capacity being considered as control. A significant increase in water use efficiency was recorded in both the varieties. This property of *Catharanthus* may be attributed to its capacity to use available water in soil more optimally possibly through decreasing the rate of evapotranspiration by partial closure of stomata or by decreasing the leaf areas, plant height and lateral stem number, i.e., tillage. Now, all these factors have greater consequences on the yield of alkaloids. Therefore, an optimal level of moisture must be maintained in order to get maximum yield.

The effect of different salinity levels and drought durations on growth and yield of *Catharanthus roseus* has been investigated also by Elfeky et al. (2007). According to their results, *C. roseus* has a capacity to tolerate 150 mM NaCl and drought up to 3 weeks and the plants appeared to be more drought resistant in comparison to their salt tolerance.

5. ***Growth Regulators and Signaling Molecules***—Various plant signaling molecules were tested for their ability to enhance alkaloid synthesis in *Catharanthus* seedlings (Aerts et al. 1996). The compounds tested include hormones, fatty acid-derived messengers and agents that can induce Systemic Acquired Resistance (SAR) in plants. Of these compounds, only methyl jasmonate enhanced the synthesis of monomeric alkaloids. Influence of growth retardants like paclobutrazol, uniconazole, and diaminozide have also been examined with this plant (Barrett and Nell 1992). Though it is reported (Cathey 1964) that these growth retardants suppress the overall growth and metabolism of plants, but it is now well established that retardants act much like phytohormones. That is at lower concentration they positively modulate metabolism but at higher concentration they are inhibitory. On the contrary, another growth-retardant Chlorocholine chloride (CCC) is reported to increase biomass and alkaloid production on *Catharanthus pusillus* (Basu 1992) and *C. roseus* (Choudhury and Gupta 1995).

Salicylic acid (SA) is also a newly recognized plant growth regulator produced under stress conditions particularly in chilling stress (Janda et al. 1999). It has been recognized to alleviate the adverse effects of salt stress by improving plant productivity through protecting photosynthetic pigments and producing antioxidant enzymes and their compounds (Larkindale and Knight 2002). Exogenous application of salicylic acid (SA) on salt-treated *Catharanthus roseus* has given positive results both in terms of growth and yield (Idrees et al. 2011).

6. ***Heavy Metals***—When we consider plants to cultivate under stress particularly on degraded lands, we must take into account the influence of heavy metals on the growth and productivity of the crop. Owing to the rampant industrial development, this has become true for normal cultivable land too. Some of the recent studies have revealed that metals have potential to increase the production of secondary metabolites in cell and tissue culture. Addition of Vanadium sulphate, for example, promoted higher content of ajmalicine and catharanthine in cell culture (Smith et al. 1987a, b). Various other earth elements like the cerium, yttrium, and neodymium are known to elicit differential response in terms of ajmalicine and catharanthine production in cell culture (Zhao et al. 2000a). Similarly, Cd treatment is reported to enhance ajmalicine secretion into the culture medium. Also addition of tetramethylammonium bromide into the culture medium enhanced the yield of ajmalicine in shake flask and bioreactor (Zhao

et al. 2000b). There are several reports regarding the effect of heavy metals or metalloids at whole plant level cultivated in the field. In this regard, effects of Cd, Pb, Ni, and Cr have been studied more extensively. According to Srivastava and Srivastava (2010), a 5 mM concentration of Cd, Pb, Mn, and Ni had been effective in increasing overall alkaloid content with two to three-fold increase in serpentine content.

2 In Vitro Cultivation through Tissue Culture

Catharanthus roseus was first time used in tissue culture as an alternative method of production way back in 1943 by Hilderbrandt and his group. After that, it was Carew 1993 who developed callus tissues from different explants of the plant. All organs like the root, stem, leaves, and anthers all have been used as explants to generate calliclones. The alkaloid status of all these calliclones has been determined. Culture media have been designed with different phytohormone combinations to enhance the yield of alkaloids (Fulzele et al. 1990).

As mentioned earlier, *Catharanthus* produces more than 125 alkaloids of various kinds. Of this, about 65 have been reported to be produced through cell culture techniques. Root is the important store of ajmalicine and serpentine, but the yield through conventional cultivation method is very low. This makes these compounds largely expensive for commercial production (Gaines 2004; Taha et al. 2008). This has led to the introduction of hairy root culture in *Catharanthus* also. *Agrobacterium rhizogenes*-mediated genetic transformation has been undertaken during the last couple of decades. An increase in the yield of alkaloids including vincristine and vinblastine has been reported for hairy roots of transgenic callus with *A. rhizogenes* (Zargar et al. 2010).

The overall yield of alkaloids in tissue culture also depends on the environment within the bioreactor and the genotype of the plant from which explants has been taken. The different nutritional, environmental, hormonal, and stress factors that govern the yield of alkaloid in vitro has been depicted in the Table 1.

A brief discussion about the optimum conditions in cell and tissue culture for maximum yield can be useful at this stage.

2.1 *Nature of Medium*

Several attempts have been made to optimize the medium composition for getting maximum yield from *Catharanthus roseus* (Zenk et al. 1977; Knobloch and Berlin 1980). The two-stage process was preferred in which in the first stage the medium composition was modulated to enhance the growth of the biomass whereas in the second stage when the appropriate biomass has been achieved the culture was induced to produce the alkaloids. In an alternative approach, a combined growth and production medium was developed which achieved both growth and production in a single stage (Morris 1986; Smith et al. 1987b).

Table 1 Factors affecting yield of *Catharanthus* alkaloid under in vitro condition

Media modulation	Culture type-cell/tissue/ organ culture	Effect on alkaloid production	References
Increase in sucrose concentration	Callus and cell suspension	Increased yield of ajmalicine, serpentine, catharanthine	Knobloch and Berlin (1980), Morris (1986), Smith et al. (1987a), Zenk et al. (1977)
Low inorganic N_2	Callus	Favored alkaloid accumulation	Knobloch and Berlin (1980), Merillon et al. (1984), Zenk et al. (1977)
Low inorganic P	Callus	Favored alkaloid accumulation	Knobloch and Berlin (1980), Zenk et al. (1977)
Supplementation of 2,4-D/NAA in the medium	Callus and Cell suspension	Inhibitory to alkaloid production	Goddijn et al. (1992), Morris (1986), Pasquali et al. (1992), Whitmer et al. (1998)
Supplementation of GA in the medium	Callus	Stimulation of alkaloid production and biomass yield	Zenk et al. (1977)
BAP supplementation singly or in combination to IAA	Callus	Stimulation of alkaloid production and biomass yield	Zenk et al. (1977)
ABA supplementation to the medium	Callus and cell suspension	Enhanced accumulation of catharanthine and ajmalicine	Smith et al. (1987a)
High or low pH	Callus	Abrupt alteration of pH causes intracellular release of alkaloids in the medium	Asada and Shuler (1989), Deus-Neumann et al. (1987),
Increased osmolarity of medium (1.7 g/l NaCl, sorbitol)	Cell suspension	Increased production of alkaloids	Smith et al. (1987a)
High concentration of selected metals in the medium, viz., 200 mM Cu, Cd	Cell suspension	Increased production of alkaloids	Tallevi and Dicosmo (1988)
Addition of jasmonate, salicylic acid, and their selected precursor in the medium	Hairy root culture	Alkaloid accumulation increased	Aerts et al. (1994), Pasquali et al. (1992)
Addition of Vanadium salt	Callus	Alkaloid accumulation increased	Smith et al. (1987a)

2.2 Phytohormones

Addition of 2,4-D in the medium is inhibitory to alkaloid production (Ouelhazi et al. 1994). Transferring the cells from a medium containing 2,4-D to a medium without 2,4-D enhances the alkaloid biosynthesis (Knobloch and Berlin 1980; Roustan et al. 1982; Merillon et al. 1984). Additions of NAA and IAA also have similar effects on alkaloid production in *Catharanthus*. This inhibitory effect of auxin-like hormones has been attributed to the down regulation of *tdc* and *sss* genes that encode tryptophan decarboxylase and strictosidine synthase enzymes respectively. These two enzymes are crucial for the biosynthesis of *Catharanthus* alkaloids (Fig. 1).

Although auxins like NAA, 2,4-D, etc. decrease overall yield of alkaloid, they can be used in the initial stage to increase the biomass of the callus tissue. In fact, a combination of low auxin and high cytokinin is better for callus proliferation, growth, and enhancement of alkaloid content in leaf callus of *Catharanthus roseus* (Kodja et al. 1989; Verma et al. 2012). According to Decendit and Liu (1992), cytokinin accomplishes its effect by removing auxin from the culture medium in non-tumorous cell lines.

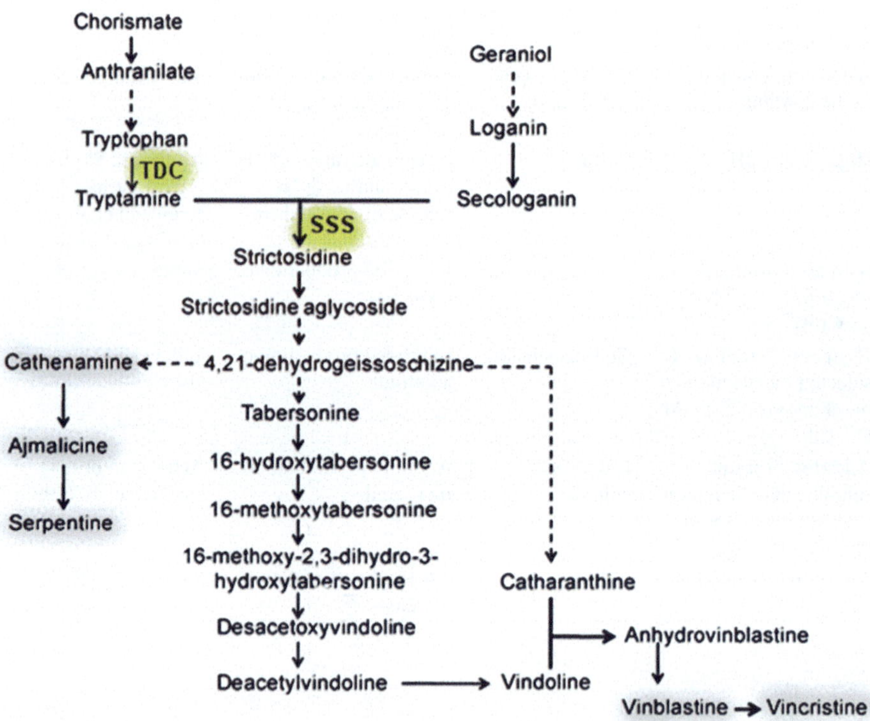

Fig. 1 A generalized pathway for biosynthesis of *Catharanthus roseus* alkaloids

2.3 Light

The serpentine to ajmalicine production ratio changes under the influence of light in tissue culture. Scragg et al. (1988) studied the serpentine production in dark grown *Catharanthus roseus* cultures during several subcultures. When the culture was exposed to light for one subculture period, a significant increase in serpentine production was observed. But continuing the exposure of light for further subcultures significantly decreases serpentine production whereas the ajmalicine production was increased.

Loyola-Vargas et al. (1992) observed a rapid greening of dark grown *C. roseus* calluses when transferred to white light. This was due to chlorophyll production which was in accordance with the increased serpentine accumulation suggesting a correlation between the two. The increase in the concentration of vindoline and catharanthine was also related to light intensity; however, high productivity could not be maintained after repeated subculturing under light.

2.4 Temperature

The effect of temperature on alkaloid production is largely dependent on the type of cell line. Scragg et al. (1988) studied the effect of temperature on growth and alkaloid production of a *Catharanthus roseus* cell line. Cultivation temperatures used were 20, 25, and 30 °C. The maximum growth was observed at 30 °C whereas the highest specific serpentine production was observed at 25 °C.

2.5 Stress Factors

Induction of stress by introducing osmotic shock, adding inorganic salts, heavy metals, fungal homogenates and exposure to UV irradiation, increases alkaloid production. As a result of stress, the enzymes of biosynthetic pathways other than the primary metabolism are induced, resulting in an accumulation of secondary products. In cultured cells of *Catharanthus roseus*, sucrose concentrations ranging from 4 to 10% (W/V) have been used to increase the alkaloid content (DiCosmo and Towers 1984). Addition of 200 nM sorbitol resulted in a 63% increase in catharanthine content. (Smith et al. 1987). When *Catharanthus* suspension cultures were exposed to an osmotic stress of 5% sucrose and 1.25% mannitol for 3 days, there was a marked increment (320%) in the total alkaloid content. The osmotic stress treatment decreased the fresh weight (7.6%) but increased the dry weight by 38% when compared to the control (Godoy-Hernandez and Loyola-vargus 1991).

The effect of UV-B exposure on the production of terpenoid indole alkaloids in *Catharanthus roseus* is remarkable in multiple shoot culture, cell suspension culture, and hairy root culture. In this regard, an increase in catharanthine content in cell suspension culture due to low dose of UV-B exposure has been reported by

Ramani and Chelliah (2007). Again in hairy root culture, an exposure of UV-B for 20 min has led to an increase in lochnericine, serpentine, and ajmalicine by 60%, 20%, and 50%, respectively (Binder et al. 2009). These terpenoid alkaloids are in fact induced to be overproduced to confer protection against UV-B radiations (Binder et al. 2009).

3 Conclusion

Catharanthus roseus, an extremely hardy plant, requires minimum amount of irrigation and water logged condition is extremely harmful for the growth and development of this plant. Propagation can be done through direct seeding but transplantation of nursery grown plants is preferable as it saves time between two crops. Ratooning rather than complete uprooting may be preferred when only stem and leaves are required for extracting vincristine and vinblastine. The plant does not require too much of nutrition; however, a mixture of nitrogen, phosphorus, and potassium can be applied. In addition, farm yard manure (FYM) is also sometimes preferred. This plant can survive without watering for more than 3 weeks and can tolerate a salt concentration of 150 mM NaCl. But for getting increased yield, life-saving intermittent watering is required. These and other kinds of stresses are known to induce alkaloid biosynthesis in *Catharanthus*. They include application of some growth inhibitors and regulators, viz., paclobutrazol, uniconazole, diaminozide, jasmonate (JA), salicylic acid (SA). Moreover, *Catharanthus roseus* has been reported to be tolerant to some of the heavy metals like Cd, Pb, Ni, and Cr (Pandey et al. 2007; Srivastava and Srivastava 2010). Therefore, cultivation of *Catharanthus* in heavy metal contaminated land could be an important consideration as it will serve two purposes, on the one hand it will help decontaminate sites through phytoremediation and on the other hand will also enhance the biosynthesis and production of alkaloids.

Despite these facts, it is matter of fact that percentage of alkaloids extracted from field grown plants is very less. Therefore, tissue culture-based cultivation methods have been developed. The stress factors that enhance yield under tissue culture conditions include osmotic shock, salinity, heavy metals, fungal homogenates, and UV-B exposure. The conditions may vary and have to be standardized according to the objective of production.

References

Aerts RJ, Gise P, Decarolis E, De L, Baumann TW (1994) Methyl Jasmonate vapours increase the developmentally controlled synthesis of alkaloids in *Catharanthus roseus* and *Cinchona* seedlings. Plant J 5:635–643

Aerts RJ, Schaefer A, Hesse M, Baumann TW, Slusarenko A (1996) Signalling molecules and synthesis of alkaloids of *Catharanthus roseus* seedlings. Phytochemistry 42(2):417–422

Amirjani MR (2013) Effects of drought stress on the alkaloid contents and growth parameters of *Catharanthus roseus*. ARPN J Agri Biol Science 8(11):745–750

Asada M, Shuler ML (1989) Stimulation of ajmalicine production and excretion from *Catharanthus roseus*: effects of adsorption *in situ*, elicitors and alginate immobilization. Appl Microbiol Biotechnol 30(5):475–481

Aslam JS, Khan H, Siddiqui ZH, Fatima Z, Maqsood M, Bhat MA, Nasim SA, Ilah A, Ahmad IZ, Khan A, Mujib A, Sharma MP (2010) Important drug: it's applications and production. Int J Compr Pharm 4(12)

Ball V (1991) Ball red book: greenhouse growing, 15th edn. Geo J Ball Publishing, West Chicago III

Barrett JE, Nell TA (1992) Efficacy of paclobutrazol and uniconazole on four bedding plant species. Hort Sci 27:896–897

Basu D (1992) Effects of growth retardant on growth flowering and alkaloid production in *Catharanthus pusillus*. Indian For 92:659–661

Binder BY, Peebles CA, Shanks JV, San KY (2009) The effects of UV-B stress on the production of terpenoid indole alkaloids in *Catharanthus roseus* hairy roots. Biotechnol Prog 25(3):861–865

Blazich FA (1995) Seed germination of annual Vinca responds to irradiation and temperature. Hort Sci 30(2):357–359

Carew DP (1993) Growth of callus tissue of *Catharanthus roseus* in suspension culture. Plant Growth Regul 13(3):287–295

Cathey HM (1964) Physiology of growth retarding chemicals. Ann Rev Plant Physiol 15:271–302

Cheruth AJ, Gopi R, Sankar B, Gomathinayagam M, Panneerselvam R (2008) Differential responses in water use efficiency in two varieties of *Catharanthus roseus* under drought stress. C R Biol 331(1):42–47

Choudhury S, Gupta K (1995) Studies on the germination mechanism of *Catharanthus roseus* (L.) G. Don. cv. alba seeds: effect of temperature and promoters. Seed Sci & Technol 23:831–842

Decendit A, Liu D (1992) Cytokinin-enhanced accumulation of indole alkaloids in *Catharanthus roseus* cell cultures: The factors affecting the cytokinin response. Plant Cell Rep 11:400–403

Deus-Neumann B, Stockijt J, Zenk MH (1987) Radioimmunoassay for the quantitative determination of catharanthine. Planta Med 53:184–188

Di Cosmo F, Towers GHN (1984) Stress and secondary metabolism in cultured plant cells. In: Timmerman BN, Steelink C, Loewus FA (eds) Phytochemical adaptation to stress. Plenum Press, New York, pp 97–175

Dobelis IN (1997) Magic and medicine of plants. The Reader's Digest Association, Pleasentville, YN

Eigste OJ, Tenney B (1943) Range concentration and number of applications of colchicines effective induction of polyploidy in *Vinca* L. Genetics 28:73–74

Elfeky SS, Osman MEH, Hamada SM, Hasan AM (2007) Effect of salinity and drought on growth criteria and biochemical analysis of *Catharanthus roseus* shoot. Int J Bot 3:202–207

Farooqi AA, Sreeram BS (2010) Cultivation of medicinal and aromatic crops. Universities Press, Hyderabad

Frischknecht PM, Batting M, Baumann TW (1987) Effect of drought and wounding stress on indole alkaloid formation in *Catharanthus roseus*. Phytochem 25:707–710

Fulzele DP, Sipahimalini AT, Meble HR (1990) *Catharanthus roseus* cell cultures: Growth and alkaloid production. Ind J Pharm Sci 52:19–22

Gaines JL (2004) Increasing alkaloid production from *Catharanthus roseus* suspensions through methyl jasmonate elicitation. Pharm Eng J 24:24–35

Gill TS, Ribeiro OK, Zentinijer GA (1977) Phytophthora blight of periwinkle in Cochella valley of California. Plant Dis Rep 61:560–561

Goddijn OJM, deKam RJ, Zanetti A, Schilperoort RA, Hoge JHC (1992) Auxin rapidly downregulates transcription of the tryptophan decarboxylase gene from *Catharanthus roseus*. Plant Mol Biol 18:1113–1120

Godoy-Hernández G, Loyola-Vargas VM (1991) Effect of fungal homogenate, enzyme inhibitors and osmotic stress on alkaloid content of Catharanthus roseus cell suspension cultures. Plant Cell Rep 10(11):537–540

Goyal R, Pathak NC (1982) A new leaf spot disease of *Catharanthus roseus* L. G. Don from India. New Bot 9:31–32

Groger D (1985) Alkaloids derived from tryptophan. In: Mother K, Schutte HR, Luckner M (eds) Biochemistry of Alkaloids. VEB Deutescher verlaq der Wessen Chaften, Berlin

Idrees M, Naeem M, Aftab T, Khan MMA, Moinuddin (2011) Salicylic acid mitigates salinity stress by improving antioxidant defence system and enhances vincristine and vinblastine alkaloids production in periwinkle [*Catharanthus roseus* (L.) G. Don]. Acta Physiol Plant 33:987–999

Janaki-Ammal EK, Bezbarauah (1963) Induced Tetraploidy in *Catharanthus roseus*. G Don Indian Acad Sci B-57:339

Janda T, Szalai G, Tari I, Paldi E (1999) Hydroponic treatment with salicylic acid decreases the effects of chilling injury in maize (*Zea mays* L.) plants. Planta 208:175–180

Jordon MA, Thrower D, Wilsen L (1991) Mechanism of inhibition of cell proliferation by *Vinca* alkaloids. Cancer Res 51:2212–2222

Kalra A, Ravindra NS, Chandrasekhar RS (1991) Influence of foliar application of fungicide on die-back disease caused by *Pythium aphanidermatum* and alkaloid yield of periwinkle *Catharanthus roseus*. Ind J Agri Sci 61:949–951

Karnick CR (1977) Studies on Indian medicinal plants—cultivation trials of Vinca rosea Linn. J Res Indian Med Yoga Homoeo 12:101–108

Knobloch KH, Berlin J (1980) Influence of the medium composition on the formation of secondary compounds in cell suspension cultures of *Catharanthus roseus* (L.) G. Don. Z Naturforsch 35c:551–556

Kodja HP, Liu D, Merillon JM, Andreau F, Raideau M, Chenieux JC (1989) Stimulation par les cytokinins de I, accumulation d' alcaloides indoliques dans des suspensions cellularies de *Catharanthus roseus* (L.) G. Don. Z Naturforsch 35c:551–556

Kulkarni RN, Baskaran K, Chandrasekhara RS, Kumar S (1999) Inheritance of morphological traits of periwinkle mutants with modified contents and yields of leaf and root alkaloid. Plant Breeding 118:71–74

Larkindale J, Knight MR (2002) Protection against heat stressinduced oxidative damage in Arabidopsis involves calcium, abscisic acid, ethylene, and salicylic acid. Plant Physiol 128:682–695

Levy A, Milo J, Aashri A, Paleswitch D (1983) Heterosis and correlation analysis of the vegetative components and ajmalicine contents in the roots of the medicinal plant- *Catharanthus roseus* L. G Don Euphytica 32:557–564

Loyola-Vargas VM, Mende-Zeel M, Monforte-Gonzales M Miranda-Ham ML (1992) Serpentine accumulation during greening in normal and tumor tissues of *Catharanthus roseus*. J Plant Physiol 140:213–217

Mastalerz JW (1976) A manual on culture of bedding plants as a greenhouse crop. In: Bedding Plants, 2nd edn. Pennsylvania Flower Growers Association, University Park, PA

Merillon JM, Rideau M, Chenieux JC (1984) Influence of sucrose on levels of ajmalicine, serpentine and tryptamine in *Catharanthus roseus* cells *in vitro*. Planta Med:497–501

Morris P (1986) Regulation of product synthesis in cell cultures of *Catharanthus roseus*. II. Comparison of production media. Planta Med 52:121–126

Ouelhazi L, Hamdi S, Chénieux JC, Rideau M (1994) Cytokinin and auxin-induced regulation of protein synthesis and poly(A)$^+$ RNA accumulation in *Catharanthus roseus* cell cultures. J Plant Physiol 144:167–174

Pandey S, Gupta K, Mukherjee AK (2007) Impact of cadmium and lead on *Catharanthus roseus* - A phytoremediation study. J Environ Biol 28(3):655–662

Pasquali G, Goddijn OJ, de Waal A, Verpoorte R, Schilperoort RA, Hoge JH, Memelink J (1992) Coordinated regulation of two indole alkaloid biosynthetic genes from *Catharanthus roseus* by auxin and elicitors. Plant Mol Biol 18(6):1121–1131

Ramani S, Chelliah J (2007) UV-B induced signalling events leading to enhanced production of Catharanthine in Catharananthus roseus cell suspension culture. BMC Plant Biol 7:61. doi:10.1186/1471-2229-7-61

Roustan JP, Ambid C, Fallot J (1982) Influence de l'acide 2,4-dichlorophénoxyacétique sur l'accumulation de certains alcaloides indoliques dans les cellules quiescentes de *Catharanthus roseus* cultivées in vitro. Physiol Vég 20:523
Schnell L (1941) The induction of polyploidy in *Vinca rosea*. Amer J Bot 28:53–55
Scragg AH, Creswell R, Ashton S et al (1988) Growth and secondary product formation of a selected *Catharanthus roseus* cell line. Enzyme Microb Technol 10:532–536
Sevestre-Rigouzzo M, Nef-Campa C, Ghesquiere A, Christin H (1993) Genetic diversity and alkaloid production in *Catharanthus roseus*. *C trichophyllous* and their hybrids. Euphytica 66:151–159
Singh SN, Vats P, Suri S et al (2001) Effect of an antidiabetic extract of *Catharanthus roseus* on enzymatic activities in Streptozotocin induced diabetic rats. J Ethnopharmacol 76(3):269–277
Smith JI, Smart NJ, Misawa M, Kurz WG, Tallevi SG, Dicosmo F (1987a) Increased accumulation of indole alkaloids by some cell lines of *Catharanthus roseus* in response to addition of vanadyl sulphate. Plant Cell Rep 6(2):142–145
Smith JI, Smart NJ, Kurz WGW, Misawa M (1987b) Stimulation of indole alkaloid production in cell suspension cultures *Catharanthus roseus* by abscisic acid. Planta Med 53:470–474
Srivastava NK, Srivastava AK (2010) Influence of Some Heavy Metals on Growth, Alkaloid Content and Composition in *Catharanthus roseus* L. Indian J Pharm Sci 72(6):275–278
Styer RC, Laffe S (1990) Tips on germinating the top ten. Greenhouse Managers 9(10):40–41
Taha HS, El-Bahr MK, Seif-El-Nasr MM (2008) *In vitro* studies on Egyptian *Catharanthus roseus* (L.) G. Don.: 1- Calli production, direc
Talha M, Radwan AS, Negm S (1975) The effect of soil moisture deficit on growth and alkaloid content of *Catharanthus roseus* G. Don. Curr Sci 44(7):614–616
Tallevi S, Dicosmo F (1988) Stimulation of Indole alkaloid content in vanadium treated *Catharanthus roseus* suspension cultures. Planta Med 54:149–152
Verma AK, Singh RR, Singh S (2012) Improved alkaloid content in callus culture of *Catharanthus roseus*. Botanica SERBICA 36(2):123–130
Walts LR (2004) The herbal encyclopedia, a practical guide to the many uses of herbs. I Universe, Inc, Lincoln
Zargar M, Farah F, Nabavi T (2010) Hairy roots production of transgenic *Catharanthus roseus* L. plants with *Agrobacterium rhizogenes* under in vitro conditions. Int J Bio-Technol Res 3(1):105–110
Zenk MH, El-Shagi H, Arens H, Stöckigt J, Weiler EW, Deus B (1977) Formation of the indole alkaloids serpentine and ajmalicine in cell suspension cultures of *Catharanthus roseus*. In: Barz W, Reinhard E, Zenk MW (eds) Plant tissue culture and its biotechnological application. Springer, Berlin, pp 27–43
Zhao J, Wei-Hua Z, Qui H (2000a) Enhanced ajmalicine production of *Catharanthus roseus* cell cultures by combined elicitor treatments: From shake flasks to 20L airlift Bioreactor. Biotechnol Lett 22:509–514
Zhao J, Zhu WH, Lu Q (2000b) Promotion of indole alkaloids production in *Catharanthus roseus* cell cultures by rare earth elements. Biotechnol Lett 22:825–828
Zhu W, Yang B, Komatsu S, Lu X, Li X, Tian J (2015) Binary stress induces an increase in indole alkaloid biosynthesis in *Catharanthus roseus*. Front Plant Sci 6(582):1–12

The Study of the Effect of Nickel Heavy Metal on Some Growth Parameters and Production of Alkaloids in *Catharanthus roseus*

Matin ArefiFard

Abstract Plants, in their life cycle, are usually exposed to various kinds of environmental stress including heavy metals. One of these heavy metals is nickel which affects many physiological activities of plants. *Catharanthus roseus* being a rich source of alkaloids (indole alkaloids) consists of more than 130 secondary metabolites. Some of these compounds have pharmacological effects and are used as medicine. Vinblastine which is the most important alkaloid found in the leaves of *Catharanthus roseus* is used to heal the diseases like: testicle and breast cancers, lymphoma, neuroblastoma, Hodgkin and non-Hodgkin's lymphoma, mycosis fungoides, histiocytosis, and Kaposi's sarcoma. It also controls mitotic activity by stopping the cells in metaphase through nonreturnable connection to tubeline. This alkaloid is the chemical analog of vincristine. In this study, the effect of 0, 2.5, 5, 10, 25, 50 mM concentrations of nickel chloride ($NiCl_2$) on the amount of germination, growth factors, and alkaloids of *Catharanthus roseus* (seeds of pink variety) were investigated. Based on the findings of the study, the nickel metal decreased the percentage of germination and the growth factors in the plant. The total amount of alkaloid noticeably increased in both root and shoot as affected by $NiCl_2$. Investigating the TLC plates/planes, an increase in the amounts of vinblastine, catharanthine, and ajmalicine alkaloids especially at higher concentrations were observed.

Keywords *Catharanthus roseus* • Nickel heavy metal • Indole alkaloids • Growth factors

M. ArefiFard (✉)
Biological Department, Arak University-Islamic Azad University of Arak, Arak, Iran
e-mail: matinarefifard@yahoo.com

M. Naeem et al. (eds.), *Catharanthus roseus*,
DOI 10.1007/978-3-319-51620-2_18

1 Introduction

Plant cells produce two sets of compounds: primary metabolites and secondary metabolites. The primary metabolites are directly involved in growth and metabolism and play an important role in making cell compounds through the process of photosynthesis (Keeling and Bohlmann 2006). Secondary metabolites are obtained from biosynthesis of primary metabolites, such as carbohydrates, lipids, proteins, and nucleic acids, and they are not involved in the process of metabolism. The most important secondary metabolites are steroids, lignins, tannins, and flavonoids. Alkaloids are the most important products of secondary metabolism in plants. They are organic compounds which have one or more nitrogen atoms in their heterocyclic chain, (a source for saving nitrogen), semi alkaline feature. Forty percent of all plant families contain at least one alkaloid-bearing species (Verpoorte et al. 2002). The plant families which fall in this category are Solanaceae, Papaveraceae, and Rubiaceae. Some alkaloids, like quinine and strychnine are very poisonous while some are used as medicines. For example, the structure of vinblastine and vincristine is used for similar synthesis of two important drugs for vinflunine and vinorelbine cancers (Lopez et al. 2011).

Secondary metabolites are chemical compounds which establishes an interaction between plant and environment and therefore, any change in environmental conditions (e.g., light, temperature, the amount of water and soil) can affect their production and quality (Kutchan 1995). The following are among the stress which can affect the amount of alkaloids production: heavy metals, drought, lack of water, and hormones. Because of the differences in alkaloids molecule construction, abundant variation, different solutions, and other features, it may be difficult to investigate the existence of alkaloids in plants.

2 *Catharanthus roseus*

The genus *Catharanthus* belongs to Apocynaceae family. It contains eight varieties and 50 different cultivars. The chromosome number of all the varieties of *Catharanthus* is 2n = 16. *Catharanthus roseus* which is known as Periwinkle in English language, however, the plant is known as Parvanesh or Parivash in Persian. *Catharanthus roseus* has three varieties: "Alba" has white flowers; "Roseus" bears pink flowers and "Ocillata" which has white flowers with purple dots in their center (Jaleel 2006). It can be propagated by putting its seed in the temperature between 22 and 25 °C where the humidity is high. The seeds are kept in a dark place until they germinate.

3 Alkaloids of *Catharanthus roseus*

Almost one fourth of medicines produced in the world are made of plants. They are either the essence of plants or plant compounds which have been synthesized. For instance, the most common pain killer (Aspirin) and vinblastine and packli taxel for cancer treatment are among the drugs which are just originated from plants and are made through in vitro regeneration. *Catharanthus roseus* is a rich source of alkaloid chemical compounds (indole alkaloids) contains more than 130 known secondary metabolites. These metabolites are secologanine, triptamine, strictozidin, serpentine, catharanthine, ajmalicine, vindoline, vincristine, vinblastine, etc. Some of these compounds have pharmacological effects (Verpoorte et al. 2002; van der Heijden et al. 2004). The alkaloids are made of two parts: indol (catharanthine) and dehydro indol (vindoline). The most important components of the plant are: vincristine, vinblastine, vineurosine, and vinrosidine. Nowadays, the derived components of these alkaloids are used for healing different cancers.

4 Vinblastine and Vincristine

Vinblastine is the most important alkaloids in the leaves of *Catharanthus roseus*. Its pre-makers are stemadenine and monomeric alkaloids. Stemadenine changes to monomeric alkaloids like catharanthine, tabersonine, and vindoline. When catharanthine and tabersonine are connected, vinblastine is produced (Aerts 1996). Vinblastine is used to heal the diseases like: testicle and breast cancers, lymphoma, neuroblastoma, Hodgkin and non-Hodgkin's lymphoma, mycosis fungoides, histiocytosis, and Kaposi's sarcoma. It also controls mitotic activity by stopping the cells in metaphase through nonreturnable connection to tuboline (Filippini et al. 2007). The mechanism of the effect of vincristine is like vinblastine.

5 Heavy Elements

Heavy metals are the metals which have the density more than 5 g/L cm^3 from copper to bismuth in the Alternative Table which have the density of more than four and are categorized in four classes: Class A, which are necessary in high density for living things (e.g., iron); Class B, which do not have any biological roles and are not poisonous in low density (e.g., Lanthanum, Stronsium); Class C, the elements which are necessary, in low density, for some kinds of living things, and are poisonous in higher densities (e.g., Zinc, Copper, Nickel, Cobalt, Molybdenum, and Chrome); and Class D, which are poisonous even in low density (Mercury, Lead, Cadmium, Uranium) (Sebastiani et al. 2004).

6 Stress Effects of Heavy Metals

The effects of heavy metals depend on many factors like environmental conditions, pH, different elements, organic materials, mediators, fertilizers, and plant variations (Eliwa 2000; Abbas and Kamel 2004). Although many metals are necessary for plants, their high densities are poisonous. Because they cause oxidative stress in plants which leads to reduction of encoding genes of antioxidant proteins (i.e., catalase, proxidase, and superoxide dismutase). In addition, when heavy metals cause stress, the leaf surface, the size of the stomatal opening, and cuticles are affected (Mehrotra et al. 2005). When high density metals are present, they take the place of necessary metals which have an important role in forming pigments and enzymes. The stress of heavy metals, like nonbiological stresses, leads to change (increase or decrease) in the synthesis of secondary metabolites of plants (Santiago et al. 2000; Mehrotra et al. 2005). Compared to the shoot part of the plant, the growth of the root is less affected by heavy metals. The root has a critical role in reserving elements and prevention of accumulation of elements in the plant body (Mazhoudia et al. 1997).

7 Nickel

A small amount of nickel plays a role in nitrogen metabolism and the plant growth (Brown et al. 1987). Nowadays, nickel has been added to the collection of the low consumption elements needed for plants (Witte-Claus et al. 2002; Benaroya et al. 2004).

Murch et al. (2003) investigating the effect of different treatments of nickel (0, 25, and 250 mM) on secondary metabolites of *Hypericum perforatum*, found that the amount of hipricine increases 15–20 times more than control group. Eliasiova (2004) reported that the amount of secondary metabolites in *Matricaria chamomlli* increased remarkably after the treatment of $NiCl_2$.

8 Purpose of the Research

Studies have shown that the change in planting conditions can increase the amount of *Catharanthus roseus* alkaloids, especially vinblastine and vincristine. In order to investigate the effect of different densities of nickel (0, 2.5, 5, 10, 25, and 50 mM of $NiCl_2$) on *Catharanthus roseus*, an experiment was conducted with factorial design, repeated three times in winter season, 2010 at the Physiology Laboratory of Faculty of Sciences, University of Arak, Iran. The purpose of the research was to investigate whether the production of amount of alkaloids and some of their growth features can be improved by applying $NiCl_2$.

9 Materials and Methods

The healthy seeds (pink variety) of *Catharanthus roseus* were sterilized using sodium hypochlorite of 1% for 10 min, ethanol 70% for 5 min, mercury chloride 0.2% for 10 min. Then, the seeds were washed three times using doubled distilled water (DDW). Ten seeds were planted, in vases, at proper distance. They were watered with Hoagland solution of ½ concentrations and kept in a planting room with photoperiods of 16 h in light and 8 h in darkness while the temperature was 28 ± 2 °C during the day, and 25 ± 2 °C during the night, and the humidity was from 30 to 40%. The surface of each vase was covered by nylon plastic till the time of germination in order to prevent vaporizing of water and change in the salt density. After 7 week, the plants of the same size were extracted unharmed and were put in microtubes, in a way that the roots were kept in Hoagland solution of ½ concentration. The planting environment was changed at every 4 days and daily aeration was done for 4 h using an air pump. After a week, the 45-day old plants were exposed to $NiCl_2$ with densities of 0 (control group), 2.5, 5, 10, 25, 50 mM for 96 h (three repetitions for each treatment).

10 Percentage of Germination and Parameters of Growth

After planting the seeds and imposing stress, the number of buds was counted and the percentage of germination was calculated every day.

11 Measurement of Water Content and Parameters of Growth

After 96 h of exposure to stress, the plants were collected. The length of the shoot (the plant height) from bottom of the body to the top of the highest leaf, and the length of the root, from its growth point to its end were measured. The percentage of water content in plant was calculated using the method offered by Sumithra et al. (2006), through the application of the following formula: WP (%) = [(FW – DW) × 100]/FW.

12 Obtaining of Alkaloids

12.1 Preparing Plant Essence

One gram of the leaves and the roots of each treatment were ground in liquid nitrogen. For each gram of tissue powder, 20 mL of methanol was added and it was kept for shaking in laboratory environment for 24 h. After filtering the solution by filter paper, the methanol essence was put in oven with the temperature of 50 °C until 2 mL remained. Then, for every gram of the plant tissue, 20 mL of 0.5 N sulfuric acid was added, and pH was adjusted between 9 and 12, by adding ammonium hydroxide of 25% density. The volume of the produced tissue was measured, and the same volume of chloroform was added to it. The solution, then, was transferred to decanter in order to separate two phases completely; the water phase at top, and the organic phase at the bottom. The organic phase was used to investigate alkaloids. The extraction was repeated two more times using chloroform (it was done based on the procedures followed by Ataei-Azimi et al. 2008). The produced essence was vaporized under the hood and in the laboratory temperature (based on the suggestions by Noori et al. 2002).

12.2 Total Amount of Alkaloids in Herbal Essence

The volume of the essence increased to 10 mL by adding 96% ethanol. 5 mL of the solution was used and its pH was adjusted between 2 and 2.5 by adding (HCL). Then, 2 mL of Dragendorf reagent and 0.8 g of $Bi(NO_3)_3{\cdot}5H_2O$ were dissolved in 40 mL of double distillation water (DDW). In addition, 8 g of potassium were separately dissolved in 20 mL of DDW. Later, the two solutions were mixed and centrifuged. In order to be sure of complete sediment of alkaloids, again some Dragendorf reagent was added to the solution and centrifuged. After taking away the solution which was over the sediment, the remaining sediment was washed by ethanol 96%, and then centrifuged. Once again the solution over the sediment was put away. 2 mL of sulfide disodium 1% (1 g sulfide disodium in 100 mL of distilled water) was added to the brownish sediment. After centrifugation, the brownish sediment was observed. The solution over the sediment was taken away and sample was centrifuged again. After excluding the surface solution, 2 mL of concentrated nitric acid was added to the sample. The volume of the produced solution was increased to 10 mL by adding DDW. 1 mL of the solution and 5 mL of thiourea solution 3% (3 g of thiourea in 100 mL of DDW) was taken from each sample. Then, the amount of attraction of the samples was read at 435 nm.

Based on the alkaloid standard curve, the total amount of alkaloid was determined: There were ten test tubes containing the stock solution of $Bi(NO_3)_3{\cdot}5H_2O$ (standard solution) (10 mg of $Bi(NO_3)_3{\cdot}5H_2O$ in 5 mL of nitric acid and adding distillation water until it reaches to the volume of 100 mL), to the ratio of 1–9, 2–8, 3–7, till 9–1 mL. In this phase, 1 mL of each test tube was taken and 5 mL of thiourea solution 3% was added to it. Then, the amount of attraction of the samples was read at 435 nm wavelength.

12.3 Thin-Layer Chromatography (TLC)

In this phase, 500 μL of ethanol (96%) was added to the remaining essence which was vaporized in the watch glass. 20 mL of the solution was used for TLC (Atael-Azimi et al. 2008). The plates of TLC (Silica gel, G60 Merck) were cut. The plates which were 10 cm × 20 cm were then spotted by capillary tube containing the essence. Spotting was done on the plates 1 cm away from the edge of the paper, and the distance among the spots was 1 cm. When the chromatograms dried, they were put in a Chromato tank containing ethyl-acetate and ethanol solutions (1/4). Later, the plates were dried under the hood and the size and color of the spots were defined using UV at a wavelength of 366 and 254 nm, respectively. Besides, the amount of RF for each spot was calculated. The amount of RF of each spot was compared with RF of vinblastine and vincristine standards, and each spot were defined (Noori 2002). After observing the TLC plates by UV, they were sprayed by Dragendorf reagent (Ataei-Azimi et al. 2008). The yellow color of the spots indicates the alkaloids, and the background of the plate changes to cream color.

12.4 Statistical Analyses

The statistical design of the experiment was random. The graphs were prepared by applying EXCEL Software. In order to analyze the measured factor, SPSS Software, 11th version was applied. One way ANOVA was used to analyze the variance and Duncan method was applied to compare the mean of the data.

13 Results

13.1 Germination

The results showed that the percentage of germination is affected by $NiCl_2$. After 5 days of planting time, germination percentage of control group reached 100% in ½ concentration of Hoalgand's solution. On the other hand, when the density of nickel was increased, the rate of germination decreased. The least percentage of germination, that is, 43.32, occurred in 50 mM of $NiCl_2$.

14 Study of the Effect of Nickel on Growth Factors

14.1 The Length of Root and Shoot Parts

The analysis of variance of data showed that $NiCl_2$ significantly increased the shoot length of plant (at the significance level of 5%). The density of $NiCl_2$, 50 mM, significantly increased the shoot length of plant by 150%, while the root length decreased 22.4% compared to the level of growth in control group. Despite the fact that $NiCl_2$ treatment decreased the root length of plant, the data analysis showed that the amount of change is not statistically significant (Fig. 1).

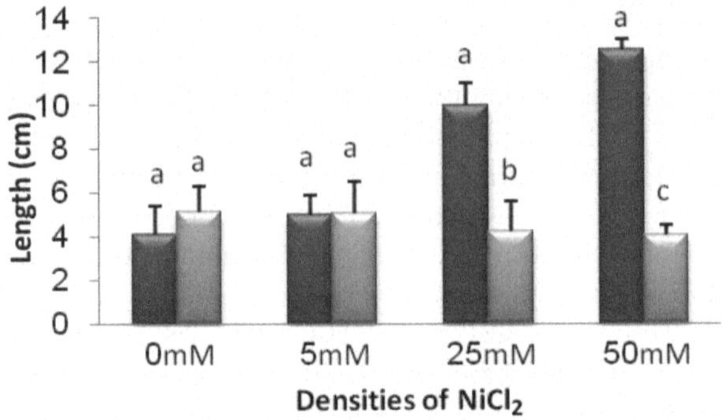

Fig. 1 Effect of different densities of $NiCl_2$ (0, 5, 25, 50 mM) on the length of root and shoot parts of the plant. The data are presented in three repetitions of the mean ± standard deviation. The means represented by different letter codes show the significant difference at the level of 5%

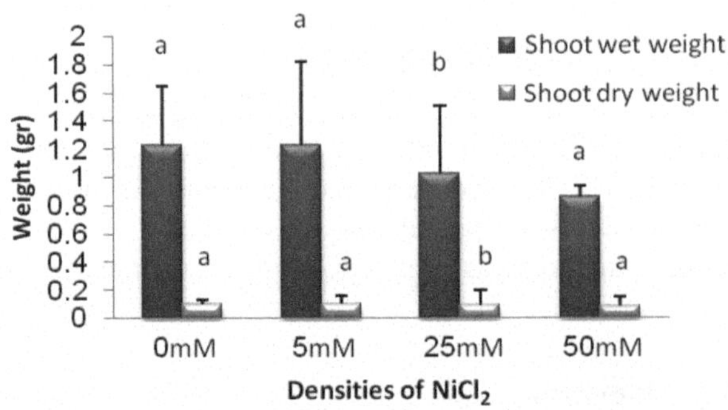

Fig. 2 Effect of different densities of $NiCl_2$ (0, 5, 25, 50 mM) on the weight of plant shoot. The data are presented in three repetitions of the mean + standard deviation. The means represented by different letter codes show the significant difference at the level of 5%

14.2 The Weight of the Fresh and Dry Shoots

Treatment with $NiCl_2$ significantly decreased the shoot fresh weight (at the level of 1%), and dry weight (at the level of 5%) in a way that when the density of the treatment increased, the shoot weight decreased (Fig. 2).

14.3 The Weight of the Fresh and Dry Roots

The $NiCl_2$ treatment significantly decreased the weight of the dry root ($p \leq 0.05$); however, the change in the weight of the plant fresh root is not statistically significant (Fig. 3).

14.4 The Water Content of Shoot

The analysis of variance of data showed that the effect of $NiCl_2$ on the water content of the shoot is not significant. Control group having highest mean (43.95%) of the water content of the shoot, while the lowest mean (32.03%) of the water content was recorded under the treatment of 50 mM $NiCl_2$ (Fig. 4).

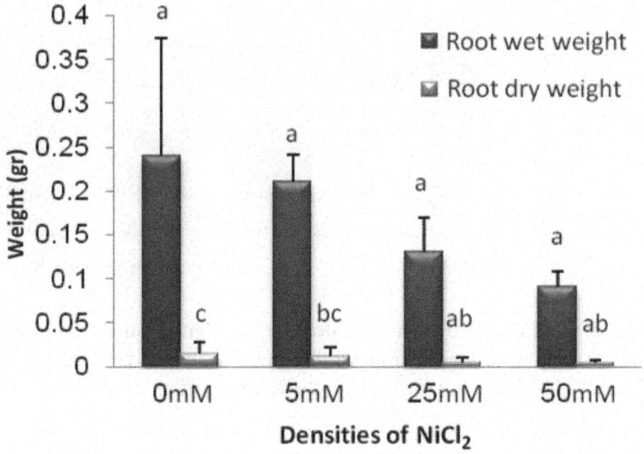

Fig. 3 Effect of different densities of $NiCl_2$ (0, 5, 25, 50 mM) on the weight of plant root. The data are presented in three repetitions of the mean ± standard deviation. The means represented by different letter codes show the significant difference at the level of 5%

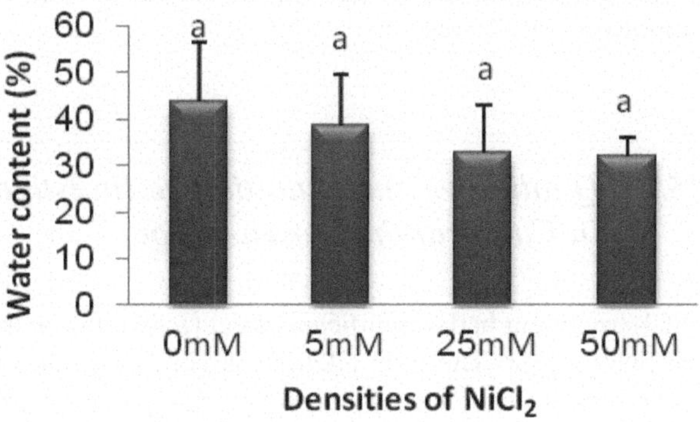

Fig. 4 Effect of different densities of $NiCl_2$ on the water content of the shoot. The data are presented in three repetitions of the mean ± standard deviation. The means represented by different letter codes show the significant difference at the level of 5%

15 The Effect of Nickel Chloride on Alkaloids of Leaf and Root

15.1 The Qualitative and Quantitative Changes of Alkaloids in Catharanthus roseus Leaf

When the density of nickel increases, the total amount of alkaloid significantly increases ($p \leq 0.05$) in *Catharanthus roseus* leaves. In this study, the total amount of alkaloid reached 1041.7 μg in control group, and when the highest treatment density was applied, the alkaloid level reached 2069.8 μg in fresh leaves (98.7% increase) (Fig. 5). Investigating qualitative changes of alkaloids following TLC method, it was observed that there was serpentine alkaloid in all densities, and ajmalicine alkaloid in all densities except in control group and the treatment of 2.5 mM. The band related to RF = 0.19 in control sample appears in a bright color; however, by increasing the density of the treatment, the color of the band gradually

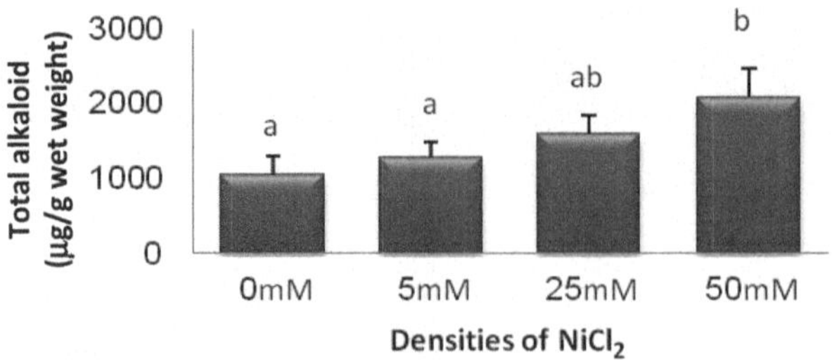

Fig. 5 The effect of different densities of $NiCl_2$ on total alkaloid in *C. roseus* leaf. The data are presented in three repetitions of the mean ± standard deviation. The means represented by different letter codes show the significant difference at the level of 5%

becomes dark. It is because of the existence of vincristine alkaloid. The band related to RF = 0.31 is observed when the density is 10 or more millimolar which refers to vinblastine (Table 1).

15.2 Qualitative and Quantitative Investigation of Alkaloids in Catharanthus roseus Root

$NiCl_2$ treatment had a significant effect ($p \leq 0.05$) on the amount of total alkaloid of the plant's root. The total alkaloid in control group reached 194.13 µg and at the highest treatment density it reached 520.84 µg in fresh root (62.72% increase) (Fig. 6). Investigating the TLC plates, it was observed that there were serpentine, vindoline, and vincristine alkaloids in all densities. There are no bands related to RF = 0.19, and RF = 0.74 (ajmalicine and catharanthine alkaloids) in control sample, and the treatment densities of 2.5 and 5 mM; however, the band is observed in the density of 10 mM, and becomes darker gradually, when the density increases. There is no band related to RF = 0.4 in control sample and treatment density of 2.5 mM; however, in density of 5 mM, the band is observed, and by increasing the density of the treatment, the color of the band gradually becomes dark. It is because of the existence of vinblastine alkaloid (Table 2).

16 Discussion

16.1 The Percentage of Germination in Nickel Stress

The results were similar with the work done by Dalal and Bairgi (1985) on *Capsular corchrus* and *Corchorus plitorius* plants. Prevention of germination due to nickel relates to the prevention of enzyme activity at the time of

Table 1 The effect of different densities of $NiCl_2$ (0, 2.5, 5, 10, 25, and 50 mM) on the type of alkaloids in *C. roseus* leaf using TLC method

Color (254 nm)	Color (366 nm)	Alkaloids	RF	Densities of $NiCl_2$
Blue	Blue	Serpentine	0.13	
Dark green		Vincristine	0.19	0 mM
Red	Pink	Catharanthine	0.72	
Blue	Blue	Serpentine	0.13	
Dark green		Vincristine	0.19	2.5 mM
Red	Pink	Catharanthine	0.75	
Dark	Yellow	Vindoline	0.81	
Blue	Blue	Serpentine	0.12	
Dark green		Vincristine	0.19	
Red	Pink	Catharanthine	0.75	5 mM
Dark	Yellow	Vindoline	0.81	
Dark green	Almond green	Ajmalicine	0.96	
Blue	Blue	Serpentine	0.14	
Dark green		Vincristine	0.18	
Dark blue		Vinblastine	0.31	10 mM
Dark	Yellow	Vindoline	0.81	
Dark green	Almond green	Ajmalicine	0.96	
Blue	Blue	Serpentine	0.14	
Dark green		Vincristine	0.21	
Dark blue		Vinblastine	0.37	25 mM
Dark green	Almond green	Ajmalicine	0.96	
Blue	Blue	Serpentine	0.13	
Dark green		Vincristine	0.21	
Dark blue		Vinblastine	0.38	50 mM
Red	Pink	Catharanthine	0.75	
Dark green	Almond green	Ajmalicine	0.96	

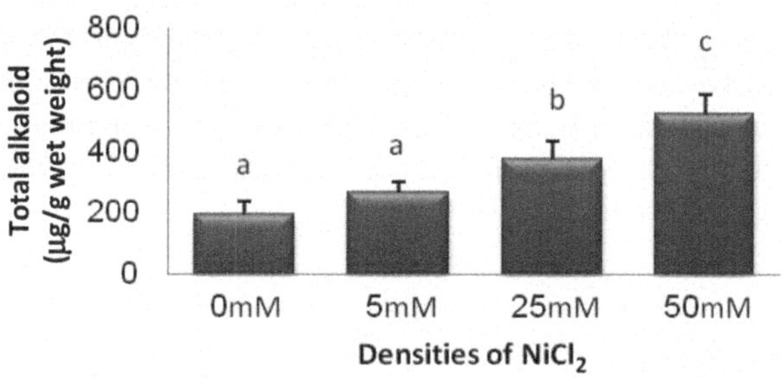

Fig. 6 The effect of different densities of $NiCl_2$ on total alkaloid in *C. roseus* root. The data are presented in three repetitions of the mean ± standard deviation. The means represented by different letter codes show the significant difference at the level of 5%

Table 2 The effect of different densities of $NiCl_2$ on the type of alkaloids in *C. roseus* root using TLC method

Color (254 nm)	Color (366 nm)	Alkaloids	RF	Densities of $NiCl_2$
Blue	Blue	Serpentine	0.11	
Dark green		Vincristine	0.18	0 mM
Dark	Yellow	Vindoline	0.82	
Blue	Blue	Serpentine	0.12	
Dark green		Vincristine	0.18	2.5 mM
Dark	Yellow	Vindoline	0.88	
Blue	Blue	Serpentine	0.11	
Dark green		Vincristine	0.19	5 mM
Dark blue		Vinblastine	0.36	
Dark	Yellow	Vindoline	0.89	
Blue	Blue	Serpentine	0.11	
Dark green		Vincristine	0.18	
Dark blue		Vinblastine	0.41	10 mM
Dark	Yellow	Vindoline	0.89	
Dark green	Almond green	Ajmalicine	0.96	
Blue	Blue	Serpentine	0.13	
Dark green		Vincristine	0.18	25 mM
Dark blue		Vinblastine	0.4	
Red	Pink	Catharanthine	0.74	
Dark	Yellow	Vindoline	0.89	
Dark green	Almond green	Ajmalicine	0.96	
Blue	Blue	Serpentine	0.11	
Dark green		Vincristine	0.18	50 mM
Dark blue		Vinblastine	0.36	
Red	Pink	Catharanthine	0.74	
Dark	Yellow	Vindoline	0.86	
Dark green	Almond green	Ajmalicine	0.96	

germination, protein synthesis, and carbohydrate metabolism (Lin and Kao 2006; Maheshwari and Dubey 2007). The high density of nickel causes change and deterrence of mineral material absorption, such as calcium and magnesium, which finally led to the decrease of germination (Gabbrielli et al. 1990; Ahmad et al. 2007).

16.2 Growth Factors

Upon treatment of heavy elements, the rate of mitotic division decreased especially in metaphase stage in meristemic cells, which leads to the decrease of the root length (Gold bold and Kettner 1991). Heavy elements stress decreases the length of the root whether it is fresh or dry which is probably because of turbulence in absorption of water, food, and low water potentiality (Azmat et al. 2006), and results in closure of stomatal opening and decrease in the turgor pressure, lowering of the sweat intensity and stabilization of CO_2 becomes limited (Ozyigit and Akinci 2009).

16.3 Alkaloid Changes

The results of Zheng and Wu 2004) showed that the production amount of ajmalicine alkaloid, and tryptophan and tryptamine aminoacids (precursor of indol alkaloids of *Catharanthus roseus* plant) increases remarkably, which leads to the noticeable increase of vincristine and vinblastine alkaloids. The obtained data showed that the degree of qualitative and quantitative changes of alkaloids is remarkable under the nickel treatment. The degree of total alkaloid in the leaf and root of the plant increased significantly by the increase in the density of $NiCl_2$ treatment.

References

Abbas SM, Kamel EA (2004) Rhizobium as a biological agent for preventing heavy metal stress. Asian J Plant Sci 3:416–424

Aerts RJ, Schäfer A, Hesse M, Baumann TW, Slusarenko A (1996) Signalling molecules and the synthesis of alkaloids in *Catharanthus roseus* seedlings. Phytochemistry 42:417–422

Ahmad MSA, Hussain M, Saddiq R, Alvi AK (2007) Mungbean: a nickel indicator, accumulator or excluder? Bull Environ Contam Toxicol 78:319–324

Ataei-Azimi A, Hashemloian BD, Ebrahimzadeh H, Majd A (2008) High in vitro production of ant-canceric indole alkaloids from periwinkle (*Catharanthus roseus*) tissue culture. Afr J Biotechnol 7:2834–2839

Azmat R, Haider S, Askari S (2006) Phyotoxicity of Pb: I. Effect of Pb on germination, growth, morphology and histomorphology of *Phaseolus mungo* and *Lens culinaris*. Pak J Biol Sci 9:979–984

Benaroya RO, Tzin V, Tel-Or E, Zamski E (2004) Lead accumulation in the aquatic fern Azolla filiculoides. Plant Physiol Biochem 42:639–645

Brown PH, Welch RM, Cary EE, Checkai RT (1987) Micronutrients: beneficial effects of nickel on plant growth. J Plant Nutr 10:2125–35

Dalal T, Bairgi P (1985) Effect of Hg, As and Pb on germination and seedling growth of two jute varieties. Environ Ecol 3:403–407

Eliasiova A, Repcak M, Pastirova A (2004) Quantitative changes of secondary metabolites of Matricaria chamomilla by abiotic stress. Z Naturforsch 59:543–548

Eliwa AM (2000) Some physiological and cytological studied on the effect of ions of heavy metals on *Pisum sativum* plant. MSc thesis. Ain Shams University, Cairo

Filippini S, Blanco A, Fernández-Marmiesse A, Álvarez-Iglesias V, Ruíz-Ponte C, Carracedo Á, Vega A (2007) Multiplex SNaPshot for detection of BRCA1/2 common mutations in Spanish and Spanish related breast/ovarian cancer families. BMC Med Genet 8:40

Gabbrielli R, Pandolfini T, Vergnano O, Palandrii MR (1990) Comparison of two serpentine species with different nickel tolerance strategies. Plant Soil 122:27–277

Goldbold DL, Kettner C (1991) Use of root elongation studies to determine aluminium and lead toxicity in *Picea abies* seedlings. J Plant Physiol138:231–235 van der Heijden R, Jacobs DI, Snoeijer W, Hallard D, Verpoorte R (2004) The *Catharanthus* alkaloids: pharmacognosy and biotechnology. Curr Med Chem 11:607–628

Jaleel CA, Gopi R, Manivannan P, Kishorekumar A, Sankar B, Panneerselvam R (2006) Paclobutrazol influences on vegetative growth and floral characteristics of *Catharanthus roseus* (L.) G. Don. Indian J Appl Pure Biol 21:369–372

Keeling CI, Bohlmann J (2006) Genes, enzymes and chemicals of terpenoid diversity in the constitutive and induced defence of conifers against insects and pathogens. New Phytol 170:657–675

Kutchan TM (1995) Alkaloid biosynthesis—the basis for metabolic engineering of medicinal plants. Plant Cell 7:1059–1070

Lin YC, Kao CH (2006) Effect of excess nickel on starch mobilization in germinating rice grains. Plant Nutr 29:1405

Lopez C, Claude B, Morin P, Max JP, Pena R, Ribet JP (2011) Synthesis and study of a molecularly imprinted polymer for the specific extraction of indole alkaloids from *Catharanthus roseus* extracts. Anal Chim Acta 683:198–205

Maheshwari R, Dubey RS (2007) Nickel toxicity inhibits ribonuclease and protease activities in rice seedlings: Protective effects of proline. Plant Growth Regul 51:231

Mazhoudia S, Chaoui A, Ghorbal MH, Ferjani EE (1997) Response of antioxidant enzymes to excess copper in tomato (*Lycopersicon esculentum*, Mill.). Plant Sci 127(1997):129–137

Mehrotra SH, Rai V, Khatoon S, Bisht S (2005) Effect of cadmium on growth, ultramorphology of leaf and secondary metabolites of *Phyllanthus amarus* Schum. and Thonn. Chemosphere 61:1644–1650

Murch SJ, Haq K, Rupasinghe HPV, Saxena PK (2003) Nickel contamination affects growth and secondary metabolite composition of St. John's wort (*Hypericum perforatum* L.). Environ Exp Bot 49:251–257

Noori M (2002) Characterization of the Iranian species of Sophora and Ammodendron (Leguminosae; Sophoreae). PhD thesis, University of London and Royal Botanic Gardens, Kew, UK

Özyigit II, Akinci S (2009) Effects of some stress factors (aluminum, cadmium and drought) on stomata of Roman nettle (*Urtica pilulifera L.*). Not Bot Horti Agrobot Cluj Napoca 37:108

Santiago LJM, Louro RP, De Oliveira DE (2000) Compartmentation of phenolic compounds and phenylalanine ammonia-lyase in leaves of Phyllanthus tenellus Roxb. and their induction by copper sulphate. Ann Bot 86:1023–1032

Sebastiani L, Scebba F, Tognetti R (2004) Heavy metal accumulation and growth responses in poplar clones Eridano (*Populus deltoides* × *maximowiczii*) and I-214 (*P.* × *euramericana*) exposed to industrial waste. Environ Exp Bot 52:79–88

Sumithra K, Jutur PP, Carmel BD, Reddy AR (2006) Salinity-induced changes in two cultivars of Vigna radiata: responses of antioxidative and proline metabolism. Plant Growth Regul 50:11–22

Verpoorte R, Contin A, Memelink J (2002) Biotechnology for the production of plant secondary metabolites. Phytochem Rev 1:13–25

Witte-Claus P, Tiller-Sarah A, Taylor-Mark A, Davies-Howard V (2002) Addition of nickel to Murashige and Skoog medium in plant tissue culture activates urease and may reduce metabolic stress. Plant Cell Tiss Org Cult 68:103–110

Zheng Z, Wu M (2004) Cadmium treatment enhances the production of alkaloid secondary metabolites in *Catharanthus roseus*. Plant sci 29:507–14

The manufacturer's authorised representative in the EU is Springer Nature Customer Service Centre GmbH, Europaplatz 3, 69115 Heidelberg, Germany. If you have any concerns regarding our products, please contact ProductSafety@springernature.com

Printed and bound by CPI Group (UK) Ltd, Croydon, CR0 4YY
25/09/2026
02222851-0005